Universitext

Universitext

Universitext is a series of textbooks that presents material from a wide variety of mathematical disciplines at master's level and beyond. The books, often well class-tested by their author, may have an informal, personal, even experimental approach to their subject matter. Some of the most successful and established books in the series have evolved through several editions, always following the evolution of teaching curricula, into very polished texts.

Thus as research topics trickle down into graduate-level teaching, first textbooks written for new, cutting-edge courses may make their way into *Universitext*.

For further volumes:
www.springer.com/series/223

Friedrich Sauvigny

Partial Differential Equations 2

Functional Analytic Methods

With Consideration of Lectures
by E. Heinz

Second Revised and Enlarged Edition

 Springer

Prof. Dr. Friedrich Sauvigny
Mathematical Institute, LS Analysis
Brandenburgian Technical University
Cottbus, Germany

ISSN 0172-5939 ISSN 2191-6675 (electronic)
Universitext
ISBN 978-1-4471-2983-7 ISBN 978-1-4471-2984-4 (eBook)
DOI 10.1007/978-1-4471-2984-4
Springer London Heidelberg New York Dordrecht

Library of Congress Control Number: 2012936042

Mathematics Subject Classification: 34L05, 35-01, 35-02, 35A01, 35A02, 35A09, 35A15, 35A16, 35A22, 35A30, 35B45, 35B65, 35D30, 35D35, 35J15, 35J25, 35J57, 35J60, 35J93, 35J96, 45-01, 46-01, 47-01, 49R05, 53A10.

Based on a previous edition of the Work:
Partial Differential Equations 2 by Friedrich Sauvigny
© Springer Heidelberg 2006

Printed on acid-free paper

Springer is part of Springer Science+Business Media (www.springer.com)

Preface – Volume 2

In this second volume we continue our textbook PARTIAL DIFFERENTIAL EQUATIONS OF GEOMETRY AND PHYSICS. Originating from both disciplines, we shall treat central questions such as eigenvalue problems, curvature estimates, and boundary value problems, for instance. With the title of our textbook we want to emphasize both the pure and applied aspects of partial differential equations. It turns out that the concepts of solutions are permanently extended in the theory of partial differential equations. Here the classical methods do not lose their significance. Besides the n-dimensional theory, we want to present the two-dimensional theory – so important to our geometric intuition.

We shall solve the differential equations by the continuity method, the variational method or the topological method. The continuity method may be preferred from a geometric point of view, since the stability of the solution is investigated there. The variational method is very attractive from the point of view in physics; however, difficult regularity questions for the weak solution appear with this method. The topological method controls the whole set of solutions during the deformation of the problem, and it does not depend on uniqueness as does the variational method.

The original version of this textbook, *Friedrich Sauvigny: Partielle Differentialgleichungen der Geometrie und der Physik 2 – Funktionalanalytische Lösungsmethoden – Unter Berücksichtigung der Vorlesungen von E. Heinz*, appeared in 2005 from *Springer-Verlag*. A translated and expanded version of this monograph followed in 2006 as *Springer-Universitext*, namely *Friedrich Sauvigny: Partial Differential Equations 2*, and we are now presenting a second edition of this textbook. We have carefully revised the original Volume 2 and added the Chapter 13 on *Boundary Value Problems from Differential Geometry*. The topics of this new Chapter 13 are listed in the table of contents, and we include a table of contents for our revised Volume 1 for the convenience of the reader.

In Chapter 7 we consider, in general, nonlinear operators in Banach spaces. With the aid of Brouwer's degree of mapping from Chapter 3 we prove Schauder's fixed point theorem in Section 1, and we supplement Banach's fixed point theorem. In Section 2 we define the Leray-Schauder degree for mappings in Banach spaces by a suitable approximation, and we prove its fundamental properties in Section 3. In this section we refer to the lecture [H4] by my academic teacher, Professor Dr. E. Heinz in Göttingen.

By transition to linear operators in Banach spaces, we prove the fundamental solution-theorem of F. Riesz via the Leray-Schauder degree. At the end of this chapter we derive the Hahn-Banach continuation theorem by Zorn's lemma (compare [HS]).

In Chapter 8 on *Linear Operators in Hilbert Spaces*, we transform the eigenvalue problems of Sturm-Liouville and of H. Weyl for differential operators into integral equations in Section 1. Then we consider weakly singular integral operators in Section 2 and prove a theorem of I. Schur on iterated kernels. In Section 3 we further develop the results from Chapter 2, Section 6 on the Hilbert space and present the abstract completion of pre-Hilbert-spaces. Bounded linear operators in Hilbert spaces are treated in Section 4: The continuation theorem, Adjoint and Hermitian operators, Hilbert-Schmidt operators, Inverse operators, Bilinear forms and the theorem of Lax-Milgram are presented. In Section 5 we study the transformation of Fourier-Plancherel as a unitary operator on the Hilbert space $L^2(\mathbb{R}^n)$.

Completely continuous, respectively compact operators are studied in Section 6 together with weak convergence. The operators with finite square norms represent an important example. The solution-theorem of Fredholm on operator equations in Hilbert spaces is deduced from the corresponding result of F. Riesz in Banach spaces. We apply these results particularly to weakly singular integral operators.

In Section 7 we prove the spectral theorem of F. Rellich on completely continuous and Hermitian operators by variational methods. Then we address the Sturm-Liouville eigenvalue problem in Section 8 and expand the relevant integral kernels into their eigenfunctions. Following ideas of H. Weyl, we treat the eigenvalue problem for the Laplacian on domains in $\mathbb{R}^n$ by the integral equation method in Section 9. In this chapter as well, we take a lecture of Professor Dr. E. Heinz into consideration (compare [H3]). For the study of eigenvalue problems we recommend the classical treatise [CH] of R. Courant and D. Hilbert.

We have been guided into functional analysis with the aid of problems concerning differential operators in mathematical physics (compare [He1] and [He2]). The usual contents of functional analysis can be taken from Chapter 2, Sections 6-8, Chapter 7, and Chapter 8. Additionally, we investigated the solvability of nonlinear operator equations in Banach spaces. For the spectral theorem of unbounded, self-adjoint operators we refer the reader to the literature.

In our compendium we shall directly construct classical solutions of boundary and initial value problems for linear and nonlinear partial differential equations with the aid of functional analytic methods. The appropriate a priori estimates with respect to the Hölder norm allow us to establish the existence of solutions in classical function spaces.

In Chapter 9, Sections 1-3, we essentially follow the book of I. N. Vekua [V] and solve the Riemann-Hilbert boundary value problem by the integral-equation method. Using the lecture [H6], we present Schauder's continuity method in the Sections 4-7 in order to solve boundary value problems for linear elliptic differential equations with n independent variables. Therefore, we completely prove the Schauder estimates.

In Chapter 10 on *Weak Solutions of Elliptic Differential Equations*, we profit from the *Grundlehren* [GT] Chapters 7 and 8 of D. Gilbarg and N. S. Trudinger. Here, we additionally recommend the textbook [Jo] of J. Jost and the compendium [E] by L. C. Evans.

We introduce Sobolev spaces in Section 1 and prove the embedding theorems in Section 2. Having established the existence of weak solutions in Section 3, we show the boundedness of weak solutions by Moser's iteration method in Section 4. Then we investigate the Hölder continuity of weak solutions in the interior and at the boundary; see the Sections 5-7. Restricting ourselves to interesting classes of equations, we can illustrate the methods of proof in a transparent way. Finally, we apply the results to equations in divergence form; see the Sections 8, 9, and 10.

In Chapter 11, Sections 1-2, we concisely lay the foundations of differential geometry (compare [BL]) and of the calculus of variations. Then we discuss the theory of characteristics for nonlinear hyperbolic differential equations in two variables (compare [CH], [G], [H5]) in Section 3 and Section 4. In particular, we solve the Cauchy initial value problem via Banach's fixed point theorem. In Section 6 we present H. Lewy's ingenious proof for the analyticity theorem of S. Bernstein. Here we would like to refer the reader to the textbook by P. Garabedian [G] as well.

On the basis of Chapter 4 from Volume 1, *Generalized Analytic Functions*, we treat *Nonlinear Elliptic Systems* in Chapter 12. Having presented Jäger's maximum principle in Section 1, we develop the general theory in the Sections 2-5 from the fundamental treatise of E. Heinz [H7] on nonlinear elliptic systems. An existence theorem for nonlinear elliptic systems is situated in the center, which is attained via the Leray-Schauder degree. In Section 6 we derive a well-known curvature estimate for minimal surfaces and obtain a Bernstein theorem. Within Section 8 we introduce conformal parameters into a nonanalytic Riemannian metric by a nonlinear continuity method. We establish the necessary a priori estimates in Section 7 which extend to the boundary.

In Chapter 13 we apply the theory of the nonlinear elliptic systems to *Boundary Value Problems from Differential Geometry*. At first, we solve the Dirichlet problem for nonparametric equations of prescribed mean curvature by the uniformization method in Section 1. Then we present in Section 2 and Section 3 a mathematical model for winding staircases: We solve a mixed boundary value problem for minimal graphs over a Riemannian surface, the so-called *etale plane*, via a nonlinear continuity method. In Section 4 we solve Plateau's problem for surfaces of constant mean curvature by the variational method.

Then we present various answers to the following central question: *In which class of closed surfaces is the sphere the only one with constant mean curvature?* In Section 5 we give an alternative proof for H. Hopf's celebrated answer to this question.

In the final Section 7, we also study the elliptic Monge-Ampère differential equation with the aid of nonlinear elliptic systems. For this fully nonlinear, elliptic p.d.e., H. Lewy and E. Heinz provided a transparent approach by controlling the associate uniformizing mapping in Section 6. We speak of Lewy-Heinz systems on account of their stupendous investigations; however, a special case of these systems already appear within the Darboux lectures [D] for the hyperbolic case. Our results on the Monge-Ampère equation will enable us to solve Weyl's embedding problem for C^3-metrics of positive Gaussian curvature on the sphere.

For further studies on *the Monge-Ampère equation* we would like to recommend the book [Sc] by F. Schulz. If our readers would like to study the theory of minimal surfaces more intensively, we recommend the three new volumes [DHS], [DHT1], and [DHT2] by U. Dierkes, S. Hildebrandt, A. Tromba and the present author within the *Springer-Grundlehren*.

We follow the Total Order Code of the Universitext series, however, adapt this to the present extensive contents. Since we see our books as an entity, we count the chapters throughout our two volumes from 1-13. In each section individually, we count the equations and refer to them by a single number; when we refer to an equation in another section of the same chapter, say the m-th section, we have to add *Section m*; when we refer to an equation in the m-th section of another chapter, say the l-th chapter, we have to add *Section m in Chapter l*.

We assemble *definitions, theorems, propositions, examples* to the expression *environment*, which is borrowed from the underlying TEX-file. Individually in each section, we count these environments consecutively by the number n, and denote the n-th environment within the m-th section by *Environment m.n*. Thus we have attributed a pair of integers $m.n$ to all definitions, theorems, propositions, and examples, which is unique within each chapter and easy to find. Referring to these environments throughout both books, we proceed as described above for the equations.

We add Figure 1.1 – Figure 1.9 to our Volume 1 and Figure 2.1 – Figure 2.11 to our Volume 2, which mostly represent portraits of mathematicians. This small photo collection of some scientists, who have contributed to the theory of partial differential equations, already shows that our area is situated in the center of modern mathematics and possesses profound interrelations with geometry and physics.

This textbook PARTIAL DIFFERENTIAL EQUATIONS has been developed from lectures that I have been giving in the Brandenburgische Technische Universität at Cottbus from the winter semester 1992/93 to the present semester. The monograph builds, in part, upon the lectures (see [H1] – [H6]) of Professor Dr. Dr.h.c. E. Heinz, whom I was fortunate to know as his student in Göttingen from 1971 to 1978 and as postdoctoral researcher in his *Oberseminar* from 1983 to 1989. As an assistant in Aachen from 1978 to 1983, I very much appreciated the elegant lecture cycles of Professor Dr. G. Hellwig (see [He1] – [He3]). Since my research fellowship at the University of Bonn in 1989/90, an intensive scientific collaboration with Professor Dr. Dr.h.c.mult. S. Hildebrandt has developed and continues to this day (see [DHS] and [DHT2] with the list of references therein). All three of these excellent representatives of mathematics will forever have my sincere gratitude and my deep respect!

Here I gratefully acknowledge the valuable and profound advice of Priv.-Doz. Dr. Frank Müller (Universität Duisburg-Essen) for the original edition *Partielle Differentialgleichungen* as well as the indispensable and excellent assistance of Dipl.-Math. Michael Hilschenz (BTU Cottbus) for the present edition *Partial Differential Equations 2*. Furthermore, my sincere thanks are devoted to Mrs. C. Prescott (Berlin) improving the English style of this second edition. Moreover, I would like to thank cordially Herr Clemens Heine (Heidelberg) and Mr. Jörg Sixt (London) as well as Mrs. Lauren Stoney (London) and all the other members of Springer for their helpful collaboration and great confidence.

Cottbus, February 2012   *Friedrich Sauvigny*

Contents – Volume 2:
Functional Analytic Methods and Differential Geometric Applications

Contents – Volume 1:
Foundations and Integral Representations

Chapter 4   Generalized Analytic Functions

Chapter 5   Potential Theory and Spherical Harmonics

Chapter 6   Linear Partial Differential Equations in $\mathbb{R}^n$

Chapter 7

Operators in Banach Spaces

We shall now present some methods from the nonlinear functional analysis. In this volume we refer to our presentation of Lebesgue's integral and the associate Lebesgue spaces from Chapter 2, especially the Sections 6-8. Furthermore, we utilize the Brouwer degree of mapping from Chapter 3 in the way E. Heinz has introduced this topological theory. The fixed point theorems of S. Banach and J. Schauder are central for the modern existence theory of partial differential equations. A detailed account of the contents for this chapter is given in the *Introduction to Volume 2* above.

1 Fixed Point Theorems

Definition 1.1. *The* Banach space $\mathcal{B}$ *is a linear normed complete (infinite-dimensional) vector space above the field of real numbers* $\mathbb{R}$.

Example 1.2. Let the set $\Omega \subset \mathbb{R}^n$ be open, $1 \le p < +\infty$, $\mathcal{B} := L^p(\Omega)$. We have $f \in L^p(\Omega)$ if and only if $f : \Omega \to \mathbb{R}$ is measurable and

$$\int_\Omega |f(x)|^p \, dx < +\infty$$

holds true. For the element $f \in \mathcal{B}$ we define the norm

$$\|f\| := \left(\int_\Omega |f(x)|^p \, dx \right)^{\frac{1}{p}},$$

and we obtain the *Lebesgue space* with $\mathcal{B}$. The case $p = 2$ reduces to the Hilbert space $L^2(\Omega)$ using the inner product

$$(f, g) := \int_\Omega f(x)g(x) \, dx.$$

F. Sauvigny, *Partial Differential Equations 2*, Universitext,
DOI 10.1007/978-1-4471-2984-4_1, © Springer-Verlag London 2012

Example 1.3. (Hilbert's sequence space ℓ^p)
For the sequence $x = (x_1, x_2, x_3, \ldots)$ we have $x \in \ell^p$ with $1 \le p < +\infty$ if and only if

$$\sum_{k=1}^{\infty} |x_k|^p < +\infty$$

is fulfilled. By the norm

$$\|x\| := \left(\sum_{k=1}^{\infty} |x_k|^p \right)^{\frac{1}{p}}$$

the set ℓ^p becomes a Banach space. From the following diagram we comprehend the relation $\ell^p \subset L^p((0, +\infty))$.

Example 1.4. (Sobolev spaces) Let the numbers $k \in \mathbb{N}$ and $1 \le p < +\infty$ be given, whereas $\Omega \subset \mathbb{R}^n$ denotes an open set. The space

$$\mathcal{B} = W^{k,p}(\Omega) := \left\{ f : \Omega \to \mathbb{R} \ : \ D^\alpha f \in L^p(\Omega) \quad \text{for all} \quad |\alpha| \le k \right\}$$

with the norm

$$\|f\|_{W^{k,p}(\Omega)} := \left(\sum_{|\alpha| \le k} \int_\Omega |D^\alpha f(x)|^p \, dx \right)^{\frac{1}{p}}, \qquad f \in \mathcal{B},$$

represents a Banach space. Here, the vector $\alpha = (\alpha_1, \ldots, \alpha_n) \in \mathbb{N}_0^n$ indicates a multi-index, and we set

$$|\alpha| := \sum_{i=1}^{n} \alpha_i \in \mathbb{N}_0 := \mathbb{N} \cup \{0\}.$$

In this context, we refer the reader to the Sections 1 and 2 of Chapter 10.

Example 1.5. Finally, we consider the *classical Banach spaces* $C^k(\overline{\Omega})$ for $k = 0, 1, 2, 3, \ldots$ on bounded domains $\Omega \subset \mathbb{R}^n$. We have $f \in C^k(\overline{\Omega})$ if and only if

$$\sup_{x \in \Omega} \left\{ \sum_{|\alpha| \le k} |D^\alpha f(x)| \right\} < +\infty$$

holds true, where $\alpha \in \mathbb{N}_0^n$ denote multi-indices. The vector space $\mathcal{B} := C^k(\overline{\Omega})$ equipped with the norm

$$\|f\|_{C^k(\overline{\Omega})} := \sum_{|\alpha| \le k} \sup_{x \in \Omega} |D^\alpha f(x)|$$

is complete and represents a Banach space. Here we abbreviate

$$D^\alpha f(x) := \frac{\partial^{|\alpha|}}{\partial x_1^{\alpha_1} \ldots \partial x_n^{\alpha_n}} f(x), \quad \alpha \in \mathbb{N}_0^n, \quad \mathbb{N}_0 := \mathbb{N} \cup \{0\}.$$

Let us start our considerations with the following

Definition 1.6. *A subset $K \subset \mathcal{B}$ of the Banach space $\mathcal{B}$ is named* convex, *if we have the inclusion $\lambda x + (1 - \lambda)y \in K$ for each two points $x, y \in K$ and each parameter $\lambda \in [0, 1]$.*

Remarks:

1. When K is closed, this set is convex if and only if the implication

$$x, y \in K \quad \Rightarrow \quad \frac{1}{2}(x + y) \in K$$

   holds true.

2. For a convex set K we have the following property: Choosing the points $x_1, \ldots, x_n \in K$ and the parameters $\lambda_k \geq 0$ for $k = 1, \ldots, n$ satisfying $\lambda_1 + \ldots + \lambda_n = 1$, we have the inclusion $\sum_{k=1}^{n} \lambda_k x_k \in K$.

Definition 1.7. *A subset $E \subset \mathcal{B}$ is called* precompact, *if each sequence*

$$\{x_n\}_{n=1,2,\ldots} \subset E$$

contains a Cauchy sequence as a subsequence. If the set E is additionally closed, which means $\{x_n\}_{n \in \mathbb{N}} \subset E$ with $x_n \to x$ for $n \to \infty$ in $\mathcal{B}$ implies $x \in E$, we call the set E compact.

Example 1.8. Let $E \subset \mathcal{B}$ be a closed and bounded subset of a finite-dimensional subspace of $\mathcal{B}$. Then the Weierstraß selection theorem yields that E is compact.

Example 1.9. For infinite-dimensional Banach spaces, bounded and closed subsets are not necessarily compact: Choosing $k \in \mathbb{N}$, we consider the set of sequences $x_k := (\delta_{kj})_{j=1,2,\ldots}$ in the space ℓ^2, with the Kronecker symbol δ_{kj}. Obviously, we have $\|x_k\| = 1$ for all $k \in \mathbb{N}$, however, $\|x_k - x_l\| = \sqrt{2}$ for all $k, l \in \mathbb{N}$ with $k \neq l$. Therefore, the set $\{x_k\}_{k=1,2,\ldots} \subset \ell^2$ is not precompact.

Example 1.10. A bounded set in $C^k(\overline{\Omega})$ is compact, if we additionally require a modulus of continuity for the k-th partial derivatives: Consider the set

$$E := \left\{ f \in C^k(\overline{\Omega}) : \begin{array}{l} \|f\|_{C^k(\overline{\Omega})} \leq M; \\ |D^\alpha f(x) - D^\alpha f(y)| \leq M'|x - y|^\vartheta \\ \text{for all } x, y \in \overline{\Omega}, \ |\alpha| = k \end{array} \right\}$$

with $k \in \mathbb{N}_0$, $M, M' \in (0, +\infty)$ and $\vartheta \in (0, 1]$. By the Theorem of Arzelà-Ascoli we easily deduce that the set

$$E \subset \mathcal{B} := C^k(\overline{\Omega})$$

is compact.

Definition 1.11. *On the subset $E \subset \mathcal{B}$ in the Banach space $\mathcal{B}$ we have defined the mapping $F : E \to \mathcal{B}$. We call F continuous, if*

$$x_n \to x \quad for \quad n \to \infty \quad in \quad E$$

implies

$$F(x_n) \to F(x) \quad for \quad n \to \infty \quad in \quad \mathcal{B}.$$

We name F completely continuous *(or* compact *as well), if additionally the set $F(E) \subset \mathcal{B}$ is precompact; this means all sequences $\{x_n\}_{n=1,2,\dots} \subset E$ contain a subsequence $\{x_{n_k}\}_k \subset \{x_n\}_n$, such that $\{F(x_{n_k})\}_{k=1,2,\dots}$ represents a Cauchy sequence in $\mathcal{B}$.*

Proposition 1.12. *Let K be a precompact subset of the Banach space $\mathcal{B}$. For all $\varepsilon > 0$ we have finitely many elements $w_1, \dots, w_N \in K$ with $N = N(\varepsilon) \in \mathbb{N}$, such that the covering property*

$$K \subset \bigcup_{j=1}^{N(\varepsilon)} \left\{ x \in \mathcal{B} : \ \|x - w_j\| \le \frac{\varepsilon}{2} \right\}$$

is fulfilled.

Proof: We choose $w_1 \in K$ and the covering property is already valid if

$$K \subset \left\{ x \in \mathcal{B} : \ \|x - w_1\| \le \frac{\varepsilon}{2} \right\}$$

holds true. When this is not the case, there exists a further point $w_2 \in K$ with $\|w_2 - w_1\| > \frac{\varepsilon}{2}$ and we consider the balls

$$\left\{ x \in \mathcal{B} : \ \|x - w_j\| \le \frac{\varepsilon}{2} \right\} \quad for \quad j = 1, 2.$$

If they do not yet cover the set K, there would exist a third point $w_3 \in K$ with $\|w_3 - w_j\| > \frac{\varepsilon}{2}$ for $j = 1, 2$. In case the procedure did not stop, we could find a sequence $\{w_j\}_{j=1,2\dots} \subset K$ of points satisfying

$$\|w_j - w_i\| > \frac{\varepsilon}{2} \quad for \quad i = 1, \dots, j-1.$$

This yields a contradiction to the precompactness of the set K.   q.e.d.

Proposition 1.13. *Let K be a precompact set in $\mathcal{B}$, and $\varepsilon > 0$ is arbitrarily given. Then we have finitely many elements $w_1, \dots, w_N \in K$ with $N = N(\varepsilon) \in \mathbb{N}$ continuous functions*

$$t_i = t_i(x) : \overline{K} \to \mathbb{R} \in C^0(\overline{K})$$

satisfying

$$t_i(x) \geq 0 \quad and \quad \sum_{i=1}^{N} t_i(x) = 1 \quad in \quad K,$$

such that the following inequality holds true:

$$\left\| \sum_{i=1}^{N} t_i(x)w_i - x \right\| \leq \varepsilon \quad for\ all \quad x \in \overline{K}.$$

Proof: We choose the points $\{w_1, \ldots, w_N\} \subset K$ according to Proposition 1.12. We define the continuous function $\varphi(\tau) : [0, +\infty) \to [0, +\infty)$ via

$$\varphi(\tau) := \begin{cases} \varepsilon - \tau, & for\ 0 \leq \tau \leq \varepsilon \\ 0, & for\ \varepsilon \leq \tau < +\infty \end{cases},$$

and obtain

$$\sum_{j=1}^{N} \varphi(\|x - w_j\|) \geq \frac{\varepsilon}{2} \quad for\ all \quad x \in \overline{K}.$$

Consequently, the functions

$$t_i(x) := \frac{\varphi(\|x - w_i\|)}{\sum\limits_{j=1}^{N} \varphi(\|x - w_j\|)}, \quad x \in \overline{K}, \quad i = 1, \ldots, N$$

are well-defined, and we note that

$$t_i \in C^0(\overline{K}, [0,1]) \quad and \quad \sum_{i=1}^{N} t_i(x) = 1 \quad for\ all \quad x \in \overline{K}.$$

Now we can estimate as follows:

$$\left\| x - \sum_{i=1}^{N} t_i(x)w_i \right\| = \left\| \sum_{i=1}^{N} t_i(x)(x - w_i) \right\|$$

$$\leq \sum_{i=1}^{N} t_i(x)\|x - w_i\|$$

$$\leq \sum_{i=1}^{N} t_i(x)\varepsilon = \varepsilon \quad for\ all \quad x \in \overline{K}.$$

This yields the stated inequality.   q.e.d.

Proposition 1.14. *Let the set $E \subset \mathcal{B}$ be closed and the function $F : E \to \mathcal{B}$ be completely continuous. To each number $\varepsilon > 0$ we then have $N = N(\varepsilon) \in \mathbb{N}$ elements $w_1, \ldots, w_N \in F(E)$ and N continuous functions $F_j : E \to \mathbb{R}$, $j = 1, \ldots, N$ satisfying*

$$F_j(x) \geq 0 \quad and \quad \sum_{j=1}^{N} F_j(x) = 1, \qquad x \in E,$$

such that the following inequality is valid:

$$\left\| F(x) - \sum_{j=1}^{N} F_j(x) w_j \right\| \leq \varepsilon \qquad \textit{for all} \quad x \in E.$$

Proof: The set $K := F(E) \subset \mathcal{B}$ is precompact and we apply Proposition 1.13. Then we have the elements

$$w_1, \ldots, w_N \in F(E)$$

and the nonnegative continuous functions

$$t_i = t_i(x), x \in \overline{K}$$

satisfying $t_1(x) + \ldots + t_N(x) = 1$ in $\overline{K}$ for each $\varepsilon > 0$, such that

$$\left\| x - \sum_{i=1}^{N} t_i(x) w_i \right\| \leq \varepsilon \qquad \text{for all} \quad x \in \overline{K}.$$

Setting $F_i(x) := t_i(F(x))$, $x \in E$, we comprehend the statement above.

q.e.d.

We now consider the *unit simplex*

$$\Sigma_{n-1} := \left\{ x \in \mathbb{R}^n \ : \ x_i \geq 0 \ \text{ for } \ i = 1, \ldots, n, \ \sum_{i=1}^{n} x_i = 1 \right\}$$

and its projection onto the plane $\mathbb{R}^{n-1} \times \{0\} \subset \mathbb{R}^n$

$$\sigma_{n-1} := \left\{ x \in \mathbb{R}^{n-1} \ : \ x_i \geq 0 \ \text{ for } \ i = 1, \ldots, n-1, \ \sum_{i=1}^{n-1} x_i \leq 1 \right\}.$$

Figure 2.1   Illustration of a Simplex

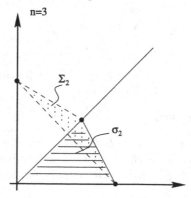

We note that

$$\Sigma_{n-1} = \left\{ (x_1, \ldots, x_n) \in \mathbb{R}^n \ : \ x_n = 1 - \sum_{i=1}^{n-1} x_i \ \text{with} \ (x_1, \ldots, x_{n-1}) \in \sigma_{n-1} \right\}.$$

Proposition 1.15. (Brouwer's fixed point theorem for the unit simplex)
Each continuous mapping $f : \Sigma_{n-1} \to \Sigma_{n-1}$ possesses a fixed point.

Proof:

1. The function $f = (f_1, \ldots, f_n) : \Sigma_{n-1} \to \Sigma_{n-1}$ being given, we define the
    mapping $g(x) = (g_1(x), \ldots, g_{n-1}(x)) : \sigma_{n-1} \to \sigma_{n-1}$ by

$$g_i(x) = g_i(x_1, \ldots, x_{n-1}) := f_i\left(x_1, \ldots, x_{n-1}, 1 - \sum_{j=1}^{n-1} x_j \right)$$

with $i = 1, \ldots, n-1$. Now the point $\eta = (\eta_1, \ldots, \eta_{n-1}) \in \sigma_{n-1}$ is a fixed
point of the mapping $g : \sigma_{n-1} \to \sigma_{n-1}$ if and only if the point

$$\left(\eta_1, \ldots, \eta_{n-1}, 1 - \sum_{i=1}^{n-1} \eta_i \right) \in \Sigma_{n-1}$$

is a fixed point of the mapping $f : \Sigma_{n-1} \to \Sigma_{n-1}$.

2. We consider the following mapping defined in part 1., namely

$$g = (g_1, \ldots, g_{n-1}) : \sigma_{n-1} \to \sigma_{n-1}.$$

The adjoint functions

$$h_i = h_i(x_1, \ldots, x_{n-1}) := \sqrt{g_i(x_1^2, \ldots, x_{n-1}^2)}, \qquad i = 1, \ldots, n-1$$

are defined on the ball

$$K := \left\{ (x_1, \ldots, x_{n-1}) \in \mathbb{R}^{n-1} \ : \ x_1^2 + \ldots + x_{n-1}^2 \leq 1 \right\}.$$

Due to Brouwer's fixed point theorem for the ball from Section 3 in Chapter
3, the continuous mapping $h = (h_1, \ldots, h_{n-1}) : K \to K$ has a fixed point
$\xi = (\xi_1, \ldots, \xi_{n-1}) \in K$, more precisely $h(\xi) = \xi$. This implies

$$g_i(\xi_1^2, \ldots, \xi_{n-1}^2) = \xi_i^2 \qquad \text{for} \quad i = 1, \ldots, n-1.$$

With the point $\eta := (\xi_1^2, \ldots, \xi_{n-1}^2) \in \sigma_{n-1}$ we finally obtain a fixed point
of the mapping $g : \sigma_{n-1} \to \sigma_{n-1}$ satisfying $g(\eta) = \eta$.
                                                                        q.e.d.

Theorem 1.16. (Schauder's fixed point theorem)
Let $A \subset \mathcal{B}$ be a closed and convex subset of the Banach space $\mathcal{B}$. Then each completely continuous mapping $F : A \to A$ possesses a fixed point $\xi \in A$, more precisely $F(\xi) = \xi$.

Proof:

1. We apply Proposition 1.14 to the completely continuous mapping F: For each $\varepsilon > 0$ there exist $N = N(\varepsilon) \in \mathbb{N}$ elements $\{w_1, \ldots, w_N\} \subset F(A) \subset A$ and N nonnegative continuous functions $F_j : A \to \mathbb{R}$, $j = 1, \ldots, N$ satisfying $F_1(x) + \ldots + F_N(x) = 1$ in A, such that

$$\left\| F(x) - \sum_{j=1}^{N(\varepsilon)} F_j(x) w_j \right\| \leq \varepsilon \qquad \text{for all} \quad x \in A.$$

We now consider the continuous function

$$g(\lambda) = (g_1(\lambda_1, \ldots, \lambda_N), \ldots, g_N(\lambda_1, \ldots, \lambda_N)) : \Sigma_{N-1} \to \Sigma_{N-1}$$

with

$$g_j(\lambda_1, \ldots, \lambda_N) := F_j\left(\sum_{i=1}^{N} \lambda_i w_i \right), \qquad j = 1, \ldots, N.$$

Due to Proposition 1.15, we have a point $\lambda \in \Sigma_{N-1}$ satisfying $g(\lambda) = \lambda$. This implies

$$F_j\left(\sum_{i=1}^{N} \lambda_i w_i \right) = \lambda_j \qquad \text{for} \quad j = 1, \ldots, N.$$

2. On account of part 1. the mapping

$$F_\varepsilon(x) := \sum_{j=1}^{N(\varepsilon)} F_j(x) w_j$$

possesses the fixed point

$$\xi_\varepsilon := \sum_{i=1}^{N} \lambda_i w_i.$$

We note that $\| F(x) - F_\varepsilon(x) \| \leq \varepsilon$ for all $x \in A$ holds true and obtain $\| F(\xi_\varepsilon) - \xi_\varepsilon \| \leq \varepsilon$. Taking the zero sequence $\varepsilon = \frac{1}{n}$, $n = 1, 2, \ldots$ as our parameter ε, we obtain a sequence of points $\{\xi_n\}_{n=1,2,\ldots}$ satisfying

$$\| F(\xi_n) - \xi_n \| \leq \frac{1}{n}, \qquad n = 1, 2, \ldots$$

Since $F(A)$ is precompact, we can select a subsequence with $F(\xi_{n_k}) \to \xi$ for $k \to \infty$. We obtain $\xi \in A$ because the set A is closed. Therefore, we deduce

$$\|\xi - \xi_{n_k}\| \leq \|F(\xi_{n_k}) - \xi_{n_k}\| + \|\xi - F(\xi_{n_k})\| \to 0 \qquad \text{for} \quad k \to \infty.$$

Together with the continuity of F, we conclude

$$\xi = \lim_{k \to \infty} F(\xi_{n_k}) = F(\lim_{k \to \infty} \xi_{n_k}) = F(\xi).$$

q.e.d.

We now provide an application of Theorem 1.16, namely

Theorem 1.17. (Leray's eigenvalue problem)
Let $K(s,t) : [a,b] \times [a,b] \to (0,+\infty)$ be a continuous and positive integral-kernel. Then the integral equation

$$\int_a^b K(s,t)x(t)\,dt = \lambda x(s), \qquad a \leq s \leq b,$$

possesses at least one positive eigenvalue λ with the adjoint nonnegative continuous eigenfunction $x(s) \not\equiv 0$.

Proof: We choose the Banach space $\mathcal{B} := C^0([a,b])$ with the norm

$$\|x\| := \max_{a \leq s \leq b} |x(s)|.$$

Then we consider the convex subset

$$A := \left\{ x = x(s) \in C^0([a,b]) \; : \; x(s) \geq 0 \text{ in } [a,b], \; \int_a^b x(s)\,ds = 1 \right\},$$

which is closed in $\mathcal{B}$. Furthermore, we study the mapping $F : A \to A$ defined by

$$F(x) := \frac{\displaystyle\int_a^b K(s,t)x(t)\,dt}{\displaystyle\int_a^b \left(\int_a^b K(s,t)x(t)\,dt \right) ds}, \qquad x \in A.$$

Via the Arzelà-Ascoli theorem one shows that the mapping $F : A \to A$ is completely continuous. Because of Schauder's fixed point theorem there exists a point $\xi \in A$ with $F(\xi) = \xi$. Consequently, we see

$$\int_a^b K(s,t)\xi(t)\,dt = \left[\int_a^b \left(\int_a^b K(s,t)\xi(t)\,dt \right) ds \right] \xi(s), \qquad s \in [a,b].$$

Therefore, the element ξ represents the desired eigenfunction for the eigenvalue

$$\lambda := \int\limits_a^b \left(\int\limits_a^b K(s,t)\xi(t)\,dt \right) ds \;\in\; (0,+\infty).$$

q.e.d.

In Brouwer's as well as Schauder's fixed point theorem, only the existence of a fixed point is established which is, in general, not uniquely determined. The subsequent fixed point theorem of S. Banach supplies both the existence and uniqueness of the fixed point. Furthermore, we shall show the continuous dependence of the fixed point from the parameter. The Picard iteration scheme, proving the existence of initial value problems with ordinary differential equations, already contains the essence of the Banach fixed point theorem within the classical spaces of continuous functions.

Definition 1.18. *The* family of operators $T_\lambda : \mathcal{B} \to \mathcal{B}$, $0 \le \lambda \le 1$, *is called* contracting, *if we have a constant* $\theta \in [0,1)$ *satisfying*

$$\|T_\lambda(x) - T_\lambda(y)\| \le \theta \|x - y\| \qquad \text{for all} \quad x,y \in \mathcal{B} \quad \text{und} \quad \lambda \in [0,1].$$

For each fixed $x \in \mathcal{B}$ *let the curve* $\{T_\lambda(x)\}_{0 \le \lambda \le 1}$ *in* $\mathcal{B}$ *be continuous. If the family* $T := T_\lambda : \mathcal{B} \to \mathcal{B}$ *for* $0 \le \lambda \le 1$ *is constant, we call the operator* T contracting.

Theorem 1.19. (Banach's fixed point theorem)
Let the family of operators

$$T_\lambda : \mathcal{B} \to \mathcal{B}, \quad 0 \le \lambda \le 1$$

be contracting on the Banach space $\mathcal{B}$. *Then there exists exactly one point* $x_\lambda \in \mathcal{B}$ *satisfying* $T_\lambda(x_\lambda) = x_\lambda$ *for each* $\lambda \in [0,1]$, *namely a fixed point of* T_λ. *Furthermore, the curve*

$$[0,1] \ni \lambda \to x_\lambda \in \mathcal{B}$$

is continuous.

Proof:

1. We define $y_\lambda := T_\lambda(0)$, $0 \le \lambda \le 1$, and set

$$\varrho := \max_{0 \le \lambda \le 1} \|y_\lambda\| \in (0,+\infty).$$

On the ball $\mathcal{B}_r := \{x \in \mathcal{B} : \|x\| \le r\}$ of radius $r := \frac{\varrho}{1-\theta} \in (0,+\infty)$ in the Banach space $\mathcal{B}$ we consider the family of mappings

$$T_\lambda : \mathcal{B}_r \to \mathcal{B}_r, \quad 0 \le \lambda \le 1.$$

Taking $x \in \mathcal{B}_r$, we have

$$\|T_\lambda(x)\| \leq \|T_\lambda(x) - T_\lambda(0)\| + \|T_\lambda(0)\|$$
$$\leq \theta\|x\| + \|y_\lambda\| \leq \theta r + \varrho$$
$$\leq \theta \frac{\varrho}{1 - \theta} + \varrho = r.$$

2. For $n = 0, 1, 2, \ldots$ we consider the *iterated points*

$$x_\lambda^{(n)} := T_\lambda^n(0) = \underbrace{T_\lambda \circ \ldots \circ T_\lambda}_{n-\text{times}}(0).$$

Evidently, we have $x_\lambda^{(0)} = 0$ and $x_\lambda^{(1)} = y_\lambda$ for $0 \leq \lambda \leq 1$. Furthermore, we observe

$$x_\lambda^{(n+1)} = x_\lambda^{(n+1)} - x_\lambda^{(0)} = \sum_{k=0}^n \left(x_\lambda^{(k+1)} - x_\lambda^{(k)} \right) = \sum_{k=0}^n \left(T_\lambda^{k+1}(0) - T_\lambda^k(0) \right).$$

Now we can estimate

$$\|T_\lambda^{k+1}(0) - T_\lambda^k(0)\| \leq \theta\|T_\lambda^k(0) - T_\lambda^{k-1}(0)\|$$
$$\leq \ldots \leq \theta^k\|T_\lambda(0) - T_\lambda^0(0)\|$$
$$= \theta^k\|y_\lambda\|, \qquad 0 \leq \lambda \leq 1, \quad k = 0, 1, 2, \ldots$$

Therefore, the series

$$\sum_{k=0}^\infty \left(T_\lambda^{k+1}(0) - T_\lambda^k(0) \right)$$

possesses the convergent majorizing function $\displaystyle\sum_{k=0}^\infty \theta^k\|y_\lambda\|$, and the following limit exists:

$$x_\lambda := \lim_{n\to\infty} x_\lambda^{(n+1)} = \sum_{k=0}^\infty \left(T_\lambda^{(k+1)}(0) - T_\lambda^{(k)}(0) \right) \in \mathcal{B}_r.$$

3. The contracting operator $T_\lambda : \mathcal{B} \to \mathcal{B}$ is continuous. Consequently, we see

$$x_\lambda = \lim_{n\to\infty} x_\lambda^{(n+1)} = \lim_{n\to\infty} T_\lambda\left(x_\lambda^{(n)}\right) = T_\lambda\left(\lim_{n\to\infty} x_\lambda^{(n)}\right) = T_\lambda(x_\lambda)$$

for $0 \leq \lambda \leq 1$. The fixed points x_λ depend continuously on the parameters $\lambda \in [0, 1]$: We choose the parameters $\lambda_1, \lambda_2 \in [a, b]$ and infer

$$\|x_{\lambda_1} - x_{\lambda_2}\| = \|T_{\lambda_1}(x_{\lambda_1}) - T_{\lambda_2}(x_{\lambda_2})\|$$
$$\leq \|T_{\lambda_1}(x_{\lambda_1}) - T_{\lambda_1}(x_{\lambda_2})\| + \|T_{\lambda_1}(x_{\lambda_2}) - T_{\lambda_2}(x_{\lambda_2})\|$$
$$\leq \theta\|x_{\lambda_1} - x_{\lambda_2}\| + \|T_{\lambda_1}(x_{\lambda_2}) - T_{\lambda_2}(x_{\lambda_2})\|$$

as well as

$$\|x_{\lambda_1} - x_{\lambda_2}\| \leq \frac{1}{1-\theta}\|T_{\lambda_1}(x_{\lambda_2}) - T_{\lambda_2}(x_{\lambda_2})\|.$$

4. Finally, we show the uniqueness of the fixed point. Therefore, we consider two elements $x_\lambda, \tilde{x}_\lambda \in \mathcal{B}$ satisfying

$$x_\lambda = T_\lambda(x_\lambda), \qquad \tilde{x}_\lambda = T_\lambda(\tilde{x}_\lambda).$$

Then the contraction inequality implies

$$\|x_\lambda - \tilde{x}_\lambda\| = \|T_\lambda(x_\lambda) - T_\lambda(\tilde{x}_\lambda)\| \leq \theta\|x_\lambda - \tilde{x}_\lambda\|$$

and $\|x_\lambda - \tilde{x}_\lambda\| = 0$ or $x_\lambda = \tilde{x}_\lambda$ for $\lambda \in [0,1]$.   q.e.d.

Remark: If the family of operators T_λ depends even differentiably on the parameter $\lambda \in [0,1]$, we can additionally deduce the differentiable dependence of the fixed point from the parameter as in part 3. of the proof above.

———

2 The Leray-Schauder Degree of Mapping

In the sequel we denote mappings between Banach spaces $\mathcal{B}$ by

$$f : \mathcal{B} \to \mathcal{B}, \quad x \mapsto f(x).$$

Let $\mathcal{B}$ be a finite-dimensional Banach space with $1 \leq \dim \mathcal{B} = n < +\infty$. Furthermore, we have the bounded open set $\Omega \subset \mathcal{B}$ and $g : \overline{\Omega} \to \mathcal{B}$ denotes a continuous mapping with the property $0 \notin g(\partial\Omega)$. At first, we shall define the degree of mapping $\delta_{\mathcal{B}}(g, \Omega)$.

Let $\{w_1, \ldots, w_n\} \subset \mathcal{B}$ constitute a basis of the linear space $\mathcal{B}$. Consider the *coordinate mapping*

$$\psi = \psi_{w_1 \ldots w_n}(x) := x_1 w_1 + \ldots + x_n w_n, \qquad x = (x_1, \ldots, x_n) \in \mathbb{R}^n.$$

Evidently, $\psi : \mathbb{R}^n \to \mathcal{B}$ holds true and the inverse mapping $\psi^{-1} : \mathcal{B} \to \mathbb{R}^n$ exists. We pull back the mapping $g : \overline{\Omega} \to \mathcal{B}$ onto the space $\mathbb{R}^n$. Therefore, we set

$$\Omega_n := \psi^{-1}(\Omega), \qquad \partial\Omega_n = \psi^{-1}(\partial\Omega), \qquad \overline{\Omega}_n = \psi^{-1}(\overline{\Omega})$$

and consider the mapping

$$g_n := \psi^{-1} \circ g \circ \psi \,|_{\overline{\Omega}_n} \qquad \text{with} \quad 0 \notin g_n(\partial\Omega_n).$$

Parallel to Chapter 3, Section 2 we can attribute the degree of mapping $d(g_n, \Omega_n)$ to the continuous mapping $g_n : \overline{\Omega}_n \to \mathbb{R}^n$.

Definition 2.1. *Let the finite-dimensional Banach space $\mathcal{B}$ be given with $n = \dim \mathcal{B} \in \mathbb{N}$, and $\Omega \subset \mathcal{B}$ denotes a bounded open set. Furthermore, the continuous mapping $g : \overline{\Omega} \to \mathcal{B}$ with $0 \notin g(\partial\Omega)$ is prescribed. Then we define the degree of mapping*

$$\delta_{\mathcal{B}}(g, \Omega) := d(g_n, \Omega_n).$$

Here we have set $g_n := \psi^{-1} \circ g \circ \psi\,|_{\overline{\Omega}_n}$ with $\Omega_n := \psi^{-1}(\Omega)$, and $\psi : \mathbb{R}^n \to \mathcal{B}$ denotes an arbitrary coordinate mapping.

We still have to show the independence of the definition above from the basis chosen: Let $\{w_1^*, \ldots, w_n^*\}$ be a further basis of $\mathcal{B}$ with the coordinate mapping

$$\psi^*(x^*) = \psi^*_{w_1^* \ldots w_n^*}(x_1^*, \ldots, x_n^*) = x_1^* w_1^* + \ldots + x_n^* w_n^* \; : \; \mathbb{R}^n \to \mathcal{B}$$

and its inverse $\psi^{*-1} : \mathcal{B} \to \mathbb{R}^n$. On $\Omega_n^* := \psi^{*-1}(\Omega)$ we define the mapping

$$g_n^* := \psi^{*-1} \circ g \circ \psi^* \,|_{\overline{\Omega}_n^*}, \qquad 0 \notin g_n^*(\partial\Omega_n^*).$$

Definition 2.1 makes sense on account of

Proposition 2.2. *We have $d(g_n^*, \Omega_n^*) = d(g_n, \Omega_n)$.*

Proof: The mapping

$$\chi := \psi^{-1} \circ \psi^* : \mathbb{R}^n \to \mathbb{R}^n$$

is linear and nonsingular, and we note that $\psi^* = \psi \circ \chi$. Furthermore, the identity $\chi(\Omega_n^*) = \Omega_n$ holds true and we calculate

$$g_n^* = \psi^{*-1} \circ g \circ \psi^* = (\psi \circ \chi)^{-1} \circ g \circ (\psi \circ \chi)$$
$$= \chi^{-1} \circ (\psi^{-1} \circ g \circ \psi) \circ \chi = \chi^{-1} \circ g_n \circ \chi \qquad \text{on} \quad \Omega_n^*.$$

Now we have a sequence of mappings

$$g_{n,\nu} : \mathbb{R}^n \to \mathbb{R}^n \in C^1(\mathbb{R}^n, \mathbb{R}^n)$$

with the following properties (compare Chapter 3, Section 4):

(a) The convergence $g_{n,\nu}(x) \to g_n(x)$ for $\nu \to \infty$ is uniformly on $\overline{\Omega}_n$;
(b) For all numbers $\nu \geq \nu_o$ the equation

$$g_{n,\nu}(x) = 0, \qquad x \in \overline{\Omega}_n$$

possesses only finitely many solutions $\{x_\nu^{(\mu)}\}_{\mu=1,\ldots,p_\nu}$ with the Jacobian

$$J_{g_{n,\nu}}(x_\nu^{(\mu)}) \neq 0 \qquad \text{for} \quad \mu = 1, \ldots, p_\nu.$$

The mapping $g_{n,\nu}^* := \chi^{-1} \circ g_{n,\nu} \circ \chi$ then satisfies

$$g_{n,\nu}^*(x) \to g_n^*(x) \qquad \text{for} \quad \nu \to \infty \quad \text{uniformly on} \quad \overline{\Omega}_n^*.$$

The zeroes $g_{n,\nu}^*(y) = 0$, $y \in \overline{\Omega}_n^*$ are evidently represented in the form

$$\chi^{-1}(x_\nu^{(\mu)}) =: y_\nu^{(\mu)} \quad \text{for} \quad \mu = 1, \ldots, p_\nu,$$

and we evaluate

$$J_{g_{n,\nu}^*}(y_\nu^{(\mu)}) = (\det \chi^{-1}) \cdot J_{g_{n,\nu}}(x_\nu^{(\mu)}) \cdot \det \chi = J_{g_{n,\nu}}(x_\nu^{(\mu)}) \quad \text{for} \quad \mu = 1, \ldots, p_\nu.$$

Theorem 4.7 from Section 4 in Chapter 3 reveals the following identity for all $\nu \geq \nu_0$:

$$d(g_{n,\nu}, \Omega_n) = \sum_{\mu=1}^{p_\nu} \operatorname{sgn} J_{g_{n,\nu}}(x_\nu^{(\mu)})$$

$$= \sum_{\mu=1}^{p_\nu} \operatorname{sgn} J_{g_{n,\nu}^*}(y_\nu^{(\mu)})$$

$$= d(g_{n,\nu}^*, \Omega_n^*).$$

Passing to the limit $\nu \to \infty$, we have proved the statement above.   q.e.d.

Via the pull-back onto the space $\mathbb{R}^n$, we immediately obtain the subsequent Propositions 2.3–2.6 from the corresponding results in Chapter 3.

Proposition 2.3. *Let* $g_\lambda : \overline{\Omega} \to \mathcal{B}$ *with* $a \leq \lambda \leq b$ *denote a family of continuous mappings, which satisfy the relation* $g_\lambda(x) \to g_{\lambda_0}(x)$ *for* $\lambda \to \lambda_0$ *uniformly on the set* $\overline{\Omega}$. *Furthermore,* $g_\lambda(x) \neq 0$ *for all* $x \in \partial\Omega$ *and* $\lambda \in [a, b]$ *holds true. Then we conclude*

$$\delta_{\mathcal{B}}(g_\lambda, \Omega) = const \qquad on \quad [a, b].$$

Proposition 2.4. *Let the mapping* $g : \overline{\Omega} \to \mathcal{B}$ *be continuous and* $g(x) \neq 0$ *for all* $x \in \partial\Omega$. *Furthermore,* $\delta_{\mathcal{B}}(g, \Omega) \neq 0$ *is valid. Then we have a point* $z \in \Omega$ *with* $g(z) = 0$.

Proposition 2.5. *Let* Ω_1 *and* Ω_2 *be bounded open disjoint subsets of* $\mathcal{B}$, *and we define their union* $\Omega := \Omega_1 \cup \Omega_2$. *Furthermore, let* $g : \overline{\Omega} \to \mathcal{B}$ *denote a continuous mapping satisfying* $0 \notin g(\partial\Omega_i)$ *for* $i = 1, 2$. *Then we have the following identity*

$$\delta_{\mathcal{B}}(g, \Omega) = \delta_{\mathcal{B}}(g, \Omega_1) + \delta_{\mathcal{B}}(g, \Omega_2).$$

Proposition 2.6. *On the open bounded subset* $\Omega \subset \mathcal{B}$ *we have defined the continuous function* $g : \overline{\Omega} \to \mathcal{B}$. *Furthermore, let* $\Omega_0 \subset \Omega$ *be an open set with the property* $g(x) \neq 0$ *for all* $x \in \overline{\Omega} \setminus \Omega_0$. *Then we have*

$$\delta_{\mathcal{B}}(g, \Omega) = \delta_{\mathcal{B}}(g, \Omega_0).$$

In the Banach space $\mathcal{B}$ we consider an open bounded subset $\Omega \subset \mathcal{B}$. Furthermore, $\mathcal{B}'$ denotes a finite-dimensional subspace of $\mathcal{B}$ with $\Omega_{\mathcal{B}'} := \Omega \cap \mathcal{B}' \neq \emptyset$. The set $\Omega_{\mathcal{B}'}$ is open and bounded in $\mathcal{B}'$, and we have

$$\partial \Omega_{\mathcal{B}'} \subset \partial \Omega \cap \mathcal{B}', \qquad \overline{\Omega}_{\mathcal{B}'} \subset \overline{\Omega} \cap \mathcal{B}'.$$

With the continuous mapping $f : \overline{\Omega} \to \mathcal{B}'$ we associate the mapping

$$\varphi_f(x) := x - f(x), \qquad x \in \overline{\Omega}.$$

For all Banach spaces $\mathcal{B}'' \supset \mathcal{B}'$ we have the inclusion

$$\varphi_f(\overline{\Omega} \cap \mathcal{B}'') \subset \mathcal{B}''.$$

Proposition 2.7. *Let the Banach spaces $\mathcal{B}' \subset \mathcal{B}'' \subset \mathcal{B}$ be given with*

$$0 < \dim \mathcal{B}' \leq \dim \mathcal{B}'' < +\infty.$$

The open bounded set $\Omega \subset \mathcal{B}$ fulfills $\Omega_{\mathcal{B}'} = \Omega \cap \mathcal{B}' \neq \emptyset$. With the continuous mapping $f : \overline{\Omega} \to \mathcal{B}'$ we associate $\varphi_f(x) := x - f(x)$, $x \in \overline{\Omega}$ satisfying

$$\varphi_f(x) \neq 0 \qquad for\ all \quad x \in \partial \Omega.$$

Then we have the equality

$$\delta_{\mathcal{B}'}(\varphi_f, \Omega_{\mathcal{B}'}) = \delta_{\mathcal{B}''}(\varphi_f, \Omega_{\mathcal{B}''}).$$

Proof: On account of $\partial \Omega_{\mathcal{B}'} \subset \partial \Omega$ and $\partial \Omega_{\mathcal{B}''} \subset \partial \Omega$ the degrees of mapping above are well-defined. Without loss of generality we can assume

$$\dim \mathcal{B}'' > \dim \mathcal{B}'.$$

We choose a basis $\{w_1, \ldots, w_n\} \subset \mathcal{B}'$ of $\mathcal{B}'$ and extend the vectors to a basis

$$\{w_1, \ldots, w_n, w_{n+1}, \ldots, w_{n+p}\} \subset \mathcal{B}''$$

of $\mathcal{B}''$; with an integer $p \in \mathbb{N}$. When we represent the mapping $\varphi_f : \mathcal{B}'' \to \mathcal{B}''$ in the coordinates belonging to the basis $\{w_1, \ldots, w_{n+p}\}$, we obtain the mapping $\varphi'' := \varphi_f \mid_{\mathcal{B}''} : \mathcal{B}'' \to \mathcal{B}''$ via

$$(x_1, \ldots, x_n, x_{n+1}, \ldots, x_{n+p}) \mapsto \begin{aligned}&(x_1 - f_1(x_1, \ldots, x_n, x_{n+1}, \ldots, x_{n+p}), \ldots \\ &x_n - f_n(x_1, \ldots, x_n, x_{n+1}, \ldots, x_{n+p}), \\ &x_{n+1}, \ldots, x_{n+p}).\end{aligned}$$

The restricted mapping $\varphi' := \varphi_f \mid_{\mathcal{B}'} : \mathcal{B}' \to \mathcal{B}'$ appears with respect to the coordinates $x_1, \ldots, x_n$ as follows:

$$(x_1, \ldots, x_n) \mapsto (x_1 - f_1(x_1, \ldots, x_n, 0, \ldots, 0), \ldots, x_n - f_n(x_1, \ldots, x_n, 0, \ldots, 0)).$$

Now the function φ'' has a zero $x'' = (\overset{\circ}{x}_1, \ldots, \overset{\circ}{x}_n, 0, \ldots, 0)$ if and only if φ' has a zero $x' = (\overset{\circ}{x}_1, \ldots, \overset{\circ}{x}_n)$. Consequently, we see $J_{\varphi''}(x'') = J_{\varphi'}(x')$ and

$$\operatorname{sgn} J_{\varphi''}(x'') = \operatorname{sgn} J_{\varphi'}(x').$$

Summing up over all zeroes, we finally obtain

$$\delta_{\mathcal{B}''}(\varphi_f, \Omega_{\mathcal{B}''}) = \delta_{\mathcal{B}'}(\varphi_f, \Omega_{\mathcal{B}'}).$$

q.e.d.

Definition 2.8. *Let Ω be a bounded open set in $\mathcal{B}$, and $\mathcal{B}'$ denotes a linear subspace of $\mathcal{B}$ with $1 \le \dim \mathcal{B}' < +\infty$ and $\Omega_{\mathcal{B}'} := \Omega \cap \mathcal{B}' \ne \emptyset$. Furthermore, let the function $f : \overline{\Omega} \to \mathcal{B}'$ be continuous, and we assume*

$$\varphi_f(x) = x - f(x) \ne 0 \qquad \text{for all} \quad x \in \partial\Omega.$$

Then we define

$$\delta_{\mathcal{B}}(\varphi_f, \Omega) := \delta_{\mathcal{B}'}(\varphi_f, \Omega_{\mathcal{B}'}).$$

We establish the independence from the choice of the finite-dimensional subspace $\mathcal{B}'$ as follows: Let $\mathcal{B}'' \subset \mathcal{B}$ with $1 \le \dim \mathcal{B}'' < +\infty$ and $\Omega \cap \mathcal{B}'' \ne \emptyset$ be an additional subspace of $\mathcal{B}$. We set $\mathcal{B}^* := \mathcal{B}' \oplus \mathcal{B}''$, such that $\mathcal{B}' \subset \mathcal{B}^*$ and $\mathcal{B}'' \subset \mathcal{B}^*$ holds true. Then Proposition 2.7 yields

$$\delta_{\mathcal{B}'}(\varphi_f, \Omega_{\mathcal{B}'}) = \delta_{\mathcal{B}^*}(\varphi_f, \Omega_{\mathcal{B}^*}) = \delta_{\mathcal{B}''}(\varphi_f, \Omega_{\mathcal{B}''}).$$

We present the transition to completely continuous mappings $f : \mathcal{B} \to \mathcal{B}$ in the sequel.

Proposition 2.9. *Let the set $A \subset \mathcal{B}$ be closed and the function $f : A \to \mathcal{B}$ be completely continuous satisfying*

$$\varphi_f(x) = x - f(x) \ne 0 \qquad \text{for all} \quad x \in A.$$

Then we have a number $\varepsilon > 0$, such that $\|\varphi_f(x)\| \ge \varepsilon$ for all $x \in A$ holds true.

Proof: If the statement were violated, we would have a sequence

$$\{x_n\}_{n=1,2,\ldots} \subset A$$

satisfying

$$\varphi_f(x_n) = x_n - f(x_n) \to 0 \qquad \text{for} \quad n \to \infty.$$

Since the set $f(A)$ is precompact, there exists a subsequence $\{x_{n_k}\}_{k=1,2,\ldots}$ with $f(x_{n_k}) \to x^* \in \mathcal{B}$ for $k \to \infty$. This implies

$$\|x_{n_k} - x^*\| \le \|x_{n_k} - f(x_{n_k})\| + \|f(x_{n_k}) - x^*\| \to 0$$

and $x_{n_k} \to x^* \in A$ for $k \to \infty$, because A is closed. Finally, we obtain

$$\varphi_f(x^*) = x^* - f(x^*) = \lim_{k \to \infty} (x_{n_k} - f(x_{n_k})) = 0$$

contradicting the assumption $\varphi_f \ne 0$ in A.   q.e.d.

Proposition 1.14 from Section 1 implies the following

Proposition 2.10. *Let $\Omega \subset \mathcal{B}$ be a bounded open set and $f : \overline{\Omega} \to \mathcal{B}$ a completely continuous function. To each number $\varepsilon > 0$ we then have a linear subspace $\mathcal{B}_\varepsilon$ with $0 < \dim \mathcal{B}_\varepsilon < +\infty$ and $\Omega \cap \mathcal{B}_\varepsilon \ne \emptyset$ as well as a continuous mapping $f_\varepsilon : \overline{\Omega} \to \mathcal{B}_\varepsilon$ with the property*

$$\|f_\varepsilon(x) - f(x)\| \le \varepsilon \qquad \text{for all} \quad x \in \overline{\Omega}.$$

Proof: With the functions $F_j(x)$, $x \in \overline{\Omega}$ for $j = 1, \ldots, N$, defined in Proposition 1.14 of Section 1, and the elements $w_1, \ldots, w_N \in \mathcal{B}$, we choose the following function

$$f_\varepsilon(x) := \sum_{j=1}^N F_j(x) w_j.$$

q.e.d.

Definition 2.11. *Let the set $\Omega \subset \mathcal{B}$ be bounded and open. Moreover, the function $f : \overline{\Omega} \to \mathcal{B}$ may be completely continuous and its associate function $\varphi_f(x) = x - f(x)$ satisfies $0 \notin \varphi_f(\partial\Omega)$. Then the function $g : \Omega \to \mathcal{B}_g \subset \mathcal{B}$ is called an* admissible approximation *of f, if the following conditions are fulfilled:*

(a) The function g is continuous;
(b) The linear subspace $\mathcal{B}_g$ satisfies $1 \le \dim \mathcal{B}_g < +\infty$ and $\Omega \cap \mathcal{B}_g \ne \emptyset$;
(c) We have the following inequality

$$\sup_{x \in \overline{\Omega}} \|g(x) - f(x)\| < \inf_{x \in \partial\Omega} \|\varphi_f(x)\|.$$

Proposition 2.12. *The mapping $f : \overline{\Omega} \to \mathcal{B}$ fulfills the assumptions of Definition 2.11, and $g : \overline{\Omega} \to \mathcal{B}_g$ as well as $h : \overline{\Omega} \to \mathcal{B}_h$ are two admissible approximations of f. Then we have*

$$\delta_{\mathcal{B}}(\varphi_g, \Omega) = \delta_{\mathcal{B}}(\varphi_h, \Omega).$$

Proof: We set $\mathcal{B}^* := \mathcal{B}_g \oplus \mathcal{B}_h$. This implies

$$\delta_{\mathcal{B}}(\varphi_g, \Omega) := \delta_{\mathcal{B}_g}(\varphi_g, \Omega_{\mathcal{B}_g}) = \delta_{\mathcal{B}^*}(\varphi_g, \Omega_{\mathcal{B}^*}),$$

and furthermore

$$\delta_{\mathcal{B}}(\varphi_h, \Omega) = \delta_{\mathcal{B}^*}(\varphi_h, \Omega_{\mathcal{B}^*}).$$

We now consider the family of mappings

$$\chi_\lambda(x) = x - \big(\lambda g(x) + (1 - \lambda)h(x)\big), \qquad x \in \overline{\Omega}, \quad \lambda \in [0,1].$$

Setting $\eta := \inf\limits_{x \in \partial\Omega} \|\varphi_f(x)\| > 0$ we can estimate as follows:

$$\|\chi_\lambda(x) - \varphi_f(x)\| = \|\lambda(g(x) - f(x)) + (1 - \lambda)(h(x) - f(x))\|$$
$$\leq \lambda\|g(x) - f(x)\| + (1 - \lambda)\|h(x) - f(x)\|$$
$$< \eta \qquad \text{for all} \quad x \in \partial\Omega.$$

Consequently, the estimate $\|\chi_\lambda(x)\| > 0$ for all $x \in \partial\Omega$ and all $\lambda \in [0,1]$ holds true, and Proposition 2.3 yields $\delta_{\mathcal{B}^*}(\chi_\lambda, \Omega_{\mathcal{B}^*}) = const$ on $[0,1]$. We then obtain

$$\delta_{\mathcal{B}}(\varphi_g, \Omega) = \delta_{\mathcal{B}^*}(\chi_1, \Omega_{\mathcal{B}^*}) = \delta_{\mathcal{B}^*}(\chi_0, \Omega_{\mathcal{B}^*}) = \delta_{\mathcal{B}}(\varphi_h, \Omega). \qquad \text{q.e.d.}$$

Definition 2.13. *The set $\Omega \subset \mathcal{B}$ is bounded and open, and we assume the function $f : \overline{\Omega} \to \mathcal{B}$ to be completely continuous such that $\varphi_f(x) = x - f(x) \neq 0$ for all $x \in \partial\Omega$ holds true. Furthermore, let $g : \overline{\Omega} \to \mathcal{B}_g$ be an admissible approximation of f. Then we call*

$$\delta_{\mathcal{B}}(\varphi_f, \Omega) := \delta_{\mathcal{B}}(\varphi_g, \Omega)$$

the Leray-Schauder degree of mapping *for (φ_f, Ω) with respect to $x = 0$.*

3 Fundamental Properties for the Degree of Mapping

At first, we collect our previous results: Let $\Omega \subset \mathcal{B}$ denote a bounded open set and $f : \overline{\Omega} \to \mathcal{B}$ a completely continuous function, such that

$$\varphi_f(x) = x - f(x) \neq 0 \qquad \text{for all} \quad x \in \partial\Omega.$$

Then we have defined our degree of mapping for φ_f by the following chain of equations:

$$\delta_{\mathcal{B}}(\varphi_f, \Omega) = \delta_{\mathcal{B}}(\varphi_g, \Omega) = \delta_{\mathcal{B}_g}(\varphi_g, \Omega \cap \mathcal{B}_g) = d(\varphi_n, \Omega_n).$$

Here g denotes an admissible approximation, $n = \dim \mathcal{B}_g$, Ω_n is the image of $\Omega \cap \mathcal{B}_g$ with respect to an arbitrary coordinate mapping ψ^{-1}, and $\varphi_n = \psi^{-1} \circ \varphi_g \circ \psi|_{\overline{\Omega}_n}$.

Theorem 3.1. (Homotopy)
Let $\Omega \subset \mathcal{B}$ be a bounded open set, and let

$$f_\lambda : \overline{\Omega} \to \mathcal{B}, \lambda \in [a, b]$$

denote a family of mappings with the following properties:

(a) For all $\lambda \in [a, b]$ the mapping $f_\lambda : \overline{\Omega} \to \mathcal{B}$ is completely continuous;
(b) To each number $\varepsilon > 0$ we have a quantity $\delta = \delta(\varepsilon) > 0$, such that

$$\|f_{\lambda_1}(x) - f_{\lambda_2}(x)\| \le \varepsilon \qquad for \ all \quad \lambda_1, \lambda_2 \in [a, b] \quad with \quad |\lambda_1 - \lambda_2| \le \delta$$

holds true for all $x \in \overline{\Omega}$;
(c) For all $x \in \partial\Omega$ and all $\lambda \in [a, b]$ we have $\varphi_{f_\lambda}(x) = x - f_\lambda(x) \ne 0$.

Then we have the identity $\delta_{\mathcal{B}}(\varphi_{f_\lambda}, \Omega) = const, \quad \lambda \in [a, b]$.

Proof: Let $\lambda_0 \in [a, b]$ be chosen arbitrarily. Then we have a number $\varepsilon > 0$ satisfying $\|\varphi_{f_{\lambda_0}}(x)\| \ge \varepsilon$ for all $x \in \partial\Omega$. We construct an admissible approximation $g : \overline{\Omega} \to \mathcal{B}_g \subset \mathcal{B}$ of f_{λ_0}, such that

$$\|g(x) - f_{\lambda_0}(x)\| \le \frac{\varepsilon}{4} \qquad for \ all \quad x \in \overline{\Omega}.$$

Therefore, we have a number $\delta = \delta(\varepsilon)$ with the following property: All parameters $\lambda \in [a, b]$ with $|\lambda - \lambda_0| \le \delta$ fulfill

$$\|g(x) - f_\lambda(x)\| \le \|g(x) - f_{\lambda_0}(x)\| + \|f_{\lambda_0}(x) - f_\lambda(x)\| \le \frac{\varepsilon}{2}, \qquad x \in \overline{\Omega}.$$

On the other hand, we have

$$\|\varphi_{f_\lambda}(x)\| \ge \|\varphi_{f_{\lambda_0}}(x)\| - \|\varphi_{f_\lambda}(x) - \varphi_{f_{\lambda_0}}(x)\| \ge \frac{3\varepsilon}{4}, \qquad x \in \partial\Omega$$

for $|\lambda - \lambda_0| \le \delta$. Consequently, the function g is an admissible approximation for all parameters $\lambda \in [a, b]$ with $|\lambda - \lambda_0| \le \delta$, and we see

$$\delta_{\mathcal{B}}(\varphi_{f_\lambda}, \Omega) = \delta_{\mathcal{B}}(g, \Omega) \quad for \ all \quad \lambda : |\lambda - \lambda_0| \le \delta.$$

A continuation argument finally yields $\delta_{\mathcal{B}}(\varphi_{f_\lambda}, \Omega) = const$ on $[a, b]$.

q.e.d.

Theorem 3.2. (Existence result)
Let the set $\Omega \subset \mathcal{B}$ be bounded and open. The mapping $f : \overline{\Omega} \to \mathcal{B}$ may be completely continuous and satisfies

$$\varphi_f(x) = x - f(x) \ne 0 \qquad for \ all \quad x \in \partial\Omega.$$

Finally, let the assumption $\delta_{\mathcal{B}}(\varphi_f, \Omega) \ne 0$ hold true. Then the equation $\varphi_f(x) = 0$ possesses a solution $x \in \Omega$, which means the mapping $x \mapsto f(x)$ has a fixed point in the set Ω.

Proof: We consider a sequence of admissible approximations $g_n : \overline{\Omega} \to \mathcal{B}_{g_n}$ for f satisfying

$$\sup_{x \in \overline{\Omega}} \|g_n(x) - f(x)\| \le \frac{1}{n}.$$

We then obtain

$$0 \ne \delta_{\mathcal{B}}(\varphi_f, \Omega) = \delta_{\mathcal{B}_{g_n}}(\varphi_{g_n}, \Omega \cap \mathcal{B}_{g_n}) \quad \text{for all} \quad n \ge n_0.$$

Because of Proposition 2.4 from Section 2 we have a sequence $x_n \in \Omega \cap \mathcal{B}_{g_n}$ for $n = n_0, n_0 + 1, \ldots$ with

$$0 = \varphi_{g_n}(x_n) = x_n - g_n(x_n).$$

This implies

$$\|x_n - f(x_n)\| = \|x_n - g_n(x_n)\| + \|g_n(x_n) - f(x_n)\| \le \frac{1}{n} \quad \text{for all} \quad n \ge n_0,$$

and therefore

$$\inf_{x \in \overline{\Omega}} \|\varphi_f(x)\| = \inf_{x \in \overline{\Omega}} \|x - f(x)\| = 0.$$

Due to Proposition 2.9 in Section 2, there exists a point $x_0 \in \Omega$ satisfying

$$\varphi_f(x_0) = x_0 - f(x_0) = 0.$$

q.e.d.

Definition 3.3. *Let $\Omega \subset \mathcal{B}$ denote a bounded open set and $f : \overline{\Omega} \to \mathcal{B}$ a completely continuous mapping with the associate mapping*

$$\varphi_f(x) = x - f(x), \quad x \in \overline{\Omega}.$$

Furthermore, let the domain $G \subset \mathcal{B} \setminus \varphi_f(\partial\Omega)$ be given. Then we define

$$\delta_{\mathcal{B}}(\varphi_f, \Omega, z) = \delta_{\mathcal{B}}(\varphi_f, \Omega, G) := \delta_{\mathcal{B}}(\varphi_{f-z}, \Omega)$$

for arbitrary points $z \in G$.

When we consider the family of mappings $f_t(x) = f(x) - z(t)$ with the continuous curve $z(t) : [0, 1] \to G$, our Theorem 3.1 guarantees the independence from the choice of the point $z \in G$. Now we could derive a product theorem as in the space $\mathbb{R}^n$, which we do not elaborate here. However, we shall generalize the index-sum formula (compare Theorem 4.4 from Chapter 3, Section 4) to completely continuous mappings between Banach spaces.

Proposition 3.4. *Let $\Omega \subset \mathcal{B}$ be a bounded open set, and the mapping $f : \overline{\Omega} \to \mathcal{B}$ is completely continuous. Furthermore, let $\Omega_0 \subset \Omega$ denote an open subset satisfying $\varphi_f(x) \ne 0$ for all $x \in \overline{\Omega} \setminus \Omega_0$. Then we have*

$$\delta_{\mathcal{B}}(\varphi_f, \Omega) = \delta_{\mathcal{B}}(\varphi_f, \Omega_0).$$

Proof: We observe $\partial \Omega \subset \overline{\Omega} \setminus \Omega_0$ and $\partial \Omega_0 \subset \overline{\Omega} \setminus \Omega_0$, which implies

$$\varphi_f(x) \neq 0 \qquad \text{for all} \quad x \in \partial \Omega \cup \partial \Omega_0.$$

Proposition 2.9 from Section 2 yields

$$\|\varphi_f(x)\| \geq \varepsilon > 0 \qquad \text{for all} \quad x \in \overline{\Omega} \setminus \Omega_0,$$

because $\overline{\Omega} \setminus \Omega_0$ is closed. Take with $g : \overline{\Omega} \to \mathcal{B}_g \subset \mathcal{B}$ an admissible approximation satisfying $\Omega_0 \cap \mathcal{B}_g \neq \emptyset$ and $\|g(x) - f(x)\| \leq \frac{\varepsilon}{2}$ for all $x \in \overline{\Omega}$. This implies

$$\|\varphi_g(x)\| \geq \|\varphi_f(x)\| - \|\varphi_f(x) - \varphi_g(x)\| \geq \frac{\varepsilon}{2} \qquad \text{for all} \quad x \in \overline{\Omega} \setminus \Omega_0.$$

Together with Proposition 2.6 from Section 2 we obtain

$$\delta_{\mathcal{B}}(\varphi_f, \Omega) = \delta_{\mathcal{B}_g}(\varphi_g, \Omega \cap \mathcal{B}_g) = \delta_{\mathcal{B}_g}(\varphi_g, \Omega_0 \cap \mathcal{B}_g) = \delta_{\mathcal{B}}(\varphi_f, \Omega_0).$$

q.e.d.

Proposition 3.5. *Let the sets $\Omega_1, \Omega_2 \subset \mathcal{B}$ be bounded, open and disjoint. Furthermore, we define $\Omega := \Omega_1 \dot\cup \Omega_2$. Then we have*

$$\delta_{\mathcal{B}}(\varphi_f, \Omega) = \delta_{\mathcal{B}}(\varphi_f, \Omega_1) + \delta_{\mathcal{B}}(\varphi_f, \Omega_2).$$

Proof: Take with $g : \overline{\Omega} \to \mathcal{B}_g \subset \mathcal{B}$ an admissible approximation of f satisfying $\Omega_i \cap \mathcal{B}_g \neq \emptyset$ for $i = 1, 2$. Then $g|_{\overline{\Omega}_i}$ are admissible approximations of $f|_{\overline{\Omega}_i}$, and Proposition 2.5 from Section 2 yields

$$\begin{aligned}
\delta_{\mathcal{B}}(\varphi_f, \Omega) &= \delta_{\mathcal{B}_g}(\varphi_g, \Omega \cap \mathcal{B}_g) \\
&= \delta_{\mathcal{B}_g}(\varphi_g, \Omega_1 \cap \mathcal{B}_g) + \delta_{\mathcal{B}_g}(\varphi_g, \Omega_2 \cap \mathcal{B}_g) \\
&= \delta_{\mathcal{B}}(\varphi_f, \Omega_1) + \delta_{\mathcal{B}}(\varphi_f, \Omega_2).
\end{aligned}$$

q.e.d.

Definition 3.6. *Let $U = U(z) \subset \mathcal{B}$ denote an open neighborhood of the point z and $f : \overline{U(z)} \to \mathcal{B}$ a completely continuous mapping. The associate mapping may fulfill*

$$\varphi_f(x) \neq 0 \quad \text{in} \quad \overline{U(z)} \setminus \{z\} \qquad \text{and} \qquad \varphi_f(z) = 0.$$

Then we define the index

$$i(\varphi_f, z) := \delta_{\mathcal{B}}(\varphi_f, K)$$

with the ball of sufficiently small radius $\epsilon > 0$ in

$$K := \{x \in \mathcal{B} : \|x - z\| < \varepsilon\} \quad \text{satisfying} \quad \overline{K} \subset U(z).$$

Theorem 3.7. (Index-sum formula)
Let the mapping $f : \overline{\Omega} \to \mathcal{B}$ be completely continuous. Furthermore, the equation $\varphi_f(x) = 0$ admits exactly p different solutions $z_1, \ldots, z_p \in \Omega$. Then we have the identity

$$\delta_{\mathcal{B}}(\varphi_f, \Omega) = \sum_{\nu=1}^{p} i(\varphi_f, z_\nu).$$

Proof: Taking $\varepsilon > 0$ sufficiently small, we consider the mutually disjoint balls

$$K_\nu := \{x \in \Omega : \|x - z_\nu\| < \varepsilon\}, \qquad \nu = 1, \ldots, p.$$

We apply Proposition 3.4 and Proposition 3.5 to $\Omega_0 := \bigcup_{\nu=1}^{p} K_\nu \subset \Omega$ as follows:

$$\delta_{\mathcal{B}}(\varphi_f, \Omega) = \delta_{\mathcal{B}}(\varphi_f, \Omega_0) = \sum_{\nu=1}^{p} \delta_{\mathcal{B}}(\varphi_f, K_\nu) = \sum_{\nu=1}^{p} i(\varphi_f, z_\nu).$$

q.e.d.

We collect our arguments to the following

Theorem 3.8. (J. Leray and J. Schauder)
Let $\Omega \subset \mathcal{B}$ be a bounded open set in the Banach space $\mathcal{B}$, and

$$f_\lambda : \overline{\Omega} \to \mathcal{B}, \quad a \leq \lambda \leq b$$

denotes a family of mappings with the following properties:

(a) For all $\lambda \in [a, b]$ the functions $f_\lambda : \overline{\Omega} \to \mathcal{B}$ are completely continuous;
(b) To each number $\varepsilon > 0$ we have a quantity $\delta = \delta(\varepsilon) > 0$, such that

$$\|f_{\lambda_1}(x) - f_{\lambda_2}(x)\| \leq \varepsilon \qquad \text{for all} \quad \lambda_1, \lambda_2 \in [a, b] \quad \text{with} \quad |\lambda_1 - \lambda_2| \leq \delta$$

holds true for all $x \in \overline{\Omega}$;
(c) For all $x \in \partial\Omega$ and all $\lambda \in [a, b]$ we have $\varphi_{f_\lambda}(x) = x - f_\lambda(x) \neq 0$;
(d) With a special $\lambda_0 \in [a, b]$ the equation

$$\varphi_{f_{\lambda_0}}(x) = x - f_{\lambda_0}(x) = 0, \quad x \in \Omega$$

has finitely many solutions $z_1, \ldots, z_p$, $p \in \mathbb{N}$ satisfying

$$\sum_{\nu=1}^{p} i(\varphi_{f_{\lambda_0}}, z_\nu) \neq 0.$$

Then the equation $\varphi_{f_\lambda}(x) = 0, \quad x \in \Omega$ *possesses at least one solution for each $\lambda \in [a, b]$.*

Remark: In Chapter 12 we shall prove the existence of solutions for nonlinear elliptic systems with the aid of Theorem 3.8.

4 Linear Operators in Banach Spaces

Let us consider two Banach spaces $\{\mathcal{B}_j, \| \ \|_j\}$ for $j = 1, 2$. Then we can define open sets in $\mathcal{B}_j$, $j = 1, 2$ with the aid of the respective norm $\| \ \|_j$. In the sequel we study linear continuous operators

$$T : \mathcal{B}_1 \to \mathcal{B}_2.$$

We call the operator T *linear* if

$$T(\alpha x + \beta y) = \alpha T(x) + \beta T(y) \qquad \text{for all} \quad x, y \in \mathcal{B}_1 \quad \text{and all} \quad \alpha, \beta \in \mathbb{R} \quad (1)$$

is valid. The operator T is *continuous* if and only if T is bounded, or equivalently

$$\|T\| := \sup_{\substack{x \in \mathcal{B}_1 \\ x \neq 0}} \frac{\|Tx\|_2}{\|x\|_1} < +\infty. \tag{2}$$

At first, we note the following

Theorem 4.1. (Open mapping principle)
The linear continuous operator $T : \mathcal{B}_1 \to \mathcal{B}_2$ is assumed surjective. Then T is an open mapping, which means the image of each open set is open.

Proof: This is achieved by methods from the set-theoretical topology. We refer the reader to [HS], pp. 39-41 (Satz 9.1) and pp. 21-22 (Lemma 4.1 and Satz 4.3).

Theorem 4.1 immediately implies

Theorem 4.2. (Inverse operator)
Let the linear continuous operator $T : \mathcal{B}_1 \to \mathcal{B}_2$ be bijective. Then the inverse operator $T^{-1} : \mathcal{B}_2 \to \mathcal{B}_1$ is continuous.

We endow the set $\mathcal{B} = \mathcal{B}_1 \times \mathcal{B}_2$ with the norm

$$\|(x, y)\| := \sqrt{\|x\|_1^2 + \|y\|_2^2}, \qquad (x, y) \in \mathcal{B} = \mathcal{B}_1 \times \mathcal{B}_2$$

and obtain a Banach space. Thus open sets are defined in $\mathcal{B}$. We now define the graph of $T : \mathcal{B}_1 \to \mathcal{B}_2$ by

$$\text{graph}\,(T) := \Big\{ (x, Tx) \in \mathcal{B}_1 \times \mathcal{B}_2 \ : \ x \in \mathcal{B}_1 \Big\}. \tag{3}$$

Theorem 4.3. (Closed graph)
For a linear operator $T : \mathcal{B}_1 \to \mathcal{B}_2$ we have the equivalence: The operator T is continuous if and only if $\text{graph}\,(T)$ in $\mathcal{B}_1 \times \mathcal{B}_2$ is closed.

Proof:

'⇒' We consider a sequence $\{x_n\}_{n=1,2,\ldots} \subset \mathcal{B}_1$ with $x_n \to x \in \mathcal{B}_1$ for $n \to \infty$. Since the operator T is continuous, we infer

$$\lim_{n \to \infty} T x_n = T(\lim_{n \to \infty} x_n) = Tx$$

and consequently

$$\lim_{n \to \infty} (x_n, T x_n) = (x, Tx) \in \text{graph}\,(T).$$

Therefore, the graph (T) is closed.

'⇐' Let the graph $(T) \subset \mathcal{B}_1 \times \mathcal{B}_2$ now be closed. Then this graph represents a Banach space. The projection

$$\pi : \text{graph}\,(T) \to \mathcal{B}_1, \quad (x, Tx) \mapsto x$$

is bijective, linear and continuous. Theorem 4.2 implies that the mapping $\pi^{-1} : \mathcal{B}_1 \to \text{graph}\,(T)$ is continuous as well. The projection

$$\varrho : \mathcal{B}_1 \times \mathcal{B}_2 \to \mathcal{B}_2, \quad (x, y) \mapsto y$$

is evidently continuous, and we finally obtain the continuity of

$$T = \varrho \circ \pi^{-1} : \mathcal{B}_1 \to \mathcal{B}_2.$$

$$\text{q.e.d.}$$

We now choose $\mathcal{B}_1 = \mathcal{B}_2 = \mathcal{B}$ and then consider linear continuous operators $T : \mathcal{B} \to \mathcal{B}$. These are injective if and only if $\ker T := T^{-1}(0)$ consists only of $\{0\}$. With the aid of the Leray-Schauder degree of mapping we now shall prove a criterion for the surjectivity of T. In the sequel we denote the open balls in $\mathcal{B}$ by

$$\mathcal{B}_r := \left\{ x \in \mathcal{B} : \|x\| < r \right\}, \qquad 0 < r < +\infty.$$

Their boundaries are described by $\partial \mathcal{B}_r = \{ x \in \mathcal{B} : \|x\| = r \}$.

Definition 4.4. *The linear operator* $K : \mathcal{B} \to \mathcal{B}$ *is called* compact *or alternatively* completely continuous, *if the condition that* $K(\partial \mathcal{B}_r)$ *is precompact holds true for a number* $r \in (0, +\infty)$.

Remarks:

1. This definition is independent of the number $r \in (0, +\infty)$.
2. A compact operator is bounded and consequently continuous. Therefore, this definition is equivalent to the Definition 1.11 from Section 1.

Definition 4.5. *With the completely continuous operator* $K : \mathcal{B} \to \mathcal{B}$ *we associate the* Fredholm operator

$$Tx := x - Kx = (Id_{\mathcal{B}} - K)(x), \qquad x \in \mathcal{B}. \tag{4}$$

Fundamentally important for the solution of linear operator equations in Banach spaces is the subsequent

Theorem 4.6. (F. Riesz)
Let $K : \mathcal{B} \to \mathcal{B}$ be a completely continuous operator on the Banach space $\mathcal{B}$ with the associate Fredholm operator

$$Tx := (Id_{\mathcal{B}} - K)(x) = x - Kx, \quad x \in \mathcal{B}.$$

Furthermore, the implication

$$Tx = 0, \quad x \in \mathcal{B} \quad \Rightarrow \quad x = 0$$

holds true, which means the kernel of T consists only of the zero element. Then the mapping $T : \mathcal{B} \to \mathcal{B}$ is bijective; the inverse operator $T^{-1} : \mathcal{B} \to \mathcal{B}$ for T exists and is bounded on $\mathcal{B}$. Especially, the operator equation

$$Tx = y, \quad x \in \mathcal{B},$$

possesses exactly one solution for all right-hand sides $y \in \mathcal{B}$.

Proof: Choosing $r \in (0, +\infty)$ arbitrarily, we study the operator

$$Tx = x - Kx, \quad x \in \mathcal{B}_r.$$

Because of the assumptions above, we have

$$Tx \neq 0 \quad \text{for all} \quad x \in \partial \mathcal{B}_r.$$

Proposition 2.9 from Section 2 yields a number $\varepsilon > 0$, such that

$$\|Tx\| \geq \varepsilon r \quad \text{for all} \quad x \in \partial \mathcal{B}_r \tag{5}$$

is correct. We prescribe $y \in \mathcal{B}$ and consider the family of operators

$$T_\lambda x := Tx - \lambda y, \quad x \in \mathcal{B}_r, \quad 0 \leq \lambda \leq 1. \tag{6}$$

Choosing r sufficiently large, we obtain

$$\|T_\lambda x\| \geq \|Tx\| - \|\lambda y\| \geq \varepsilon r - \|y\| > 0$$

for all $x \in \partial \mathcal{B}_r$ and all $\lambda \in [0,1]$. At the initial value $\lambda = 0$ the equation $T_\lambda x = 0$, $x \in \mathcal{B}_r$ admits exactly one solution, namely the element $x = 0$ with the index $i(T, 0) \neq 0$. Due to the Leray-Schauder theorem, our equation $T_\lambda x = 0$, $x \in \mathcal{B}_r$ possesses at least one solution for each $\lambda \in [0,1]$. For the value $\lambda = 1$ especially, we find a solution $x \in \mathcal{B}_r$ satisfying

$$Tx = y.$$

Since the point y has been chosen arbitrarily, the mapping $T : \mathcal{B} \to \mathcal{B}$ is surjective. The injectivity of T immediately follows from ker $T = \{0\}$. Finally, the inequality

$$\|Tx\| \geq \varepsilon \|x\| \qquad \text{for all} \quad x \in \mathcal{B}$$

implies the boundedness of the operator T^{-1}, namely

$$\|T^{-1}y\| \leq \frac{1}{\varepsilon}\|y\| \qquad \text{for all} \quad y \in \mathcal{B}.$$

q.e.d.

We call the linear operator $\quad F : \mathcal{B} \to \mathbb{R} \quad$ a *linear functional* on the Banach space $\mathcal{B}$. Concluding this chapter we prove the well-known

Theorem 4.7. (Extension theorem of Hahn-Banach)
Let $\mathcal{L}$ be a subspace of the Banach space $\mathcal{B}$, and $f : \mathcal{L} \to \mathbb{R}$ denotes a linear continuous mapping with

$$\|f\| := \sup_{\substack{x \in \mathcal{L} \\ x \neq 0}} \frac{|f(x)|}{\|x\|}.$$

Then we have a continuous linear functional $F : \mathcal{B} \to \mathbb{R}$ satisfying

$$F(x) = f(x) \quad \text{for all} \quad x \in \mathcal{L}$$

and $\|F\| = \|f\|$.

Definition 4.8. *Let $\mathcal{L} \subset \mathcal{B}$ denote a subspace of the Banach space $\mathcal{B}$. We call the function $p = p(x) : \mathcal{L} \to \mathbb{R}$ superlinear (on $\mathcal{L}$) if*

$$p(\lambda x) = \lambda p(x) \qquad \text{for all} \quad x \in \mathcal{L} \quad \text{and all} \quad \lambda \in [0, +\infty) \tag{7}$$

and

$$p(x + y) \leq p(x) + p(y) \qquad \text{for all} \quad x, y \in \mathcal{L} \tag{8}$$

holds true.

Proposition 4.9. *With the assumptions of Theorem 4.7 the function*

$$p(x) := \inf_{y \in \mathcal{L}} \left\{ \|f\| \, \|x - y\| + f(y) \right\}, \qquad x \in \mathcal{B}, \tag{9}$$

is superlinear in $\mathcal{B}$, and we have

$$p(x) \leq \|f\| \, \|x\|, \quad x \in \mathcal{B}; \qquad p(x) \leq f(x), \quad x \in \mathcal{L}. \tag{10}$$

Proof: At first, we note that

$$p(x) := \inf_{y \in \mathcal{L}} \left\{ \|f\| \, \|x - y\| + f(y) \right\}$$

$$\geq \inf_{y \in \mathcal{L}} \left\{ f(y) + \|f\| \, \|y\| - \|f\| \, \|x\| \right\}$$

$$\geq -\|f\| \, \|x\| > -\infty, \qquad x \in \mathcal{B}.$$

We now deduce (7): For $\lambda = 0$, we have

$$p(0\,x) = \inf_{y \in \mathcal{L}} \left\{ \|f\| \, \|y\| + f(y) \right\} = 0 = 0\,p(x) \qquad \text{for all} \quad x \in \mathcal{B}.$$

For $\lambda \in (0, +\infty)$, we calculate

$$p(\lambda x) = \inf_{y \in \mathcal{L}} \left\{ \|f\| \, \|\lambda x - y\| + f(y) \right\}$$

$$= \inf_{y \in \mathcal{L}} \left\{ \|f\| \, \|\lambda x - \lambda y\| + f(\lambda y) \right\}$$

$$= \lambda \inf_{y \in \mathcal{L}} \left\{ \|f\| \, \|x - y\| + f(y) \right\}$$

$$= \lambda p(x) \qquad \text{for all} \quad x \in \mathcal{B}.$$

We then deduce (8): Let the elements $x, z \in \mathcal{B}$ be chosen arbitrarily. The number $\varepsilon > 0$ given, there exist elements $y_1, y_2 \in \mathcal{L}$ satisfying

$$p(x) \geq \|f\| \, \|x - y_1\| + f(y_1) - \varepsilon,$$

$$p(z) \geq \|f\| \, \|z - y_2\| + f(y_2) - \varepsilon.$$

Therefore, we can estimate as follows

$$p(x + z) = \inf_{y \in \mathcal{L}} \left\{ \|f\| \, \|x + z - y\| + f(y) \right\}$$

$$\leq \|f\| \, \|x + z - (y_1 + y_2)\| + f(y_1 + y_2)$$

$$\leq \|f\| \left\{ \|x - y_1\| + \|z - y_2\| \right\} + f(y_1 + y_2)$$

$$= \left\{ \|f\| \, \|x - y_1\| + f(y_1) \right\} + \left\{ \|f\| \, \|z - y_2\| + f(y_2) \right\}$$

$$\leq p(x) + p(z) + 2\varepsilon.$$

The transition to the limit $\varepsilon \to 0$ yields the superlinearity of $p(x)$.
We additionally show (10): Choosing $y = 0$ in the definition of $p(x)$, we obtain

$$p(x) = \inf_{y \in \mathcal{L}} \left\{ \|f\| \, \|x - y\| + f(y) \right\} \leq \|f\| \, \|x\| + f(0) = \|f\| \, \|x\|, \qquad x \in \mathcal{B}.$$

Correspondingly, the choice $y = x \in \mathcal{L}$ implies the inequality

$$p(x) = \inf_{y \in \mathcal{L}} \Big\{ \|f\| \, \|x - y\| + f(y) \Big\} \leq f(x), \qquad x \in \mathcal{L}.$$

This completes the proof.   q.e.d.

We now consider the set of functions which are superlinear in $\mathcal{L}$, namely

$$\mathcal{F} := \mathcal{S}(\mathcal{L}) := \Big\{ p : \mathcal{L} \to \mathbb{R} \; : \; p \text{ is superlinear in } \mathcal{L} \Big\}.$$

With respect to the relation

$$p, \tilde{p} \in \mathcal{S}(\mathcal{L}) : \qquad p \leq \tilde{p} \quad \Leftrightarrow \quad p(x) \leq \tilde{p}(x) \text{ for all } x \in \mathcal{L} \qquad (11)$$

this set is *partially ordered* in the following sense:

$$p \leq p; \qquad p \leq \tilde{p}, \; \tilde{p} \leq \hat{p} \; \Rightarrow \; p \leq \hat{p}; \qquad p \leq \tilde{p}, \; \tilde{p} \leq p \; \Rightarrow \; p = \tilde{p}. \qquad (12)$$

A subset $\mathcal{E} \subset \mathcal{F}$ is called *totally ordered*, if each two elements $p, \tilde{p} \in \mathcal{E}$ satisfy at least one of the alternatives $p \leq \tilde{p}$ or $\tilde{p} \leq p$. The element $p_* \in \mathcal{F}$ is called a *lower bound* of $\mathcal{E}$ if

$$p_* \leq p \qquad \text{for all} \quad p \in \mathcal{E} \qquad (13)$$

is correct.

Proposition 4.10. *Each totally ordered subset $\mathcal{E} \subset \mathcal{S}(\mathcal{L})$ possesses a lower bound $p_* = p_*(\mathcal{E}) \in \mathcal{S}(\mathcal{L})$.*

Proof: Let $\mathcal{E} = \{p_i\}_{i \in I} \subset \mathcal{S}(\mathcal{L})$ be a totally ordered subset. We choose

$$p_*(x) := \inf_{i \in I} p_i(x), \qquad x \in \mathcal{L}$$

as a lower bound and show that p_* represents a superlinear function. Here it suffices to prove the inequality (8). Let $x, y \in \mathcal{L}$ be chosen arbitrarily. For each number $\varepsilon > 0$ we then find an index $j \in I$ such that

$$p_*(x) \geq p_j(x) - \varepsilon.$$

Similarly, we find an index $k \in I$ satisfying

$$p_*(y) \geq p_k(y) - \varepsilon.$$

Since we have alternatively the inequalities $p_j \geq p_k$ or $p_k \geq p_j$ in $\mathcal{L}$, both are even valid with the same index - say $j \in I$. Therefore, we obtain

$$p_*(x + y) = \inf_{i \in I} p_i(x + y) \leq p_j(x + y)$$

$$\leq p_j(x) + p_j(y) \leq p_*(x) + p_*(y) + 2\varepsilon.$$

The transition to the limit $\varepsilon \to 0$ yields the statement above.   q.e.d.

Definition 4.11. *In a partially ordered set $\mathcal{F}$ we call $p \in \mathcal{F}$ a minimal element of $\mathcal{F}$ if the implication*

$$\tilde{p} \in \mathcal{F} \quad \text{with} \quad \tilde{p} \leq p \quad \Rightarrow \quad \tilde{p} = p \tag{14}$$

is correct. Therefore, strictly smaller elements for p do not exist!

Proposition 4.12. *The function $p \in \mathcal{S}(\mathcal{L})$ is a minimal element of $\mathcal{S}(\mathcal{L})$ if and only if $p : \mathcal{L} \to \mathbb{R}$ is linear.*

Proof:

'$\Leftarrow$' Let $p(x) : \mathcal{L} \to \mathbb{R}$ be linear. Furthermore, we choose $\tilde{p}(x) \in \mathcal{S}(\mathcal{L})$ with $\tilde{p} \leq p$, which implies $\tilde{p}(x) \leq p(x)$ for all $x \in \mathcal{L}$. We then infer $\tilde{p} = p$ immediately. If there existed a point $y \in \mathcal{L}$ with $\tilde{p}(y) < p(y)$, we could deduce

$$0 = \tilde{p}(y - y) \leq \tilde{p}(y) + \tilde{p}(-y) < p(y) + p(-y) = p(y - y) = 0.$$

'$\Rightarrow$' The point $a \in \mathcal{L}$ being fixed, we consider the function

$$p_a(x) := \inf_{t \geq 0} \left\{ p(x + ta) - tp(a) \right\}, \qquad x \in \mathcal{L}. \tag{15}$$

We easily see $p_a(x) \leq p(x)$, $x \in \mathcal{L}$. Furthermore, we calculate

$$p_a(\lambda x) = \inf_{t \geq 0} \left\{ p(\lambda x + ta) - tp(a) \right\}$$

$$= \inf_{t \geq 0} \left\{ p(\lambda x + \lambda ta) - \lambda tp(a) \right\}$$

$$= \lambda \inf_{t \geq 0} \left\{ p(x + ta) - tp(a) \right\}$$

$$= \lambda p_a(x), \qquad x \in \mathcal{L}$$

for $\lambda > 0$. In the case $\lambda = 0$ this identity is trivially fulfilled.
We now show that $p_a(x)$ also is subject to the inequality (8): Let the points $x, y \in \mathcal{L}$ be chosen. As in the proof of Proposition 4.9 we select values $t_1 \geq 0$ and $t_2 \geq 0$, where the respective infima $p_a(x)$ and $p_a(y)$ can be approximated up to the quantity $\varepsilon > 0$. This implies

$$p_a(x + y) = \inf_{t \geq 0} \left\{ p(x + y + ta) - tp(a) \right\}$$

$$\leq p(x + y + (t_1 + t_2)a) - (t_1 + t_2)p(a)$$

$$\leq \left\{ p(x + t_1 a) - t_1 p(a) \right\} + \left\{ p(y + t_2 a) - t_2 p(a) \right\}$$

$$\leq p_a(x) + p_a(y) + 2\varepsilon,$$

and the passage to the limit $\varepsilon \to 0$ yields (8). Consequently, the function $p_a(x)$, $x \in \mathcal{L}$, is superlinear. Since $p(x)$ is a minimal element in $\mathcal{S}(\mathcal{L})$ we infer

$$p(x) \le p_a(x) = \inf_{t \ge 0} \left\{ p(x + ta) - tp(a) \right\} \le p(x + a) - p(a)$$

or equivalently

$$p(x) + p(a) \le p(x + a) \le p(x) + p(a) \qquad \text{for all} \quad x, a \in \mathcal{L}.$$

Therefore, the function $p : \mathcal{L} \to \mathbb{R}$ is linear.   q.e.d.

From the set theory we need the following

Proposition 4.13. (Lemma of Zorn)
In the partially ordered set $\mathcal{F}$ we assume that each totally ordered subset $\mathcal{E} \subset \mathcal{F}$ possesses a lower bound. Then a minimal element exists in $\mathcal{F}$.

We now arrive at the

Proof of Theorem 4.7: With the assumptions of Theorem 4.7 we consider the super linear function $p(x)$ from Proposition 4.9 and define the partially ordered set

$$\mathcal{F} := \left\{ \tilde{p} \in \mathcal{S}(\mathcal{B}) \,:\, \tilde{p} \le p \right\}.$$

On account of Proposition 4.10 each totally ordered subset $\mathcal{E} \subset \mathcal{F}$ possesses a lower bound. From Proposition 4.13 we infer the existence of a minimal element in $\mathcal{F}$, namely $F : \mathcal{B} \to \mathbb{R}$. Due to Proposition 4.12 the latter represents a linear function. We note (10) and obtain the following inequalities

$$F(x) \le p(x) \le \|f\| \, \|x\|, \quad x \in \mathcal{B}$$

and

$$-F(x) = F(-x) \le \|f\| \, \| - x\| = \|f\| \, \|x\|, \quad x \in \mathcal{B}.$$

This implies $|F(x)| \le \|f\| \, \|x\|$ for all $x \in \mathcal{B}$ and consequently $\|F\| = \|f\|$.

All $x \in \mathcal{L}$ satisfy the inequality $F(x) \le p(x) \le f(x)$ and $f : \mathcal{L} \to \mathbb{R}$ is linear. Therefore, we infer

$$F(x) = f(x) \qquad \text{for all} \quad x \in \mathcal{L}$$

from Proposition 4.12.   q.e.d.

5 Some Historical Notices to the Chapters 3 and 7

By L. Kronecker, H. Poincaré, and L. Brouwer about 1900, the degree of mapping in Euclidean spaces has been developed in the framework of combinatorial topology. Here we refer the reader to the textbook *Topologie* by P. Alexandroff and H. Hopf from 1935.

The analytical definition for the degree of mapping was invented by E. Heinz in 1959, utilizing A.Sard's lemma on the critical values of differentiable mappings from 1942. Our representation in Chapter 3 contains the Jordan-Brouwer theorem in $\mathbb{R}^n$, whose proof was given by L. Bers via the product formula. A beautiful approach to Jordan's curve theorem was already invented by E. Schmidt. We would like to recommend B. von Kérékjartó's monograph *Flächentopologie* from 1923 in this context.

The first definition for the degree of mapping in Banach spaces was given by J. Leray and J. Schauder in their joint paper on functional equations from 1934. They discovered that the existence question is independent of the answer to the uniqueness problem. Finally, we would like to mention S.Banach's influential book *Théorie des Opérations Linéaires* from 1932 in connection with Section 4 of Chapter 7.

Figure 2.2   PORTRAIT OF D. HILBERT (1862–1943)
taken from page 244 of the biography by *C. Reid: Hilbert*, Springer-Verlag, Berlin... (1970).

Chapter 8

Linear Operators in Hilbert Spaces

Motivated by the eigenvalue problems for ordinary and partial differential operators, we shall develop the spectral theory for linear operators in Hilbert spaces. Here we transform the unbounded differential operators into singular integral operators which are completely continuous. With his study of integral equations D. Hilbert, together with his students E. Schmidt, I. Schur, and H. Weyl, opened a new era for the analysis.

1 Various Eigenvalue Problems

At first, we consider the resolution of linear systems of equations: For the given matrix $A = (a_{ij})_{i,j=1,\ldots,n} \in \mathbb{R}^{n \times n}$ we associate the mapping

$$x \mapsto Ax : \mathbb{R}^n \to \mathbb{R}^n$$

and the system of equations

$$\sum_{k=1}^{n} a_{ik}x_k = y_i, \quad i = 1,\ldots,n, \qquad \text{or equivalently} \qquad Ax = y$$

with the right-hand side $y = (y_1,\ldots,y_n)^t$. The system $Ax = y$ has a solution for all $y \in \mathbb{R}^n$ if and only if the homogeneous equation $Ax = 0$ possesses only the trivial solution $x = 0$ and we have $x = A^{-1}y$. We remark that the concept of determinants is not necessary in this context.

In Theorem 4.6 from Chapter 7, Section 4 by F. Riesz, we have transferred this solvability theory to linear operators in Banach spaces: Let $\mathcal{B}$ be a real Banach space and $K : \mathcal{B} \to \mathcal{B}$ a linear completely continuous operator with the associate operator $Tx := x - Kx$, $x \in \mathcal{B}$. If the implication

$$Tx = 0 \quad \Rightarrow \quad x = 0$$

F. Sauvigny, *Partial Differential Equations 2*, Universitext,
DOI 10.1007/978-1-4471-2984-4_2, © Springer-Verlag London 2012

holds true, the equation

$$Tx = y, \qquad x \in \mathcal{B}$$

possesses exactly one solution for all $y \in \mathcal{B}$.

We now consider the *principal axes transformation of Hermitian matrices*: Let $A = (a_{ik})_{i,k=1,\ldots,n} \in \mathbb{C}^{n \times n}$ denote a Hermitian matrix, which means $a_{ik} = \overline{a}_{ki}$ for all $i, k = 1, \ldots, n$. Then A possesses a complete orthonormal system of eigenvectors $\varphi_1, \ldots, \varphi_n \in \mathbb{C}^n$ with the real eigenvalues $\lambda_1, \ldots, \lambda_n \in \mathbb{R}$, more precisely

$$A\varphi_i = \lambda_i \varphi_i, \qquad i = 1, \ldots, n,$$

and we have

$$(\varphi_i, \varphi_k) = \delta_{ik}, \qquad i, k = 1, \ldots, n.$$

Here we have used the inner product $(x, y) := \overline{x}^t \cdot y$. By $\xi_i := (\varphi_i, x)$ we denote the i-th component of $x \in \mathbb{C}^n$ with respect to $(\varphi_1, \ldots, \varphi_n)$ which implies

$$x = \sum_{i=1}^{n} \xi_i \varphi_i.$$

Then we note that

$$Ax = \sum_{i=1}^{n} \xi_i A\varphi_i = \sum_{i=1}^{n} \lambda_i \xi_i \varphi_i.$$

We define the diagonal matrix

$$\Lambda := \begin{pmatrix} \lambda_1 & 0 & \cdots & 0 \\ 0 & \ddots & \ddots & \vdots \\ \vdots & \ddots & \ddots & 0 \\ 0 & \cdots & 0 & \lambda_n \end{pmatrix} \in \mathbb{R}^{n \times n}$$

and the unitary matrix $U^{-1} := (\varphi_1, \ldots, \varphi_n) \in \mathbb{C}^{n \times n}$. Now we obtain the representation $x = U^{-1}\xi$ with $\xi = (\xi_1, \ldots, \xi_n)^t$ and consequently $\xi = Ux$. This implies the equation

$$U \circ A \circ U^{-1} \circ \xi = U \circ A \circ x = U \circ U^{-1} \circ \Lambda \circ \xi = \Lambda \circ \xi$$

and consequently the *unitary transformation*

$$\Lambda = U \circ A \circ U^{-1}.$$

Now we observe $U^{-1} = U^* = \overline{U}^t$, and we calculate the transformation of the associate Hermitian form

$$\sum_{i,k=1}^{n} a_{ik}\overline{x}_i x_k = (x, Ax) = \left(\sum_{i=1}^{n}(\varphi_i, x)\varphi_i, \sum_{j=1}^{n}(\varphi_j, x)\lambda_j\varphi_j \right)$$

$$= \sum_{i,j=1}^{n} \overline{(\varphi_i, x)}(\varphi_j, x)\lambda_j\delta_{ij}$$

$$= \sum_{i=1}^{n} |(\varphi_i, x)|^2\lambda_i = \sum_{i=1}^{n} \lambda_i|\xi_i|^2.$$

In the present chapter we intend to deduce the corresponding theorems for operators in Hilbert spaces. Specializing them to integral operators we shall treat eigenvalue problems for ordinary and partial differential equations.

Example 1.1. Let the domain of definition

$$\mathcal{D} := \left\{ u = u(x) \in C^2[0, \pi] \; : \; u(0) = 0 = u(\pi) \right\}$$

and the differential operator

$$Lu(x) := -u''(x), \quad x \in [0, \pi], \quad \text{for} \quad u \in \mathcal{D}$$

be given. Which numbers $\lambda \in \mathbb{R}$ admit a nontrivial solution of the eigenvalue problem

$$Lu(x) = \lambda u(x), \quad 0 \le x \le \pi, \quad u \in \mathcal{D}, \tag{1}$$

satisfying $u \in \mathcal{D}$ and $u \not\equiv 0$?

$\lambda = 0$: We have $u''(x) = 0$ for $x \in [0, \pi]$ and consequently $u(x) = ax + b$ with the constants $a, b \in \mathbb{R}$. The boundary conditions for u yield $0 = u(0) = b$ and $0 = u(\pi) = a\pi + b = a\pi$, and we obtain

$$u(x) \equiv 0, \quad x \in [0, \pi].$$

Therefore, $\lambda = 0$ is not an eigenvalue of (1).

$\lambda < 0$: Setting $\lambda = -k^2$ with $k \in (0, +\infty)$, we rewrite (1) into the form

$$u''(x) - k^2 u(x) = 0, \quad x \in [0, \pi].$$

Evidently, the functions $\{e^{kx}, e^{-kx}\}$ constitute a fundamental system of the differential equation. Taking $u(0) = 0$ into account, we infer

$$u(x) = Ae^{kx} + Be^{-kx} = A(e^{kx} - e^{-kx}) = 2A\sinh(kx).$$

Noting $u(\pi) = 0$, we finally obtain

$$u(x) \equiv 0, \quad x \in [0, \pi].$$

Therefore, negative eigenvalues $\lambda < 0$ of (1) do not exist.

$\lambda > 0$: We now consider $\lambda = k^2$ with a number $k \in (0, +\infty)$. Then (1) is written in the form

$$u''(x) + k^2 u(x) = 0, \qquad x \in [0, \pi].$$

The fundamental system is given by $\{\cos(kx), \sin(kx)\}$ and the general solution by

$$u(x) = A\cos(kx) + B\sin(kx), \qquad x \in [0, \pi].$$

From $0 = u(0) = A$ we infer $u(x) = B\sin(kx)$ for $x \in [0, \pi]$, and the boundary condition $0 = u(\pi) = B\sin(k\pi)$ implies $k \in \mathbb{N}$. In this way we obtain the eigenvalues $\lambda = k^2$ of (1) with the eigenfunctions

$$u_k(x) = \sin(kx), \quad x \in [0, \pi], \qquad k = 1, 2, \ldots$$

Example 1.2. Let the domain

$$G := \Big\{ x = (x_1, \ldots, x_n) \in \mathbb{R}^n \ : \ x_i \in (0, \pi), \ i = 1, \ldots, n \Big\} = (0, \pi)^n \subset \mathbb{R}^n$$

be given. On the domain of definition

$$\mathcal{D} := \Big\{ u \in C^2(G) \cap C^0(\overline{G}) \ : \ u|_{\partial G} = 0 \Big\}$$

we define the differential operator

$$Lu(x) := -\Delta u(x) = -\sum_{i=1}^{n} \frac{\partial^2}{\partial x_i^2} u(x), \qquad x \in G.$$

We consider the eigenvalue problem

$$Lu(x) = \lambda u(x), \qquad x \in G, \tag{2}$$

for $u \in \mathcal{D}$ and $\lambda \in \mathbb{R}$. In order to solve this problem we propose the *ansatz of separation*

$$u(x) = u(x_1, \ldots, x_n) := u_1(x_1) \cdot u_2(x_2) \cdot \ldots \cdot u_n(x_n), \qquad x \in G.$$

The differential equation (2) becomes

$$-\sum_{i=1}^{n} u_1(x_1) \cdot \ldots \cdot u_{i-1}(x_{i-1}) u_i''(x_i) u_{i+1}(x_{i+1}) \cdot \ldots \cdot u_n(x_n) = \lambda u_1(x_1) \cdot \ldots \cdot u_n(x_n)$$

and consequently

$$-\sum_{i=1}^{n} \frac{u_i''(x_i)}{u_i(x_i)} = \lambda, \qquad x \in G.$$

We now choose $u_i(x_i) := \sin(k_i x_i)$ with $k_i \in \mathbb{N}$ for $i = 1, \ldots, n$, and we obtain

$$-\sum_{i=1}^{n} \frac{u_i''(x_i)}{u_i(x_i)} = \sum_{i=1}^{n} k_i^2 = \lambda \in (0, \infty).$$

The solutions of the eigenvalue problem (2) appear as follows:

$$u(x_1, \ldots, x_n) := \sin(k_1 x_1) \cdot \ldots \cdot \sin(k_n x_n) \quad \text{and} \quad \lambda = k_1^2 + \ldots + k_n^2$$

for $k_1, \ldots, k_n \in \mathbb{N}$. Normalizing

$$u_{k_1, \ldots, k_n}(x_1, \ldots, x_n) := \left(\frac{2}{\pi}\right)^{\frac{n}{2}} \sin(k_1 x_1) \cdot \ldots \cdot \sin(k_n x_n), \qquad x \in G \qquad (3)$$

and using the inner product

$$(u, v) := \int_G u(x_1, \ldots, x_n) v(x_1, \ldots, x_n) \, dx_1 \ldots dx_n \qquad u, v \in \mathcal{D},$$

we obtain the orthonormal system of functions

$$(u_{k_1 \ldots k_n}, u_{l_1 \ldots l_n}) = \delta_{k_1 l_1} \cdot \ldots \cdot \delta_{k_n l_n} \qquad \text{for} \quad k_1, \ldots, k_n, l_1, \ldots, l_n \in \mathbb{N}. \qquad (4)$$

On account of

$$L u_{k_1 \ldots k_n} = (k_1^2 + \ldots + k_n^2) u_{k_1 \ldots k_n}$$

and

$$\|u_{k_1 \ldots k_n}\|_{L^2(G)}^2 := (u_{k_1 \ldots k_n}, u_{k_1 \ldots k_n}) = 1$$

for all $k_1 \ldots k_n \in \mathbb{N}$ we infer

$$\sup_{u \in \mathcal{D}, \|u\|=1} \|Lu\| \geq \sup_{k_1 \ldots k_n \in \mathbb{N}} \|L u_{k_1 \ldots k_n}\| = \sup_{k_1 \ldots k_n \in \mathbb{N}} (k_1^2 + \ldots + k_n^2) = +\infty. \quad (5)$$

Consequently $L = -\Delta : L^2(G) \to L^2(G)$ represents an unbounded operator on the Hilbert space $L^2(G)$.

The following question is of central interest: Do the given functions

$$\{u_{k_1 \ldots k_n}\}_{k_1 \ldots k_n = 1, 2, \ldots}$$

constitute a complete system? Can we expand an arbitrary function into such a series of functions?

We now consider the *Sturm-Liouville eigenvalue problem:*
Let the numbers $c_1, c_2, d_1, d_2 \in \mathbb{R}$ satisfying $c_1^2 + c_2^2 > 0$ and $d_1^2 + d_2^2 > 0$ be prescribed. We choose the linear space

$$\mathcal{D} := \left\{ f \in C^2([a, b], \mathbb{R}) \ : \ c_1 f(a) + c_2 f'(a) = 0 = d_1 f(b) + d_2 f'(b) \right\}$$

as our domain of definition, where the numbers $-\infty < a < b < +\infty$ are fixed. With the functions $p = p(x) \in C^1([a, b], (0, +\infty))$ and $q = q(x) \in C^0([a, b], \mathbb{R})$ we define the *Sturm-Liouville operator*

$$Lu(x) := -\left(p(x) u'(x)\right)' + q(x) u(x), \quad x \in [a, b], \qquad \text{for} \quad u \in \mathcal{D}.$$

Proposition 1.3. *The operator* $L : \mathcal{D} \to \mathbb{C}^0([a,b], \mathbb{R})$ *is linear and symmetric satisfying*

$$L(\alpha u + \beta v) = \alpha L u + \beta L v \qquad \textit{for all} \quad u, v \in \mathcal{D}, \quad \alpha, \beta \in \mathbb{R}$$

and

$$\int_a^b u(x) \Big(L v(x) \Big) \, dx = \int_a^b \Big(L u(x) \Big) v(x) \, dx \qquad \textit{for all} \quad u, v \in \mathcal{D}.$$

Proof: The linearity is evident, and we calculate for $u, v \in \mathcal{D}$ as follows:

$$vLu - uLv = v(-(pu')' + qu) - u(-(pv')' + qv)$$

$$= \frac{d}{dx} \Big\{ p(x)(u(x)v'(x) - u'(x)v(x)) \Big\}, \qquad x \in [a, b]. \tag{6}$$

This implies

$$\int_a^b (vLu - uLv) \, dx = \Big[-p(x)(u(x), u'(x)) \cdot (-v'(x), v(x))^t \Big]_a^b = 0,$$

since the vectors $(u(x), u'(x))$ and $(v(x), v'(x))$ are parallel for $x = a$ and $x = b$, respectively; here we take $u, v \in \mathcal{D}$ into account.

q.e.d.

We now investigate the eigenvalue problem

$$Lu = \lambda u, \qquad u \in \mathcal{D}. \tag{7}$$

Setting

$$(u, v) := \int_a^b u(x)v(x) \, dx \qquad \text{for} \quad u, v \in \mathcal{D},$$

we obtain an orthonormal system of eigenfunctions

$$u_k(x) \in \mathcal{D} \qquad \text{with} \quad (u_k, u_l) = \delta_{kl}, \qquad k, l \in \mathbb{N}$$

in Section 8 satisfying

$$Lu_k = \lambda_k u_k, \qquad k = 1, 2, \ldots$$

According to the Example 1.1 we expect the asymptotic behavior

$$-\infty < \lambda_1 \le \lambda_2 \le \lambda_3 \le \ldots \to +\infty \tag{8}$$

for the eigenvalues. Consequently, the operator L is unbounded. We shall derive the expansion into the series

$$f(x) = \sum_{k=1}^{\infty} c_k u_k(x) \qquad \text{with} \quad c_k = (u_k, f)$$

for all functions $f \in \mathcal{D}$. At first, we require the

Assumption 0: The equation $Lu = 0$ with $u \in \mathcal{D}$ admits only the trivial solution $u \equiv 0$.

The domain $\mathcal{D}$ is not complete with respect to the norm $\|u\| := \sqrt{(u,u)}$ for $u \in \mathcal{D}$, and the operator L is unbounded in general. Therefore, we cannot prove the existence of the inverse L^{-1} by the Theorem of F. Riesz. With the *Assumption 0*, however, we shall construct the inverse with the aid of *Green's function for the Sturm-Liouville operator* $K = K(x, y)$. Having achieved this, we shall transform (7) equivalently into an eigenvalue problem for the bounded operator L^{-1}, namely

$$L^{-1}u = \frac{1}{\lambda}u, \qquad u \in \mathcal{D}. \tag{9}$$

For the construction of the inverse we consider the ordinary differential equation

$$
\begin{aligned}
Lu(x) &= -(p(x)u'(x))' + q(x)u(x) \\
&= -p(x)u''(x) - p'(x)u'(x) + q(x)u(x) \tag{10} \\
&= f(x), \qquad a \le x \le b.
\end{aligned}
$$

The homogeneous equation $Lu = 0$ possesses a fundamental system $\alpha = \alpha(x)$, $\beta = \beta(x)$ satisfying

$$L\alpha(x) \equiv 0 \equiv L\beta(x) \qquad \text{in} \quad [a, b].$$

We construct a solution of (10) by the method *variation of the constants*

$$u(x) = A(x)\alpha(x) + B(x)\beta(x), \qquad a \le x \le b, \tag{11}$$

under the subsidiary condition

$$A'(x)\alpha(x) + B'(x)\beta(x) = 0. \tag{12}$$

With the aid of (12) we calculate

$$u'(x) = A(x)\alpha'(x) + B(x)\beta'(x)$$

and

$$u''(x) = A(x)\alpha''(x) + B(x)\beta''(x) + A'(x)\alpha'(x) + B'(x)\beta'(x).$$

Together with the formula (11) we obtain

$$Lu(x) = A(x)L\alpha(x) + B(x)L\beta(x) - p(x)\Big\{A'(x)\alpha'(x) + B'(x)\beta'(x)\Big\} = f(x),$$

and therefore

$$- p(x)\Big\{ A'(x)\alpha'(x) + B'(x)\beta'(x) \Big\} = f(x), \qquad a \leq x \leq b. \qquad (13)$$

By the ansatz

$$A'(x) = \beta(x)k(x), \quad B'(x) = -\alpha(x)k(x), \qquad a \leq x \leq b, \qquad (14)$$

with a continuous function $k = k(x)$, $x \in [a, b]$, the relation (12) is fulfilled and (13) becomes

$$- p(x) \left\{ \beta(x)\alpha'(x) - \alpha(x)\beta'(x) \right\} k(x) = f(x), \qquad a \leq x \leq b. \qquad (15)$$

Proposition 1.4. *The relation $p(x)\{\alpha(x)\beta'(x) - \alpha'(x)\beta(x)\} =$const in $[a, b]$ holds true.*

Proof: Applying (6) to $u = \alpha(x)$ and $v = \beta(x)$ we infer

$$0 = \frac{d}{dx}\Big\{ p(x)\left(\alpha(x)\beta'(x) - \alpha'(x)\beta(x)\right)\Big\} \qquad \text{in} \quad [a, b].$$

$$\text{q.e.d.}$$

We now choose $\alpha = \alpha(x)$ and $\beta = \beta(x)$ to solve the homogeneous equation $Lu = 0$ satisfying

$$p(x)\Big\{ \alpha(x)\beta'(x) - \alpha'(x)\beta(x) \Big\} \equiv 1 \qquad \text{in} \quad [a, b] \qquad (16)$$

and

$$c_1\beta(a) + c_2\beta'(a) = 0 = d_1\alpha(b) + d_2\alpha'(b). \qquad (17)$$

Here we solve the initial value problems

$$L\alpha = 0 \quad \text{in} \ [a, b], \qquad \alpha(b) = d_2, \qquad \alpha'(b) = -d_1$$

and

$$L\beta = 0 \quad \text{in} \ [a, b], \qquad \beta(a) = \frac{1}{M}c_2, \qquad \beta'(a) = -\frac{1}{M}c_1.$$

Thereby, we determine $M \neq 0$ such that

$$p(a)\Big\{ \alpha(a)\beta'(a) - \alpha'(a)\beta(a) \Big\} = -\frac{1}{M}p(a)\Big\{ c_1\alpha(a) + c_2\alpha'(a) \Big\} = 1$$

is fulfilled choosing

$$M = -p(a)\Big\{ c_1\alpha(a) + c_2\alpha'(a) \Big\}.$$

The statement $M \neq 0$ is contained in the following

Proposition 1.5. *The functions $\{\alpha, \beta\}$ constitute a fundamental system.*

Proof: If the statement were violated, we have a number $\mu \neq 0$ with the property

$$\alpha(x) = \mu\beta(x), \qquad a \leq x \leq b.$$

We deduce $\alpha \in \mathcal{D}$ from (17), and the *Assumption 0* yields a contradiction with $\alpha \equiv 0$.

q.e.d.

The relations (15) and (16) imply

$$k(x) = f(x) \qquad \text{in} \quad [a, b], \tag{18}$$

and (14) yields

$$A(x) = \int_a^x \beta(y)f(y)\,dy + \text{const}, \qquad B(x) = \int_x^b \alpha(y)f(y)\,dy + \text{const}. \tag{19}$$

We summarize our considerations to the following

Theorem 1.6. *The Sturm-Liouville equation $Lu = f$ for $u \in \mathcal{D}$ with the right-hand side $f \in C^0([a,b])$ is solved by the function*

$$u(x) = \alpha(x) \int_a^x \beta(y)f(y)\,dy + \beta(x) \int_x^b \alpha(y)f(y)\,dy = \int_a^b K(x,y)f(y)\,dy. \tag{20}$$

With the aid of the fundamental system $\{\alpha, \beta\}$ of $Lu = 0$ satisfying (16) and (17), we here define the Green's function of the Sturm-Liouville operator as follows:

$$K(x,y) = \begin{cases} \alpha(x)\beta(y), & a \leq y \leq x \\ \beta(x)\alpha(y), & x \leq y \leq b. \end{cases} \tag{21}$$

Proof: The derivation above implies that the function $u(x)$ from (20) satisfies the differential equation $Lu = f$. Furthermore, we see

$$u(a) = \beta(a) \int_a^b \alpha(y)f(y)\,dy, \qquad u'(a) = \beta'(a) \int_a^b \alpha(y)f(y)\,dy,$$

and (17) gives us

$$c_1 u(a) + c_2 u'(a) = (c_1\beta(a) + c_2\beta'(a)) \int_a^b \alpha(y)f(y)\,dy = 0.$$

In the same way we determine

$$u(b) = \alpha(b) \int_a^b \beta(y)f(y)\,dy, \qquad u'(b) = \alpha'(b) \int_a^b \beta(y)f(y)\,dy$$

and

$$d_1 u(b) + d_2 u'(b) = (d_1 \alpha(b) + d_2 \alpha'(b)) \int_a^b \beta(y)f(y)\,dy = 0.$$

q.e.d.

Theorem 1.6 directly implies the following

Theorem 1.7. *With the Assumption 0, the subsequent statements are equivalent:*

I.   *The function $u \in \mathcal{D}$ with $u \not\equiv 0$ satisfies $Lu = \lambda u$.*

II. *The function $u \in \mathcal{D}$ with $u \not\equiv 0$ satisfies $\displaystyle\int_a^b K(x,y)u(y)\,dy = \frac{1}{\lambda}u(x)$ for $a \leq x \leq b$.*

We shall now address the *eigenvalue problem of the n-dimensional oscillation equation* considered by H. von Helmholtz : Let $G \subset \mathbb{R}^n$ be a bounded Dirichlet domain, which means that all continuous functions $g = g(x) : \partial G \to \mathbb{R}$ possess a solution of the Dirichlet problem

$$u = u(x) \in C^2(G) \cap C^0(\overline{G}),$$

$$\Delta u(x) = 0 \quad \text{in} \quad G, \tag{22}$$

$$u(x) = g(x) \quad \text{on} \quad \partial G$$

(compare Chapter 5, Section 3). The further assumption will be eliminated by Proposition 9.1 in Section 9, namely that G satisfies the conditions of the Gaussian integral theorem from Section 5 in Chapter 1. Then we can specify the *Green's function of the Laplace operator* for the domain G as follows:

$$H(x,y) = \begin{cases} -\dfrac{1}{2\pi} \log|y - x| + h(x,y), & n = 2 \\[2mm] \dfrac{1}{(n-2)\omega_n} \dfrac{1}{|y-x|^{n-2}} + h(x,y), & n \geq 3 \end{cases} \tag{23}$$

for $(x,y) \in G \otimes G := \{(\xi, \eta) \in G \times G : \xi \neq \eta\}$. Here we have $\Delta_y h(x,y) = 0$ in G and

$$h(x,y) = \begin{cases} \dfrac{1}{2\pi} \log|y - x|, & n = 2 \\[2mm] -\dfrac{1}{(n-2)\omega_n} \dfrac{1}{|y-x|^{n-2}}, & n \geq 3 \end{cases} \tag{24}$$

for $x \in G$ and $y \in \partial G$. Furthermore, ω_n denotes the area of the unit sphere in $\mathbb{R}^n$. Due to Chapter 5, Sections 1 and 2, we can represent a solution of the problem

$$u = u(x) \in C^2(G) \cap C^0(\overline{G}),$$

$$-\Delta u(x) = f(x) \quad \text{in} \quad G, \tag{25}$$

$$u(x) = 0 \quad \text{on} \quad \partial G$$

in the following form

$$u(x) = \int_G H(x,y) f(y)\, dy, \qquad x \in G. \tag{26}$$

For the deduction of (26) we consider the domain $G_\varepsilon := \{y \in G : |y - x| > \varepsilon\}$ with a small $\varepsilon > 0$. The Gaussian integral theorem implies

$$\int_{G_\varepsilon} \Big(H(x,y) \Delta u(y) - u(y) \Delta_y H(x,y) \Big)\, dy$$

$$= \int_{\partial G} \Big(H(x,y) \frac{\partial u}{\partial \nu}(y) - u(y) \frac{\partial H}{\partial \nu}(x,y) \Big)\, d\sigma(y)$$

$$- \int_{y:|y-x|=\varepsilon} \Big(H(x,y) \frac{\partial u}{\partial \nu}(y) - u(y) \frac{\partial H}{\partial \nu}(x,y) \Big)\, d\sigma(y).$$

Observing $\varepsilon \downarrow 0$ we obtain

$$- \int_G H(x,y) f(y)\, dy = \lim_{\varepsilon \downarrow 0} \int_{r=|y-x|=\varepsilon} u(y) \Big(\frac{1}{(n-2)\omega_n}(2-n) r^{1-n} \Big)\, d\sigma(y)$$

$$= -\lim_{\varepsilon \downarrow 0} \Big(\frac{1}{\varepsilon^{n-1}\omega_n} \int_{r=|y-x|=\varepsilon} u(y)\, d\sigma(y) \Big)$$

$$= -u(x) \qquad \text{for all} \quad x \in G$$

in the case $n \geq 3$, and similarly in the case $n = 2$.

We now derive the *symmetry of Green's function*, namely

$$H(x,y) = H(y,x) \qquad \text{for all} \quad (x,y) \in G \otimes G. \tag{27}$$

Here we choose the points $x, y \in G$ satisfying $x \neq y$, and on the domain

$$G_\varepsilon := \Big\{ z \in G : |z - x| > \varepsilon \text{ and } |z - y| > \varepsilon \Big\}$$

we consider the functions $p(z) := H(x,z)$ and $q(z) := H(y,z)$, $z \in G_\varepsilon$. For $\varepsilon \downarrow 0$ the Gaussian integral theorem implies

$$0 = \lim_{\varepsilon \downarrow 0} \int_{G_\varepsilon} (q\Delta p - p\Delta q)\, dz = \lim_{\varepsilon \downarrow 0} \int_{\partial G_\varepsilon} \left(q\frac{\partial p}{\partial \nu} - p\frac{\partial q}{\partial \nu} \right) d\sigma(z)$$

$$= -\lim_{\varepsilon \downarrow 0} \int_{|z-x|=\varepsilon} \left(q\frac{\partial p}{\partial \nu} - p\frac{\partial q}{\partial \nu} \right) d\sigma(z) - \lim_{\varepsilon \downarrow 0} \int_{|z-y|=\varepsilon} \left(q\frac{\partial p}{\partial \nu} - p\frac{\partial q}{\partial \nu} \right) d\sigma(z)$$

$$= q(x) - p(y) \; = \; H(y,x) - H(x,y) \qquad \text{for all} \quad x,y \in G \quad \text{with} \quad x \neq y.$$

We now show a *growth condition for Green's function* $H(x,y)$ as follows: With $\varepsilon > 0$ given we define the harmonic function

$$W_\varepsilon(x,y) := \begin{cases} -\dfrac{1}{2\pi}(1+\varepsilon) \log \dfrac{|y-x|}{d}, & n = 2 \\[2mm] \dfrac{1+\varepsilon}{(n-2)\omega_n} |y-x|^{2-n}, & n \geq 3 \end{cases}$$

setting $d := \operatorname{diam} G$. We consider the function $\Phi_\varepsilon(x,y) := W_\varepsilon(x,y) - H(x,y)$ and choose $\delta > 0$ so small that

$$\Phi_\varepsilon(x,y) \geq 0 \qquad \text{for all} \quad y : |y-x| = \delta \quad \text{and all} \quad y \in \partial G$$

is satisfied. Applying the maximum principle to the harmonic function $\Phi_\varepsilon(x,.)$ on the domain $G_\delta := \{ y \in G : |x-y| > \delta \}$ we infer $\Phi_\varepsilon(x,y) \geq 0$ in G_δ and consequently

$$H(x,y) \leq W_\varepsilon(x,y) \qquad \text{for all} \quad \varepsilon > 0.$$

Therefore, we obtain

$$0 \leq H(x,y) \leq \begin{cases} -\dfrac{1}{2\pi} \log \dfrac{|y-x|}{d}, & n = 2 \\[2mm] \dfrac{1}{(n-2)\omega_n} |y-x|^{2-n}, & n \geq 3 \end{cases}$$

for all $(x,y) \in G \otimes G$ and finally the growth condition

$$|H(x,y)| \leq \frac{\text{const}}{|x-y|^\alpha} \qquad \text{for all} \quad (x,y) \in G \otimes G \tag{28}$$

with $\alpha := n - 2 < n$.

Definition 1.8. *Let* $G \subset \mathbb{R}^n$ *denote a bounded domain where* $n \in \mathbb{N}$ *holds true, and let the number* $\alpha \in [0,n)$ *be chosen arbitrarily. A function* $K = K(x,y) \in C^0(G \otimes G, \mathbb{C})$ *is called a* singular kernel of the order α - *briefly* $K \in S_\alpha(G, \mathbb{C})$ - *if we have a constant* $c \in [0, +\infty)$ *satisfying*

$$|K(x,y)| \leq \frac{c}{|x-y|^\alpha} \qquad \text{for all} \quad (x,y) \in G \otimes G. \tag{29}$$

We name the kernel $K \in S_\alpha(G, \mathbb{C})$ Hermitian, *if*

$$K(x, y) = \overline{K(y, x)} \qquad \text{for all} \quad (x, y) \in G \otimes G \tag{30}$$

is valid. The real kernels belong to the class $\mathcal{S}_\alpha(G) := \mathcal{S}_\alpha(G, \mathbb{R})$, *and these kernels* $K \in \mathcal{S}_\alpha(G)$ *are Hermitian if and only if they are symmetric in the following sense:*

$$K(x, y) = K(y, x) \qquad \text{for all} \quad (x, y) \in G \otimes G. \tag{31}$$

We summarize our considerations about the n-dimensional oscillation equation to the following

Theorem 1.9. *Let* $G \subset \mathbb{R}^n$, $n = 2, 3, \ldots$ *denote a Dirichlet domain satisfying the assumptions for the Gaussian integral theorem. Furthermore, we fix the domain of definition*

$$\mathcal{D} := \Big\{ u = u(x) \in C^2(G) \cap C^0(\overline{G}) \; : \; u(x) = 0 \text{ for all } x \in \partial G \Big\}.$$

Then the following two statements are equivalent:

I. *The function* $u \in \mathcal{D}$ *with* $u \not\equiv 0$ *solves the differential equation*

$$-\Delta u(x) = \lambda u(x) \qquad in \quad G$$

for a number $\lambda \in \mathbb{R}$.

II. *The function* $u \in \mathcal{D}$ *with* $u \not\equiv 0$ *solves the integral equation*

$$\int\limits_G H(x, y) u(y)\, dy = \frac{1}{\lambda} u(x) \qquad in \quad G$$

for a number $\lambda \in \mathbb{R} \setminus \{0\}$.

Here Green's function $H(x, y)$ *of the Laplace operator for the domain* G *represents a symmetric real singular kernel of the regularity class* $\mathcal{S}_{n-2}(G)$.

We finally consider *singular integral operators*: On the bounded domain $G \subset \mathbb{R}^n$ with $n \in \mathbb{N}$, the singular kernel $K = K(x, y) \in \mathcal{S}_\alpha(G, \mathbb{C})$ of the order $\alpha \in [0, n)$ is defined. On the domain of definition

$$\mathcal{D} := \left\{ u(x) : G \to \mathbb{C} \in C^0(G, \mathbb{C}) \; : \; \begin{array}{c} \text{There exists a number } c \in [0, +\infty) \\ \text{satisfying } |u(x)| \le c \text{ for all } x \in G \end{array} \right\}$$

$$=: C_b^0(G, \mathbb{C}) = C^0(G, \mathbb{C}) \cap L^\infty(G, \mathbb{C})$$

we consider the integral operator $\mathbb{K} : \mathcal{D} \to C^0(G, \mathbb{C})$ given by

$$\mathbb{K}u(x) := \int\limits_G K(x, y) u(y)\, dy, \quad x \in G, \qquad \text{with} \quad u \in \mathcal{D}.$$

Evidently, we obtain with $\mathbb{K} : \mathcal{D} \to C^0(G, \mathbb{C})$ a linear operator.

Theorem 1.10. *Let the kernel* $K = K(x,y) \in \mathcal{S}_\alpha(G,\mathbb{C})$ *with* $\alpha \in [0,n)$ *be Hermitian. Then we have the following statements:*

a) *If* $u \in \mathcal{D}$ *is an eigenfunction of the associate integral operator, more precisely* $u \not\equiv 0$ *and* $\mathbb{K}u = \lambda u$ *with* $\lambda \in \mathbb{C}$, *we then infer* $\lambda \in \mathbb{R}$.

b) *Given the two eigenfunctions* $u_i \in \mathcal{D}$ *with* $\mathbb{K}u_i = \lambda_i u_i$ *and* $i = 1, 2$ *for the eigenvalues* $\lambda_1 \neq \lambda_2$, *we infer* $(u_1, u_2) = 0$. *Here we used the inner product*

$$(u, v) := \int_G \overline{u(x)} v(x) \, dx \qquad for \quad u, v \in \mathcal{D}.$$

Proof:

a) Let $u \in \mathcal{D} \setminus \{0\}$ be a solution of the problem $\mathbb{K}u = \lambda u$ with a number $\lambda \in \mathbb{C}$. This implies

$$\lambda u(x) = \int_G K(x,y) u(y) \, dy, \qquad x \in G.$$

We multiply the equation by $\overline{u(x)}$ and afterwards integrate over the domain G with respect to x, and we obtain

$$\lambda(u, u) = \int_G \int_G K(x,y) \overline{u(x)} u(y) \, dx \, dy \ \in \mathbb{R}.$$

Since the inner product (u, u) is a real expression, the number λ has to be real.

b) Let the eigenfunctions $u_i \in \mathcal{D}$ satisfying $\mathbb{K}u_i = \lambda_i u_i$ with $i = 1, 2$ and the eigenvalues $\lambda_1 \neq \lambda_2$ be given. On account of $\lambda_1, \lambda_2 \in \mathbb{R}$ we infer

$$\lambda_1(u_1, u_2) = (\lambda_1 u_1, u_2) = (\mathbb{K}u_1, u_2) = (u_1, \mathbb{K}u_2) = (u_1, \lambda_2 u_2) = \lambda_2(u_1, u_2)$$

or $(\lambda_1 - \lambda_2)(u_1, u_2) = 0$ and consequently $(u_1, u_2) = 0$. We namely deduce for all $u, v \in \mathcal{D}$:

$$(\mathbb{K}u, v) = \int_G \overline{\left(\int_G K(x,y) u(y) \, dy \right)} v(x) \, dx = \int_G \int_G \overline{K(x,y) u(y)} v(x) \, dx \, dy$$

$$= \int_G \int_G K(y,x) v(x) \overline{u(y)} \, dx \, dy = \int_G \overline{u(y)} \left(\int_G K(y,x) v(x) \, dx \right) dy$$

$$= (u, \mathbb{K}v).$$

$$\text{q.e.d.}$$

2 Singular Integral Equations

In Section 1 we have equivalently transformed eigenvalue problems for differential equations into so-called *integral equations of the first kind*

$$\int_G K(x,y)u(y)\,dy = \mu u(x), \qquad x \in G, \tag{1}$$

with the singular kernels $K = K(x,y)$. Parallel to the swinging equation we take a bounded Dirichlet domain $G \subset \mathbb{R}^n$ satisfying the assumptions of the Gaussian integral theorem with the associate Green's function $H = H(x,y) \in \mathcal{S}_{n-2}(G)$ for the Laplace operator. Especially for the unit ball $B := \{x \in \mathbb{R}^n : |x| < 1\}$ we obtain as Green's function in the case $n = 2$:

$$G(\zeta, z) := \frac{1}{2\pi} \log \left| \frac{1 - \overline{z}\zeta}{\zeta - z} \right|, \qquad (\zeta, z) \in B \otimes B, \tag{2}$$

and in the case $n \geq 3$:

$$G(x,y) := \frac{1}{(n-2)\omega_n} \left\{ \frac{1}{|y-x|^{n-2}} - \frac{1}{|x|^{n-2}\left|y - \frac{x}{|x|^2}\right|^{n-2}} \right\}, \qquad (x,y) \in B \otimes B. \tag{3}$$

We now consider the Dirichlet problem

$$u = u(x) \in C^2(G) \cap C^0(\overline{G}),$$

$$\Delta u(x) + \sum_{i=1}^{n} b_i(x)u_{x_i}(x) + c(x)u(x) = f(x), \qquad x \in G, \tag{4}$$

$$u(x) = 0, \qquad x \in \partial G.$$

Here we assume the functions $b_i(x)$, $i = 1, \ldots, n$, $c(x)$ and $f(x)$ to be Hölder continuous in $\overline{G}$. We transfer the equation (4) into an integral equation as follows: With the representation

$$-\Delta u(x) = \sum_{i=1}^{n} b_i(x)u_{x_i}(x) + c(x)u(x) - f(x) =: g(x), \qquad x \in G, \tag{5}$$

we deduce (similarly to the oscillation equation)

$$u(x) = \int_G H(x,y) \left\{ \sum_{i=1}^{n} b_i(y)u_{x_i}(y) + c(y)u(y) - f(y) \right\} dy$$

and

$$u(x) - \int\limits_G \left\{ (H(x,y)c(y))\,u(y) + \sum_{i=1}^n (H(x,y)b_i(y))\,u_{x_i}(y) \right\} dy$$

$$= - \int\limits_G H(x,y)f(y)\,dy \qquad \text{for all} \quad x \in G. \tag{6}$$

We differentiate (6) with respect to x_j for $j = 1, \ldots, n$ and obtain the additional n equations

$$u_{x_j}(x) - \int\limits_G \left\{ (H_{x_j}(x,y)c(y))\,u(y) + \sum_{i=1}^n \left(H_{x_j}(x,y)b_i(y)\right)\,u_{x_i}(y) \right\} dy$$

$$= - \int\limits_G H_{x_j}(x,y)f(y)\,dy, \qquad x \in G, \quad j = 1, \ldots, n. \tag{7}$$

Setting

$$K_{00}(x,y) := H(x,y)c(y), \qquad K_{0i}(x,y) := H(x,y)b_i(y),$$

$$K_{j0}(x,y) := H_{x_j}(x,y)c(y), \qquad K_{ji}(x,y) := H_{x_j}(x,y)b_i(y)$$

for $i, j = 1, \ldots, n$ and

$$f_0(x) := - \int\limits_G H(x,y)f(y)\,dy, \qquad f_j(x) := - \int\limits_G H_{x_j}(x,y)f(y)\,dy$$

for $j = 1, \ldots, n$, we arrive at the following

Theorem 2.1. *The solution $u = u(x)$ of (4) is transferred into the system of Fredholm's integral equations*

$$u_j(x) - \int\limits_G \sum_{i=0}^n K_{ji}(x,y)u_i(y)\,dy = f_j(x), \qquad x \in G, \quad j = 0, \ldots, n \tag{8}$$

with the functions $u_0(x) := u(x)$ and $u_i(x) := u_{x_i}(x)$ for $i = 1, \ldots, n$. Here the singular kernels $K_{ji}(x,y) \in \mathcal{S}_{n-1}(G)$ are real for $i, j = 0, \ldots, n$. However, they are not symmetric in general.

Remark: In the special case $n = 2$, $G = B$, $b_1(x) \equiv 0 \equiv b_2(x)$ in B we can transfer the problem

$$u = u(z) \in C^2(B) \cap C^0(\overline{B}),$$

$$\Delta u(z) + c(z)u(z) = f(z), \qquad z \in B, \tag{9}$$

$$u(z) = 0, \qquad z \in \partial B,$$

into Fredholm's integral equation

$$u(z) - \int_B \frac{1}{2\pi} \log\left|\frac{1 - \bar{z}\zeta}{\zeta - z}\right| c(\zeta) u(\zeta) \, d\zeta = -\int_B \frac{1}{2\pi} \log\left|\frac{1 - \bar{z}\zeta}{\zeta - z}\right| f(\zeta) \, d\zeta, \qquad z \in B.$$

(10)

Sometimes (10) is called an *integral equation of the second kind*. We remark that the integral kernel which appears is not symmetric in general.

For the L^p-spaces used in the following we refer the reader to Chapter 2, Section 7. On the bounded domain $G \subset \mathbb{R}^n$ we choose a singular kernel $K = K(x, y) \in \mathcal{S}_\alpha(G, \mathbb{C})$ with $\alpha \in [0, n)$. On the domain of definition

$$\mathcal{D} := \left\{ f : G \to \mathbb{C} \in C^0(G) : \sup_{x \in G} |f(x)| < +\infty \right\}$$

we consider the associate integral operator

$$\mathbb{K}f(x) := \int_G K(x, y) f(y) \, dy, \quad x \in G, \qquad \text{for} \quad f \in \mathcal{D}.$$

(11)

Choosing an exponent $p \in (1, \frac{n}{\alpha})$, we obtain a constant $C = C(c, \alpha, n, p) \in (0, +\infty)$ with the following property

$$\int_G |K(x, y)|^p dy \le C \qquad \text{for all} \quad x \in G,$$

(12)

because of Section 1, Definition 1.8. When $q \in (\frac{n}{n-\alpha}, +\infty)$ denotes the conjugate exponent to p satisfying $\frac{1}{p} + \frac{1}{q} = 1$, Hölder's inequality from Theorem 7.4 in Chapter 2, Section 7 yields the following estimate:

$$|\mathbb{K}f(x)| \le \int_G |K(x, y)| |f(y)| \, dy$$

$$\le \left(\int_G |K(x, y)|^p \, dy \right)^{\frac{1}{p}} \left(\int_G |f(y)|^q \, dy \right)^{\frac{1}{q}}$$

(13)

$$\le C^{\frac{1}{p}} \|f\|_{L^q(G)}, \qquad x \in G$$

for all $f \in \mathcal{D}$. Here the symbol

$$\|f\|_p = \|f\|_{L^p(G)} := \left(\int_G |f(x)|^p dx \right)^{\frac{1}{p}}, \qquad 1 \le p < +\infty$$

denotes the L^p-norm on the Banach space

$$L^p(G) := \left\{ f : G \to \mathbb{C} \text{ measurable} : \|f\|_{L^p(G)} < +\infty \right\}$$

(compare Chapter 2, Sections 6 and 7). Furthermore, we introduce the C^0-norm

$$\|f\|_{C^0(G)} := \sup_{x \in G} |f(x)|, \qquad f \in \mathcal{D},$$

and (13) yields the estimate

$$\|\mathbb{K}f\|_{C^0(G)} \leq C\|f\|_{L^q(G)} \qquad \text{for all } f \in \mathcal{D} \tag{14}$$

with a constant $C \in (0, +\infty)$. Therefore, $\mathbb{K} : \mathcal{D} \to C^0(G)$ represents a bounded linear operator, where $\mathcal{D}$ is endowed with the $L^q(G)$-norm (see Chapter 2, Section 6). Parallel to Theorem 8.1 from Section 8 in Chapter 2, we can now continue $\mathbb{K}$ to the operator

$$\mathbb{K} : L^q(G) \to C^0(G) \tag{15}$$

on the Banach space $L^q(G)$. The set $C_0^\infty(G) \subset \mathcal{D}$ is dense in the space $L^q(G)$, and for each $f \in L^q(G)$ we have a sequence

$$\{f_j\}_{j=1,2,\ldots} \subset C_0^\infty(G) \qquad \text{satisfying} \qquad \|f - f_j\|_{L^q(G)} \to 0 \; (j \to \infty).$$

We then define

$$\mathbb{K}f := \lim_{j \to \infty} \mathbb{K}f_j \qquad \text{in } C^0(G). \tag{16}$$

We summarize our considerations to the following

Theorem 2.2. *The integral operator $\mathbb{K} : \mathcal{D} \to C^0(G)$ with the singular kernel $K \in \mathcal{S}_\alpha(G)$ and $\alpha \in [0, n)$ can be uniquely continued to the bounded linear operator $\mathbb{K} : L^q(G) \to C^0(G)$ satisfying*

$$\|\mathbb{K}f\|_{C^0(G)} \leq C(q)\|f\|_{L^q(G)}, \qquad f \in L^q(G) \tag{17}$$

for each $q \in (\frac{n}{n-\alpha}, +\infty)$, according to (16). Here we have chosen the constant $C = C(q) \in (0, +\infty)$ appropriately.

Remark: In the case $n \geq 3$, Green's function of the Laplace operator $H = H(x, y)$ belongs to the class $\mathcal{S}_{n-2}(G)$ which means $\alpha = n-2$ and $q \in (\frac{n}{2}, +\infty)$. Therefore, the associate singular integral operator

$$\mathbb{H} : L^q(G) \to C^0(G)$$

is even defined on the Hilbert space $L^2(G)$ for $n = 3$. In the case $n > 3$, Green's function $\mathbb{H}$ cannot be continued onto the Hilbert space $L^2(G)$.

For the orders $\alpha \in [0, n)$ and $\beta \in [0, n)$ let $K = K(x, y) \in \mathcal{S}_\alpha(G, \mathbb{C})$ and $L = L(y, z) \in \mathcal{S}_\beta(G, \mathbb{C})$ denote two singular kernels with the associate integral operators

$$\mathbb{K}f(x) := \int_G K(x, y)f(y)\, dy, \quad x \in G; \qquad f \in \mathcal{D},$$

$$\mathbb{L}f(y) := \int_G L(y, z)f(z)\, dz, \quad y \in G; \qquad f \in \mathcal{D}. \tag{18}$$

With the aid of Fubini's theorem from Chapter 2, Section 5 we now calculate for all $f \in \mathcal{D}$ and all $x \in G$ as follows:

$$
\begin{aligned}
\mathbb{K} \circ \mathbb{L} f(x) &= \mathbb{K} \left(\int_G L(y, z) f(z) \, dz \right) \bigg|_x \\
&= \int_G K(x, y) \left(\int_G L(y, z) f(z) \, dz \right) dy \\
&= \int_G \int_G K(x, y) L(y, z) f(z) \, dz \, dy \qquad\qquad (19) \\
&= \int_G \left(\int_G K(x, y) L(y, z) \, dy \right) f(z) \, dz \\
&= \int_G M(x, z) f(z) \, dz = \mathbb{M} f(x), \qquad x \in G.
\end{aligned}
$$

Here we have as the product kernel

$$
M(x, z) = \int_G K(x, y) L(y, z) \, dy, \qquad (x, z) \in G \otimes G. \qquad (20)
$$

Proposition 2.3. *We have the regularity result* $M = M(x, z) \in C^0(G \otimes G, \mathbb{C})$.

Proof: We take the point $(x^0, z^0) \in G \otimes G$ such that $x^0, z^0 \in G$ and $x^0 \neq z^0$ holds true. Then we choose the number $0 < \delta < \frac{1}{4}|x^0 - z^0|$ sufficiently small and define the sets

$$
B_\delta := \left\{ y \in G : |y - x^0| \leq 2\delta \text{ or } |y - z^0| \leq 2\delta \right\},
$$

$$
G_\delta := G \backslash B_\delta = \left\{ y \in G : |y - x^0| > 2\delta \text{ and } |y - z^0| > 2\delta \right\}.
$$

Given the quantity $\varepsilon > 0$, we find a number $\delta = \delta(\varepsilon) > 0$ with the property

$$
\int_{B_\delta} |K(x, y) L(y, z)| \, dy \leq \varepsilon \qquad (21)
$$

for all $x, z \in G$ with $|x - x^0| \leq \delta$ and $|z - z^0| \leq \delta$, taking $K \in \mathcal{S}_\alpha$ and $L \in \mathcal{S}_\beta$ into account. Furthermore, we have a number $\eta \in (0, \delta]$ such that

$$
\left| K(x, y) L(y, z) - K(x^0, y) L(y, z^0) \right| \leq \varepsilon \qquad (22)
$$

holds true for all $y \in G_\delta$ and $x, z \in G$ with $|x - x^0| \leq \eta$ and $|z - z^0| \leq \eta$. Finally, we obtain the following estimate

$$|M(x,z) - M(x^0, z^0)| \leq \int\limits_{G_\delta} \Big| K(x,y)L(y,z) - K(x^0,y)L(y,z^0) \Big| \, dy$$

$$+ \int\limits_{B_\delta} \Big| K(x,y)L(y,z) - K(x^0,y)L(y,z^0) \Big| \, dy$$

$$\leq \varepsilon|G| + 2\varepsilon$$

for all $x, z \in G$ with $|x - x^0| \leq \eta$ and $|z - z^0| \leq \eta$. Therefore, the regularity result $M = M(x,z) \in C^0(G \otimes G)$ is correct.

q.e.d.

Proposition 2.4. *If $\alpha + \beta < n$ holds true, we have $M \in \mathcal{S}_0(G, \mathbb{C})$.*

Proof: We have to prove only that the kernel M is bounded. Without loss of generality, we can assume $\alpha > 0$ and $\beta > 0$. Taking $(x,z) \in G \otimes G$, we estimate with the aid of Hölder's inequality as follows:

$$|M(x,z)| \leq \int\limits_G |K(x,y)||L(y,z)| \, dy \leq c_1 c_2 \int\limits_G \frac{1}{|x-y|^\alpha} \frac{1}{|y-z|^\beta} \, dy$$

$$\leq c_1 c_2 \left(\int\limits_G \frac{1}{|x-y|^{\alpha+\beta}} \, dy \right)^{\frac{\alpha}{\alpha+\beta}} \left(\int\limits_G \frac{1}{|y-z|^{\alpha+\beta}} \, dy \right)^{\frac{\beta}{\alpha+\beta}}$$

$$\leq c_1 c_2 C \qquad \text{for all} \quad (x,z) \in G \otimes G.$$

Here we observe $C := \sup\limits_{x \in G} \int\limits_G \frac{1}{|x-y|^{\alpha+\beta}} \, dy < +\infty$, since $\alpha + \beta < n$ holds true.

q.e.d.

Proposition 2.5. *In the case $\alpha + \beta > n$, we have the regularity result $M \in \mathcal{S}_{\alpha+\beta-n}(G, \mathbb{C})$.*

Proof: We set $R := \operatorname{diam} G \in (0, +\infty)$, and for the points $x, z \in G$ satisfying $x \neq z$ we define the quantity $\delta := |x - z| \in (0, R)$. Then we calculate

$$|M(x,z)| \leq \int\limits_G |K(x,y)||L(y,z)| \, dy \leq c \int\limits_G \frac{1}{|x-y|^\alpha} \cdot \frac{1}{|y-z|^\beta} \, dy$$

$$= c \int\limits_{\substack{y \in G \\ |y-x| \leq \frac{1}{2}\delta}} \frac{1}{|x-y|^\alpha} \frac{1}{|y-z|^\beta} \, dy + c \int\limits_{\substack{y \in G \\ \frac{1}{2}\delta \leq |y-x| \leq 2\delta}} \frac{1}{|x-y|^\alpha} \frac{1}{|y-z|^\beta} \, dy$$

$$+ c \int\limits_{\substack{y \in G \\ |y-x| \geq 2\delta}} \frac{1}{|x-y|^\alpha} \frac{1}{|y-z|^\beta} \, dy$$

with a constant $c \in (0, +\infty)$.

Taking the point $y \in G$ with $|y - x| \leq \frac{1}{2}\delta$, we estimate as follows:

$$|y - z| \geq |z - x| - |x - y| \geq \delta - \frac{1}{2}\delta = \frac{1}{2}\delta.$$

Taking the point $y \in G$ with $|y - x| \geq 2\delta$, we obtain

$$|y - z| \geq |y - x| - |x - z| = |y - x| - \delta \geq |y - x| - \frac{1}{2}|y - x| = \frac{1}{2}|y - x|.$$

Consequently, we see

$$|M(x, z)| \leq \frac{c}{(\frac{1}{2}\delta)^\beta} \int\limits_{y: |y-x| \leq \frac{1}{2}\delta} \frac{1}{|y - x|^\alpha} \, dy + \frac{c}{(\frac{1}{2}\delta)^\alpha} \int\limits_{y: |y-x| \leq 2\delta} \frac{1}{|y - z|^\beta} \, dy$$

$$+ \frac{c}{(\frac{1}{2})^\beta} \int\limits_{y: |y-x| \geq 2\delta} \frac{1}{|y - x|^{\alpha+\beta}} \, dy$$

$$\leq \frac{c}{(\frac{1}{2}\delta)^\beta} \int\limits_{y: |y-x| \leq \frac{1}{2}\delta} \frac{1}{|y - x|^\alpha} \, dy + \frac{c}{(\frac{1}{2}\delta)^\alpha} \int\limits_{y: |y-z| \leq 3\delta} \frac{1}{|y - z|^\beta} \, dy$$

$$+ \frac{c}{(\frac{1}{2})^\beta} \int\limits_{y: |y-x| \geq 2\delta} \frac{1}{|y - x|^{\alpha+\beta}} \, dy. \tag{23}$$

We now substitute

$$y = x + \varrho\xi, \quad dy = \omega_n \varrho^{n-1} \, d\varrho, \quad \varrho \in (0, \frac{1}{2}\delta), \quad \xi \in S^{n-1},$$

and calculate

$$\int\limits_{y: |y-x| \leq \frac{1}{2}\delta} \frac{1}{|y - x|^\alpha} \, dy = \int\limits_0^{\frac{1}{2}\delta} \varrho^{-\alpha} \varrho^{n-1} \omega_n \, d\varrho = \omega_n \int\limits_0^{\frac{1}{2}\delta} \varrho^{n-\alpha-1} \, d\varrho$$

$$= \frac{\omega_n}{n - \alpha} \left[\varrho^{n-\alpha} \right]_0^{\frac{1}{2}\delta} = \frac{\omega_n}{n - \alpha} \left(\frac{1}{2}\delta \right)^{n-\alpha}. \tag{24}$$

Analogously, we get

$$\int\limits_{y: |y-z| \leq 3\delta} \frac{1}{|y - z|^\beta} \, dy = \frac{\omega_n}{n - \beta} (3\delta)^{n-\beta}. \tag{25}$$

With the aid of the substitution above, we deduce

$$\int\limits_{y:|y-x|\geq 2\delta} \frac{1}{|y-x|^{\alpha+\beta}}\, dy = \omega_n \int\limits_{2\delta}^{+\infty} \varrho^{-\alpha-\beta}\varrho^{n-1}d\varrho$$

$$= \omega_n \frac{1}{n-(\alpha+\beta)}\Big[\varrho^{n-\alpha-\beta}\Big]_{2\delta}^{+\infty} \tag{26}$$

$$= \frac{\omega_n}{\alpha+\beta-n}(2\delta)^{n-\alpha-\beta}.$$

Combining (23), (24), (25), and (26) we finally obtain the estimate

$$|M(x,z)| \leq c\Big\{\Big(\frac{1}{2}\Big)^{n-\alpha-\beta}\frac{\omega_n}{n-\alpha} + 3^{n-\beta}\Big(\frac{1}{2}\Big)^{-\alpha}\frac{\omega_n}{n-\beta} + \frac{2^{n-\alpha}\omega_n}{\alpha+\beta-n}\Big\}\delta^{n-\alpha-\beta}$$

$$= \frac{C(n,\alpha,\beta)}{|x-z|^{\alpha+\beta-n}} \qquad \text{for all} \quad x,z \in G \quad \text{with} \quad x \neq z.$$

$$\tag{27}$$

Therefore, the statement $M \in \mathcal{S}_{\alpha+\beta-n}(G,\mathbb{C})$ follows.   q.e.d.

We summarize our arguments to the subsequent

Theorem 2.6. (I. Schur)
To the given orders $\alpha \in [0,n)$, $\beta \in [0,n)$ let $K = K(x,y) \in \mathcal{S}_\alpha(G,\mathbb{C})$, $L = L(y,z) \in \mathcal{S}_\beta(G,\mathbb{C})$ denote singular kernels with the associate integral operators $\mathbb{K}$, $\mathbb{L}$. Then the composition

$$\mathbb{K}\circ\mathbb{L}f(x) = \int\limits_G M(x,z)f(z)\,dz, \quad x \in G, \qquad f \in \mathcal{D}$$

represents a singular integral operator as well, where its product kernel

$$M(x,z) = \int\limits_G K(x,y)L(y,z)\,dy, \qquad (x,z) \in G \otimes G$$

satisfies the following regularity properties:

$$M = M(x,y) \in \begin{cases} \mathcal{S}_0(G,\mathbb{C}), & \text{if } \alpha+\beta < n \\ \mathcal{S}_{\alpha+\beta-n}(G,\mathbb{C}), & \text{if } \alpha+\beta > n \end{cases}.$$

Theorem 2.7. (Iterated kernels)
Let $K = K(x,y) \in \mathcal{S}_\alpha(G,\mathbb{C})$ denote a singular kernel of the order $0 < \alpha < n$ with the associate integral operator $\mathbb{K}$. Then we have a positive integer $k = k(K) \in \mathbb{N}$ and a kernel $L = L(x,y) \in \mathcal{S}_0(G,\mathbb{C})$ with the associate integral operator $\mathbb{L}$ such that

$$\mathbb{K}^k f = \mathbb{L}f \qquad \text{for all} \quad f \in \mathcal{D}.$$

Proof: We choose $\beta \in (\alpha, n)$ satisfying

$$\beta \neq \frac{m}{m+1} n \qquad \text{for all} \quad m \in \mathbb{N}.$$

This implies

$$\beta + m(\beta - n) \neq 0 \qquad \text{for all} \quad m \in \mathbb{N}.$$

With the aid of the theorem by I. Schur we now consider the iterated kernels:

$$K \in \mathcal{S}_\alpha \subset \mathcal{S}_\beta, \qquad K^2 = K \circ K \in \mathcal{S}_{\beta+\beta-n} = \mathcal{S}_{\beta+1(\beta-n)},$$

$$K^3 = K \circ K \circ K \in \mathcal{S}_{\beta+2(\beta-n)}, \quad \ldots, \quad K^k = \underbrace{K \circ \ldots \circ K}_{k} \in \mathcal{S}_{\beta+(k-1)(\beta-n)}.$$

We now determine the number $k \in \mathbb{N}$ such that

$$\beta + (k-2)(\beta - n) > 0 \qquad \text{and} \qquad \beta + (k-1)(\beta - n) < 0$$

is satisfied, and we infer

$$\{\beta + (k-2)(\beta - n)\} + \beta = \beta + (k-1)(\beta - n) + n < n.$$

Theorem 2.6 finally yields $K^k \in \mathcal{S}_0(G, \mathbb{C})$.   q.e.d.

An outlook on the treatment of the eigenvalue problem for the n-dimensional oscillation equation (Weyl's eigenvalue problem): Parallel to Theorem 1.9 from Section 1 we use the domain of definition

$$\mathcal{D}_0 := \left\{ u = u(x) \in C^2(G) \cap C^0(\overline{G}) \,:\, u(x) = 0 \text{ on } \partial G \right\}$$

and consider the eigenvalue problem for the n-dimensional oscillation equation

$$- \Delta u(x) = \lambda u(x), \quad x \in G; \qquad u \in \mathcal{D}_0 \setminus \{0\}, \quad \lambda \in \mathbb{R}. \tag{28}$$

In Section 9 we show the property $\lambda > 0$. Then the differential equation (28) can be transferred into the singular integral equation

$$\mathbb{H}u(x) := \int_G H(x, y)u(y)\, dy = \frac{1}{\lambda} u(x), \qquad x \in G, \tag{29}$$

with the singular kernel $H = H(x, y) \in \mathcal{S}_{n-2}(G)$ which is symmetric. Due to Theorem 2.7 we now choose a number $k \in \mathbb{N}$ satisfying

$$\mathbb{H}^k u = \mathbb{K}u = \int_G K(x, y)u(y)\, dy, \qquad u \in \mathcal{D}_0,$$

with a kernel $K = K(x, y) \in \mathcal{S}_0(G)$. The eigenvalue problem (29) is transferred into the equation

$$\mathbb{K}u = \mathbb{H}^k u = \frac{1}{\lambda^k} u(x), \qquad x \in G. \tag{30}$$

Now we can continue the operator $\mathbb{K} : L^q(G) \to C^0(G)$ for each exponent $q > 1$, due to Theorem 2.2. With (30) we obtain an eigenvalue problem on the Hilbert space $L^2(G) \subset L^q(G)$, if $q \in (1,2]$ holds true. Therefore, it suffices in the following considerations to investigate eigenvalue problems for operators in Hilbert spaces.

3 The Abstract Hilbert Space

We now continue the considerations from Section 6 in Chapter 2.

Postulate (A): $\mathcal{H}$ is a linear space. This means $\mathcal{H}$ is an additive Abelian group with 0 as its neutral element:

$$x, y \in \mathcal{H} \;\Rightarrow\; x + y \in \mathcal{H}, \qquad x = 0 \in \mathcal{H}.$$

Furthermore, we have a scalar multiplication in $\mathcal{H}$: With the number $\lambda \in \mathbb{C}$ and the element $x \in \mathcal{H}$ the statement $\lambda x \in \mathcal{H}$ is correct, and the axioms for vector spaces are valid.

Postulate (B): In $\mathcal{H}$ we have defined the inner product

$$\mathcal{H} \times \mathcal{H} \to \mathbb{C}$$
$$(x, y) \mapsto (x, y)_{\mathcal{H}}$$

with the following properties:

(a) $(x, \alpha y)_{\mathcal{H}} = \alpha (x, y)_{\mathcal{H}}$ for all $x, y \in \mathcal{H}$ and $\alpha \in \mathbb{C}$
(b) $(x, y)_{\mathcal{H}} = \overline{(y, x)_{\mathcal{H}}}$ for all $x, y \in \mathcal{H}$ (Hermitian character)
(c) $(x_1 + x_2, y)_{\mathcal{H}} = (x_1, y)_{\mathcal{H}} + (x_2, y)_{\mathcal{H}}$ for all $x_1, x_2, y \in \mathcal{H}$
(d) $(x, x)_{\mathcal{H}} \geq 0$ for all $x \in \mathcal{H}$ and $(x, x)_{\mathcal{H}} = 0 \Leftrightarrow x = 0$ (positive-definiteness)

Postulate (C): For each positive integer $n \in \mathbb{N}$ we have n linear independent elements $x_1, \ldots, x_n \in \mathcal{H}$, which means

$$\alpha_1, \ldots \alpha_n \in \mathbb{C}, \quad \sum_{i=1}^{n} \alpha_i x_i = 0 \quad \Rightarrow \quad \alpha_1 = \ldots = \alpha_n = 0.$$

Definition 3.1. *If the set $\mathcal{H}'$ satisfies the Postulates (A), (B), and (C) then $\mathcal{H}'$ is named a* pre-Hilbert-space.

Example 3.2. Let $G \subset \mathbb{R}^n$ denote a bounded open set and define

$$\mathcal{H}' := \left\{ f : G \to \mathbb{C} \in C^0(G) \; : \; \sup_{x \in G} |f(x)| < +\infty \right\}.$$

With the inner product

$$(f, g) := \int\limits_{G} \overline{f(x)} g(x) \, dx, \qquad f, g \in \mathcal{H}', \tag{1}$$

the vector space $\mathcal{H}'$ becomes a pre-Hilbert-space.

Theorem 3.3. *In pre-Hilbert-spaces $\mathcal{H}'$ we have the following calculus rules for the inner product $(.\,,.)$:*

a) *For all $x, y, y_1, y_2 \in \mathcal{H}'$, $\alpha \in \mathbb{C}$ we have*

$$(\alpha x, y) = \overline{\alpha}(x, y), \qquad (x, y_1 + y_2) = (x, y_1) + (x, y_2).$$

Consequently, the bilinear form $(.\,,.)$ is antilinear in the first and linear in the second component.

b) *The Cauchy-Schwarz inequality is satisfied:*

$$|(x, y)| \leq \sqrt{(x, x)} \sqrt{(y, y)} \qquad \text{for all} \quad x, y \in \mathcal{H}'.$$

c) *Setting $\|x\| := \sqrt{(x, x)}$ with $x \in \mathcal{H}'$, the pre-Hilbert-space $\mathcal{H}'$ becomes a normed space. This means*

$$\|x\| = 0 \iff x = 0,$$
$$\|x + y\| \leq \|x\| + \|y\| \qquad \text{for all} \quad x, y \in \mathcal{H}',$$
$$\|\lambda x\| = |\lambda| \|x\| \qquad \text{for all} \quad x \in \mathcal{H}', \, \lambda \in \mathbb{C},$$
$$\|x - y\| \geq |\, \|x\| - \|y\| \,| \qquad \text{for all} \quad x, y \in \mathcal{H}'.$$

d) *The inner product is continuous on $\mathcal{H}'$ in the following sense: From the assumptions*

$$x_n \to x \; (n \to \infty) \qquad \text{with} \quad \{x_n\}_{n=1,2,\dots} \subset \mathcal{H}' \quad \text{and} \quad x \in \mathcal{H}'$$

and

$$y_n \to y \; (n \to \infty) \qquad \text{with} \quad \{y_n\}_{n=1,2,\dots} \subset \mathcal{H}' \quad \text{and} \quad y \in \mathcal{H}'$$

we infer

$$(x_n, y_n) \to (x, y) \; (n \to \infty).$$

Here the symbol $x_n \to x \; (n \to \infty)$ indicates that $\|x_n - x\| \to 0 \; (n \to \infty)$ is satisfied.

Proof:

a) We calculate

$$(\alpha x, y) = \overline{(y, \alpha x)} = \overline{\alpha(y, x)} = \overline{\alpha}(x, y)$$

and

$$(x, y_1 + y_2) = \overline{(y_1 + y_2, x)} = \overline{(y_1, x) + (y_2, x)}$$
$$= \overline{(y_1, x)} + \overline{(y_2, x)} = (x, y_1) + (x, y_2).$$

b) and c) are contained in Section 6 of Chapter 2; more precisely in Theorem 6.7 with its proof and the Remark following Definition 6.1.

d) The subsequent estimate yields this statement:

$$|(x_n, y_n) - (x, y)| \leq |(x_n, y_n) - (x_n, y)| + |(x_n, y) - (x, y)|$$
$$= |(x_n, y_n - y)| + |(x_n - x, y)|$$
$$\leq \|x_n\|\|y_n - y\| + \|x_n - x\|\|y\| \to 0 \ (n \to \infty).$$

$$\text{q.e.d.}$$

Postulate (D): $\mathcal{H}$ is complete. This means each sequence $\{x_n\}_{n=1,2,\ldots} \subset \mathcal{H}$ satisfying $\|x_n - x_m\| \to 0 \ (n, m \to \infty)$ possesses a limit element $x \in \mathcal{H}$ such that

$$\lim_{n \to \infty} \|x_n - x\| = 0.$$

Definition 3.4. *If $\mathcal{H}$ satifies the Postulates (A), (B), (C), and (D) we name $\mathcal{H}$ a Hilbert space.*

Remark: The Hilbert space $\mathcal{H}$ becomes a Banach space via the norm given in Theorem 3.3,c).

Definition 3.5. *The Hilbert space $\mathcal{H}$ is* separable, *if the following Postulate (E) holds true additionally:*

Postulate (E): *There exists a sequence $\{x_n\}_{n=1,2,\ldots} \subset \mathcal{H}$ which is dense in $\mathcal{H}$: This means for all $x \in \mathcal{H}$ and every $\varepsilon > 0$ we have an index $n \in \mathbb{N}$ satisfying $\|x - x_n\| < \varepsilon$.*

Example 3.6. Hilbert's sequential space
We endow the set of sequences

$$l_2 := \left\{ x = (x_1, x_2, \ldots) \in \mathbb{C} \times \mathbb{C} \times \ldots : \sum_{k=1}^{\infty} |x_k|^2 < +\infty \right\}$$

with the inner product

$$(x, y) := \sum_{k=1}^{\infty} \overline{x_k} y_k \ \in \mathbb{C}$$

and we obtain a separable Hilbert space.

Example 3.7. Let $G \subset \mathbb{R}^n$ be a bounded open set, and by

$$L^2(G) := \left\{ f : G \to \mathbb{C} \text{ measurable} : \int\limits_G |f(x)|^2 \, dx < +\infty \right\}$$

we denote the Lebesgue space of the square-integrable functions with the inner product

$$(f, g) := \int\limits_G \overline{f(x)} g(x) \, dx \qquad \text{for} \quad f, g \in L^2(G).$$

Then $\mathcal{H} = L^2(G)$ represents a separable Hilbert space. The pre-Hilbert-space $\mathcal{H}'$ described in Example 3.2 lies dense in $\mathcal{H}$ (compare Chapter 2, Section 7).

Parallel to the transition from rational numbers $\mathbb{Q}$ to real numbers $\mathbb{R}$ we prove the following result using the ideas of D. Hilbert, which were presented in his famous book on *The Foundations of Geometry:*

Theorem 3.8. (Hilbert's fundamental theorem)
Each pre-Hilbert-space $\mathcal{H}'$ can be completed to a Hilbert space $\mathcal{H}$ such that $\mathcal{H}'$ lies dense in $\mathcal{H}$. We name $\mathcal{H}$ the abstract completion *of $\mathcal{H}'$. When $\mathcal{H}'$ satisfies the Postulate (E), then the abstract completion $\mathcal{H}$ is a separable Hilbert space.*

Proof: Let $\mathcal{H}'$ be a pre-Hilbert-space. We then consider the Cauchy sequences $\{f'_n\}_{n=1,2,\ldots} \subset \mathcal{H}'$ and $\{g'_n\}_{n=1,2,\ldots} \subset \mathcal{H}'$. We call them equivalent if

$$f'_n - g'_n \to 0 \ (n \to \infty)$$

is satisfied. Now we set

$$\mathcal{H} := \left\{ f = [f'_n]_{n=1,2,\ldots} : \begin{array}{l} [f'_n] \text{ is the equivalence class} \\ \text{of the Cauchy sequences } \{f'_n\}_{n=1,2,\ldots} \subset \mathcal{H}' \end{array} \right\}.$$

For $f = [f'_n]_n \in \mathcal{H}$ and $g = [g'_n]_n \in \mathcal{H}$ we evidently have the statement

$$[f'_n]_n = [g'_n]_n \quad \Leftrightarrow \quad \|f'_n - g'_n\| \to 0 \quad (n \to \infty).$$

To Postulate (A): On $\mathcal{H}$ we define a vector space structure as follows: For $\alpha, \beta \in \mathbb{C}$ and $f = [f'_n]_n \in \mathcal{H}$, $g = [g'_n]_n \in \mathcal{H}$ we set

$$\alpha f + \beta g := [\alpha f'_n + \beta g'_n]_n.$$

The null element is the equivalence class of all zero sequences in $\mathcal{H}'$:

$$0 = [f'_n]_n \quad \text{with} \quad \{f'_n\}_{n=1,2,\ldots} \subset \mathcal{H}' \quad \text{and} \quad \|f'_n\| \to 0 \ (n \to \infty).$$

To Postulate (B): For the elements $f = [f'_n]_n \in \mathcal{H}$ and $g = [g'_n]_n \in \mathcal{H}$ we define the inner product

$$(f, g) := \lim_{n \to \infty} (f'_n, g'_n).$$

On account of

$$|(f'_n, g'_n) - (f'_m, g'_m)| \leq |(f'_n - f'_m, g'_n)| + |(f'_m, g'_n - g'_m)|$$

$$\leq \|f'_n - f'_m\|\|g'_n\| + \|f'_m\|\|g'_n - g'_m\| \to 0 \ (n, m \to \infty),$$

the limit given above exists. One easily verifies that the so-defined inner product satisfies the Postulate (B).

To the Postulates (C) and (E): Let $\widetilde{\mathcal{H}}$ denote the set of all $f \in \mathcal{H}$ satisfying $f = [f', f', \ldots]$ and $f' \in \mathcal{H}'$. Then the vector spaces $\mathcal{H}'$ and $\widetilde{\mathcal{H}}$ are isomorphic, and consequently $\mathcal{H}'$ is embedded into $\mathcal{H}$. Now $\widetilde{\mathcal{H}}$ is dense in $\mathcal{H}$: Taking $f = [f'_n]_n \in \mathcal{H}$ we set $\tilde{f}_m = [f'_m, f'_m, \ldots] \in \widetilde{\mathcal{H}}$ and see

$$\|f - \tilde{f}_m\| = \lim_{n \to \infty} \|f'_n - f'_m\| \to 0 \ (m \to \infty).$$

Evidently, the Postulate (C) remains valid for $\mathcal{H}$. In the case that $\mathcal{H}'$ additionally satisfies the Postulate (E), this holds true for $\mathcal{H}$ as well.

To Postulate (D): Let $\{f_n\}_{n=1,2,\ldots} \subset \mathcal{H}$ with

$$\|f_n - f_m\| \to 0 \ (n, m \to \infty)$$

be chosen. Since $\widetilde{\mathcal{H}}$ lies dense in $\mathcal{H}$ we have a sequence $\{\tilde{f}_n\}_{n=1,2,\ldots} \subset \widetilde{\mathcal{H}}$ satisfying

$$\|f_n - \tilde{f}_n\| \leq \frac{1}{n}, \quad n = 1, 2, \ldots$$

Here we have $\tilde{f}_n = [f'_n, f'_n, \ldots]$ with $f'_n \in \mathcal{H}'$. We now set $f := [f'_n]_{n=1,2,\ldots}$ and show that $f \in \mathcal{H}$ and $\|f - f_n\| \to 0 \ (n \to \infty)$ are correct. At first, we estimate

$$\|f'_n - f'_m\| = \|\tilde{f}_n - \tilde{f}_m\| \leq \|\tilde{f}_n - f_n\| + \|f_n - f_m\| + \|f_m - \tilde{f}_m\|$$

$$\leq \frac{1}{n} + \|f_n - f_m\| + \frac{1}{m} \ \to \ 0 \ (n, m \to \infty).$$

Now we have

$$\|f - f_m\| \leq \|f - \tilde{f}_m\| + \|\tilde{f}_m - f_m\| \leq \|f - \tilde{f}_m\| + \frac{1}{m},$$

and note that $\tilde{f}_m = [f'_m, f'_m, \ldots]$ and $f = [f'_1, f'_2, \ldots]$. Then we infer

$$\|f - \tilde{f}_m\| = \lim_{n \to \infty} \|f'_n - f'_m\| \leq \varepsilon_m$$

with the numbers $\varepsilon_m > 0$ and $m \in \mathbb{N}$ satisfying $\varepsilon_m \to 0\,(m \to \infty)$. We summarize our considerations to

$$\|f - f_m\| \le \varepsilon_m + \frac{1}{m} \to 0\ (m \to \infty).$$

q.e.d.

Remark: We can complete the pre-Hilbert-space $\mathcal{H}'$ from the Example 3.2 abstractly to a Hilbert space $\mathcal{H}$ with the aid of Theorem 3.8. Alternatively, we can concretely complete $\mathcal{H}'$ to the Hilbert space

$$L^2(G, \mathbb{C}) := \Big\{ f : G \to \mathbb{C} \text{ measurable} : \int_G |f(x)|^2 dx < +\infty \Big\},$$

whose inner product is given in (1).

Definition 3.9. *A sequence of elements $\{\varphi_1, \varphi_2, \ldots\} \subset \mathcal{H}'$ in a pre-Hilbert-space $\mathcal{H}'$ is called* orthonormal *if and only if*

$$(\varphi_i, \varphi_j) = \delta_{ij} \qquad \text{for all} \quad i, j \in \mathbb{N}$$

is correct. We name the orthonormal system $\{\varphi_k\}_{k=1,2,\ldots}$ complete - briefly we speak of a c.o.n.s. - if each element $f \in \mathcal{H}'$ satisfies the completeness *relation*

$$\|f\|^2 = \sum_{k=1}^{\infty} |(\varphi_k, f)|^2. \tag{2}$$

This definition is justified by the following theorem, whose proof is contained in the Propositions 6.17, 6.18 and Theorem 6.19 at the end of Section 6 in Chapter 2.

Theorem 3.10. *Let $\{\varphi_k\}_{k=1,2,\ldots} \subset \mathcal{H}'$ represent an orthonormal system. For all $f \in \mathcal{H}'$ we then have Bessel's inequality*

$$\sum_{k=1}^{\infty} |(\varphi_k, f)|^2 \le \|f\|^2. \tag{3}$$

An element $f \in \mathcal{H}'$ satisfies the equation

$$\sum_{k=1}^{\infty} |(\varphi_k, f)|^2 = \|f\|^2 \tag{4}$$

if and only if

$$\lim_{N \to \infty} \Big\| f - \sum_{k=1}^{N} (\varphi_k, f) \varphi_k \Big\| = 0 \tag{5}$$

holds true. The latter statement means that $f \in \mathcal{H}'$ can be represented by the Fourier series

$$\sum_{k=1}^{\infty} (\varphi_k, f)\varphi_k$$

converging with respect to the norm $\| \cdot \|$ *in the Hilbert space.*

Example 3.11. With the Fourier series and the spherical harmonic functions in Section 4 and Section 5 from Chapter 5, respectively, we obtain two c.o.n.s. in the adequate Hilbert spaces.

Theorem 3.12. *An orthonormal system* $\{\varphi_k\}_{k=1,2,...}$ *in the pre-Hilbert-space* $\mathcal{H}'$ *is complete if and only if the relation* $(\varphi_k, x) = 0$, $k = 1, 2, \ldots$ *with* $x \in \mathcal{H}'$ *implies the identity* $x = 0$.

Proof:

'$\Rightarrow$' Let $x \in \mathcal{H}'$ and $(\varphi_k, x) = 0$ hold true for all $k \in \mathbb{N}$. Then the completeness relation yields

$$\|x\|^2 = \sum_{k=1}^{\infty} |(\varphi_k, x)|^2 = 0$$

as well as $\|x\| = 0$ and consequently $x = 0 \in \mathcal{H}'$.

'$\Leftarrow$' Let $\{\varphi_k\}_{k=1,2,...}$ be an orthonormal system such that the statement $(\varphi_k, x) = 0$ for all $k \in \mathbb{N}$ implies $x = 0$. For arbitrary $y \in \mathcal{H}$ we set

$$x := y - \sum_{k=1}^{\infty} (\varphi_k, y)\varphi_k$$

and we calculate

$$(\varphi_l, x) = (\varphi_l, y) - \left(\varphi_l, \sum_{k=1}^{\infty} (\varphi_k, y)\varphi_k\right) = (\varphi_l, y) - (\varphi_l, y) = 0$$

for all $l \in \mathbb{N}$. This implies $x = 0$ and consequently

$$y = \sum_{k=1}^{\infty} (\varphi_k, y)\varphi_k.$$

Therefore, the system $\{\varphi_k\}_{k=1,2,...} \subset \mathcal{H}'$ is complete due to Theorem 3.10.

q.e.d.

Theorem 3.13. *Let* $\mathcal{H}$ *denote a separable Hilbert space.*

a) *Then there exists a c.o.n.s.* $\{\varphi_k\}_{k=1,2,...} \subset \mathcal{H}$.

b) *For two arbitrary elements $x, y \in \mathcal{H}$ we have the* Parseval equation

$$(x, y) = \sum_{k=1}^{\infty} \overline{(\varphi_k, x)}(\varphi_k, y). \tag{6}$$

c) *The Hilbert space $\mathcal{H}$ is isomorphic to the Hilbert sequential space l_2 via the mapping*

$$\Phi : \mathcal{H} \rightarrow l_2, \quad x \mapsto (x_1, x_2, \ldots) \quad \text{with} \quad x_k := (\varphi_k, x).$$

By the prescription

$$x = \sum_{k=1}^{\infty} x_k \varphi_k \quad \text{with} \quad (x_1, x_2, \ldots) \in l_2$$

the mapping inverse to Φ is given.

Proof:

a) Since $\mathcal{H}$ is separable, we have a sequence $\{g_1, g_2, \ldots\} \subset \mathcal{H}$ which is dense in $\mathcal{H}$. We eliminate the linear dependent functions from $\{g_1, g_2, \ldots\}$, and construct a system of linear independent functions $\{f_1, f_2, \ldots\}$ in $\mathcal{H}$ with the following property:

$$[g_1, \ldots, g_n] \subset [f_1, \ldots, f_p] \quad \text{for all} \quad p \geq n \geq 1, \tag{7}$$

denoting with $[g_1, \ldots, g_n]$ and $[f_1, \ldots, f_p]$ the $\mathbb{C}$-linear spaces spanned by the elements $g_1, \ldots, g_n$ and $f_1, \ldots, f_p$, respectively; here $n, p \in \mathbb{N}$ holds true. Now we apply the *orthonormalizing procedure of E. Schmidt* to the system of functions $\{f_k\}_{k=1,2\ldots}$:

$$\varphi_1 := \frac{1}{\|f_1\|} f_1, \quad \varphi_2 := \frac{f_2 - (\varphi_1, f_2)\varphi_1}{\|f_2 - (\varphi_1, f_2)\varphi_1\|}, \quad \ldots,$$

$$\varphi_n := \frac{f_n - \sum_{j=1}^{n-1} (\varphi_j, f_n)\varphi_j}{\left\| f_n - \sum_{j=1}^{n-1} (\varphi_j, f_n)\varphi_j \right\|}, \quad n = 1, 2, \ldots$$

We evidently have $(\varphi_j, \varphi_k) = \delta_{jk}$ for $j, k = 1, 2, \ldots$ and

$$[g_1, \ldots, g_n] \subset [\varphi_1, \ldots, \varphi_p] \quad \text{for all} \quad p \geq n \geq 1. \tag{8}$$

When $f \in \mathcal{H}$ and $\varepsilon > 0$ are given, we have an index $n \in \mathbb{N}$ satisfying $\|f - g_n\| \leq \varepsilon$. Due to (8) we find $p \geq n$ numbers $c_1, \ldots, c_p \in \mathbb{C}$ with

$$\left\| f - \sum_{k=1}^{p} c_k \varphi_k \right\| \leq \varepsilon.$$

Because of the minimal property of the Fourier coefficients (compare Chapter 2, Section 6, Corollary to Proposition 6.17) we can still choose n and p such that

$$\left\| f - \sum_{k=1}^{p} (\varphi_k, f)\varphi_k \right\| \leq \varepsilon \qquad \text{for all} \quad \varepsilon > 0.$$

Observing the limit process $\varepsilon \downarrow 0$ we deduce

$$f = \sum_{k=1}^{\infty} (\varphi_k, f)\varphi_k,$$

and $\{\varphi_k\}_{k=1,2,\ldots}$ is a c.o.n.s.

b) For two elements $x, y \in \mathcal{H}$ with the representations

$$x = \sum_{k=1}^{\infty} (\varphi_k, x)\varphi_k, \qquad y = \sum_{l=1}^{\infty} (\varphi_l, y)\varphi_l$$

we evaluate the inner product

$$(x, y) = \lim_{n \to \infty} \left(\sum_{k=1}^{n} (\varphi_k, x)\varphi_k, \sum_{l=1}^{n} (\varphi_l, y)\varphi_l \right)$$

$$= \lim_{n \to \infty} \sum_{k,l=1}^{n} \overline{(\varphi_k, x)}(\varphi_l, y)(\varphi_k, \varphi_l)$$

$$= \lim_{n \to \infty} \sum_{k=1}^{n} \overline{(\varphi_k, x)}(\varphi_k, y)$$

$$= \sum_{k=1}^{\infty} \overline{(\varphi_k, x)}(\varphi_k, y).$$

c) Here nothing has to be shown any more.   q.e.d.

Definition 3.14. *We name $M \subset \mathcal{H}$ a linear subspace of the Hilbert space $\mathcal{H}$, if for arbitrary elements $f, g \in M$ and all numbers $\alpha, \beta \in \mathbb{C}$ we obtain the inclusion*

$$\alpha f + \beta g \in M.$$

A linear subspace $M \subset \mathcal{H}$ is called closed, *if each Cauchy sequence*

$$\{f_n\}_{n=1,2,\ldots} \subset M$$

fulfills

$$f := \lim_{n \to \infty} f_n \in M.$$

With a linear subspace $M \subset \mathcal{H}$ we denote by

$$\overline{M} := \left\{ f \in \mathcal{H} : \begin{array}{l} \text{There exists a Cauchy sequence } \{f_n\}_{n=1,2,\ldots} \subset M \\ \text{satisfying } f = \lim_{n \to \infty} f_n \end{array} \right\}$$

the closure *of M.*

Example 3.15. The space $C_0^\infty(G) =: \mathcal{M} \subset \mathcal{H} := L^2(G)$ is a nonclosed linear subspace, and we have $\overline{\mathcal{M}} = \mathcal{H}$.

Definition 3.16. *We call $\mathcal{H}$ a unitary space, if the following Postulate (C') is satisfied in addition to the Postulates (A) and (B).*

Postulate (C'): *With an integer $n \in \mathbb{N}$ we have $\dim \mathcal{H} = n$.*

Remarks:

1. In an n-dimensional unitary space $\mathcal{H}$ we have n linearly independent elements $\{f_1, \ldots, f_n\}$, and each $g \in \mathcal{H}$ can be represented in the form

$$g = \sum_{k=1}^{n} c_k f_k \quad \text{with} \quad c_1, \ldots, c_n \in \mathbb{C}.$$

2. A unitary space $\mathcal{H}$ possesses an orthonormal basis $\{\varphi_1, \ldots, \varphi_n\}$ with $n = \dim \mathcal{H}$ satisfying

$$f = \sum_{k=1}^{n} (\varphi_k, f)\varphi_k \quad \text{for all} \quad f \in \mathcal{H}.$$

3. Each unitary space $\mathcal{H}$ endowed with the inner product

$$(x, y) := \sum_{k=1}^{n} \overline{x_k} y_k, \qquad x = (x_1, \ldots, x_n), \ y = (y_1, \ldots, y_n) \in \mathbb{C}$$

   is isomorphic to $\mathbb{C}^n$, where $n = \dim \mathcal{H}$ holds true.
4. Each unitary space is complete.

Noticing the Definitions 3.14 and 3.16, we easily prove the following

Theorem 3.17. *Let $\mathcal{H}$ denote a Hilbert space, and $\mathcal{M}$ is a closed linear subspace of $\mathcal{H}$. Then $\mathcal{M}$ represents either a Hilbert space or a unitary space. In the case that $\mathcal{H}$ is separable the same holds true for $\mathcal{M}$.*

Definition 3.18. *Let $\mathcal{M}$ be a linear subspace of $\mathcal{H}$. Then we define the orthogonal space of $\mathcal{M}$ in $\mathcal{H}$ setting*

$$\mathcal{M}^\perp := \left\{ g \in \mathcal{H} \ : \ (g, f) = 0 \ \text{for all} \ f \in \mathcal{M} \right\}.$$

Remark: On account of the continuity for the inner product, the orthogonal space $\mathcal{M}^\perp \subset \mathcal{H}$ is closed.

From Theorem 6.9 on the Orthogonal projection in Chapter 2, Section 6 we take over the proof of the following

Theorem 3.19. (Projection theorem)
Let M denote a closed linear subspace in $\mathcal{H}$. Then each element $x \in \mathcal{H}$ can be uniquely represented in the form $x = x_1 + x_2$ with $x_1 \in M$ and $x_2 \in M^\perp$. We then obtain the decomposition

$$\mathcal{H} = M \oplus M^\perp.$$

We still note the subsequent

Theorem 3.20. *Let M be a linear subspace in a Hilbert space $\mathcal{H}$. Then M lies dense in $\mathcal{H}$ if and only if the following implication is correct:*

$$\varphi \in \mathcal{H}: \quad (f, \varphi) = 0 \text{ for all } f \in M \quad \Rightarrow \quad \varphi = 0. \tag{9}$$

Proof: We have the orthogonal decomposition $\mathcal{H} = \overline{M} \oplus M^\perp$ via the Projection theorem. Now the subspace M lies dense in $\mathcal{H}$ if the statement $\overline{M} = \mathcal{H}$ and consequently $M^\perp = \{0\}$ is correct. The latter statement coincides with the implication (9).

q.e.d.

4 Bounded Linear Operators in Hilbert Spaces

We begin with the fundamental

Definition 4.1. *On the Hilbert space $\mathcal{H}$ the mapping $A : \mathcal{H} \to \mathbb{C}$ is a bounded linear functional, if the following conditions are fulfilled:*

a) $A(\alpha f + \beta g) = \alpha A f + \beta A g$ *for all* $f, g \in \mathcal{H}$ *and* $\alpha, \beta \in \mathbb{C}$,
b) $|Af| \le c\|f\|$ *for all* $f \in \mathcal{H}$ *, with a constant* $c \in [0, +\infty)$.

Due to Theorem 6.11 of Section 6 in Chapter 2, the following three statements are equivalent for a linear functional:

(i) A is bounded,
(ii) A is continuous at one point,
(iii) A is continuous at all points of the Hilbert space.

We define the *norm of the bounded linear functional A* by

$$\|A\| := \sup_{x \in \mathcal{H}, \|x\| \le 1} |Ax| = \sup_{x \in \mathcal{H}, \|x\| = 1} |Ax| < +\infty.$$

In Chapter 2, Section 6, Theorem 6.14 we have proved the following statement:

Theorem 4.2. (Representation theorem of Fréchet and Riesz)
Each bounded linear functional $A : \mathcal{H} \to \mathbb{C}$ on the Hilbert space $\mathcal{H}$ can be represented in the form

$$Af = (g, f) \qquad \text{for all} \quad f \in \mathcal{H} \tag{1}$$

with a uniquely determined generating element $g \in \mathcal{H}$.

Definition 4.3. *Let $\mathcal{D}$ denote a linear subspace of the Hilbert space $\mathcal{H}$. A linear operator T consists of a function $T : \mathcal{D} \to \mathcal{H}$ with the following property*

$$T(c_1 u_1 + c_2 u_2) = c_1 T(u_1) + c_2 T(u_2) \qquad \text{for all} \quad u_1, u_2 \in \mathcal{D} \quad \text{and} \quad c_1, c_2 \in \mathbb{C}.$$

Definition 4.4. *A linear operator $T : \mathcal{D} \to \mathcal{H}$ is called* bounded, *if we have a number $c \in [0, +\infty)$ such that*

$$\|Tu\| \leq c\|u\| \qquad \text{for all} \quad u \in \mathcal{D} \tag{2}$$

holds true. Then the norm of T is defined by

$$\|T\| := \sup_{u \in \mathcal{D},\, u \neq 0} \frac{\|Tu\|}{\|u\|} = \sup_{u \in \mathcal{D},\, \|u\| \leq 1} \|Tu\| = \sup_{u \in \mathcal{D},\, \|u\| = 1} \|Tu\|. \tag{3}$$

Remark: The Example 1.2 in Section 1 presents an unbounded operator with $T := -\Delta$.

Definition 4.5. *Let $\mathcal{D}_T, \mathcal{D}_{\widetilde{T}}$ denote two linear subspaces of the Hilbert space $\mathcal{H}$. Then the mapping*

$$\widetilde{T} : \mathcal{D}_{\widetilde{T}} \to \mathcal{H}$$

is called the extension *of the bounded linear operator $T : \mathcal{D}_T \to \mathcal{H}$, if the following properties are satisfied*

a) $\mathcal{D}_T \subset \mathcal{D}_{\widetilde{T}}$,
b) $\widetilde{T}u = Tu \quad$ for all $\quad u \in \mathcal{D}_T$.

We then write $T \subset \widetilde{T}$.

For bounded operators it suffices to define them on dense subspaces of Hilbert spaces due to the following

Theorem 4.6. (Extension theorem)
Let $T : \mathcal{D} \to \mathcal{H}$ denote a bounded linear operator, and the linear subspace $\mathcal{D} \subset \mathcal{H}$ lies dense in the Hilbert space $\mathcal{H}$. Then there exists a uniquely determined bounded extension $\widetilde{T} \supset T$ satisfying $\mathcal{D}_{\widetilde{T}} = \mathcal{H}$ and $\|\widetilde{T}\| = \|T\|$.

Proof:

1. We define $\widetilde{T} : \mathcal{H} \to \mathcal{H}$ as follows: Taking $f \in \mathcal{H}$ we have a sequence $\{f_n\}_{n=1,2,\dots} \subset \mathcal{D}_T$ satisfying $f_n \to f$ $(n \to \infty)$ in $\mathcal{H}$. Now $\{Tf_n\}_{n=1,2\dots}$ gives a Cauchy sequence in $\mathcal{H}$ on account of

$$\|Tf_n - Tf_m\| = \|T(f_n - f_m)\| \leq \|T\| \|f_n - f_m\| \to 0 \quad (n, m \to \infty).$$

Therefore the limit $\lim_{n \to \infty} Tf_n$ exists, and we set

$$\widetilde{T}f := \lim_{n \to \infty} Tf_n, \qquad f \in \mathcal{H}.$$

This notion is uniquely determined: Taking namely a further sequence $\{f_n'\}_{n=1,2,\ldots} \subset \mathcal{D}_T$ satisfying $f_n' \to f$ $(n \to \infty)$ in $\mathcal{H}$, we observe

$$\|Tf_n - Tf_n'\| \le \|T\|\|f_n - f_n'\| \le \|T\| \left(\|f_n - f\| + \|f - f_n'\|\right) \to 0 \ (n \to \infty).$$

Finally, we note that $\widetilde{T}f = Tf$ for all $f \in \mathcal{D}_T$.

2. Now the relation

$$\|\widetilde{T}\| := \sup_{f \in \mathcal{H},\, \|f\| \le 1} \|\widetilde{T}f\| = \sup_{f \in \mathcal{D}_T,\, \|f\| \le 1} \|\widetilde{T}f\| = \sup_{f \in \mathcal{D}_T,\, \|f\| \le 1} \|Tf\| = \|T\|$$

is correct. Furthermore, the operator $\widetilde{T} : \mathcal{H} \to \mathcal{H}$ is linear: For two elements

$$f = \lim_{n \to \infty} f_n, \quad g = \lim_{n \to \infty} g_n \qquad \text{from} \quad \mathcal{H}$$

with $\{f_n\}_n \subset \mathcal{D}_T$ and $\{g_n\}_n \subset \mathcal{D}_T$, we have the equation

$$\widetilde{T}(\alpha f + \beta g) = \widetilde{T}\left(\lim_{n \to \infty}(\alpha f_n + \beta g_n)\right) = \lim_{n \to \infty}(\alpha Tf_n + \beta Tg_n)$$

$$= \alpha \lim_{n \to \infty} Tf_n + \beta \lim_{n \to \infty} Tg_n = \alpha \widetilde{T}f + \beta \widetilde{T}g$$

for arbitrary $\alpha, \beta \in \mathbb{C}$, on account of the continuity of $\widetilde{T}$.

3. If $\widehat{T}, \widetilde{T} : \mathcal{H} \to \mathcal{H}$ are two extensions of $\mathcal{D}_T$ on $\mathcal{H}$, we have

$$(\widetilde{T} - \widehat{T})(f) = 0 \qquad \text{for all} \quad f \in \mathcal{D}_T.$$

Since $\mathcal{D}_T \subset \mathcal{H}$ lies dense and $(\widetilde{T} - \widehat{T}) : \mathcal{H} \to \mathcal{H}$ is continuous, we infer

$$(\widetilde{T} - \widehat{T})(f) = 0 \qquad \text{for all} \quad f \in \mathcal{H}$$

and consequently $\widetilde{T} = \widehat{T}$.   q.e.d.

Theorem 4.7. *Let $T : \mathcal{H} \to \mathcal{H}$ denote a bounded linear operator in the Hilbert space $\mathcal{H}$. Then we have a uniquely determined linear operator $T^* : \mathcal{H} \to \mathcal{H}$ such that*

$$(Tf, g) = (f, T^*g) \qquad \text{for all} \quad f, g \in \mathcal{H} \tag{4}$$

is correct. Furthermore, we have

$$\|T^*\| = \|T\| \qquad \text{and} \qquad T^{**} = T,$$

which means that the operation $*$ is an involution.

Definition 4.8. *The operator T^* is named the* adjoint operator *of T.*

Definition 4.9. *A bounded linear operator H is called Hermitian if $H^* = H$ holds true, which means*

$$(Hx, y) = (x, Hy) \qquad \text{for all} \quad x, y \in \mathcal{H}.$$

Proof of Theorem 4.7:
-*Uniqueness:* Let T_1 and T_2 be two adjoint operators to T: Then (4) yields

$$(f, T_1 g) = (Tf, g) = (f, T_2 g) \qquad \text{for all} \quad f, g \in \mathcal{H}$$

as well as $(f, (T_1 - T_2)g) = 0$ for all $f, g \in \mathcal{H}$, and consequently $T_1 = T_2$.
-*Existence:* For a fixed $g \in \mathcal{H}$ we consider the bounded linear functional

$$A_g(f) := (g, Tf), \qquad f \in \mathcal{H}.$$

This functional is bounded according to

$$|A_g(f)| \le \|g\| \|Tf\| \le (\|g\| \|T\|) \|f\| \qquad \text{for all} \quad f \in \mathcal{H}.$$

We apply the Representation theorem of Fréchet-Riesz: For each $g \in \mathcal{H}$ we have an element $g^* \in \mathcal{H}$ with the property

$$(g, Tf) = A_g(f) = (g^*, f) \qquad \text{for all} \quad f \in \mathcal{H}.$$

We now set $T^* g := g^*$, and the so-defined mapping $T^* : \mathcal{H} \to \mathcal{H}$ has the property (4).
-*Linearity:* We take the elements $g_1, g_2 \in \mathcal{H}$ and the numbers $c_1, c_2 \in \mathbb{C}$. With the aid of (4) we then calculate

$$
\begin{aligned}
(T^*(c_1 g_1 + c_2 g_2), f) &= (c_1 g_1 + c_2 g_2, Tf) \\
&= \bar{c}_1 (g_1, Tf) + \bar{c}_2 (g_2, Tf) \\
&= \bar{c}_1 (T^* g_1, f) + \bar{c}_2 (T^* g_2, f) \\
&= (c_1 T^* g_1 + c_2 T^* g_2, f) \qquad \text{for all} \quad f \in \mathcal{H}.
\end{aligned}
$$

-*Boundedness:* At first, we note that

$$(Tf, g) = (f, T^* g) = (T^{**} f, g) \qquad \text{for all} \quad f, g \in \mathcal{H}.$$

Therefore, T is an involution. With the aid of (4) we obtain the following estimate

$$\|(f, T^* g)\| \le \|Tf\| \|g\| \le \|T\| \|f\| \|g\| \qquad \text{for all} \quad f, g \in \mathcal{H}.$$

Inserting $f = T^* g$, we obtain

$$\|T^* g\|^2 \le \|T\| \|T^* g\| \|g\|$$

and consequently

$$\|T^* g\| \le \|T\| \|g\|,$$

which means

$$\|T^*\| \le \|T\|.$$

Since T is an involution, we infer

$$\|T\| = \|T^{**}\| \le \|T^*\|.$$

Consequently, the relation $\|T\| = \|T^*\|$ is correct.   q.e.d.

Example 4.10. Let the following cube in $\mathbb{R}^n$ be given, namely

$$Q := \Big\{ x = (x_1, \ldots, x_n) \in \mathbb{R}^n \ : \ |x_j| \leq R \text{ for } j = 1, \ldots, n \Big\}$$

whose sides have the length $2R \in (0, +\infty)$. Then we consider a *Hilbert-Schmidt integral kernel*

$$K = K(x,y) : Q \times Q \to \mathbb{C} \in L^2(Q \times Q, \mathbb{C}). \tag{5}$$

On account of

$$\int\limits_{Q \times Q} |K(x,y)|^2 dx\, dy < \infty$$

we have a null-set $N \subset Q$ with the following property: For all $x \in Q \setminus N$ the function $y \mapsto K(x,y)$ is measurable on Q and $\int\limits_Q |K(x,y)|^2\, dy < \infty$ is satisfied.

Furthermore, we have

$$\int\limits_Q \Big(\int\limits_Q |K(x,y)|^2\, dy \Big) dx = \int\limits_{Q \times Q} |K(x,y)|^2\, dx\, dy =: \|K\|^2 < \infty$$

because of the Fubini-Tonelli theorem. Taking $f \in L^2(Q, \mathbb{C})$ we define the *Hilbert-Schmidt operator*

$$\mathbb{K}f(x) = \begin{cases} \displaystyle\int\limits_Q K(x,y)f(y)\, dy, & x \in Q \setminus N \\ 0, & x \in N \end{cases}. \tag{6}$$

The Hölder inequality yields

$$|\mathbb{K}f(x)|^2 \leq \Big(\int\limits_Q |K(x,y)|^2\, dy \Big) \|f\|^2$$

for all $x \in Q \setminus N$ and integration with respect to $x \in Q$ gives

$$\int\limits_Q |\mathbb{K}f(x)|^2 dx \leq \Big(\int\limits_{Q \times Q} |K(x,y)|^2\, dx\, dy \Big) \|f\|^2 = \|K\|^2 \|f\|^2,$$

and finally

$$\|\mathbb{K}f\| \leq \|K\| \|f\| \qquad \text{for all} \quad f \in L^2(Q, \mathbb{C}). \tag{7}$$

Therefore, we obtain the following

Theorem 4.11. *The Hilbert-Schmidt operator* $\mathbb{K} : \mathcal{H} \to \mathcal{H}$ *from (6) with the integral kernel (5) represents a bounded linear operator on the Hilbert space* $\mathcal{H} = L^2(Q, \mathbb{C})$, *and we have the estimate*

$$\|\mathbb{K}\| \leq \|K\|.$$

Remarks:

1. The singular kernels

$$K = K(x, y) \in \mathcal{S}_\alpha(G, \mathbb{C}) \qquad \text{with} \quad \alpha \in [0, n)$$

   generate special Hilbert-Schmidt operators. The statements from Section 2, Theorem 2.6 and Theorem 2.7, valid for these special operators, will later be utilized to obtain regularity results concerning the solutions of the integral equation.

2. The kernel

$$K^*(x, y) := \overline{K(y, x)} \in L^2(Q \times Q, \mathbb{C})$$

   generates the adjoint operator $\mathbb{K}^*$ belonging to the Hilbert-Schmidt operator $\mathbb{K}$.

3. The operator $\mathbb{K}$ is Hermitian if and only if the following identity is satisfied:

$$K(x, y) = \overline{K(y, x)} \quad \text{a. e. in} \quad Q \times Q.$$

We now shall investigate the inverse of a linear operator.

Definition 4.12. *Let $T : \mathcal{D}_T \to \mathcal{H}$ denote a linear operator on the subset $\mathcal{D}_T \subset \mathcal{H}$ of the Hilbert space $\mathcal{H}$ with the range $\mathcal{W}_T := T(\mathcal{D}_T) \subset \mathcal{H}$. Furthermore, let the mapping*

$$x \mapsto Tx, \qquad x \in \mathcal{D}_T$$

be injective. Setting $f := T^{-1}g$ the inverse $T^{-1} : \mathcal{W}_T \to \mathcal{D}_T \subset \mathcal{H}$ of the operator T is then defined if $Tf = g$ holds true. We note that

$$\mathcal{D}_{T^{-1}} = \mathcal{W}_T, \qquad \mathcal{W}_{T^{-1}} = \mathcal{D}_T.$$

We immediately obtain the following

Theorem 4.13. *The operator $T^{-1} : \mathcal{W}_T \to \mathcal{D}_T$ is linear and does exist if and only if the equation*

$$Tx = 0, \qquad x \in \mathcal{D}_T$$

possesses only the trivial solution $x = 0$.

Theorem 4.14. (O. Toeplitz)
Let $T : \mathcal{H} \to \mathcal{H}$ denote a bounded linear operator in the Hilbert space $\mathcal{H}$. Then the operator T possesses a bounded inverse in $\mathcal{H}$ - namely $T^{-1} : \mathcal{H} \to \mathcal{H}$ - if and only if the following conditions are satisfied:

a) For all $x \in \mathcal{H}$ we have $\|Tx\| \geq d\|x\|$ with a bound $d \in (0, +\infty)$.
b) The homogeneous equation $T^*x = 0$ admits only the trivial solution $x = 0$.

Proof:

'$\Rightarrow$' We assume that the bounded inverse $T^{-1} : \mathcal{H} \to \mathcal{H}$ exists. Then we have a number $c > 0$ satisfying

$$\|T^{-1}x\| \le c\|x\| \qquad \text{for all} \quad x \in \mathcal{H}.$$

With $x := Tf$ we infer

$$\|Tf\| \ge \frac{1}{c}\|f\|.$$

Therefore, the condition a) is fulfilled with $d := \frac{1}{c}$.
If $z \in \mathcal{H}$ is a solution of $T^*z = 0$, we deduce

$$(Tx, z) = (x, T^*z) = (x, 0) = 0 \qquad \text{for all} \quad x \in \mathcal{H}.$$

Inserting the element $x = T^{-1}z$, we obtain $z = 0$.
'$\Leftarrow$' At first, we show that $\mathcal{W}_T \subset \mathcal{H}$ is closed: Let $\{y_n\}_{n=1,2,\dots} \subset \mathcal{W}_T$ denote an arbitrary sequence with $y_n \to y$ $(n \to \infty)$ in $\mathcal{H}$. We set $y_n = Tx_n$, $n = 1, 2, \dots$, and with the aid of a) we get the inequality

$$\|x_n - x_m\| \le \frac{1}{d}\|y_n - y_m\| \to 0 \quad (m, n \to \infty).$$

This implies $x_n \to x$ $(n \to \infty)$, and the continuity of T yields

$$Tx = T(\lim_{n\to\infty} x_n) = \lim_{n\to\infty} Tx_n = \lim_{n\to\infty} y_n = y \ \in \mathcal{W}_T.$$

Consequently, $\mathcal{W}_T$ is closed in $\mathcal{H}$, and we have the orthogonal decomposition

$$\mathcal{H} = \mathcal{W}_T \oplus \mathcal{W}_T^{\perp}.$$

Now we take $z \in \mathcal{W}_T^{\perp}$ and obtain

$$0 = (z, Tx) = (T^*z, x) \qquad \text{for all} \quad x \in \mathcal{H}$$

and finally $T^*z = 0$. The condition b) therefore yields $z = 0$ and consequently $\mathcal{H} = \mathcal{W}_T$, which means T is surjective. The injectivity of T follows immediately from a). Consequently, T^{-1} exists and is bounded due to a) with

$$\|T^{-1}\| \le \frac{1}{d}.$$

q.e.d.

Remark: If $H : \mathcal{H} \to \mathcal{H}$ is a bounded Hermitian operator satisfying

$$\|Hx\| \ge d\|x\| \qquad \text{for all} \quad x \in \mathcal{H}$$

with the number $d \in (0, +\infty)$, Theorem 4.14 implies the existence of the bounded inverse $H^{-1} : \mathcal{H} \to \mathcal{H}$.

Theorem 4.15. *Let $T : \mathcal{H} \to \mathcal{H}$ denote a bounded linear operator in the Hilbert space $\mathcal{H}$. Furthermore, let the bounded inverse $T^{-1} : \mathcal{H} \to \mathcal{H}$ be defined. Then the operator T^* possesses an inverse $(T^*)^{-1}$, which is defined and bounded in $\mathcal{H}$. Furthermore, we have $(T^{-1})^* = (T^*)^{-1}$.*

Proof: Because of the assumptions we see

$$\|T^{-1}x\| \leq c\|x\| \qquad \text{for all} \quad x \in \mathcal{H}.$$

When we insert the element $x = T^{-1}y$ into the relation

$$(Tx, y) = (x, T^*y) \qquad \text{for all} \quad y \in \mathcal{H}$$

we obtain

$$\|y\|^2 \leq \|T^{-1}y\|\|T^*y\| \leq c\|y\|\|T^*y\|$$

and consequently

$$\|T^*y\| \geq \frac{1}{c}\|y\| \qquad \text{for all} \quad y \in \mathcal{H}.$$

Theorem 4.7 and the relation

$$(T^*)^*f = T^{**}f = Tf$$

imply, with $(T^*)^*f = 0$ then $f = 0$ holds true. Theorem 4.14 shows that the inverse

$$(T^*)^{-1} : \mathcal{H} \to \mathcal{H}$$

exists, and we have

$$\|(T^*)^{-1}\| \leq \|T^{-1}\|.$$

Let the elements $f, g \in \mathcal{H}$ be chosen arbitrarily. With $x = T^{-1}f$ and $y = (T^*)^{-1}g$ we then obtain the relation

$$(f, (T^*)^{-1}g) = (Tx, y) = (x, T^*y) = (T^{-1}f, g) = (f, (T^{-1})^*g).$$

Consequently, the identity $(T^*)^{-1} = (T^{-1})^*$ is correct.   q.e.d.

In the Hilbert space $\mathcal{H}$ we consider a closed linear nonvoid subspace $\mathcal{M} \subset \mathcal{H}$, and the Projection theorem yields the decomposition $\mathcal{H} = \mathcal{M} \oplus \mathcal{M}^\perp$. Noting that

$$f = f_1 + f_2 \in \mathcal{H} \qquad \text{with} \quad f_1 \in \mathcal{M}, \ f_2 \in \mathcal{M}^\perp$$

holds true, the following definition for a projector P is sensible:

$$P : \mathcal{H} \to \mathcal{M} \qquad \text{via} \quad f = f_1 + f_2 \mapsto Pf := f_1.$$

We consider

$$\|Pf\|^2 = \|f_1\|^2 \leq \|f\|^2 \qquad \text{for all} \quad f \in \mathcal{H}$$

and

$$\|Pf\| = \|f\| \qquad \text{for all} \quad f \in \mathcal{M}.$$

The norm of the projector consequently satisfies

$$\|P\| = 1.$$

Furthermore, we observe

$$P^2 f = P \circ P f = P f \quad \text{and} \quad P^2 = P \qquad \text{in} \quad \mathcal{H},$$

and we conclude

$$(Pf, g) = (f_1, g_1 + g_2) = (f_1, g_1)$$
$$= (f_1 + f_2, g_1) = (f, Pg) \qquad \text{for all} \quad f, g \in \mathcal{H},$$

which means $P = P^*$.

Definition 4.16. *A bounded linear operator* $P : \mathcal{H} \to \mathcal{H}$ *is a projection operator or a projector if the following holds true:*

a) P *is Hermitian, which means* $P = P^*$;
b) $P^2 = P$.

Theorem 4.17. *Let* $P : \mathcal{H} \to \mathcal{H}$ *denote a projector. Then the set*

$$\mathcal{M} := \Big\{ g \in \mathcal{H} \, : \, g = Pf \text{ with } f \in \mathcal{H} \Big\}$$

is a closed linear subspace in $\mathcal{H}$. *Furthermore, we have*

$$f = Pf + (f - Pf) \ \in \ \mathcal{M} \oplus \mathcal{M}^\perp.$$

Proof:

1. We show that $\mathcal{M} = P(\mathcal{H})$ is closed: Let $\{g_n\}_{n=1,2,\dots} \subset \mathcal{M}$ be a sequence with $g_n \to g$ $(n \to \infty)$ in $\mathcal{H}$. On account of $g_n = Pf_n$ with $f_n \in \mathcal{H}$ we infer

$$P g_n = P^2 f_n = P f_n = g_n.$$

   Since P is continuous, $Pg = g$ follows and consequently $g \in \mathcal{M}$.
2. We take $f \in \mathcal{H}$, set $f_1 := Pf$ and $f_2 := f - Pf$, and observe $f_1 \in \mathcal{M}$. Furthermore, all $h \in \mathcal{M}$ satisfy

$$(f_2, h) = (f - Pf, h) = (f - Pf, Ph) = (Pf - P^2 f, h) = 0.$$

Consequently, $f_2 = f - Pf \in \mathcal{M}^\perp$ is correct. \qquad q.e.d.

Remark: In the Hilbert space $\mathcal{H}$ let the linear subspace $\mathcal{M} \subset \mathcal{H}$ be given. The sequence $\{\varphi_j\}_{j=1,2,\ldots}$ is assumed to constitute a c.o.n.s. in $\mathcal{M}$. Then we have

$$P_{\mathcal{M}} f = \sum_{j=1}^{j_0} (\varphi_j, f) \varphi_j, \qquad f \in \mathcal{H}$$

with $j_0 \in \mathbb{N} \cup \{\infty\}$.

In physics the energy of a system is measured with the aid of *bilinear forms*. Linear operators are then attributed to the latter.

Definition 4.18. *A complex-valued function*

$$B(.,.) : \mathcal{H} \times \mathcal{H} \to \mathbb{C}$$

is named a bilinear form *if*

$$B(f, c_1 g_1 + c_2 g_2) = c_1 B(f, g_1) + c_2 B(f, g_2) \qquad \textit{for all} \quad f, g_1, g_2 \in \mathcal{H}$$
$$B(c_1 f_1 + c_2 f_2, g) = \bar{c}_1 B(f_1, g) + \bar{c}_2 B(f_2, g) \qquad \textit{for all} \quad f_1, f_2, g \in \mathcal{H} \tag{8}$$

and all $c_1, c_2 \in \mathbb{C}$ holds true. The bilinear form is Hermitian *if*

$$B(f, g) = \overline{B(g, f)} \qquad \textit{for all} \quad f, g \in \mathcal{H} \tag{9}$$

is correct, and we name B symmetric *if*

$$B(f, g) = B(g, f) \qquad \textit{for all} \quad f, g \in \mathcal{H} \tag{10}$$

holds true. For real-valued bilinear forms the conditions (9) and (10) are equivalent. The bilinear form B is bounded *if we have a constant $c \in [0, +\infty)$ with the property*

$$|B(f, g)| \le c \|f\| \|g\| \qquad \textit{for all} \quad f, g \in \mathcal{H}. \tag{11}$$

A Hermitian bilinear form is strictly positive-definite *if we have a constant $c \in (0, +\infty)$ such that*

$$B(f, f) \ge c \|f\|^2 \qquad \textit{for all} \quad f \in \mathcal{H} \tag{12}$$

is satisfied.

Remarks:

1. Alternatively, one calls (12) the *coercivity condition*.
2. For a given bounded linear operator $T : \mathcal{H} \to \mathcal{H}$, we obtain with

$$B(f, g) := (Tf, g), \qquad f, g \in \mathcal{H}$$

a bilinear form. This is bounded on account of

$$|B(f, g)| \le \|Tf\| \|g\| \le \|T\| \|f\| \|g\|, \qquad f, g \in \mathcal{H}.$$

If T is Hermitian, the bilinear form is Hermitian as well since we have

$$B(f, g) = (Tf, g) = (f, Tg) = \overline{(Tg, f)} = \overline{B(g, f)}, \qquad f, g \in \mathcal{H}.$$

We now address the inverse question.

Theorem 4.19. (Representation theorem for bilinear forms)
For each bounded bilinear form $B = B(f,g)$ with $f,g \in \mathcal{H}$, there exists a uniquely determined bounded linear operator $T : \mathcal{H} \to \mathcal{H}$ satisfying

$$B(f,g) = (Tf,g) \qquad \text{for all} \quad f,g \in \mathcal{H}. \tag{13}$$

If B is Hermitian then T is Hermitian as well.

Proof: For a fixed element $f \in \mathcal{H}$ we obtain with

$$L_f(g) := B(f,g), \qquad g \in \mathcal{H}$$

a bounded linear functional on $\mathcal{H}$. Due to the representation theorem of Fréchet-Riesz, we have an element $f^* \in \mathcal{H}$ with the property

$$(f^*,g) = B(f,g) = L_f(g) \qquad \text{for all} \quad g \in \mathcal{H}. \tag{14}$$

Now f^* is uniquely determined by f, and we set

$$Tf := f^*, \qquad f \in \mathcal{H}.$$

1. The operator $T : \mathcal{H} \to \mathcal{H}$ is linear: We calculate

$$(T(c_1 f_1 + c_2 f_2), g) = B(c_1 f_1 + c_2 f_2, g) = \overline{c}_1 B(f_1, g) + \overline{c}_2 B(f_2, g)$$

$$= \overline{c}_1 (Tf_1, g) + \overline{c}_2 (Tf_2, g) = (c_1 Tf_1 + c_2 Tf_2, g)$$

for all $f_1, f_2, g \in \mathcal{H}$ and all $c_1, c_2 \in \mathbb{C}$.

2. Since the bilinear form $B(f,g) = (Tf,g)$ is bounded, we have

$$|(Tf,g)| \le c\|f\|\|g\| \qquad \text{for all} \quad f,g \in \mathcal{H}.$$

With $g = \frac{Tf}{\|Tf\|}$ we easily comprehend the inequality

$$\|Tf\| \le c\|f\| \qquad \text{for all} \quad f \in \mathcal{H},$$

and we conclude

$$\|T\| \le c < +\infty.$$

3. If B is Hermitian, we see

$$(Tf,g) = B(f,g) = \overline{B(g,f)} = \overline{(Tg,f)} = (f,Tg) \qquad \text{for all} \quad f,g \in \mathcal{H}.$$

Therefore, the operator T is Hermitian.   q.e.d.

Theorem 4.20. (Lax, Milgram)
Let $B : \mathcal{H} \times \mathcal{H} \to \mathbb{C}$ denote a Hermitian bilinear form, which is bounded due to

$$|B(f,g)| \leq c^+ \|f\| \|g\| \qquad \text{for all} \quad f,g \in \mathcal{H} \tag{15}$$

and satisfies the following coercivity condition

$$B(f,f) \geq c^- \|f\|^2 \qquad \text{for all} \quad f \in \mathcal{H}. \tag{16}$$

Here the constants $0 < c^- \leq c^+ < +\infty$ have been chosen adequately. Then we have a bounded Hermitian operator $T : \mathcal{H} \to \mathcal{H}$ satisfying $\|T\| \leq c^+$ and

$$B(f,g) = (Tf,g) \qquad \text{for all} \quad f,g \in \mathcal{H}. \tag{17}$$

This operator possesses a bounded inverse $T^{-1} : \mathcal{H} \to \mathcal{H}$ which is Hermitian and subject to

$$\|T^{-1}\| \leq \frac{1}{c^-}.$$

Proof: From Theorem 4.19 we obtain a Hermitian operator $T : \mathcal{H} \to \mathcal{H}$ with $\|T\| \leq c^+$ and the property (17). Together with (8) we arrive at

$$c^- \|f\|^2 \leq B(f,f) = (Tf,f) \leq \|Tf\| \|f\| \qquad \text{for all} \quad f \in \mathcal{H}$$

and consequently

$$\|Tf\| \geq c^- \|f\| \qquad \text{for all} \quad f \in \mathcal{H}. \tag{18}$$

Due to the Theorem of Toeplitz, the operator T possesses a bounded inverse $T^{-1} : \mathcal{H} \to \mathcal{H}$, which is Hermitian because of Theorem 4.15. Finally, the relation (18) implies

$$\|T^{-1}\| \leq \frac{1}{c^-}.$$

q.e.d.

Remarks:

1. The Theorems 4.19 and 4.20 remain valid for real bilinear forms, if we replace *Hermitian* with *symmetric*.
2. Theorem 4.20 provides the basic result for the weak solvability of elliptic differential equations, as we shall see in Section 3 of Chapter 10.
3. In its full strength, the Theorem of Lax-Milgram even holds true for real, not necessarily symmetric, bilinear forms. However, we only need the symmetric situation, when we apply our Theorem 4.20 to elliptic p.d.e.s – without first order terms.

5 Unitary Operators

Definition 5.1. *Let $\mathcal{H}$ and $\mathcal{H}'$ denote two Hilbert spaces with the inner products (x, y) and $(x, y)'$. Then the linear operator $V : \mathcal{H} \to \mathcal{H}'$ is called* isometric *if the following holds true:*

$$(Vf, Vg)' = (f, g) \qquad \text{for all} \quad f, g \in \mathcal{H}. \tag{1}$$

Remarks:

1. With the isometric operator $V : \mathcal{H} \to \mathcal{H}'$ we calculate

$$\|Vf - Vg\|'^2 = \|V(f - g)\|'^2 = \Big(V(f - g), V(f - g) \Big)'$$
$$= (f - g, f - g) = \|f - g\|^2 \qquad \text{for all} \quad f, g \in \mathcal{H}. \tag{2}$$

   Therefore, the relation $f \neq g$ implies $Vf \neq Vg$, and consequently the operator V is injective.

2. The operator V is bounded. Noting that

$$\|Vf\|' = \sqrt{(Vf, Vf)'} = \sqrt{(f, f)} = \|f\| \qquad \text{for all} \quad f \in \mathcal{H},$$

   we infer

$$\|V\| = 1. \tag{3}$$

3. We have $\mathcal{D}_V = \mathcal{H}$ for the domain of definition of an isometric operator V, and the range $\mathcal{W}_V \subset \mathcal{H}'$ is closed. We take a sequence $g_n = Vf_n \in \mathcal{W}_V$, $n = 1, 2, \ldots$ satisfying $g_n \to g$ $(n \to \infty)$ and observe that $\{f_n\}_{n=1,2,\ldots} \subset \mathcal{H}$ is a Cauchy sequence due to (2), namely

$$\|f_n - f_m\|^2 = \|g_n - g_m\|'^2 \to 0 \ (n, m \to \infty).$$

   This implies $f_n \to f \in \mathcal{H}$ $(n \to \infty)$ and furthermore

$$g = \lim_{n \to \infty} g_n = \lim_{n \to \infty} Vf_n = V(\lim_{n \to \infty} f_n) = Vf \ \in \mathcal{W}_V,$$

   since V is continuous. Consequently, $\mathcal{W}_V \subset \mathcal{H}$ is closed.

4. In the case $\dim \mathcal{H} = \dim \mathcal{H}' < +\infty$, the injectivity implies the surjectivity. For infinite-dimensional Hilbert spaces $\mathcal{H}$ and $\mathcal{H}'$ this is not true, as illustrated by the following example:

Example 5.2. We consider the so-called *shift-operator* in Hilbert's sequential space $\mathcal{H} := l_2 =: \mathcal{H}'$:

$$V : \mathcal{H} \to \mathcal{H}',$$
$$(x_1, x_2, \ldots) \mapsto (0, \ldots, 0, x_1, x_2, \ldots).$$

The operator V is evidently isometric, however, V is not surjective.

Definition 5.3. *An isometric operator* $V : \mathcal{H} \to \mathcal{H}'$ *is called* unitary *if* $V : \mathcal{H} \to \mathcal{H}'$ *is surjective, more precisely* $V(\mathcal{H}) = \mathcal{H}'$.

Remark: For a unitary operator $U : \mathcal{H} \to \mathcal{H}'$ there exists its inverse $U^{-1} : \mathcal{H}' \to \mathcal{H}$, and we have the identity

$$(U^{-1}f, U^{-1}g) = (U \circ U^{-1}f, U \circ U^{-1}g)' = (f, g)' \qquad \text{for all} \quad f, g \in \mathcal{H}' \quad (4)$$

on account of (1). Therefore, the inverse U^{-1} is unitary as well.

Definition 5.4. *Let* $\mathcal{H}$ *and* $\mathcal{H}'$ *denote two Hilbert spaces with the inner products* (x, y) *and* $(x, y)'$. *Furthermore,* T *and* T' *are two linear operators in* $\mathcal{H}$ *and* $\mathcal{H}'$, *respectively. Then the operators* T *and* T' *are named* unitary equivalent, *if there exists a unitary operator* $U : \mathcal{H} \to \mathcal{H}'$ *satisfying*

$$T' = U \circ T \circ U^{-1}. \qquad (5)$$

Theorem 5.5. *A bounded linear operator* $V : \mathcal{H} \to \mathcal{H}$ *is unitary if and only if*

$$V^* \circ V = V \circ V^* = \mathbb{E} \qquad (6)$$

is correct. Here the symbol $\mathbb{E} : \mathcal{H} \to \mathcal{H}$ *denotes the identity operator.*

Proof:

'$\Rightarrow$' At first, we remark that $V : \mathcal{H} \to \mathcal{H}$ is isometric if and only if

$$(V^* \circ Vf, g) = (Vf, Vg) = (f, g) = (\mathbb{E}f, g) \qquad \text{for all} \quad f, g \in \mathcal{H}$$

or equivalently

$$V^* \circ V = \mathbb{E}$$

is valid.

If V is unitary, we have the existence of $V^{-1} : \mathcal{H} \to \mathcal{H}$. From the last relation we infer $V^* = V^{-1}$ and therefore $V \circ V^* = \mathbb{E}$.

'$\Leftarrow$' Now let the identity (6) be satisfied for $V : \mathcal{H} \to \mathcal{H}$. Then we infer $V^{-1} = V^*$. In particular, the operator V is surjective and isometric as well according to the following relation:

$$(f, g) = (V^* \circ Vf, g) = (Vf, Vg) \qquad \text{for all} \quad f, g \in \mathcal{H}.$$

q.e.d.

Now we shall prove the Theorem of Fourier-Plancherel (see Theorem 3.1 in Section 3 of Chapter 6), which we have already applied there. At first, we present the transition from Fourier series to the Fourier integral: Taking $c > 0$ arbitrarily, the functions

$$\left\{\frac{1}{\sqrt{2c}}\,e^{-\frac{\pi}{c}ikx}\right\}_{k\in\mathbb{Z}}$$

constitute a complete orthonormal system of functions on the interval $[-c,+c]$. For all

$$f \in L^2([-c,+c],\mathbb{R}) \cap C_0^0((-c,+c),\mathbb{R})$$

the completeness relation yields the following identity:

$$\int\limits_{-c}^{+c}|f(x)|^2\,dx = \sum_{k=-\infty}^{+\infty}\left|\int\limits_{-c}^{+c}\frac{1}{\sqrt{2c}}\,e^{\frac{\pi}{c}ikx}f(x)\,dx\right|^2$$

$$= \sum_{k=-\infty}^{+\infty}\frac{1}{2c}\left|\int\limits_{-c}^{+c}e^{\frac{\pi}{c}iky}f(y)\,dy\right|^2.$$

We set

$$g(x) := \frac{1}{\sqrt{2\pi}}\int\limits_{-c}^{+c}e^{ixy}f(y)\,dy, \qquad x \in \mathbb{R},$$

and $x_k = \frac{\pi}{c}k$ for $k \in \mathbb{Z}$. Then we obtain $x_k - x_{k-1} = \frac{\pi}{c}$ and

$$\frac{1}{2c}\left|\int\limits_{-c}^{+c}e^{\frac{\pi}{c}iky}f(y)\,dy\right|^2 = \frac{1}{2c}\left|\sqrt{2\pi}g(x_k)\right|^2 = \frac{\pi}{c}|g(x_k)|^2 = |g(x_k)|^2(x_k - x_{k-1})$$

for all $k \in \mathbb{Z}$. For all $c > 0$, the following identity holds true:

$$\int\limits_{-c}^{+c}|f(x)|^2\,dx = \sum_{k=-\infty}^{+\infty}|g(x_k)|^2(x_k - x_{k-1}).$$

The transition to the limit $c \to +\infty$ yields

$$\int\limits_{-\infty}^{+\infty}|f(x)|^2\,dx = \int\limits_{-\infty}^{+\infty}|g(x)|^2\,dx. \qquad (7)$$

We expect the operator

$$Tf(x) := \frac{1}{\sqrt{2\pi}}\int\limits_{-\infty}^{+\infty}e^{ixy}f(y)\,dy, \qquad x \in \mathbb{R} \qquad (8)$$

to be unitary on the space $L^2(\mathbb{R})$.

More generally, we define *Fourier's integral operator* on the Euclidean space $\mathbb{R}^n$ as follows:

$$Tf(x) = \frac{1}{\sqrt{2\pi}^n} \int\limits_{\mathbb{R}^n} e^{i(x \cdot y)} f(y) \, dy, \qquad x \in \mathbb{R}^n. \tag{9}$$

We shall prove that $T : L^2(\mathbb{R}^n) \to L^2(\mathbb{R}^n)$ is unitary.

At first, we determine T from (8) explicitly for the *characteristic function*

$$\varphi_{a,b}(x) = \varphi(a, b, x) = \begin{cases} 1, & a \le x \le b \\ 0, & x < a \text{ or } x > b \end{cases}. \tag{10}$$

We calculate

$$T\varphi_{a,b}(0) = \frac{1}{\sqrt{2\pi}} \int\limits_{-\infty}^{+\infty} e^{i0y} \varphi_{a,b}(y) \, dy = \frac{b-a}{\sqrt{2\pi}}, \tag{11}$$

and for $x \ne 0$ we evaluate

$$T\varphi_{a,b}(x) = \frac{1}{\sqrt{2\pi}} \int\limits_{-\infty}^{+\infty} e^{ixy} \varphi_{a,b}(y) \, dy = \frac{1}{\sqrt{2\pi}} \int\limits_{a}^{b} e^{ixy} \, dy$$
$$= \frac{1}{\sqrt{2\pi}} \left[\frac{e^{ixy}}{ix} \right]_{y=a}^{y=b} = \frac{1}{\sqrt{2\pi}} \frac{e^{ibx} - e^{iax}}{ix}. \tag{12}$$

Proposition 5.6. *For Cauchy's principal values*

$$\psi(h) = \frac{1}{\pi} \int\limits_{-\infty}^{+\infty} \frac{e^{ihx} - 1}{x^2} \, dx := \lim_{\varepsilon \downarrow 0} \left\{ \frac{1}{\pi} \int\limits_{-\infty}^{-\varepsilon} \frac{e^{ihx} - 1}{x^2} \, dx + \frac{1}{\pi} \int\limits_{+\varepsilon}^{+\infty} \frac{e^{ihx} - 1}{x^2} \, dx \right\}$$

we have

$$\psi(h) = -|h|, \qquad h \in \mathbb{R}.$$

Proof: Taking an arbitrary $h \ge 0$, we consider the holomorphic function

$$f(z) := \frac{e^{ihz} - 1}{z^2} = \frac{1 + ihz + \frac{1}{2}(ihz)^2 + \ldots - 1}{z^2}$$
$$= \frac{ih}{z} + \ldots, \qquad z \in \mathbb{C} \setminus \{0\}. \tag{13}$$

For $0 < \varepsilon < R < +\infty$ we utilize the domain

$$G_{\varepsilon,R} := \left\{ z \in \mathbb{C} : \varepsilon < |z| < R, \ \operatorname{Re} z > 0 \right\}$$

with the positive-oriented boundary

$$[-R, -\varepsilon] \cup K_\varepsilon \cup [+\varepsilon, +R] \cup K_R = \partial G_{\varepsilon,R}.$$

Here we have defined the semicircle

$$K_R : \quad z = Re^{i\varphi}, \; 0 \leq \varphi \leq \pi,$$

and the following semicircle

$$-K_\varepsilon : \quad z = \varepsilon e^{i\varphi}, \; 0 \leq \varphi \leq \pi$$

negatively traversed. Since f is a holomorphic function in $\overline{G_{\varepsilon,R}}$, Cauchy's integral theorem yields the following identity:

$$0 = \int_{\partial G_{\varepsilon,R}} f(z)\, dz = \int_{-R}^{-\varepsilon} \frac{e^{ihx} - 1}{x^2}\, dx + \int_{+\varepsilon}^{+R} \frac{e^{ihx} - 1}{x^2}\, dx$$
$$+ \int_{K_R} \frac{e^{ihz} - 1}{z^2}\, dz - \int_{-K_\varepsilon} \frac{e^{ihz} - 1}{z^2}\, dz \tag{14}$$

for all $0 < \varepsilon < R < +\infty$. With the aid of (13), let us calculate

$$\lim_{\varepsilon \downarrow 0} \int_{-K_\varepsilon} \frac{e^{ihz} - 1}{z^2}\, dz = \lim_{\varepsilon \downarrow 0} \int_{-K_\varepsilon} \left(\frac{ih}{z} + \ldots \right) dz = \lim_{\varepsilon \downarrow 0} \int_{-K_\varepsilon} \frac{ih}{z}\, dz$$
$$= \lim_{\varepsilon \downarrow 0} \int_0^\pi \frac{ih}{\varepsilon e^{i\varphi}} i\varepsilon e^{i\varphi}\, d\varphi = -h\pi. \tag{15}$$

Futhermore, we deduce

$$\int_{K_R} \frac{e^{ihz} - 1}{z^2}\, dz = \int_0^\pi \frac{\exp\left\{ ih(\cos\varphi + i\sin\varphi)R \right\} - 1}{R^2 e^{2i\varphi}} iRe^{i\varphi}\, d\varphi$$
$$= \frac{i}{R} \int_0^\pi e^{-i\varphi} \left\{ e^{ihR\cos\varphi} e^{-hR\sin\varphi} - 1 \right\} d\varphi,$$

and we estimate for all $R > 0$ as follows:

$$\left| \int_{K_R} \frac{e^{ihz} - 1}{z^2}\, dz \right| \leq \frac{1}{R} \int_0^\pi 1 \cdot \left\{ 1 \cdot e^{-hR\sin\varphi} + 1 \right\} d\varphi$$
$$\leq \frac{2\pi}{R} \to 0 \; (R \to +\infty).$$

This implies

$$\lim_{R \to +\infty} \int_{K_R} \frac{e^{ihz} - 1}{z^2} \, dz = 0 \qquad \text{for all} \quad h \geq 0. \tag{16}$$

In (14) we observe the transition $\varepsilon \downarrow 0$ and $R \uparrow +\infty$. With the aid of (15) and (16), we obtain the identity

$$0 = \int_{-\infty}^{+\infty} \frac{e^{ihx} - 1}{x^2} \, dx + h\pi$$

and consequently

$$\psi(h) = -h \qquad \text{for all} \quad h \geq 0. \tag{17}$$

Via the substitution $y = -x$, we evaluate

$$\psi(-h) = \lim_{\varepsilon \downarrow 0} \left\{ \frac{1}{\pi} \int_{-\infty}^{-\varepsilon} \frac{e^{ih(-x)} - 1}{x^2} \, dx + \frac{1}{\pi} \int_{\varepsilon}^{+\infty} \frac{e^{ih(-x)} - 1}{x^2} \, dx \right\}$$

$$= \lim_{\varepsilon \downarrow 0} \left\{ \frac{1}{\pi} \int_{\varepsilon}^{+\infty} \frac{e^{ihy} - 1}{y^2} \, dy + \frac{1}{\pi} \int_{-\infty}^{-\varepsilon} \frac{e^{ihy} - 1}{y^2} \, dy \right\}$$

$$= \psi(h) \qquad \text{for all} \quad h \in \mathbb{R}.$$

Finally, we obtain the following identity from (17):

$$\psi(h) = -|h| \qquad \text{for all} \quad h \in \mathbb{R}.$$
q.e.d.

Proposition 5.7. *With respect to the inner product in $L^2(\mathbb{R}, \mathbb{C})$, we see*

$$(T\varphi_{a,b}, T\varphi_{c,d}) = \begin{cases} 0, & \text{if } -\infty < a < b \leq c < d < +\infty \\ b - a, & \text{if } -\infty < a = c < b = d < +\infty \end{cases}. \tag{18}$$

Proof: We utilize (12) and calculate

$$(T\varphi_{a,b}, T\varphi_{c,d}) = \frac{1}{2\pi} \int_{-\infty}^{+\infty} \frac{(e^{-ibx} - e^{-iax})(e^{idx} - e^{icx})}{x^2} \, dx$$

$$= \frac{1}{2\pi} \int_{-\infty}^{+\infty} \frac{e^{i(d-b)x} - e^{i(c-b)x} - e^{i(d-a)x} + e^{i(c-a)x}}{x^2} \, dx.$$

If $-\infty < a < b \leq c < d < +\infty$ is fulfilled, Proposition 5.6 implies

$$(T\varphi_{a,b}, T\varphi_{c,d}) = \frac{1}{2}\Big\{\psi(d-b) - \psi(c-b) - \psi(d-a) + \psi(c-a)\Big\}$$

$$= \frac{1}{2}\Big\{b - d + c - b + d - a + a - c\Big\} = 0.$$

If $-\infty < a = c < b = d < +\infty$ holds true, we obtain

$$(T\varphi_{a,b}, T\varphi_{c,d}) = \frac{1}{2\pi} \int\limits_{-\infty}^{+\infty} \frac{1 - e^{i(c-b)x} - e^{i(d-a)x} + 1}{x^2}\, dx$$

$$= -\frac{1}{2}\Big\{\psi(c-b) + \psi(d-a)\Big\} = -\frac{1}{2}\{c - b + a - d\}$$

$$= -\frac{1}{2}\{2a - 2b\} = b - a.$$

q.e.d.

Let the rectangle

$$Q := \Big\{x = (x_1, \ldots, x_n) \in \mathbb{R}^n \ : \ a_\alpha \le x_\alpha \le b_\alpha \text{ for } \alpha = 1, \ldots, n\Big\}$$

in $\mathbb{R}^n$ be given. We decompose the interval $[a_\alpha, b_\alpha]$ into the parts

$$a_\alpha = x_\alpha^{(0)} < x_\alpha^{(1)} < \ldots < x_\alpha^{(m_\alpha)} = b_\alpha \qquad \text{for} \quad \alpha = 1, \ldots, n$$

and set

$$I_\alpha^{(k_\alpha)} = [x_\alpha^{(k_\alpha - 1)}, x_\alpha^{(k_\alpha)}]$$

for $1 \le k_\alpha \le m_\alpha$ and $\alpha = 1, \ldots, n$. Finally, we define the following rectangles for $k = (k_1, \ldots, k_n) \in \mathbb{N}^n$ with $1 \le k_\alpha \le m_\alpha$:

$$I^{(k)} = I_1^{(k_1)} \times \ldots \times I_n^{(k_n)} \subset \mathbb{R}^n.$$

We define the characteristic function of the set $I^{(k)}$ by

$$\varphi_{I^{(k)}}(x) = \begin{cases} 1, & x \in I^{(k)} \\ 0, & x \in \mathbb{R}^n \setminus I^{(k)} \end{cases}$$

and similarly

$$\varphi_{I_\alpha^{(k_\alpha)}}(x) = \begin{cases} 1, & x \in I_\alpha^{(k_\alpha)} \\ 0, & x \in \mathbb{R} \setminus I_\alpha^{(k_\alpha)} \end{cases} \qquad \text{for} \quad \alpha = 1, \ldots, n.$$

Then we see

$$\varphi_{I^{(k)}}(x) = \varphi_{I_1^{(k_1)}}(x_1) \cdot \ldots \cdot \varphi_{I_n^{(k_n)}}(x_n), \qquad x \in \mathbb{R}^n. \tag{19}$$

Now we calculate

$$T\varphi_{I^{(k)}}(x) = \frac{1}{\sqrt{2\pi}^n} \int_{\mathbb{R}^n} e^{i(x\cdot y)} \varphi_{I^{(k)}}(y)\, dy$$

$$= \left(\int_{-\infty}^{+\infty} \frac{e^{ix_1 y_1}}{\sqrt{2\pi}} \varphi_{I_1^{(k_1)}}(y_1)\, dy_1 \right) \cdot \ldots \cdot \left(\int_{-\infty}^{+\infty} \frac{e^{ix_n y_n}}{\sqrt{2\pi}} \varphi_{I_n^{(k_n)}}(y_n)\, dy_n \right) \quad (20)$$

$$= T\varphi_{I_1^{(k_1)}}(x_1) \cdot \ldots \cdot T\varphi_{I_n^{(k_n)}}(x_n).$$

For the admissible multi-indices $k = (k_1,\ldots,k_n)$ and $l = (l_1,\ldots,l_n)$ we deduce

$$(T\varphi_{I^{(k)}}, T\varphi_{I^{(l)}})$$

$$= \int_{\mathbb{R}^n} \overline{\left(T\varphi_{I_1^{(k_1)}}(x_1) \cdot \ldots \cdot T\varphi_{I_n^{(k_n)}}(x_n) \right)} \left(T\varphi_{I_1^{(l_1)}}(x_1) \cdot \ldots \cdot T\varphi_{I_n^{(l_n)}}(x_n) \right) dx$$

$$= (T\varphi_{I_1^{(k_1)}}, T\varphi_{I_1^{(l_1)}}) \cdot \ldots \cdot (T\varphi_{I_n^{(k_n)}}, T\varphi_{I_n^{(l_n)}})$$

$$= |I_1^{(k_1)}| \cdot \ldots \cdot |I_n^{(k_n)}| \,\delta_{k_1 l_1} \cdot \ldots \cdot \delta_{k_n l_n}$$

$$= |I^{(k)}| \,\delta_{k_1 l_1} \cdot \ldots \cdot \delta_{k_n l_n}$$

and consequently

$$(T\varphi_{I^{(k)}}, T\varphi_{I^{(l)}}) = \begin{cases} |I^{(k)}|, & k = l \\ 0, & k \neq l \end{cases}. \quad (21)$$

Here we have set

$$|I_\alpha^{(k_\alpha)}| = x_\alpha^{(k_\alpha)} - x_\alpha^{(k_\alpha - 1)} \quad \text{and} \quad |I^{(k)}| = |I_1^{(k_1)}| \cdot \ldots \cdot |I_n^{(k_n)}|.$$

We summarize our considerations to the following

Proposition 5.8. *Let $\varphi_k := \varphi_{I^{(k)}}$ with $k = (k_1,\ldots,k_n)$ and $1 \le k_\alpha \le m_\alpha$ for $\alpha = 1,\ldots,n$ be chosen. Then we have the inclusion*

$$T\varphi_k \in L^2(\mathbb{R}^n)$$

and furthermore the equation

$$(T\varphi_k, T\varphi_l) = (\varphi_k, \varphi_l) \qquad \text{holds true for all admissible } k,l. \quad (22)$$

We now consider the linear subspace $\mathcal{D} \subset L^2(\mathbb{R}^n)$ of the step functions in $\mathbb{R}^n$. They consist of all those functions f satisfying the following conditions:

1. Outside of a rectangle $Q \subset \mathbb{R}^n$ the relation $f(x) = 0$ is correct.

2. There exists a decomposition $Q = \bigcup_k I^{(k)}$ of the rectangle as above, and we have the representation

$$f(x) = \sum_k c_k \varphi_k$$

with the coefficients $c_k \in \mathbb{C}$ and the characteristic functions $\varphi_k := \varphi_{I^{(k)}}$.

Proposition 5.9. *For all step functions* $f, g \in \mathcal{D}$ *we have*

$$(Tf, Tg) = (f, g).$$

Proof: We choose a rectangle $Q \supset \mathrm{supp}\,(f) \cup \mathrm{supp}\,(g)$ and find a canonical decomposition of Q, such that

$$f = \sum_k c_k \varphi_k, \qquad g = \sum_l d_l \varphi_l.$$

Therefore, we obtain

$$(Tf, Tg) = \left(\sum_k c_k T\varphi_k, \sum_l d_l T\varphi_l \right) = \sum_{k,l} \overline{c_k} d_l (T\varphi_k, T\varphi_l)$$

$$= \sum_{k,l} \overline{c_k} d_l (\varphi_k, \varphi_l) = (f, g).$$

q.e.d.

We now consider the integral operator

$$Sf := \frac{1}{\sqrt{2\pi}^n} \int\limits_{\mathbb{R}^n} e^{-i(x \cdot y)} f(y)\, dy, \qquad f \in \mathcal{D},$$

and note that

$$Sf = \overline{T\overline{f}}, \qquad f \in \mathcal{D}. \tag{23}$$

Since $T : \mathcal{D} \to L^2(\mathbb{R}^n)$ represents a linear bounded operator, this is the case for S as well. Furthermore, the operator S is isometric according to

$$(Sf, Sg) = (\overline{T\overline{f}}, \overline{T\overline{g}}) = \overline{(T\overline{f}, T\overline{g})} = \overline{(\overline{f}, \overline{g})} = (f, g)$$

for all $f, g \in \mathcal{D}$.

Intermediate statement: We have the identity

$$(Tf, g) = (f, Sg) \qquad \text{for all} \quad f, g \in \mathcal{D}.$$

Proof: We now calculate

$$(Tf, g) = \int\limits_Q \Big(\frac{1}{\sqrt{2\pi}^n} \int\limits_Q e^{-i(x \cdot y)} \overline{f}(y)\, dy \Big) g(x)\, dx$$

$$= \frac{1}{\sqrt{2\pi}^n} \int\limits_{Q \times Q} e^{-i(x \cdot y)} \overline{f}(y) g(x)\, dy\, dx$$

$$= \int\limits_Q \overline{f}(y) \Big(\frac{1}{\sqrt{2\pi}^n} \int\limits_Q e^{-i(x \cdot y)} g(x)\, dx \Big)\, dy$$

$$= (f, Sg).$$

We sum up our considerations to the following

Proposition 5.10. *Let $\mathcal{D}$ denote the set of all step functions in $\mathbb{R}^n$. Then the operators*

$$T, S : \mathcal{D} \to L^2(\mathbb{R}^n)$$

are isometric, and S is adjoint to T which means

$$(Tf, Tg) = (f, g) = (Sf, Sg) \tag{24}$$

as well as

$$(Tf, g) = (f, Sg) \tag{25}$$

for all $f, g \in \mathcal{D}$.

Now the linear space $\mathcal{D} \subset L^2(\mathbb{R}^n)$ lies dense, and consequently for each $f \in L^2(\mathbb{R}^n)$ we have a sequence

$$\{f_k\}_{k=1,2,\dots} \subset \mathcal{D} \qquad \text{with} \qquad \|f_k - f\|_{L^2(\mathbb{R}^n)} \to 0 \ (k \to \infty).$$

Therefore, we can uniquely extend the bounded operators T, S of $\mathcal{D}$ onto $L^2(\mathbb{R}^n)$ as follows:

$$Tf := \lim_{k \to \infty} Tf_k, \qquad Sf := \lim_{k \to \infty} Sf_k. \tag{26}$$

The relations (24) and (25) yield

$$S \circ T = T^* \circ T = E = S^* \circ S = T \circ S \qquad \text{on} \quad \mathcal{D},$$

and we infer

$$S \circ T = E = T \circ S \qquad \text{on} \quad L^2(\mathbb{R}^n). \tag{27}$$

Consequently, the operator $T : L^2(\mathbb{R}^n) \to L^2(\mathbb{R}^n)$ is unitary and satisfies

$$T^* = S = T^{-1}.$$

From (26) we shall derive a direct representation of T and S as follows: We choose

$$f \in L_0^2(\mathbb{R}^n) := \Big\{ g \in L^2(\mathbb{R}^n) \ : \ \operatorname{supp}(g) \subset \mathbb{R}^n \text{ is compact} \Big\}.$$

For $f \in L_0^2(\mathbb{R}^n)$ we have a sequence $\{f_k\}_{k=1,2,\dots} \subset \mathcal{D}$ such that

$$\operatorname{supp}(f), \operatorname{supp}(f_k) \subset Q, \qquad k = 1, 2, \dots$$

- where $Q \subset \mathbb{R}^n$ is a fixed rectangle - and

$$\|f - f_k\|_{L^2(Q)} \to 0 \ (k \to \infty)$$

holds true. For all $x \in \mathbb{R}^n$ we obtain the estimate

$$\left| Tf_k(x) - \frac{1}{\sqrt{2\pi}^n} \int_Q e^{i(x \cdot y)} f(y) \, dy \right| = \frac{1}{\sqrt{2\pi}^n} \left| \int_Q e^{i(x \cdot y)} (f_k(y) - f(y)) \, dy \right|$$

$$\leq \frac{1}{\sqrt{2\pi}^n} \int_Q |f_k(y) - f(y)| \, dy \tag{28}$$

$$\leq \frac{1}{\sqrt{2\pi}^n} \sqrt{|Q|} \|f_k - f\|_{L^2(Q)} \to 0 \ (k \to \infty)$$

and consequently

$$\left\| Tf_k(x) - \frac{1}{\sqrt{2\pi}^n} \int_Q e^{i(x \cdot y)} f(y) \, dy \right\|_{L^2(\mathbb{R}^n)} \to 0 \ (k \to \infty). \tag{29}$$

Together with (26) we obtain

$$\left\| Tf(x) - \frac{1}{\sqrt{2\pi}^n} \int_Q e^{i(x \cdot y)} f(y) \, dy \right\|_{L^2(\mathbb{R}^n)}$$

$$\leq \|Tf - Tf_k\|_{L^2(\mathbb{R}^n)} + \left\| Tf_k(x) - \frac{1}{\sqrt{2\pi}^n} \int_Q e^{i(x \cdot y)} f(y) \, dy \right\|_{L^2(\mathbb{R}^n)}$$

$$\to 0 \ (k \to \infty).$$

For all $f \in L_0^2(\mathbb{R}^n)$ therefore the relation

$$Tf(x) = \frac{1}{\sqrt{2\pi}^n} \int_Q e^{i(x \cdot y)} f(y) \, dy \qquad \text{a. e. in} \quad \mathbb{R}^n \tag{30}$$

is correct. Taking $f \in L^2(\mathbb{R}^n)$ arbitrarily, we choose a sequence of rectangles

$$Q_1 \subset Q_2 \subset \dots \qquad \text{with} \quad \bigcup_{n=1}^{\infty} Q_k = \mathbb{R}^n$$

and set

$$f_k(x) = \begin{cases} f(x), & x \in Q_k \\ 0, & x \in \mathbb{R}^n \setminus Q_k \end{cases}.$$

Then the relation $\|f_k - f\|_{L^2(\mathbb{R}^n)} \to 0$ for $(k \to \infty)$ is valid, and (26) yields

$$Tf = \lim_{k \to \infty} Tf_k = \text{l.i.m.} \frac{1}{\sqrt{2\pi}^n} \int_{Q_k} e^{i(x \cdot y)} f(y) \, dy, \qquad (31)$$

$$Sf = \lim_{k \to \infty} Sf_k = \text{l.i.m.} \frac{1}{\sqrt{2\pi}^n} \int_{Q_k} e^{-i(x \cdot y)} f(y) \, dy. \qquad (32)$$

Here the symbol l.i.m. denotes the limit for $k \to \infty$ in the quadratic means, more precisely in the $L^2(\mathbb{R}^n)$-norm.

We summarize our considerations to the following

Theorem 5.11. (Fourier, Plancherel)
The Fourier integral operator $T : L^2(\mathbb{R}^n) \to L^2(\mathbb{R}^n)$ exists according to (31) and is unitary. Moreover, the adjoint integral operator $S : L^2(\mathbb{R}^n) \to L^2(\mathbb{R}^n)$ from (32) is unitary as well, and we have the identities

$$S \circ T = T \circ S = \mathbb{E} \qquad on \quad L^2(\mathbb{R}^n).$$

6 Completely Continuous Operators in Hilbert Spaces

We owe the following notion of convergence to David Hilbert:

Definition 6.1. *In the Hilbert space $\mathcal{H}$ a sequence $\{x_n\}_{n=1,2,\ldots} \subset \mathcal{H}$ is called weakly convergent towards an element $x \in \mathcal{H}$, symbolically $x_n \rightharpoonup x \ (n \to \infty)$, if the relation*

$$\lim_{n \to \infty} (x_n, y) = (x, y) \qquad for \ all \quad y \in \mathcal{H}$$

is satisfied.

Example 6.2. Let $\{\varphi_i\}_{i=1,2,\ldots}$ denote an orthonormal system in the Hilbert space $\mathcal{H}$, and we observe

$$\|\varphi_i - \varphi_j\| = \sqrt{(\varphi_i - \varphi_j, \varphi_i - \varphi_j)} = \sqrt{(\varphi_i, \varphi_i) + (\varphi_j, \varphi_j)} = \sqrt{2}$$

for all $i, j \in \mathbb{N}$ with $i \neq j$. Consequently, $\{\varphi_i\}_{i=1,2,\ldots}$ does not contain a subsequence which represents a Cauchy sequence with respect to the Hilbert space norm. Because of Bessel's inequality, all $f \in \mathcal{H}$ are subject to the estimate

$$\sum_{i=1}^{\infty} |(\varphi_i, f)|^2 \leq \|f\|^2 < +\infty,$$

and we infer

$$\lim_{i \to \infty} (\varphi_i, f) = 0 = (0, f) \qquad \text{for all} \quad f \in \mathcal{H}.$$

Thus we obtain $\varphi_i \rightharpoonup 0$ $(i \to \infty)$ and note

$$\|0\| \leq \liminf_{i \to \infty} \|\varphi_i\| = 1.$$

Theorem 6.3. (Principle of uniform boundedness)
On the Hilbert space $\mathcal{H}$ let a sequence of the following bounded linear functionals $A_n : \mathcal{H} \to \mathbb{C}$ with $n \in \mathbb{N}$ be given, such that each element $f \in \mathcal{H}$ possesses a constant $c_f \in [0, +\infty)$ with the property

$$|A_n f| \leq c_f, \qquad n = 1, 2, \dots \tag{1}$$

Then we have a constant $\alpha \in [0, +\infty)$ satisfying

$$\|A_n\| \leq \alpha \qquad \text{for all} \quad n \in \mathbb{N}. \tag{2}$$

Proof:

1. Let $A : \mathcal{H} \to \mathbb{C}$ denote a bounded linear functional such that

$$|Af| \leq c \qquad \text{for all} \quad f \in \mathcal{H} \quad \text{with} \quad \|f - f_0\| \leq \varepsilon.$$

Here we have chosen an element $f_0 \in \mathcal{H}$ as well as a quantity $\varepsilon > 0$ and a constant $c \geq 0$. Then we have the estimate

$$\|A\| \leq \frac{2c}{\varepsilon}.$$

Setting $x := \frac{1}{\varepsilon}(f - f_0)$, we infer $\|x\| \leq 1$ and

$$|Ax| = \left| \frac{1}{\varepsilon} Af - \frac{1}{\varepsilon} Af_0 \right| \leq \frac{1}{\varepsilon} \left(|Af| + |Af_0| \right) \leq \frac{2c}{\varepsilon}$$

and finally $\|A\| \leq \frac{2c}{\varepsilon}$.

2. If the statement (2) does not hold true, part 1 of this proof together with the continuity of the functionals $\{A_n\}_n$ enables us to construct a sequence of balls

$$\Sigma_n := \left\{ f \in \mathcal{H} : \|f - f_n\| \leq \varepsilon_n \right\}, \qquad n \in \mathbb{N}$$

satisfying $\Sigma_1 \supset \Sigma_2 \supset \dots$ with $\varepsilon_n \downarrow 0$ $(n \to \infty)$ and a subsequence $1 \leq n_1 < n_2 < \dots$ such that

$$|A_{n_j} x| \geq j \qquad \text{for all} \quad x \in \Sigma_j \quad \text{and} \quad j = 1, 2, \dots \tag{3}$$

is correct. Evidently, the relation (3) yields a contradiction to (1).

$$\text{q.e.d.}$$

Theorem 6.4. (Weak convergence criterion)
Let the sequence $\{x_n\}_{n=1,2,\ldots} \subset \mathcal{H}$ be given in a Hilbert space such that all elements $y \in \mathcal{H}$ possess the limit

$$\lim_{n\to\infty} (x_n, y).$$

Then the sequence $\{x_n\}_n$ is bounded in $\mathcal{H}$ and weakly convergent towards an element $x \in \mathcal{H}$, which means $x_n \rightharpoonup x \ (n \to \infty)$.

Proof: We consider the bounded linear functionals

$$A_n(y) := (x_n, y), \qquad y \in \mathcal{H}$$

with the norms $\|A_n\| = \|x_n\|$ for $n = 1, 2, \ldots$. Since the limits

$$\lim_{n\to\infty} A_n(y) =: A(y)$$

exist for all $y \in \mathcal{H}$ by assumption, Theorem 6.3 gives us a constant $c \in [0, +\infty)$ with $\|x_n\| = \|A_n\| \le c$ for all $n \in \mathbb{N}$. This implies $\|A\| \le c$, and the representation theorem of Fréchet-Riesz yields the existence of exactly one element $x \in \mathcal{H}$ satisfying

$$A(y) = (x, y), \qquad y \in \mathcal{H}$$

for the bounded linear functional A. We obtain

$$\lim_{n\to\infty} (x_n, y) = \lim_{n\to\infty} A_n(y) = A(y) = (x, y) \qquad \text{for all} \quad y \in \mathcal{H}$$

which means $x_n \rightharpoonup x \ (n \to \infty)$.   q.e.d.

Though it is not possible in general to select a subsequence convergent with respect to the norm out of a bounded sequence in a Hilbert space (compare the Example 6.2 above), we can prove the following fundamental result (compare Theorem 8.9 in Chapter 2, Section 8 for the special case $\mathcal{H} = L^2(X)$):

Theorem 6.5. (Hilbert's selection theorem)
Each bounded sequence $\{x_n\}_{n=1,2,\ldots} \subset \mathcal{H}$ in a Hilbert space $\mathcal{H}$ contains a weakly convergent subsequence $\{x_{n_k}\}_{k=1,2,\ldots}$.

Proof:

1. The sequence $\{x_n\}_{n=1,2,\ldots}$ is bounded and we have a constant $c \in [0, +\infty)$, such that

$$\|x_n\| \le c, \qquad n = 1, 2, \ldots \tag{4}$$

   is correct. On account of

$$|(x_1, x_n)| \le c\|x_1\| \qquad \text{for all} \quad n \in \mathbb{N}$$

we find a subsequence $\{x_n^{(1)}\}_n \subset \{x_n\}_n$, such that $\lim_{n\to\infty} (x_1, x_n^{(1)})$ exists. Noting that

$$|(x_2, x_n^{(1)})| \leq c\|x_2\| \qquad \text{for all} \quad n \in \mathbb{N}$$

we select a further subsequence $\{x_n^{(2)}\}_n \subset \{x_n^{(1)}\}_n$ whose limit exists, namely $\lim_{n\to\infty} (x_2, x_n^{(2)})$. The continuation of this procedure gives us a chain of subsequences

$$\{x_n\}_n \supset \{x_n^{(1)}\}_n \supset \{x_n^{(2)}\}_n \supset \ldots \supset \{x_n^{(k)}\}_n,$$

such that the limits

$$\lim_{n\to\infty} (x_i, x_n^{(k)})$$

exist for $i = 1, \ldots, k$. With the aid of Cantor's diagonal procedure we get the sequence $x_k' := x_k^{(k)}$. Then the sequence $\{(x_i, x_k')\}_{k=1,2,\ldots}$ is convergent for all $i \in \mathbb{N}$. Denoting the linear subspace of all finite linear combinations by $\mathcal{M}$, namely

$$x = \sum_{i=1}^{N(x)} \alpha_i x_i, \qquad \alpha_i \in \mathbb{C}, \quad N(x) \in \mathbb{N},$$

the following limits exist:

$$\lim_{k\to\infty} (x, x_k') \qquad \text{for all} \quad x \in \mathcal{M}. \tag{5}$$

2. Now we make the transition to the linear closed subspace $\mathcal{M} \subset \overline{\mathcal{M}} \subset \mathcal{H}$, and the following limits exist as well:

$$\lim_{k\to\infty} (y, x_k') \qquad \text{for all} \quad y \in \overline{\mathcal{M}}. \tag{6}$$

Here we note that we can extend the bounded linear functional

$$A(y) := \lim_{k\to\infty} (x_k', y) = \overline{\lim_{k\to\infty} (y, x_k')}, \qquad y \in \mathcal{M}$$

continuously onto the closure $\overline{\mathcal{M}}$. Due to the Projection theorem each element $y \in \mathcal{H}$ can be represented in the form $y = y_1 + y_2$ with $y_1 \in \overline{\mathcal{M}}$ and $y_2 \in \overline{\mathcal{M}}^\perp$. This implies the existence of

$$\lim_{k\to\infty} (y, x_k') = \lim_{k\to\infty} (y_1 + y_2, x_k') = \lim_{k\to\infty} (y_1, x_k')$$

for all $y \in \mathcal{H}$. Consequently, the sequence $\{x_k'\}_{k=1,2,\ldots}$ in the Hilbert space $\mathcal{H}$ converges weakly.

$$\text{q.e.d.}$$

Remarks to the weak convergence:

1. If the sequence $x_n \to x$ $(n \to \infty)$ converges strongly, which means

$$\lim_{n\to\infty} \|x_n - x\| = 0,$$

then it converges weakly $x_n \rightharpoonup x$ $(n \to \infty)$ as well. For arbitrary elements $y \in \mathcal{H}$ we namely deduce

$$|(x_n, y) - (x, y)| = |(x_n - x, y)| \leq \|x_n - x\|\|y\| \to 0 \ (n \to \infty).$$

2. The norm is lower-semi-continuous with respect to weak convergence. This means

$$x_n \rightharpoonup x \ (n \to \infty) \quad \Rightarrow \quad \liminf_{n\to\infty} \|x_n\| \geq \|x\|, \qquad x_n, x \in \mathcal{H}$$

for a real Hilbert space $\mathcal{H}$. We namely observe

$$\|x_n\|^2 - \|x\|^2 = (x_n, x_n) - (x, x) = (x_n - x, x_n + x)$$
$$= (x_n - x, x_n - x) + 2(x - x_n, x), \qquad n = 1, 2, \ldots,$$

and consequently

$$\liminf_{n\to\infty} \|x_n\|^2 - \|x\|^2 = \liminf_{n\to\infty} \|x_n - x\|^2 + 2\liminf_{n\to\infty}(x - x_n, x) \geq 0,$$

and finally

$$\liminf_{n\to\infty} \|x_n\| \geq \|x\|.$$

3. From $x_n \rightharpoonup x$ $(n \to \infty)$ and $y_n \to y$ $(n \to \infty)$ we infer $(x_n, y_n) \to (x, y)$ $(n \to \infty)$. Here we utilize the estimate

$$|(x_n, y_n) - (x, y)| = |(x_n, y_n) - (x_n, y) + (x_n, y) - (x, y)|$$
$$\leq |(x_n, y_n - y)| + |(x_n - x, y)|$$
$$\leq \|y_n - y\|\|x_n\| + |(x_n - x, y)| \to 0 \ (n \to \infty).$$

Definition 6.6. *A subset $\Sigma \subset \mathcal{H}$ of a Hilbert space is named* precompact, *if each sequence $\{y_n\}_{n=1,2,\ldots} \subset \Sigma$ contains a strongly convergent subsequence $\{y_{n_k}\}_{k=1,2,\ldots} \subset \{y_n\}_n$, which means*

$$\lim_{k,l\to\infty} \|y_{n_k} - y_{n_l}\| = 0.$$

Definition 6.7. *A linear operator $K : \mathcal{H}_1 \to \mathcal{H}_2$ is called* completely continuous *or alternatively* compact, *if the following set*

$$\Sigma := \Big\{ y = Kx \ : \ x \in \mathcal{H}_1 \text{ with } \|x\|_1 \leq r \Big\} \subset \mathcal{H}_2$$

is precompact, with a certain radius $r \in (0, +\infty)$ given. This means that each sequence $\{x_n\}_{n=1,2,\ldots} \subset \mathcal{H}_1$ with $\|x_n\|_1 \leq r$ for $n \in \mathbb{N}$ contains a subsequence $\{x_{n_k}\}_{k=1,2,\ldots}$ such that $\{Kx_{n_k}\}_{k=1,2,\ldots} \subset \mathcal{H}_2$ converges strongly.

Remarks:

1. It suffices to choose $r = 1$ in Definition 6.7.
2. A completely continuous linear operator $K : \mathcal{H}_1 \to \mathcal{H}_2$ is bounded. If this were not the case, there would exist a sequence $\{x_n\}_{n=1,2,...} \subset \mathcal{H}_1$ with $\|x_n\|_1 = 1$ for all $n \in \mathbb{N}$ and $\|Kx_n\|_2 \to +\infty$ $(n \to \infty)$. Therefore, we cannot select a convergent subsequence from $\{Kx_n\}_{n=1,2,...}$ in the Hilbert space $\mathcal{H}_2$. This yields a contradiction to Definition 6.7.

Theorem 6.8. *Let $K : \mathcal{H}_1 \to \mathcal{H}_2$ denote a linear operator between the Hilbert spaces $\mathcal{H}_1$ and $\mathcal{H}_2$. The operator K is completely continuous if and only if for each weakly convergent sequence $x_n \rightharpoonup x$ $(n \to \infty)$ in $\mathcal{H}_1$ the statement*

$$Kx_n \to Kx \ (n \to \infty) \quad in \quad \mathcal{H}_2$$

follows. Consequently, the operator K is completely continuous if and only if each weakly convergent sequence in $\mathcal{H}_1$ is transferred into a strongly convergent sequence in $\mathcal{H}_2$.

Proof:

'$\Leftarrow$' Let $\{x_n\}_{n=1,2,...} \subset \mathcal{H}_1$ be a sequence with $\|x_n\|_1 \le 1$, $n \in \mathbb{N}$. Hilbert's selection theorem yields a subsequence $\{x_{n_k}\}_{k=1,2,...} \subset \{x_n\}_n$ satisfying

$$x_{n_k} \rightharpoonup x \ (k \to \infty) \quad in \quad \mathcal{H}_1.$$

By assumption

$$y_{n_k} \to Kx \ (k \to \infty)$$

is fulfilled for the subsequence $y_{n_k} := Kx_{n_k}$, $k = 1, 2, \ldots$. Consequently, the operator $K : \mathcal{H}_1 \to \mathcal{H}_2$ is completely continuous.

'$\Rightarrow$' Now let K be completely continuous, and $\{x_n\}_{n=1,2,...} \subset \mathcal{H}_1$ denotes a sequence satisfying $x_n \rightharpoonup x = 0$ $(n \to \infty)$. We then have to prove: $Kx_n \to Kx = 0$ $(n \to \infty)$ in $\mathcal{H}_2$. If the latter statement were false, there would exist a number $d > 0$ and a subsequence $\{x_n{}'\}_n \subset \{x_n\}_n$ satisfying

$$\|Kx_n{}'\| \ge d > 0 \qquad \text{for all} \quad n \in \mathbb{N}.$$

Since the operator K is completely continuous, we have a further subsequence

$$\{x_n{}''\}_n \subset \{x_n{}'\} \quad \text{with} \quad Kx_n{}'' \to y \ (n \to \infty).$$

Therefore, we obtain with the statement

$$0 < d^2 \le (y, y) = \lim_{n \to \infty} (y, Kx_n{}'') = \lim_{n \to \infty} (K^*y, x_n{}'') = (K^*y, 0) = 0$$

a contradiction.   q.e.d.

Remarks about completely continuous operators:

1. If $K : \mathcal{H} \to \mathcal{H}$ is a bounded linear operator with a finite-dimensional range $\mathcal{W}_K := K(\mathcal{H})$, then K is completely continuous.
2. If $T_1 : \mathcal{H}_1 \to \mathcal{H}_2$ and $T_2 : \mathcal{H}_2 \to \mathcal{H}_3$ are two bounded linear operators, and T_1 or T_2 is completely continuous, then the operator

$$T := T_2 \circ T_1 : \mathcal{H}_1 \to \mathcal{H}_3$$

is completely continuous as well. If the operator T_1 is completely continuous for instance, then the weakly convergent sequence $x_n \rightharpoonup x \ (n \to \infty)$ in $\mathcal{H}_1$ is transferred into the strongly convergent sequence $T_1 x_n \to T_1 x$ $(n \to \infty)$ in $\mathcal{H}_2$. Since T_2 is continuous, we infer

$$T x_n = T_2 \circ T_1 x_n \to T_2 \circ T_1 x = T x \ (n \to \infty) \qquad \text{in} \quad \mathcal{H}_3.$$

3. The operator $K : \mathcal{H} \to \mathcal{H}$ is completely continuous on the Hilbert space $\mathcal{H}$ if and only if the adjoint operator $K^* : \mathcal{H} \to \mathcal{H}$ is completely continuous. *Proof:* Let $K : \mathcal{H} \to \mathcal{H}$ be completely continuous, then the operator $K \circ K^*$ is completely continuous as well. From an arbitrary sequence

$$\{x_n\}_{n=1,2,\ldots} \subset \mathcal{H} \qquad \text{with} \quad \|x_n\| \leq 1, \ n \in \mathbb{N}$$

we can extract a subsequence $\{x_{n_k}\}_{k=1,2,\ldots}$ such that $\{K \circ K^* x_{n_k}\}_k$ converges strongly in $\mathcal{H}$. We infer

$$
\begin{aligned}
\|K^* x_{n_k} - K^* x_{n_l}\|^2 &= \|K^*(x_{n_k} - x_{n_l})\|^2 \\
&= \left(K^*(x_{n_k} - x_{n_l}), K^*(x_{n_k} - x_{n_l}) \right) \\
&= \left(K \circ K^*(x_{n_k} - x_{n_l}), x_{n_k} - x_{n_l} \right) \\
&\leq \|K \circ K^*(x_{n_k} - x_{n_l})\| \|x_{n_k} - x_{n_l}\| \to 0 \ (k, l \to \infty).
\end{aligned}
$$

Consequently, the sequence $\{K^* x_{n_k}\}_{k=1,2,\ldots}$ converges in $\mathcal{H}$, and the operator K^* is completely continuous.
The inverse direction can be seen via the identity $K = (K^*)^*$.
4. Let $A : \mathcal{H} \to \mathcal{H}$ be a completely continuous Hermitian linear operator on the Hilbert space $\mathcal{H}$. Then the associate *bilinear form*

$$\alpha(x, y) := (Ax, y) = (x, Ay), \qquad (x, y) \in \mathcal{H} \times \mathcal{H}$$

is continuous with respect to weak convergence. This means, with $x_n \rightharpoonup x$ $(n \to \infty)$ and $y_n \rightharpoonup y \ (n \to \infty)$ in $\mathcal{H}$ we have the limit relation

$$\alpha(x_n, y_n) \to \alpha(x, y) \ (n \to \infty).$$

Proof: Remark 3 concerning the weak convergence in combination with Theorem 6.8 yield this statement.

Definition 6.9. *Let $\mathcal{H}$ denote a separable Hilbert space with two c.o.n.s.*

$$\varphi = \{\varphi_i\}_{i=1,2,\dots} \quad and \quad \psi = \{\psi_i\}_{i=1,2,\dots}.$$

The linear operator $T : \mathcal{H} \to \mathcal{H}$ has a finite square-norm *if*

$$N(T;\varphi,\psi) := \sqrt{\sum_{i,k=1}^{\infty} |(T\varphi_i,\psi_k)|^2} < +\infty$$

holds true.

Proposition 6.10. *Let $T : \mathcal{H} \to \mathcal{H}$ denote a linear operator as in Definition 6.9 with $N(T;\varphi,\psi) < +\infty$. Then we have*

$$\|T\| \le N(T;\varphi,\psi) = \sqrt{\sum_{i=1}^{\infty} \|T\varphi_i\|^2}. \tag{7}$$

Proof: At first, we observe

$$N(T;\varphi,\psi)^2 = \sum_{i,k=1}^{\infty} |(T\varphi_i,\psi_k)|^2 = \sum_{i=1}^{\infty}\left(\sum_{k=1}^{\infty} |(T\varphi_i,\psi_k)|^2\right) = \sum_{i=1}^{\infty} \|T\varphi_i\|^2.$$

With the series

$$f = \sum_{i=1}^{\infty} c_i\,\varphi_i \in \mathcal{H}$$

we calculate

$$Tf = \sum_{i=1}^{\infty} c_i T\varphi_i$$

and consequently

$$\|Tf\| \le \sum_{i=1}^{\infty} |c_i| \|T\varphi_i\|.$$

This implies

$$\|Tf\| \le \sqrt{\sum_{i=1}^{\infty} |c_i|^2}\sqrt{\sum_{i=1}^{\infty} \|T\varphi_i\|^2} = \|f\| N(T;\varphi,\psi) \qquad \text{for all} \quad f \in \mathcal{H}$$

and therefore $\|T\| \le N(T;\varphi,\psi)$.   q.e.d.

Proposition 6.11. *We consider with*

$$\varphi = \{\varphi_i\}_{i=1,2,\dots}, \quad \varphi' = \{\varphi_i'\}_{i=1,2,\dots}, \quad \psi = \{\psi_i\}_{i=1,2,\dots}, \quad \psi' = \{\psi_i'\}_{i=1,2,\dots}$$

four complete orthonormal systems in $\mathcal{H}$. Then the identity

$$N(T;\varphi,\psi) = N(T;\varphi',\psi') =: N(T)$$

holds true. Furthermore, the equality $N(T) = N(T^*)$ is correct.

Proof: We calculate as follows:

$$
N(T; \varphi, \psi)^2 = \sum_{i,k=1}^{\infty} |(T\varphi_i, \psi_k)|^2 = \sum_{i=1}^{\infty} \|T\varphi_i\|^2
$$

$$
= \sum_{i,k=1}^{\infty} |(\psi_k', T\varphi_i)|^2 = \sum_{i,k=1}^{\infty} |(T^*\psi_k', \varphi_i)|^2
$$

$$
= \sum_{k=1}^{\infty} \|T^*\psi_k'\|^2 = \sum_{i,k=1}^{\infty} |(T^*\psi_k', \varphi_i')|^2
$$

$$
= \sum_{i,k=1}^{\infty} |(\psi_k', T\varphi_i')|^2 = \sum_{i,k=1}^{\infty} |(T\varphi_i', \psi_k')|^2 = N(T; \varphi', \psi').
$$

From the identity above we also infer $N(T) = N(T^*)$.   q.e.d.

Remark: Proposition 6.11 implies that the square-norm is independent of the chosen c.o.n.s.

Example 6.12. On the rectangle

$$
Q := \left\{ x = (x_1, \ldots, x_n) \in \mathbb{R}^n \; : \; a_i \le x_i \le b_i \text{ for } i = 1, \ldots, n \right\}
$$

let the kernel $K = K(x,y) : Q \times Q \to \mathbb{C} \in L^2(Q \times Q, \mathbb{C})$ with

$$
\int_{Q \times Q} |K(x,y)|^2 \, dx \, dy < +\infty \tag{8}
$$

be given. As in the Example 4.10 from Section 4 we define the Hilbert-Schmidt operator

$$
\mathbb{K}f(x) := \int_Q K(x,y) f(y) \, dy \qquad \text{for almost all} \quad x \in Q.
$$

On account of Theorem 4.11 from Section 4, the linear operator $\mathbb{K} : L^2(Q) \to L^2(Q)$ is bounded by $\|\mathbb{K}\| \le \|K\|_{L^2(Q \times Q)}$.

Statement: The Hilbert-Schmidt operator $\mathbb{K}$ has the finite square-norm

$$
N(\mathbb{K}) = \sqrt{\int_{Q \times Q} |K(x,y)|^2 \, dx \, dy} < +\infty. \tag{9}
$$

Proof: Let $\{\varphi_i(x)\}_{i=1,2,\ldots}$ constitute a c.o.n.s. in $L^2(Q)$. Then we set

$$\psi_i(x) = \int_Q K(x,y)\varphi_i(y)\,dy = \mathbb{K}\varphi_i(x) \qquad \text{a. e. in} \quad Q \quad \text{for} \quad i = 1, 2, \ldots$$

We calculate

$$\sum_{i=1}^{\infty} |\psi_i(x)|^2 = \sum_{i=1}^{\infty} \left| \int_Q K(x,y)\varphi_i(y)\,dy \right|^2$$

$$= \sum_{i=1}^{\infty} \left| (\overline{K(x,\cdot)}, \varphi_i) \right|^2 = \int_Q |K(x,y)|^2\,dy,$$

and Fubini's theorem yields

$$\int_{Q \times Q} |K(x,y)|^2\,dx\,dy = \int_Q \sum_{i=1}^{\infty} |\psi_i(x)|^2\,dx = \sum_{i=1}^{\infty} \int_Q |\psi_i(x)|^2\,dx$$

$$= \sum_{i=1}^{\infty} \|\mathbb{K}\varphi_i\|^2 = N(\mathbb{K})^2.$$

q.e.d.

Central significance possesses the subsequent

Theorem 6.13. *On the separable Hilbert space $\mathcal{H}$ let $T : \mathcal{H} \to \mathcal{H}$ denote a linear operator with finite square-norm $N(T) < +\infty$. Then the operator $T : \mathcal{H} \to \mathcal{H}$ is completely continuous.*

Proof: Let the sequence $f_n \rightharpoonup f = 0$ $(n \to \infty)$ be weakly convergent. If $\{\varphi_i\}_{i=1,2,\ldots}$ represents a c.o.n.s. in $\mathcal{H}$, we have the expansion

$$f_n = \sum_{i=1}^{\infty} c_n^i \varphi_i$$

with

$$\lim_{n \to \infty} c_n^i = 0 \qquad \text{for} \quad i = 1, 2, \ldots \tag{10}$$

and

$$\sum_{i=1}^{\infty} |c_n^i|^2 \le M^2 \qquad \text{for} \quad n = 1, 2, \ldots \tag{11}$$

Due to Proposition 6.10, the operator $T : \mathcal{H} \to \mathcal{H}$ is continuous. Thus we infer

$$Tf_n = \sum_{i=1}^{\infty} c_n^i T\varphi_i$$

and

$$\|Tf_n\| \le \left\| \sum_{i=1}^{N} c_n^i T\varphi_i \right\| + \left\| \sum_{i=N+1}^{\infty} c_n^i T\varphi_i \right\|$$

$$\le \left\| \sum_{i=1}^{N} c_n^i T\varphi_i \right\| + \sqrt{\sum_{i=1}^{\infty} |c_n^i|^2} \sqrt{\sum_{i=N+1}^{\infty} \|T\varphi_i\|^2}.$$

With the aid of (11), we obtain

$$\|Tf_n\| \le \left\| \sum_{i=1}^{N} c_n^i T\varphi_i \right\| + M \sqrt{\sum_{i=N+1}^{\infty} \|T\varphi_i\|^2}, \qquad n = 1, 2, \ldots \qquad (12)$$

Given the quantity $\varepsilon > 0$, we choose an integer $N = N(\varepsilon) \in \mathbb{N}$ so large that

$$M \sqrt{\sum_{i=N+1}^{\infty} \|T\varphi_i\|^2} \le \varepsilon$$

is attained. Observing (10), we then can choose a number $n_0 = n_0(\varepsilon) \in \mathbb{N}$ such that

$$\left\| \sum_{i=1}^{N(\varepsilon)} c_n^i T\varphi_i \right\| \le \varepsilon \qquad \text{for all} \quad n \ge n_0$$

is correct. Altogether, we obtain the estimate

$$\|Tf_n\| \le 2\varepsilon \qquad \text{for all} \quad n \ge n_0$$

with the quantity $\varepsilon > 0$ given. Therefore, $Tf_n \to 0$ $(n \to \infty)$ holds true. q.e.d.

Remark: Due to Theorem 6.13, the Hilbert-Schmidt operators are completely continuous.

Definition 6.14. *With the completely continuous operator $K : \mathcal{H} \to \mathcal{H}$ on the Hilbert space $\mathcal{H}$ we associate the so-called* Fredholm operator

$$T := \mathbb{E} - K : \mathcal{H} \to \mathcal{H} \quad via \quad Tx := \mathbb{E}x - Kx = x - Kx, \quad x \in \mathcal{H}.$$

Using the theorem of F. Riesz we now prove the important

Theorem 6.15. (Fredholm)
Let $K : \mathcal{H} \to \mathcal{H}$ denote a completely continuous linear operator on the Hilbert space $\mathcal{H}$ with the associate Fredholm operator $T := \mathbb{E} - K$. Then we have the following statements:

i) The null-spaces

$$\mathcal{N}_T := \Big\{ x \in \mathcal{H} \ : \ Tx = 0 \Big\}$$

of the operator T and

$$\mathcal{N}_{T^*} := \left\{ x \in \mathcal{H} \; : \; T^* x = 0 \right\}$$

of $T^* = \mathbb{E} - K^*$ fulfill the identity

$$\omega := \dim \mathcal{N}_T = \dim \mathcal{N}_{T^*} \; \in \; \mathbb{N}_0 = \mathbb{N} \cup \{0\}. \tag{13}$$

ii) The operator equation

$$x - Kx = Tx = y, \qquad x \in \mathcal{H} \tag{14}$$

is solvable for the right-hand side $y \in \mathcal{H}$ if and only if $y \in \mathcal{N}_{T^*}^{\perp}$ is correct, which means

$$(y, z) = 0 \qquad \text{for all} \quad z \in \mathcal{N}_{T^*} \tag{15}$$

is satisfied.

iii) If $\omega = 0$ holds true, the bounded inverse operator $T^{-1} : \mathcal{H} \to \mathcal{H}$ exists.

Proof:

1. At first, we show $\dim \mathcal{N}_T < +\infty$. If this were violated, there would exist an orthonormal system $\{\varphi_i\}_{i=1,2,\ldots}$ satisfying

$$0 = T\varphi_i = \varphi_i - K\varphi_i, \qquad i = 1, 2, \ldots$$

Since the operator K is completely continuous, we can select a subsequence $\{\varphi_{i_j}\}_{j=1,2,\ldots} \subset \{\varphi_i\}_i$ such that $\varphi_{i_j} \to \varphi$ $(j \to \infty)$ in $\mathcal{H}$. This contradicts the statement

$$\| \varphi_i - \varphi_j \| = \sqrt{2} \qquad \text{for all} \quad i, j \in \mathbb{N} \quad \text{with} \quad i \neq j.$$

Therefore, we see $\dim \mathcal{N}_T \in \mathbb{N}_0$. With K the operator K^* is completely continuous as well, and we comprehend $\dim \mathcal{N}_{T^*} \in \mathbb{N}_0$.
We now decompose $\mathcal{H}$ into the closed linear subspaces

$$\mathcal{H} = \mathcal{N}_T^{\perp} \oplus \mathcal{N}_T. \tag{16}$$

Furthermore, we assume

$$\dim \mathcal{N}_T \leq \dim \mathcal{N}_{T^*}.$$

If this were not the case, we could replace T by T^* and T^* by $T^{**} = T$. Finally, we set

$$\mathcal{W}_T := T(\mathcal{H}) \quad \text{and} \quad \mathcal{W}_{T^*} := T^*(\mathcal{H}).$$

2. We now have $y \in \mathcal{W}_T^{\perp}$ if and only if

$$0 = (y, Tx) = (T^*y, x) \qquad \text{for all} \quad x \in \mathcal{H}$$

is correct, and therefore $T^*y = 0$ or equivalently $y \in \mathcal{N}_{T^*}$ holds true. This implies

$$\mathcal{N}_{T^*} = \mathcal{W}_T^{\perp} \quad \text{or equivalently} \quad \mathcal{W}_T = \mathcal{N}_{T^*}^{\perp}. \tag{17}$$

In particular, the range of T is closed in $\mathcal{H}$. Let us consider the orthonormal basis

$$\{\varphi_1, \ldots, \varphi_d\} \subset \mathcal{N}_T$$

in $\mathcal{N}_T$ and the orthonormal basis

$$\{\psi_1, \ldots, \psi_{d^*}\} \subset \mathcal{N}_{T^*}$$

in $\mathcal{N}_{T^*}$ satisfying $0 \le d \le d^* < +\infty$. We modify the operator T and obtain a Fredholm operator

$$Sx := Tx - \sum_{i=1}^{d} (\varphi_i, x)\psi_i, \qquad x \in \mathcal{H}. \tag{18}$$

On account of (16) and (17), the null-space of the operator S evidently satisfies

$$\mathcal{N}_S = \Big\{ x \in \mathcal{H} \,:\, Sx = 0 \Big\} = \{0\}.$$

Theorem 4.6 from Chapter 7, Section 4 of F. Riesz implies that the mapping $S : \mathcal{H} \to \mathcal{H}$ is surjective. Consequently, $d^* = d$ and $\dim \mathcal{N}_T = \dim \mathcal{N}_{T^*}$ are correct. Furthermore, the mapping $T : \mathcal{N}_T^{\perp} \to \mathcal{N}_{T^*}^{\perp}$ is invertible. In the special case

$$\omega = \dim \mathcal{N}_T = \dim \mathcal{N}_{T^*} = 0$$

the theorem of F. Riesz quoted above guarantees the existence of the bounded inverse operator on the entire Hilbert space $\mathcal{H}$.

q.e.d.

Remark: Theorem 6.15 is especially valid for linear operators $K : \mathcal{H} \to \mathcal{H}$ on the separable Hilbert space $\mathcal{H}$ with finite square-norm $N(K) < +\infty$. We can estimate the dimension of the null-space for the operator T according to

$$\dim \mathcal{N}_T \le N(K)^2. \tag{19}$$

If $\{\varphi_1, \ldots, \varphi_d\}$ namely denotes an orthonormal system in $\mathcal{N}_T$, we enlarge it to a c.o.n.s. $\{\varphi_i\}_{i=1,2,\ldots}$ in $\mathcal{H}$ and obtain

$$N(K)^2 = \sum_{i=1}^{\infty} \|K\varphi_i\|^2 \geq \sum_{i=1}^{d} \|K\varphi_i\|^2$$

$$= \sum_{i=1}^{d} \|\varphi_i\|^2 = d = \dim \mathcal{N}_T.$$

We collect our results for Hilbert-Schmidt operators in the following

Theorem 6.16. (D. Hilbert, E. Schmidt)
On the rectangle $Q = [a_1, b_1] \times \ldots \times [a_n, b_n]$ let the integral kernel

$$K = K(x, y) : Q \times Q \to \mathbb{C} \in L^2(Q \times Q)$$

be given. We consider the linear subspaces of $L^2(Q)$ satisfying

$$\mathcal{N} : \quad \int_Q K(x, y) f(y) \, dy = f(x), \quad f \in L^2(Q),$$

$$\mathcal{N}^* : \quad \int_Q K^*(x, y) \psi(y) \, dy = \psi(x), \quad \psi \in L^2(Q),$$

with $K^*(x, y) := \overline{K(y, x)}$ for $(x, y) \in Q \times Q$, and the following statement

$$\dim \mathcal{N} = \dim \mathcal{N}^* \leq \int_{Q \times Q} |K(x, y)|^2 \, dx \, dy < +\infty \tag{20}$$

holds true. The right-hand side $f(x) \in L^2(Q)$ given, the integral equation

$$u(x) - \int_Q K(x, y) u(y) \, dy = f(x), \quad u \in L^2(Q),$$

can be solved if and only if

$$\int_Q \overline{f(x)} \psi(x) \, dx = 0 \quad \text{for all} \quad \psi \in \mathcal{N}^*$$

is satisfied.

Proof: The Hilbert-Schmidt operator

$$\mathbb{K} f(x) := \int_Q K(x, y) f(y) \, dy, \quad f \in L^2(Q)$$

has the finite square-norm

$$N(\mathbb{K}) = \sqrt{\int\limits_{Q \times Q} |K(x,y)|^2 dx\,dy} < +\infty.$$

Due to Theorem 6.13, the operator is completely continuous with the adjoint operator

$$\mathbb{K}^* f(x) := \int\limits_{Q} K^*(x,y) f(y)\,dy, \qquad f \in L^2(Q).$$

From Theorem 6.15 and the subsequent remark we infer the statements given. q.e.d.

In the bounded domain $G \subset \mathbb{R}^n$ we consider the weakly singular kernels

$$K = K(x,y) \in \mathcal{S}_\alpha(G, \mathbb{C})$$

from Section 1, Definition 1.8 with $\alpha \in [0,n)$ and their associate integral operators

$$\mathbb{K}f(x) := \int\limits_{G} K(x,y) f(y)\,dy, \qquad x \in G,$$

$$\text{for} \qquad f \in \mathcal{D} := C^0(G, \mathbb{C}) \cap L^\infty(G, \mathbb{C}). \tag{21}$$

Proposition 6.17. *Let the kernel* $K = K(x,y) \in \mathcal{S}_\alpha(G, \mathbb{C})$ *with the properties*

$$\int\limits_{G} |K(x,y)|\,dy \le M, \qquad x \in G,$$

$$\int\limits_{G} |K(x,y)|\,dx \le N, \qquad y \in G, \tag{22}$$

be given. Then the operator $\mathbb{K} : \mathcal{H} \to \mathcal{H}$ *can be extended from* $\mathcal{D}$ *onto the Hilbert space* $\mathcal{H} = L^2(G, \mathbb{C})$, *and the following estimate holds true:*

$$\|\mathbb{K}\| \le \sqrt{MN}.$$

Proof: We estimate for arbitrary functions $f, g \in \mathcal{D}$ as follows:

$$|(g, \mathbb{K}f)| = \left| \int_G \overline{g(x)} \left(\int_G K(x,y) f(y) \, dy \right) dx \right|$$

$$\leq \int_{G \otimes G} |K(x,y)| |g(x)| |f(y)| \, dx \, dy$$

$$= \int_{G \otimes G} \left(|K(x,y)|^{\frac{1}{2}} |g(x)| \right) \left(|K(x,y)|^{\frac{1}{2}} |f(y)| \right) dx \, dy$$

$$\leq \left(\int_{G \otimes G} |K(x,y)| |g(x)|^2 \, dx \, dy \right)^{\frac{1}{2}} \left(\int_{G \otimes G} |K(x,y)| |f(y)|^2 \, dx \, dy \right)^{\frac{1}{2}}$$

$$= \left(\int_G |g(x)|^2 \left(\int_G |K(x,y)| \, dy \right) dx \right)^{\frac{1}{2}} \left(\int_G |f(y)|^2 \left(\int_G |K(x,y)| \, dx \right) dy \right)^{\frac{1}{2}}$$

$$\leq \sqrt{MN} \sqrt{\int_G |g(x)|^2 \, dx} \sqrt{\int_G |f(y)|^2 \, dy} = \sqrt{MN} \|g\| \|f\|.$$

Consequently, the operator $\mathbb{K} : \mathcal{H} \to \mathcal{H}$ is defined with $\|\mathbb{K}\| \leq \sqrt{MN}$.   q.e.d.

We define the auxiliary function

$$\Theta(t) := \begin{cases} 0, & 0 \leq t \leq 1 \\ t - 1, & 1 \leq t \leq 2 \\ 1, & 2 \leq t \end{cases}.$$

With $K = K(x,y) \in \mathcal{S}_\alpha(G, \mathbb{C})$ and $\delta \in (0, \delta_0)$, let us consider the continuous kernels

$$K_\delta(x,y) := K(x,y) \Theta\left(\frac{|x-y|}{\delta} \right), \qquad (x,y) \in G \times G, \tag{23}$$

together with their associate integral operators $\mathbb{K}_\delta$. For all $\delta \in (0, \delta_0)$ the operator $\mathbb{K}_\delta : \mathcal{H} \to \mathcal{H}$ is completely continuous, due to Theorem 6.13. With the aid of Proposition 6.17, we easily deduce the limit relation

$$\|\mathbb{K}_\delta - \mathbb{K}\| \to 0 \; (\delta \to 0). \tag{24}$$

The complete continuity of $\mathbb{K}$ is inferred from the following

Proposition 6.18. *On the Hilbert space $\mathcal{H}$ let the sequence of completely continuous operators $A_j : \mathcal{H} \to \mathcal{H}$ for $j = 1, 2, \ldots$ be given, converging due to $\|A_j - A\| \to 0 \; (j \to \infty)$ towards the bounded linear operator $A : \mathcal{H} \to \mathcal{H}$. Then the limit operator $A : \mathcal{H} \to \mathcal{H}$ is completely continuous.*

Proof: Taking the sequence $\{x_k\}_{k=1,2,\ldots} \subset \mathcal{H}$ with $\|x_k\| \leq 1$ for all $k \in \mathbb{N}$, we can select a subsequence $\{x_k^{(1)}\}_k \subset \{x_k\}_k$ such that $\{A_1 x_k^{(1)}\}_{k=1,2,\ldots} \subset \mathcal{H}$

converges. Furthermore, we have a subsequence $\{x_k^{(2)}\}_k \subset \{x_k^{(1)}\}_k$ such that $\{A_2 x_k^{(2)}\}_k \subset \mathcal{H}$ converges. In this way we successively select subsequences

$$\{x_k^{(1)}\} \supset \{x_k^{(2)}\} \supset \ldots$$

and go over to the diagonal sequence $x_k' := x_k^{(k)}$ for $k = 1, 2, \ldots$. We then show the sequence $\{A x_k'\}_{k=1,2,\ldots}$ to be convergent in $\mathcal{H}$ as well: At first, we estimate

$$\|A x_k' - A x_l'\| \leq \|A x_k' - A_j x_k'\| + \|A_j x_k' - A_j x_l'\| + \|A_j x_l' - A x_l'\|$$
$$\leq \|A - A_j\| \|x_k'\| + \|A_j x_k' - A_j x_l'\| + \|A_j - A\| \|x_l'\|. \tag{25}$$

With a given quantity $\varepsilon > 0$, we now choose an index $j \in \mathbb{N}$ so large that $\|A - A_j\| \leq \varepsilon$ is correct. Furthermore, we chose an integer $N = N(\varepsilon) \in \mathbb{N}$ satisfying

$$\|A_j x_k' - A_j x_l'\| \leq \varepsilon \qquad \text{for all} \quad k, l \geq N.$$

From the relation (25) we obtain the inequality

$$\|A x_k' - A x_l'\| \leq 2\varepsilon + \varepsilon = 3\varepsilon \qquad \text{for all} \quad k, l \geq N(\varepsilon).$$

Therefore, the sequence $\{A x_k'\}_{k=1,2,\ldots}$ converges in $\mathcal{H}$.   q.e.d.

Theorem 6.19. (Weakly singular integral equations)
On the bounded domain $G \subset \mathbb{R}^n$ let the weakly singular kernel $K = K(x, y)$ of the class $\mathcal{S}_\alpha(G, \mathbb{C})$ with $\alpha \in [0, n)$ and the integral operator $\mathbb{K}$ be given. Then the null-spaces

$$\mathcal{N} : \quad \int_G K(x, y) \varphi(y) \, dy = \varphi(x), \quad x \in G; \quad \varphi \in \mathcal{D}$$

$$\mathcal{N}^* : \quad \int_G \overline{K(x, y)} \psi(x) \, dx = \psi(y), \quad y \in G; \quad \psi \in \mathcal{D}$$

satisfy the identity $\dim \mathcal{N} = \dim \mathcal{N}^* < +\infty$. *The integral equation*

$$u(x) - \int_G K(x, y) u(y) \, dy = f(x), \quad x \in G; \quad u \in \mathcal{D}, \tag{26}$$

can be solved for the given right-hand side $f \in \mathcal{D}$ if and only if the following condition holds true:

$$\int_G \overline{\psi(x)} f(x) = 0 \qquad \text{for all} \quad \psi \in \mathcal{N}^*. \tag{27}$$

Proof: Because of Proposition 6.18 the integral operator $\mathbb{K} : \mathcal{H} \to \mathcal{H}$ is completely continuous, and Fredholm's Theorem 6.15 can be applied in the Hilbert space $\mathcal{H} = L^2(G, \mathbb{C})$. We still have to show that (26) is solvable in $\mathcal{D}$. Therefore, let $u \in \mathcal{H}$ be a solution of the integral equation

$$\mathbb{E}u - \mathbb{K}u = f \tag{28}$$

with the continuous right-hand side $f \in \mathcal{D}$. Because of the Theorem by I. Schur on iterated kernels (compare Section 2, Theorem 2.7) there exists an integer $k \in \mathbb{N}$, such that $\mathbb{K}^k = \mathbb{L}$ with a bounded kernel $L = L(x, y) \in \mathcal{S}_0(G, \mathbb{C})$ is correct. From Theorem 2.2 in Section 2 the property $\mathbb{L}u \in \mathcal{D}$ is satisfied. Via (28) we now obtain the following identity:

$$\mathbb{E}u - \mathbb{L}u = \mathbb{E}u - \mathbb{K}^k u = (\mathbb{E} + \mathbb{K} + \ldots + \mathbb{K}^{k-1})f =: g. \tag{29}$$

Due to Section 2, Theorem 2.2 we have $\mathbb{E} + \mathbb{K} + \ldots + \mathbb{K}^{k-1} : \mathcal{D} \to \mathcal{D}$ and consequently $g \in \mathcal{D}$ holds true. Finally, we obtain

$$u = g + \mathbb{L}u \in \mathcal{D}.$$

q.e.d.

7 Spectral Theory for Completely Continuous Hermitian Operators

At first, we consider the following

Example 7.1. On the Hilbert space $\mathcal{H} = L^2((0, 1), \mathbb{C})$ with the inner product

$$(f, g) = \int_0^1 \overline{f(x)} g(x) \, dx, \qquad f, g \in \mathcal{H}$$

we define the linear operator

$$Af(x) := xf(x), \quad x \in (0, 1); \qquad f = f(x) \in \mathcal{H}.$$

On account of

$$\|Af\|^2 = \int_0^1 x^2 \overline{f(x)} f(x) \, dx \le \int_0^1 |f(x)|^2 \, dx = \|f\|^2,$$

the operator A is bounded by $\|A\| \le 1$. Furthermore, the operator A is symmetric according to

$$(Af, g) = \int\limits_0^1 x\overline{f(x)}g(x)\, dx = (f, Ag) \qquad \text{for all} \quad f, g \in \mathcal{H}.$$

We now claim that A does not possess eigenvalues. From the identity $Af = \lambda f$ we namely infer

$$(x - \lambda)f(x) = 0 \qquad \text{a. e. in} \quad (0, 1)$$

and consequently $f(x) = 0$ a. e. in $(0, 1)$ which implies $f = 0 \in \mathcal{H}$.

Theorem 7.2. *Let $A : \mathcal{H} \to \mathcal{H}$ denote a completely continuous Hermitian operator on the Hilbert space $\mathcal{H}$. Then we have an element $\varphi \in \mathcal{H}$ with $\|\varphi\| = 1$ and a number $\lambda \in \mathbb{R}$ with $|\lambda| = \|A\|$ such that*

$$A\varphi = \lambda\varphi.$$

Consequently, the numbers $+\|A\|$ or $-\|A\|$ are eigenvalues of the operator A. Furthermore, we have the following estimate:

$$|(x, Ax)| \le |\lambda|(x, x) \qquad \text{for all} \quad x \in \mathcal{H}. \tag{1}$$

Proof:

1. At first, we show

$$\|A\| = \sup_{x \in \mathcal{H},\, \|x\|=1} |(Ax, x)|. \tag{2}$$

From the estimate

$$|(Ax, x)| \le \|Ax\|\,\|x\| \le \|A\|\,\|x\|^2 = \|A\|$$

for all $x \in \mathcal{H}$ with $\|x\| = 1$ we infer

$$\sup_{x \in \mathcal{H},\, \|x\|=1} |(Ax, x)| \le \|A\|.$$

In order to show the reverse inequality, we choose an arbitrary $\alpha \in [0, +\infty)$ satisfying

$$|(Ax, x)| \le \alpha\|x\|^2 \qquad \text{for all} \quad x \in \mathcal{H}.$$

With arbitrary elements $f, g \in \mathcal{H}$ we calculate

$$(A(f + g), f + g) - (A(f - g), f - g) = 2\{(Af, g) + (Ag, f)\} = 4\operatorname{Re}(Af, g)$$

and consequently

$$4|\operatorname{Re}(Af, g)| \le |(A(f + g), f + g)| + |(A(f - g), f - g)|$$
$$\le \alpha\{\|f + g\|^2 + \|f - g\|^2\} = 2\alpha\{\|f\|^2 + \|g\|^2\}.$$

We now replace

$$f = \sqrt{\frac{\|y\|}{\|x\|}}\, x, \qquad g = e^{i\varphi}\sqrt{\frac{\|x\|}{\|y\|}}\, y$$

with a suitable angle $\varphi \in [0, 2\pi)$, such that the inequality

$$4|(Ax, y)| \le 2\alpha\left\{\frac{\|y\|}{\|x\|}\|x\|^2 + \frac{\|x\|}{\|y\|}\|y\|^2\right\} = 4\alpha\|x\|\,\|y\|$$

follows and equivalently

$$|(Ax, y)| \le \alpha\|x\|\,\|y\| \qquad \text{for all} \quad x, y \in \mathcal{H}.$$

Inserting the element $y = Ax$, we obtain

$$\|Ax\|^2 \le \alpha\|x\|\,\|Ax\| \qquad \text{or equivalently} \qquad \|Ax\| \le \alpha\|x\|$$

for all $x \in \mathcal{H}$, and therefore $\|A\| \le \alpha$. Finally, we see

$$\sup_{x\in\mathcal{H},\,\|x\|=1} |(Ax, x)| = \inf\left\{\alpha \in [0, +\infty) : \begin{array}{l} |(Ax, x)| \le \alpha\|x\|^2 \\ \text{for all } x \in \mathcal{H} \end{array}\right\} \ge \|A\|.$$

2. We now consider the variational problem

$$\|A\| = \sup_{x\in\mathcal{H}\setminus\{0\}} \frac{|(Ax, x)|}{\|x\|^2} = \sup_{x\in\mathcal{H},\,\|x\|=1} |(Ax, x)|, \tag{3}$$

and without loss of generality we assume $A \ne 0$. Let $\{x_n\}_{n=1,2,\ldots} \subset \mathcal{H}$ denote a sequence with $\|x_n\| = 1$ for all $n \in \mathbb{N}$ satisfying

$$|(Ax_n, x_n)| \to \|A\| \ (n \to \infty).$$

Then we have a subsequence $\{x_n'\}_{n=1,2,\ldots} \subset \{x_n\}_{n=1,2,\ldots}$ and an element $x \in \mathcal{H}$ with $\|x\| \le 1$, such that $x_n' \rightharpoonup x \ (n \to \infty)$ and

$$(Ax_n', x_n') \to \lambda \ \in \ \{-\|A\|, \|A\|\}$$

hold true. Since the bilinear form $(y, z) \mapsto (Ay, z)$ is weakly continuous, we infer

$$0 \ne \lambda = \lim_{n\to\infty} (Ax_n', x_n') = (Ax, x)$$

and therefore $x \ne 0$. Now the condition $\|x\| = 1$ holds true: If $\|x\| < 1$ were correct, we would obtain

$$\frac{|(Ax, x)|}{\|x\|^2} > \frac{|\lambda|}{1} = \|A\|$$

contradicting (3).

3. Without loss of generality we now assume $\lambda = +\|A\|$; and the element $x \in \mathcal{H}$ satisfying $\|x\| = 1$ may solve the variational problem (3) from part 2 of our proof. We therefore have

$$(Ax, x) = \lambda \|x\|^2.$$

Taking an arbitrary element $y \in \mathcal{H}$, there exists a quantity $\varepsilon_0 = \varepsilon_0(y) > 0$ such that all $\varepsilon \in (-\varepsilon_0, \varepsilon_0)$ fulfill

$$(A(x + \varepsilon y), x + \varepsilon y) \leq \lambda(x + \varepsilon y, x + \varepsilon y)$$

and consequently

$$(Ax, x) + \varepsilon\{(Ax, y) + (Ay, x)\} \leq \lambda\|x\|^2 + \varepsilon\lambda\{(x, y) + (y, x)\} + o(\varepsilon).$$

This implies

$$\varepsilon \operatorname{Re}(Ax - \lambda x, y) \leq o(\varepsilon),$$

and consequently $\operatorname{Re}(Ax - \lambda x, y) \leq o(1)$ for all $y \in \mathcal{H}$. Therefore, the relation

$$\operatorname{Re}(Ax - \lambda x, y) = 0 \qquad \text{for all} \quad y \in \mathcal{H}$$

has to be fulfilled and especially

$$Ax = \lambda x.$$

$$\text{q.e.d.}$$

Theorem 7.3. (Spectral theorem of F. Rellich)
Let the completely continuous Hermitian operator $A : \mathcal{H} \to \mathcal{H}$ be given on the Hilbert space $\mathcal{H}$ satisfying $A \neq 0$. Then we have a finite or countably infinite system of orthonormal elements $\{\varphi_i\}_{i=1,2,\ldots}$ in $\mathcal{H}$ such that

a) The elements φ_i are eigenfunctions to the eigenvalues $\lambda_i \in \mathbb{R}$ ordered as follows:

$$\|A\| = |\lambda_1| \geq |\lambda_2| \geq |\lambda_3| \geq \ldots > 0,$$

more precisely

$$A\varphi_i = \lambda_i \varphi_i, \qquad i = 1, 2, \ldots$$

If the set $\{\varphi_i\}_i$ is infinite, we have the asymptotic behavior

$$\lim_{i \to \infty} \lambda_i = 0.$$

b) For all $x \in \mathcal{H}$ we have the representations

$$Ax = \sum_{i=1,2,\ldots} \lambda_i(\varphi_i, x)\varphi_i \qquad and \qquad (x, Ax) = \sum_{i=1,2,\ldots} \lambda_i |(\varphi_i, x)|^2.$$

Remark: This theorem remains true for not separable Hilbert spaces. If the system $\{\varphi_i\}_{i=1,\dots,N}$ is finite, the series above reduce to sums.

Proof of Theorem 7.3: On account of $\|A\| > 0$, Theorem 7.2 yields the existence of an element $\varphi_1 \in \mathcal{H}$ with $\|\varphi_1\| = 1$ satisfying

$$A\varphi_1 = \lambda_1\varphi_1, \qquad \lambda_1 \in \{-\|A\|, +\|A\|\}.$$

Furthermore, we have

$$|(Ax, x)| \leq |\lambda_1|(x, x) \qquad \text{for all} \quad x \in \mathcal{H}.$$

We now assume that we have already found $m \geq 1$ orthonormal eigenelements $\varphi_1, \dots, \varphi_m$ with the associate eigenvalues $\lambda_1, \dots, \lambda_m \in \mathbb{R}$ satisfying the property a). Then we consider the completely continuous Hermitian operator

$$B_m x = Ax - \sum_{i=1}^{m} \lambda_i(\varphi_i, x)\varphi_i.$$

Case 1: We have $B_m = 0$. Then the following representation holds true:

$$Ax = \sum_{i=1}^{m} \lambda_i(\varphi_i, x)\varphi_i.$$

Case 2: We have $B_m \neq 0$. Due to Theorem 7.2 we have an element $\varphi \in \mathcal{H}$ with $\|\varphi\| = 1$, such that $B_m\varphi = \lambda\varphi$ and consequently

$$A\varphi - \sum_{i=1}^{m} \lambda_i(\varphi_i, \varphi)\varphi_i = \lambda\varphi$$

is satisfied with $|\lambda| = \|B_m\| > 0$. Multiplication by φ_k, with an index $k \in \{1, \dots, m\}$, from the left yields

$$\lambda(\varphi_k, \varphi) = (\varphi_k, A\varphi) - \lambda_k(\varphi_k, \varphi) = (A\varphi_k, \varphi) - \lambda_k(\varphi_k, \varphi)$$
$$= \lambda_k(\varphi_k, \varphi) - \lambda_k(\varphi_k, \varphi) = 0, \qquad k = 1, \dots m.$$

Therefore, the system $\{\varphi_1, \dots, \varphi_m, \varphi\}$ is orthonormal as well; we set $\varphi_{m+1} := \varphi$ and $\lambda_{m+1} := \lambda \neq 0$. Now we deduce $|\lambda_{m+1}| \leq |\lambda_m|$:
By construction the following estimate

$$|(x, B_m x)| \leq |\lambda_m|(x, x) \qquad \text{for all} \quad x \in \mathcal{H}$$

holds true, and for $x = \varphi_{m+1}$ we obtain

$$|\lambda_m| \geq |(\varphi_{m+1}, B_m\varphi_{m+1})| = |(\varphi_{m+1}, \lambda_{m+1}\varphi_{m+1})| = |\lambda_{m+1}|.$$

We now assume that the procedure above does not end. Since the elements $\{\varphi_i\}_i$ are orthonormal, we infer $\varphi_i \rightharpoonup 0$ $(i \to \infty)$ and the complete continuity of the operator A yields

$$|\lambda_i| = \|A\varphi_i\| \to 0 \ (i \to \infty).$$

On account of $\|B_m\| = |\lambda_{m+1}|$ we obtain the statement

$$\left\| A - \sum_{i=1}^{m} \lambda_i(\varphi_i, \cdot)\varphi_i \right\| = |\lambda_{m+1}| \to 0 \quad (m \to \infty) \tag{4}$$

and consequently

$$Ax = \sum_{i=1}^{\infty} \lambda_i(\varphi_i, x)\varphi_i, \qquad x \in \mathcal{H}.$$

Therefore, all $y = Ax$ with $x \in \mathcal{H}$ can be represented in the form

$$y = \sum_{i=1}^{\infty} (\varphi_i, y)\varphi_i,$$

and the system $\{\varphi_i\}_{i=1,2,\ldots}$ is complete in $\overline{\mathcal{W}_A} = \overline{A(\mathcal{H})}$.   q.e.d.

Theorem 7.4. *The Hermitian operator $A : \mathcal{H} \to \mathcal{H}$ with finite square-norm $N(A) < +\infty$ satisfying $A \neq 0$ is defined on the separable Hilbert space $\mathcal{H}$. The operator A may possess a countably infinite system of orthonormal eigenelements $\{\varphi_i\}_{i=1,2,\ldots}$ and associate eigenvalues $\{\lambda_i\}_{i=1,2,\ldots}$ with the properties a) and b) from Theorem 7.3. We set*

$$A_n f := Af - \sum_{i=1}^{n} \lambda_i(\varphi_i, f)\varphi_i, \qquad n = 1, 2, \ldots$$

Then the sequence of square-norms

$$N(A_n)^2 = \sum_{i=n+1}^{\infty} \lambda_i^2, \qquad n = 1, 2, \ldots$$

is a zero sequence.

Proof: Noting that $N(A) < +\infty$, the operator $A : \mathcal{H} \to \mathcal{H}$ is completely continuous and Theorem 7.3 gives us the representation

$$y = \sum_{i=1}^{\infty} (\varphi_i, y)\varphi_i \qquad \text{for all} \quad y \in \mathcal{W}_A.$$

We observe the decomposition $\mathcal{H} = \overline{\mathcal{W}_A} \oplus \mathcal{N}_A$. The relation $y \in \mathcal{N}_A$ or equivalently $Ay = 0$ holds true if and only if

$$0 = (Ay, x) = (y, Ax) \qquad \text{for all} \quad x \in \mathcal{H}$$

is satisfied, which means $\mathcal{N}_A = \mathcal{W}_A^{\perp}$.

Let now $\{\psi_i\}_{i=1,2,\ldots}$ represent a c.o.n.s. in $\mathcal{N}_A$. Then the set $\{\varphi_i\}_i \cup \{\psi_i\}_i$ constitutes a c.o.n.s. in $\mathcal{H}$. This allows us to evaluate

$$N(A)^2 = \sum_{i=1}^{\infty} \|A\varphi_i\|^2 + \sum_{i=1}^{\infty} \|A\psi_i\|^2 = \sum_{i=1}^{\infty} \lambda_i^2 < +\infty$$

and finally

$$N(A_n)^2 = \sum_{i=1}^{\infty} \|A_n\varphi_i\|^2 + \sum_{i=1}^{\infty} \|A_n\psi_i\|^2 = \sum_{i=n+1}^{\infty} \lambda_i^2 \to 0 \ (n \to \infty).$$

$$\text{q.e.d.}$$

We specialize Theorem 7.3 and Theorem 7.4 to Hilbert-Schmidt operators and obtain the following

Theorem 7.5. (Spectral theorem of D. Hilbert and E. Schmidt)
*On the rectangle $Q \subset \mathbb{R}^n$ with $n \in \mathbb{N}$ let the integral kernel
$K = K(x,y) : Q \times Q \to \mathbb{C} \in L^2(Q \times Q)$ be given satisfying*

$$\int\limits_{Q \times Q} |K(x,y)|^2 \, dx \, dy > 0$$

and

$$\overline{K(y,x)} = K(x,y) \qquad \text{for almost all} \quad (x,y) \in Q \times Q. \tag{5}$$

Then we have a finite or countably infinite system of eigenfunctions

$$\{\varphi_i(x)\}_{i=1,2,\ldots} \subset L^2(Q, \mathbb{C})$$

with the associate eigenvalues $\{\lambda_i\}_{i=1,2,\ldots} \subset \mathbb{R}$, such that the following integral-eigenvalue-equation

$$\int\limits_Q K(x,y)\varphi_i(y) \, dy = \lambda_i \varphi_i(x) \qquad \text{for almost all} \quad x \in Q \tag{6}$$

is satisfied with $i = 1, 2, \ldots$. The eigenvalues have the properties

$$|\lambda_1| \geq |\lambda_2| \geq \ldots > 0 \qquad \text{and} \qquad \lim_{i \to \infty} \lambda_i = 0. \tag{7}$$

Furthermore, we have the eigenvalue expansions

$$\int\limits_{Q \times Q} |K(x,y)|^2 \, dx \, dy = \sum_{i=1}^{\infty} \lambda_i^2 < +\infty \tag{8}$$

and

$$\int\limits_{Q \times Q} \left| K(x,y) - \sum_{i=1}^{n} \lambda_i \varphi_i(x)\overline{\varphi_i(y)} \right|^2 dx \, dy = \sum_{i=n+1}^{\infty} \lambda_i^2 \to 0 \ (n \to \infty). \tag{9}$$

8 The Sturm-Liouville Eigenvalue Problem

We need the following result:

Theorem 8.1. (Eigenvalue problem for weakly singular integral operators)
Let the weakly singular kernel $K = K(x,y) \in S_\alpha(G, \mathbb{C})$ with $\alpha \in [0, n)$ be given on the bounded domain $G \subset \mathbb{R}^n$, and we have $K(x,y) \not\equiv 0$ and

$$K(x,y) = \overline{K(y,x)} \qquad \text{for all} \quad (x,y) \in G \otimes G.$$

We denote the associate integral operator by $\mathbb{K}$, and define as our domain of definition

$$\mathcal{D} := \left\{ f \in C^0(G, \mathbb{C}) \ : \ \sup_{x \in G} |f(x)| < +\infty \right\}.$$

Statements: *Then we have a finite or countably infinite orthonormal system of eigenfunctions $\{\varphi_i\}_{i \in I} \subset \mathcal{D}$ with their eigenvalues $\lambda_i \in \mathbb{R} \setminus \{0\}$ for $i \in I$ satisfying*

$$\int_G K(x,y)\varphi_i(y)\,dy = \lambda_i \varphi_i(x), \quad x \in G, \quad i \in I. \tag{1}$$

If $I = \{1, 2, \ldots\}$ is countably infinite, we have

$$\lim_{i \to \infty} \lambda_i = 0, \tag{2}$$

and each function $g = \mathbb{K}f$ with $f \in \mathcal{D}$ can be approximated in the square-means according to

$$\lim_{n \to \infty} \int_G \left| g(x) - \sum_{i=1}^{n} g_i \varphi_i(x) \right|^2 dx = 0. \tag{3}$$

Here we have set

$$g_i := \int_G \overline{\varphi_i(x)} g(x)\,dx, \qquad i \in I \tag{4}$$

for the Fourier coefficients. When we assume $\alpha \in [0, \frac{n}{2})$ in addition, the function $g = \mathbb{K}f$ with $f \in \mathcal{D}$ can be expanded into the following uniformly convergent series:

$$g(x) = \sum_{i=1}^{\infty} g_i \varphi_i(x), \qquad x \in G. \tag{5}$$

Proof: As it has been elaborated in the proof of Theorem 6.19 from Section 6, the operator $\mathbb{K} : \mathcal{H} \to \mathcal{H}$ is completely continuous on the Hilbert space $\mathcal{H} = L^2(G, \mathbb{C})$. Furthermore, we have the regularity result

$$\mathbb{E}u - \mathbb{K}u = v \quad \text{with} \quad u \in \mathcal{H} \quad \text{and} \quad v \in \mathcal{D} \quad \Rightarrow \quad u \in \mathcal{D}. \qquad (6)$$

Due to Rellich's spectral theorem from Section 7, Theorem 7.3 the operator $\mathbb{K}$ possesses a finite or countably infinite system of orthonormal eigenfunctions $\{\varphi_i\}_{i=1,2,\dots} \subset \mathcal{H}$. Then we have the identities

$$\int_G K(x,y)\varphi_i(y)\,dy = \lambda_i \varphi_i(x), \qquad x \in G$$

for all $i = 1, 2, \dots$ with $|\lambda_1| \geq |\lambda_2| \geq \dots > 0$ and $\lambda_i \to 0$ $(i \to \infty)$, in the case that infinitely many eigenfunctions exist. Because of the regularity result (6) we see $\varphi_i \in \mathcal{D}$ for $i = 1, 2, \dots$. Furthermore, the function $g = \mathbb{K}f$ with an arbitrary element $f \in \mathcal{D}$ satisfies the following identity in the Hilbert space:

$$g = \mathbb{K}f = \sum_{i=1,2,\dots} \lambda_i(\varphi_i, f)\varphi_i = \sum_{i=1,2,\dots} (\mathbb{K}\varphi_i, f)\varphi_i$$

$$= \sum_{i=1,2,\dots} (\varphi_i, \mathbb{K}f)\varphi_i = \sum_{i=1,2,\dots} (\varphi_i, g)\varphi_i$$

or equivalently (3), where the Fourier coefficients g_i are defined in (4). When we assume $\alpha \in [0, \frac{n}{2})$ in addition, the linear operator $\mathbb{K} : L^2(G) \to C^0(G)$ is bounded by

$$\|\mathbb{K}f\|_{C^0(G)} \leq C\|f\|_{L^2(G)} \qquad \text{for all} \quad f \in \mathcal{D}$$

due to Theorem 2.2 from Section 2. The series $\displaystyle\sum_{i=1,2,\dots} (\varphi_i, f)\varphi_i$, convergent in the Hilbert space $\mathcal{H} = L^2(G, \mathbb{C})$, is consequently transferred by the operator $\mathbb{K}$ into the uniformly convergent series

$$\sum_{i=1,2,\dots} \lambda_i(\varphi_i, f)\varphi_i = \mathbb{K}\Big(\sum_{i=1,2,\dots} (\varphi_i, f)\varphi_i \Big) = g.$$

q.e.d.

Theorem 8.2. (Expansion theorem for kernels)
Let $K = K(x,y) : G \times G \to \mathbb{C}$ denote a Hermitian integral kernel of the class $S_0(G, \mathbb{C})$, which is continuous on $G \times G$. For the associate integral operator $\mathbb{K}$ we assume

$$(f, \mathbb{K}f) \geq 0 \qquad \text{for all} \quad f \in \mathcal{D}. \qquad (7)$$

Then we have a representation by the uniformly convergent series in each compact set $\Gamma \subset G$ as follows:

$$K(x,y) = \sum_{i=1}^{\infty} \lambda_i \varphi_i(x)\overline{\varphi_i(y)}, \qquad (x, y) \in \Gamma \times \Gamma. \qquad (8)$$

Proof:

1. We show firstly that

$$K(x,x) \geq 0 \qquad \text{for all} \quad x \in G \qquad (9)$$

is fulfilled. Here we utilize the function $\varphi = \varphi(y) : \mathbb{R}^n \to [0, +\infty) \in C^0(\mathbb{R}^n)$ satisfying

$$\varphi(y) = 0, \quad |y| \geq 1, \qquad \text{and} \qquad \int\limits_{\mathbb{R}^n} \varphi(y)\, dy = 1.$$

The quantity $\delta > 0$ given arbitrarily, we consider the approximate point distributions about $x \in G$, namely

$$f_\delta(y) := \frac{1}{\delta^n} \varphi\Big(\frac{1}{\delta}(y - x)\Big), \qquad y \in \mathbb{R}^n.$$

We insert the function $f_\delta \in \mathcal{D}$ into (7) and obtain

$$0 \leq (f_\delta, \mathbb{K}f_\delta) = \int\limits_{\mathbb{R}^n} \int\limits_{\mathbb{R}^n} f_\delta(y) K(y,z) f_\delta(z)\, dy\, dz$$

$$= \int\limits_{\mathbb{R}^n} \int\limits_{\mathbb{R}^n} K(x,x) f_\delta(y) f_\delta(z)\, dy\, dz$$

$$+ \int\limits_{\mathbb{R}^n} \int\limits_{\mathbb{R}^n} (K(y,z) - K(x,x)) f_\delta(y) f_\delta(z)\, dy\, dz$$

$$= K(x,x) + \int\limits_{\mathbb{R}^n} \int\limits_{\mathbb{R}^n} (K(y,z) - K(x,x)) f_\delta(y) f_\delta(z)\, dy\, dz.$$

Since the second term on the right-hand side vanishes for $\delta \to 0$, we infer the inequality (9).

2. We now show

$$0 \leq \sum_{i=1}^{\infty} \lambda_i |\varphi_i(x)|^2 \leq K(x,x) < +\infty \qquad \text{for all} \quad x \in G. \qquad (10)$$

Let us define the integral kernel

$$K_N(x,y) := K(x,y) - \sum_{i=1}^{N} \lambda_i \varphi_i(x) \overline{\varphi_i(y)}$$

with the associate integral operator $\mathbb{K}_N$. The latter satisfies

$$(f, \mathbb{K}_N f) = \sum_{i=N+1}^{\infty} \lambda_i |(\varphi_i, f)|^2 \geq 0 \qquad \text{for all} \quad f \in \mathcal{D}.$$

From part 1 of our proof we obtain

$$K_N(x,x) \geq 0 \quad \text{for all} \quad x \in G$$

or equivalently

$$K(x,x) \geq \sum_{i=1}^{N} \lambda_i \varphi_i(x) \overline{\varphi_i(x)}, \qquad x \in G$$

for all $N \in \mathbb{N}$, which implies (10).

3. Let the point $x \in G$ be chosen as fixed. With an arbitrary quantity $\varepsilon > 0$, we then can estimate

$$\sum_{i=N+1}^{\infty} |\lambda_i \varphi_i(x) \overline{\varphi_i(y)}| \leq \sqrt{\sum_{i=N+1}^{\infty} \lambda_i |\varphi_i(x)|^2} \sqrt{\sum_{i=N+1}^{\infty} \lambda_i |\varphi_i(y)|^2}$$

$$\leq \varepsilon \sqrt{K(y,y)} \leq \varepsilon \cdot \text{const}, \qquad y \in G,$$

for all $N \geq N_0(\varepsilon)$. Therefore, we have the following statement for each fixed $x \in G$:

The series $\quad \Phi(y) := \sum_{i=1}^{\infty} \lambda_i \varphi_i(x) \overline{\varphi_i(y)} \quad$ converges uniformly in $\quad G$. (11)

4. Due to Theorem 7.5 from Section 7, we have the relation

$$K(x,y) = \sum_{i=1}^{\infty} \lambda_i \varphi_i(x) \overline{\varphi_i(y)} \tag{12}$$

in the $L^2(G \times G, \mathbb{C})$-sense. Choosing the point $x \in G$ and the function $f \in C_0^0(G)$ arbitrarily, we obtain the following identity via (11) and (12):

$$\int_G K(x,y) f(y) \, dy = \lim_{N \to \infty} \int_G \left(\sum_{i=1}^{N} \lambda_i \varphi_i(x) \overline{\varphi_i(y)} \right) f(y) \, dy$$

$$= \int_G \left(\sum_{i=1}^{\infty} \lambda_i \varphi_i(x) \overline{\varphi_i(y)} \right) f(y) \, dy.$$

This implies

$$\int_G \left(K(x,y) - \sum_{i=1}^{\infty} \lambda_i \varphi_i(x) \overline{\varphi_i(y)} \right) f(y) \, dy = 0 \qquad \text{for all} \quad f \in C_0^0(G).$$

Since $K \in C^0(G \times G)$ holds true, and the series in the integrand is continuous with respect to $y \in G$, we deduce the pointwise identity

$$K(x,y) = \sum_{i=1}^{\infty} \lambda_i \varphi_i(x) \overline{\varphi_i(y)} \qquad \text{for all} \quad x, y \in G. \tag{13}$$

5. Especially for $x = y$, we infer the following identity from (13):

$$K(x,x) = \sum_{i=1}^{\infty} \lambda_i |\varphi_i(x)|^2, \qquad x \in G.$$

Because of Dini's theorem, the series converges uniformly on each compact set $\Gamma \subset G$. Finally, we obtain the following inequality for arbitrary $\varepsilon > 0$ and suitable $N \geq N_0(\varepsilon)$, namely

$$\left| \sum_{i=N+1}^{\infty} \lambda_i \varphi_i(x) \overline{\varphi_i(y)} \right| \leq \sqrt{\sum_{i=N+1}^{\infty} \lambda_i |\varphi_i(x)|^2} \sqrt{\sum_{i=N+1}^{\infty} \lambda_i |\varphi_i(y)|^2} \leq \varepsilon^2$$

for all $(x,y) \in \Gamma \times \Gamma$. 	q.e.d.

Theorem 8.3. (The Sturm-Liouville eigenvalue problem)
We prescribe $a, b \in \mathbb{R}$ with $a < b$ and the coefficient functions

$$p = p(x) \in C^1([a,b],(0,+\infty)), \qquad q = q(x) \in C^0([a,b],\mathbb{R}),$$

and consider the Sturm-Liouville operator $\mathbb{L} : C^2([a,b],\mathbb{C}) \to C^0([a,b],\mathbb{C})$ *defined by*

$$\mathbb{L}u(x) := -(p(x)u'(x))' + q(x)u(x), \qquad x \in [a,b].$$

Furthermore, we use the real boundary operators $\mathbb{B}_j : C^2([a,b],\mathbb{C}) \to \mathbb{C}$ *for $j = 1, 2$ defined by*

$$\mathbb{B}_1 u := c_1 u(a) + c_2 u'(a) \qquad with \quad c_1^2 + c_2^2 > 0$$

and

$$\mathbb{B}_2 u := d_1 u(b) + d_2 u'(b) \qquad with \quad d_1^2 + d_2^2 > 0.$$

Finally, we fix the domain of definition

$$\mathcal{D} := \left\{ u \in C^2([a,b],\mathbb{C}) : \mathbb{B}_1 u = 0 = \mathbb{B}_2 u \right\}.$$

Statements: *Then we have a sequence $\{\lambda_i\}_{i=1,2,\ldots} \subset \mathbb{R}$ of eigenvalues satisfying*

$$-\infty < \lambda_1 \leq \lambda_2 \leq \ldots \qquad and \qquad \lim_{i \to \infty} \lambda_i = +\infty$$

and the associate eigenfunctions $\{\varphi_i\}_{i=1,2,\ldots} \subset \mathcal{D}$ with the following properties:

a) We have

$$\mathbb{L}\varphi_i = \lambda_i \varphi_i, \qquad \int_a^b \overline{\varphi_i(x)} \varphi_j(x) \, dx = \delta_{ij} \qquad for \ all \quad i, j \in \mathbb{N},$$

and the identity

$$\sum_{i=1}^{\infty} \left| \int_a^b \overline{\varphi_i(x)} f(x)\, dx \right|^2 = \int_a^b |f(x)|^2\, dx \qquad \text{for all} \quad f \in \mathcal{D}$$

is satisfied.

b) Each function $g \in \mathcal{D}$ can be expanded into the uniformly convergent series on the interval $[a, b]$ as follows:

$$g(x) = \sum_{i=1}^{\infty} g_i \varphi_i(x), \quad x \in [a, b], \qquad \text{with} \quad g_i := \int_a^b \overline{\varphi_i(x)} g(x)\, dx, \quad i \in \mathbb{N}.$$

c) If the property $\lambda_i \neq 0$ for all $i \in \mathbb{N}$ is satisfied, the following series

$$\sum_{i=1}^{\infty} \frac{1}{\lambda_i} \varphi_i(x) \overline{\varphi_i(y)}, \qquad (x, y) \in [a, b] \times [a, b]$$

converges uniformly towards the Green function K of $\mathbb{L}$ under the boundary conditions $\mathbb{B}_1 = 0 = \mathbb{B}_2$.

Proof: We continue our considerations from Section 1 concerning the Sturm-Liouville eigenvalue problem.

1. All eigenvalues of $\mathbb{L}$ are real. Since the coefficient functions p and q are real, we obtain the following statement from Proposition 1.3 in Section 1 by separation into the real and imaginary parts:

$$\int_a^b \overline{\mathbb{L}u(x)} v(x)\, dx = \int_a^b u(x) \overline{\mathbb{L}v(x)}\, dx \qquad \text{for all} \quad u, v \in \mathcal{D}. \qquad (14)$$

We calculate

$$\overline{\lambda_i} \int_a^b |\varphi_i(x)|^2\, dx = \int_a^b \overline{\lambda_i \varphi_i(x)} \varphi_i(x)\, dx = \int_a^b \overline{\mathbb{L}\varphi_i}\, \varphi_i\, dx$$

$$= \int_a^b \overline{\varphi_i} \mathbb{L}\varphi_i\, dx = \int_a^b \overline{\varphi_i} \lambda_i \varphi_i\, dx$$

$$= \lambda_i \int_a^b |\varphi_i(x)|^2\, dx, \qquad i=1,2,\ldots$$

This implies

$$\lambda_i = \overline{\lambda_i} \qquad \text{for all} \quad i \in \mathbb{N}.$$

2. We now prove that the sequence of eigenvalues is bounded from below. Here we consider the class of admissible functions

$$\mathcal{D}_0 := \left\{ u = u(x) \in C^2([a,b], \mathbb{C}) \; : \; u(a) = 0 = u(b) \right\}.$$

If $u \in \mathcal{D}_0$ is a solution of $\mathbb{L}u = \lambda u$, we infer

$$\lambda \geq q_* := \inf_{a \leq x \leq b} q(x). \tag{15}$$

With the aid of partial integration, we evaluate

$$\lambda \int_a^b |u(x)|^2 \, dx = \lambda \int_a^b \overline{u(x)} u(x) \, dx = \int_a^b \overline{\mathbb{L}u(x)} u(x) \, dx$$

$$= \int_a^b \left\{ p(x)|u'(x)|^2 + q(x)|u(x)|^2 \right\} dx$$

$$\geq q_* \int_a^b |u(x)|^2 \, dx.$$

We now show indirectly that the operator $\mathbb{L}$ on $\mathcal{D}$ possesses at most two eigenvalues smaller than q_*. On the contrary, we assume that we had three eigenfunctions $\varphi_1, \varphi_2, \varphi_3 \in \mathcal{D}$ satisfying

$$\mathbb{L}\varphi_i = \lambda_i \varphi_i, \quad i = 1, 2, 3 \quad \text{and} \quad \lambda_1 \leq \lambda_2 \leq \lambda_3 < q_*.$$

Then we can find numbers $\alpha_1, \alpha_2, \alpha_3 \in \mathbb{C}$ with $|\alpha_1|^2 + |\alpha_2|^2 + |\alpha_3|^2 = 1$ such that

$$v := \sum_{i=1}^3 \alpha_i \varphi_i \in \mathcal{D}_0$$

is correct. On account of (15) we see

$$q_* \int_a^b |v(x)|^2 \, dx \leq \int_a^b \overline{\mathbb{L}v(x)} \, v(x) \, dx = \int_a^b \left(\overline{\sum_{i=1}^3 \lambda_i \alpha_i \varphi_i} \right) \left(\sum_{j=1}^3 \alpha_j \varphi_j \right) dx$$

$$= \int_a^b \sum_{i=1}^3 \lambda_i |\alpha_i|^2 |\varphi_i(x)|^2 \, dx \leq \lambda_3 \int_a^b \sum_{i=1}^3 |\alpha_i|^2 |\varphi_i(x)|^2 \, dx$$

$$= \lambda_3 \int_a^b |v(x)|^2 \, dx.$$

We obtain $\lambda_3 \geq q_*$ in contradiction to $\lambda_3 < q_*$.

3. We name $\lambda_1 \in \mathbb{R}$ the least eigenvalue of $\mathbb{L}$ on $\mathcal{D}$, existing due to part 2 of our proof. Then we obtain in

$$\tilde{\mathbb{L}} := \mathbb{L} - \lambda_1 \mathbb{E} + \mathbb{E}$$

a Sturm-Liouville operator with the eigenvalues $\tilde{\lambda}_k \geq 1$, $k = 1, 2, \ldots$
Due to Theorem 1.6 in Section 1, the operator $\mathbb{L}$ on $\mathcal{D}$ possesses a symmetric Green's function

$$K = K(x, y), \; (x, y) \in [a, b] \times [a, b] \quad \text{of the class} \quad C^0([a, b] \times [a, b], \mathbb{R}).$$

We now take Theorem 1.7 from Section 1 into account and utilize Theorem 8.1 for the given integral equation. Then we obtain a sequence of eigenfunctions

$$\{\varphi_i\}_{i=1,2,\ldots} \subset \mathcal{D} \text{ satisfying } \mathbb{L}\varphi_i = \lambda_i \varphi_i \, (i \in \mathbb{N}) \text{ and } \lambda_1 \leq \lambda_2 \leq \ldots \to +\infty.$$

From Theorem 8.1 and Theorem 8.2 we immediately infer all the statements above.   q.e.d.

9 Weyl's Eigenvalue Problem for the Laplace Operator

We need the following generalization of the Gaussian integral theorem, which does not require regularity assumptions for the boundary of the domain with vanishing boundary values:

Proposition 9.1. (Giesecke, E. Heinz)

I. *Let the bounded domain $G \subset \mathbb{R}^n$ be given, in which $N \in \mathbb{N}_0$ mutually disjoint balls*

$$K_j := \left\{ x \in \mathbb{R}^n \, : \, |x - x^{(j)}| \leq r_j \right\}, \qquad j = 1, 2, \ldots, N$$

with their radii $r_j > 0$ and their centers $x^{(j)}$ are contained. We set

$$G' := G \setminus \{x^{(1)}, \ldots, x^{(N)}\} \qquad \text{and} \qquad G'' := G \setminus \bigcup_{j=1}^{N} K_j.$$

The topological closure of the set G'' is denoted by $\overline{G''}$.
II. *For the two functions $u, v \in C^2(G') \cap C^0(\overline{G''})$ we assume*

$$u|_{\partial G} = 0 = v|_{\partial G}; \qquad \int_{G''} \left\{ |\Delta u(x)| + |\Delta v(x)| \right\} dx < +\infty.$$

Statement: *Then the following identity*

$$\int_{G''} (v\Delta u + \nabla v \cdot \nabla u)\, dx = -\sum_{j=1}^{N} \int_{|x-x^{(j)}|=r_j} v\frac{\partial u}{\partial \nu_j}\, d\Omega_j \qquad (1)$$

holds true. Here the symbol ν_j denotes the exterior normal to K_j and $d\Omega_j$ the surface element on the spheres $\{x : |x - x^{(j)}| = r_j\} = \partial K_j$ for $j = 1, \ldots, N$.

From Proposition 9.1 we immediately infer the subsequent

Proposition 9.2. *With the assumptions from Proposition 9.1 we have Green's identity*

$$\int_{G''} (v\Delta u - u\Delta v)\, dx = -\sum_{j=1}^{N} \int_{|x-x^{(j)}|=r_j} \left(v\frac{\partial u}{\partial \nu_j} - u\frac{\partial v}{\partial \nu_j}\right) d\Omega_j. \qquad (2)$$

Proof of Proposition 9.1:

1. At first, we assume that $v \in C_0^0(G)$ is satisfied in addition to the assumptions above, and we consider the vector-field $f = v\nabla u$. Then the Gaussian integral theorem yields the identity

$$\int_{G''} (v\Delta u + \nabla v \cdot \nabla u)\, dx = -\sum_{j=1}^{N} \int_{|x-x^{(j)}|=r_j} v\frac{\partial u}{\partial \nu_j}\, d\Omega_j. \qquad (3)$$

We approximate an arbitrary function $v \in C^2(G') \cap C^0(\overline{G''})$ by a sequence $\{v_k\}_{k=1,2,\ldots}$ as follows: Let $\{w_k(t)\}_{k=1,2,\ldots} \subset C^\infty(\mathbb{R}, [0,1]))$ denote a sequence of functions with the properties

$$w_k(t) = \begin{cases} 1, & |t| \geq \dfrac{1}{k} \\[2mm] 0, & |t| \leq \dfrac{1}{2k} \end{cases}, \qquad k = 1, 2, \ldots$$

The functions

$$\varphi_k(t) := \int_0^t w_k(s)\, ds, \qquad t \in \mathbb{R}$$

then satisfy

$$\varphi_k(0) = 0, \quad \varphi_k'(t) = w_k(t), \qquad k = 1, 2, \ldots,$$

and we estimate

$$|\varphi_k(t) - t| = \left| \int_0^t (w_k(s) - 1)\, ds \right| \leq \frac{2}{k}, \qquad k = 1, 2, \ldots$$

Now we define the sequence

$$v_k(x) := \varphi_k(v(x)), \quad x \in \overline{G''}, \qquad k = 1, 2, \ldots \tag{4}$$

and consider the relation

$$|v_k(x) - v(x)| = |\varphi_k(v(x)) - v(x)| \leq \frac{2}{k} \to 0 \ (k \to \infty)$$

for all $x \in G''$, which implies

$$v_k(x) \to v(x) \ (k \to \infty) \qquad \text{uniformly in} \quad \overline{G''}. \tag{5}$$

2. We prove that

$$E := \Big\{ x \in G'' \ : \ v(x) = 0, \ \nabla v(x) \neq 0 \Big\}$$

represents a Lebesgue null-set: Here we choose the point $z \in E$ arbitrarily. Taking the quantity $\varepsilon > 0$ sufficiently small, the set

$$E \cap \Big\{ x \in G'' \ : \ |x - z| < \varepsilon \Big\}$$

constitutes a graph, and is consequently a Lebesgue null-set due to the theorem on implicit functions. We exhaust the set G'' with the aid of the cube decomposition. For each point $z \in E$ we consider a sufficiently small cube $z \in W \subset G''$ such that $W \cap E$ is a Lebesgue null-set. Now the set E consists of a countable union of those sets $W \cap E$, and the σ-additivity of the Lebesgue measure yields the statement above.

3. For all points $x \in G'' \setminus E$ we deduce

$$\nabla v_k(x) = \varphi_k'(v(x))\nabla v(x) = w_k(v(x))\nabla v(x) \to \nabla v(x) \quad (k \to \infty),$$

which holds true a. e. in G'', due to part 2 of our proof. We insert $v = v_k$ into (3) and obtain

$$\int\limits_{G''} (v_k \Delta u + \nabla v_k \cdot \nabla u)\, dx = -\sum_{j=1}^{N} \int\limits_{|x-x^{(j)}|=r_j} v_k \frac{\partial u}{\partial \nu_j}\, d\Omega_j, \qquad k \in \mathbb{N}.$$

Then we observe the passage to the limit $k \to \infty$ and see

$$\int\limits_{G''} v(x)\Delta u(x)\, dx + \lim_{k \to \infty} \int\limits_{G''} (\nabla v_k(x) \cdot \nabla u(x))\, dx$$

$$= -\sum_{j=1}^{N} \int\limits_{|x-x^{(j)}|=r_j} v(x)\frac{\partial u(x)}{\partial \nu_j}\, d\Omega_j. \tag{6}$$

Inserting $v(x) = u(x)$ into (6) and noting that

$$\nabla v_k(x) = w_k(u(x))\nabla u(x)$$

holds true, we see

$$\int_{G''} u(x)\Delta u(x)\,dx + \lim_{k\to\infty} \int_{G''} w_k(u(x))|\nabla u(x)|^2\,dx$$

$$= -\sum_{j=1}^{N} \int_{|x-x^{(j)}|=r_j} u(x)\frac{\partial u(x)}{\partial \nu_j}\,d\Omega_j. \tag{7}$$

Fatou's theorem yields

$$\int_{G''} |\nabla u(x)|^2\,dx < +\infty \quad \text{and} \quad \int_{G''} |\nabla v(x)|^2\,dx < +\infty. \tag{8}$$

On account of

$$|\nabla v_k(x) \cdot \nabla u(x)| = |w_k(v(x))|\,|\nabla v(x) \cdot \nabla u(x)|$$

$$\le \frac{1}{2}(|\nabla u(x)|^2 + |\nabla v(x)|^2), \qquad x \in G'',$$

we have an integrable majorant for the limit in (6). The identity (1) then follows by Lebesgue's convergence theorem.   q.e.d.

We now continue the considerations from Section 1 concerning the eigenvalue problem of the n-dimensional oscillation equation : Let $G \subset \mathbb{R}^n$ denote a bounded Dirichlet domain. On the linear space

$$\mathcal{E} := \left\{ u = u(x) \in C^2(G) \cap C^0(\overline{G}) \,:\, u|_{\partial G} = 0 \right\}$$

we consider *Weyl's eigenvalue problem*

$$-\Delta u(x) = \lambda u(x), \quad x \in G \quad \text{with} \quad u \in \mathcal{E} \setminus \{0\} \quad \text{and} \quad \lambda \in \mathbb{R}. \tag{9}$$

Proposition 9.3. *All eigenvalues λ of (9) have the property $\lambda > 0$.*

Proof: With $u \in \mathcal{E} \setminus \{0\}$ we consider a solution of (9) belonging to the eigenvalue $\lambda \in \mathbb{R}$. Then we infer

$$\int_G |\Delta u(x)|\,dx = |\lambda| \int_G |u(x)|\,dx < +\infty.$$

We apply Proposition 9.1 with $v = u$ and obtain

$$\int_G |\nabla u(x)|^2\,dx = -\int_G u(x)\Delta u(x)\,dx = \lambda \int_G |u(x)|^2\,dx$$

or equivalently

$$\lambda = \frac{\displaystyle\int_G |\nabla u(x)|^2 \, dx}{\displaystyle\int_G |u(x)|^2 \, dx} > 0.$$

q.e.d.

Remark: The *Rayleigh quotient* appears in the last formula.

We do not mention the case $n = 2$ particularly, and we shall utilize Green's function in the dimensions $n = 3, 4, \ldots$ as follows:

$$H(x, y) = \frac{1}{(n-2)\omega_n} \frac{1}{|y - x|^{n-2}} + h(x, y), \qquad (x, y) \in G \otimes G. \tag{10}$$

A solution u of (9) evidently belongs to the space

$$\mathcal{D} := \left\{ u = u(x) \in C^0(G) \ : \ \sup_{x \in G} |u(x)| < +\infty \right\} = C^0(G) \cap L^\infty(G)$$

and satisfies the integral-equation problem

$$u(x) = \lambda \int_G H(x, y) u(y) \, dy, \quad x \in G \ \text{ with } \ u \in \mathcal{D} \setminus \{0\} \text{ and } \lambda \in \mathbb{R}. \tag{11}$$

We deduce the latter statements as in Section 1 (compare Theorem 1.9 there), using the Propositions 9.1 and 9.2 above. We have already shown the symmetry of Green's function and controlled the growth condition:

$$\begin{aligned} H &= H(x, y) \in \mathcal{S}_{n-2}(G), \\ 0 &\le H(x, y) = H(y, x), \qquad (x, y) \in G \otimes G. \end{aligned} \tag{12}$$

Now we shall prove that a solution u of (11) solves (9) as well.

Proposition 9.4. *Let the function $u = u(x) \in \mathcal{D}$ be given, and the parameter integral*

$$v(x) := \int_G H(x, y) u(y) \, dy, \qquad x \in G$$

be defined. Then we have the properties $v \in C^0(\overline{G})$ and $v|_{\partial G} = 0$.

Proof: With the aid of the auxiliary function

$$\Theta(t) = \begin{cases} 0, & 0 \le t \le 1 \\ t - 1, & 1 \le t \le 2 \\ 1, & 2 \le t \end{cases}$$

we define the continuous integral kernel

$$H_\delta(x, y) := H(x, y) \Theta\left(\frac{|x - y|}{\delta}\right)$$

$$= \Theta\left(\frac{|x - y|}{\delta}\right)\left(\frac{1}{(n - 2)\omega_n}|y - x|^{2-n} + h(x, y)\right), \qquad \delta > 0.$$

For all $\delta > 0$, the parameter integral

$$v_\delta(x) := \int_G H_\delta(x, y) u(y)\, dy, \quad x \in \overline{G}$$

is continuous on $\overline{G}$, and we observe the boundary condition

$$v_\delta|_{\partial G} = 0.$$

Furthermore, we have the following inequality

$$|v_\delta(x) - v(x)| \leq \int_G \left|\Theta\left(\frac{|x - y|}{\delta}\right) - 1\right| |H(x, y)|\, |u(y)|\, dy$$

$$\leq \int_{y:|y-x|\leq 2\delta} \frac{c}{|y - x|^{n-2}}\, dy \leq \gamma(\delta) \to 0 \ (\delta \downarrow 0) \qquad (13)$$

for all $x \in G$. This implies

$$v_\delta(x) \to v(x) \ (\delta \downarrow 0) \qquad \text{uniformly in} \quad G,$$

and therefore $v \in C^0(\overline{G})$ and $v|_{\partial G} = 0$ holds true.   q.e.d.

Proposition 9.5. *Each solution u of (11) satisfies $u \in C^2(G)$ and fulfills the eigenvalue equation*

$$-\Delta u(x) = \lambda u(x), \qquad x \in G.$$

Proof: We take an arbitrary point $z \in G$, and then choose a quantity $\varepsilon > 0$ so small that the inclusion

$$K_\varepsilon(z) := \left\{x \in \mathbb{R}^n \ : \ |x - z| \leq \varepsilon\right\} \subset G$$

is valid. From the integral equation (11) we infer

$$u(x) = \lambda \int_{K_\varepsilon(z)} \frac{1}{(n - 2)\omega_n} \frac{1}{|y - x|^{n-2}} u(y)\, dy$$

$$+ \lambda \int_{G\backslash K_\varepsilon(z)} \frac{1}{(n - 2)\omega_n} \frac{1}{|y - x|^{n-2}} u(y)\, dy + \int_G h(x, y) u(y)\, dy \qquad (14)$$

$$= \lambda \int_{K_\varepsilon(z)} \frac{1}{(n - 2)\omega_n} \frac{1}{|y - x|^{n-2}} u(y)\, dy + \psi_{z,\varepsilon}(x), \qquad x \in \overset{\circ}{K}_\varepsilon(z).$$

Here the function $\psi_{z,\varepsilon}(x)$ is harmonic in $\overset{\circ}{K_\varepsilon}(z)$. We can differentiate the equation (14) once (but not twice!) and obtain, via the Gaussian integral theorem, the following identity

$$
\begin{aligned}
\nabla u(x) &= \lambda \int\limits_{K_\varepsilon(z)} \frac{1}{(n-2)\omega_n} \Big(\nabla_x \frac{1}{|y-x|^{n-2}} \Big) u(y)\, dy + \nabla \psi_{z,\varepsilon}(x) \\
&= -\lambda \int\limits_{K_\varepsilon(z)} \frac{1}{(n-2)\omega_n} \Big(\nabla_y \frac{1}{|y-x|^{n-2}} \Big) u(y)\, dy + \nabla \psi_{z,\varepsilon}(x) \\
&= -\lambda \int\limits_{K_\varepsilon(z)} \frac{1}{(n-2)\omega_n} \nabla_y \Big(\frac{u(y)}{|y-x|^{n-2}} \Big)\, dy \\
&\quad +\lambda \int\limits_{K_\varepsilon(z)} \frac{1}{(n-2)\omega_n} \frac{\nabla u(y)}{|y-x|^{n-2}} u(y)\, dy + \nabla \psi_{z,\varepsilon}(x) \\
&= -\lambda \int\limits_{\partial K_\varepsilon(z)} \frac{1}{(n-2)\omega_n} \frac{u(y)}{|y-x|^{n-2}} \nu(y)\, d\Omega(y) \\
&\quad +\lambda \int\limits_{K_\varepsilon(z)} \frac{1}{(n-2)\omega_n} \frac{\nabla u(y)}{|y-x|^{n-2}}\, dy + \nabla \psi_{z,\varepsilon}(x)
\end{aligned}
\tag{15}
$$

for all $x \in \overset{\circ}{K_\varepsilon}(z)$. Here the symbol $\nu(y)$ denotes the exterior normal to the ball $K_\varepsilon(z)$ and $d\Omega(y)$ the surface element on the sphere $\partial K_\varepsilon(z)$.
From (15) we infer the statement

$$
u \in C^2(G),
\tag{16}
$$

since the point $z \in G$ could be chosen arbitrarily. We differentiate (15) once more, choose $x = z$, and evaluate the limit $\varepsilon \downarrow 0$. Thus we obtain

$$
\begin{aligned}
\Delta u(z) &= \lim_{\varepsilon \downarrow 0} \Big\{ -\lambda \int\limits_{\partial K_\varepsilon(z)} \frac{1}{(n-2)\omega_n} \Big(\nabla_x \frac{u(y)}{|y-x|^{n-2}} \Big)\Big|_{x=z} \nu(y)\, d\Omega(y) \Big\} \\
&\quad + \lim_{\varepsilon \downarrow 0} \Big\{ \lambda \int\limits_{K_\varepsilon(z)} \frac{1}{(n-2)\omega_n} \Big(\nabla_x \frac{\nabla u(y)}{|y-x|^{n-2}} \Big)\Big|_{x=z}\, dy \Big\} \\
&= -\lambda u(z) + 0 = -\lambda u(z) \qquad \text{for all} \quad z \in G.
\end{aligned}
\tag{17}
$$

q.e.d.

We summarize our considerations to the following

Proposition 9.6. *The function u solves the eigenvalue problem (9) if and only if the function u solves the eigenvalue problem (11).*

Theorem 9.7. (H. Weyl)
On each bounded Dirichlet domain $G \subset \mathbb{R}^n$ with $n = 2, 3, \ldots$, the Laplace

operator possesses a c.o.n.s. of eigenfunctions $\varphi_k \in \mathcal{E}$, $k = 1, 2, \ldots$. This means

$$- \Delta\varphi_k(x) = \lambda_k\varphi_k(x), \quad x \in G, \qquad k = 1, 2, \ldots, \tag{18}$$

and the eigenvalues have the properties

$$0 < \lambda_1 \leq \lambda_2 \leq \lambda_3 \leq \ldots \to +\infty. \tag{19}$$

Proof: Equivalently to (9) we consider the integral eigenvalue problem (11)

$$\int\limits_G H(x, y)u(y)\, dy = \mu u(x), \quad x \in G, \qquad \mu = \frac{1}{\lambda}$$

with the symmetric weakly singular kernel $H(x, y)$ from (10) and (12). The statements of the theorem can now be inferred from Section 8, Theorem 8.1. q.e.d.

Remarks:

1. In the spaces $\mathbb{R}^2$ and $\mathbb{R}^3$, we can even uniformly expand each function $f \in \mathcal{E}$ into the series of eigenfunctions for the Laplace operator.
2. The least eigenvalue λ_1 for the Laplacian on the bounded domain $G \subset \mathbb{R}^n$ satisfies

$$\lambda_1(G) = \inf_{\varphi\in W_0^{1,2}(G)\cap G^0(\overline{G}),\, \varphi\neq 0} \frac{\displaystyle\int\limits_G |\nabla\varphi(x)|^2\, dx}{\displaystyle\int\limits_G |\varphi(x)|^2\, dx}. \tag{20}$$

Here we refer the reader to the Sobolev spaces in the Sections 1 and 2 of Chapter 10. From the relation (20) we immediately infer the *monotonicity property of the least eigenvalue:*

$$G \subset G_* \quad \Rightarrow \quad \lambda_1(G) \geq \lambda_1(G_*). \tag{21}$$

With the aid of a regularity theorem for weak solutions of the Laplace equation one proves the *strict monotonicity property:*

$$G \subset\subset G_* \quad \Rightarrow \quad \lambda_1(G) > \lambda_1(G_*). \tag{22}$$

In this context we refer the reader to [CH], Band II, Kapitel VI.

3. Comparing sufficiently regular domains $G \subset \mathbb{R}^n$ with the ball of the same volume $K \subset \mathbb{R}^n$ - which means $|K| = |G|$ - we have the estimate

$$\lambda_1(G) \geq \lambda_1(K). \tag{23}$$

Here the equality is attained only in the case that G is already a ball in $\mathbb{R}^n$. This *Theorem of Faber and Krahn* is based on the isoperimetric inequality in $\mathbb{R}^n$ and had already been conjectured by Rayleigh in his book *Theory of the Sound.* In the case $n = 2$ we recommend the following paper by E. KRAHN: *Über eine von Rayleigh formulierte Minimaleigenschaft des Kreises.* Mathematische Annalen **94** (1924), 97-100.

4. If the function $\varphi \in \mathcal{E}$ is a solution of (9) for the eigenvalue $\lambda \in \mathbb{R}$, we infer

$$\lambda = \lambda_1 \qquad \Leftrightarrow \qquad \varphi(x) \neq 0 \quad \text{for all } x \in G. \tag{24}$$

   Therefore, the eigenfunction to the least eigenvalue λ_1 has no zeroes in G.
5. About the eigenfunctions for higher eigenvalues and their nodal domains almost no results are available (compare [CH]).
6. Endowing the domain $G \subset \mathbb{R}^n$ with the elliptic Riemannian metric

$$ds^2 = \sum_{i,j=1}^{n} g_{ij}(x) \, dx_i \, dx_j,$$

   we propose the integral equation method in order to treat the eigenvalue problem of the Laplace-Beltrami operator

$$\Delta = \frac{1}{\sqrt{g(x)}} \sum_{i=1}^{n} \frac{\partial}{\partial x_i} \Big(\sqrt{g(x)} \sum_{j=1}^{n} g^{ij}(x) \frac{\partial}{\partial x_j} \Big) \tag{25}$$

   with $((g^{kl})_{kl} = (g_{ij})_{ij}^{-1}$ and $g = \det(g_{ij})_{ij}$. In this context we need the generalized Green's function for elliptic operators in divergence form which is weakly singular again. Here we refer the reader to our approach in Chapter 10, Sections 9 and 10, or to the following original paper by M. GRÜTER, K. O. WIDMAN: *The Green function for uniformly elliptic equations*. Manuscripta mathematica **37** (1982), 303-342.
7. Theorems of Faber-Krahn type are valid even for the operators (25). Here we recommend the monographs by G. POLYA: *Isoperimetric Inequalities*, Princeton University Press, Princeton N.J., 1944 and by C. BANDLE: *Isoperimetric Inequalities in Mathematical Physics*, Pitman, 1984.
8. The spectral theory for unbounded operators is presented e.g. in the Kapitel IV: *Selbstadjungierte Operatoren im Hilbertraum* of the monograph by H. TRIEBEL: *Höhere Analysis*. Verlag der Wissenschaften, Berlin, 1972.
9. A simple proof of the spectral theorem for selfadjoint operators has been discovered by H. LEINFELDER: *A geometric proof of the spectral theorem for unbounded selfadjoint operators*. Mathematische Annalen **242** (1979), 85-96.

10 Some Historical Notices to Chapter 8

The investigation of eigenvalue problems for ordinary differential operators started in 1837; then C.F. Sturm invented his well-known comparison theorem, essential for the stability question of geodesics. C.G. Jacobi (1804–1851) created the general stability theory for one-dimensional variational problems. In order to study the stability question for parametric minimal surfaces, H.A. Schwarz investigated eigenvalue problems for the two-dimensional Laplacian already in the *Festschrift dedicated to his academic teacher Karl Weierstrass* from 1885.

D. Hilbert created the theory of integral equations in the years 1904–1910, solving linear systems of infinitely many equations. This theory may be seen as one of Hilbert's greatest achievements, and it was substantially further developed by his students H. Weyl and E. Schmidt.

We have presented H. Weyl's approach to the eigenvalue problem of the n-dimensional Laplacian via the integral equation method in this chapter. In his famous textbook together with Hilbert, R. Courant solved eigenvalue problems for partial differential equations alternatively by direct variational methods. His student F. Rellich then created a spectral theory for abstract operators in Hilbert spaces, as well as K. Friedrichs.

In the meantime, physicists became intensively interested in eigenvalue problems for partial differential equations; these are situated in the center of *Quantum Mechanics* – evolving in the 1930s. Their source of information were mainly the textbooks *Methoden der Mathematischen Physik* [CH] by R. Courant and D. Hilbert.

Figure 2.3   PORTRAIT OF R. COURANT (1888–1972)
taken from page 240 of the biography by *C. Reid: Hilbert*, Springer-Verlag, Berlin... (1970).

Chapter 9

Linear Elliptic Differential Equations

At first, we transform boundary value problems for elliptic differential equations with two independent variables into a Riemann-Hilbert boundary value problem in Section 1. The latter can be solved by the integral equation method due to I. N. Vekua in Section 2 and Section 3. Then, we derive potential-theoretic estimates for the solution of Poisson's equation in Section 4. For use in Chapter 12 we prove corresponding inequalities for solutions of the inhomogeneous Cauchy-Riemann equation. For elliptic differential equations in n variables we solve the Dirichlet problem by the continuity method in the classical function space $C^{2+\alpha}(\overline{\Omega})$; see Section 5 and Section 6. The necessary Schauder estimates are completely derived in the last paragraph.

1 The Differential Equation
$\Delta\phi + p(x,y)\phi_x + q(x,y)\phi_y = r(x,y)$

In the simply connected domain $\Omega \subset \mathbb{C}$ we take the bounded coefficient functions

$$p = p(x,y), \ q = q(x,y), \ r = r(x,y) \in C^0(\Omega, \mathbb{R}),$$

and consider the differential operator

$$\mathcal{L} := \Delta + p(x,y)\frac{\partial}{\partial x} + q(x,y)\frac{\partial}{\partial y}. \tag{1}$$

We define the complex-valued function

$$a = a(z) := -\frac{1}{4}\Big(p(x,y) + iq(x,y)\Big), \quad z = x + iy \in \Omega, \tag{2}$$

and remark that

$$\frac{\partial}{\partial z} = \frac{1}{2}\Big(\frac{\partial}{\partial x} - i\frac{\partial}{\partial y}\Big), \qquad \frac{\partial}{\partial \overline{z}} = \frac{1}{2}\Big(\frac{\partial}{\partial x} + i\frac{\partial}{\partial y}\Big).$$

F. Sauvigny, *Partial Differential Equations 2*, Universitext,
DOI 10.1007/978-1-4471-2984-4_3, © Springer-Verlag London 2012

With arbitrary functions $\phi = \phi(x, y) \in C^2(\Omega, \mathbb{R})$ we calculate

$$
\begin{aligned}
\frac{1}{4}\mathcal{L}\phi(x, y) &= \frac{1}{4}\left(\Delta\phi(x, y) + p\phi_x + q\phi_y\right) \\
&= \phi_{z\bar{z}} + \frac{1}{2}\operatorname{Re}\left\{(p + iq)\frac{1}{2}\left(\phi_x - i\phi_y\right)\right\} \\
&= \phi_{z\bar{z}} - 2\operatorname{Re}\{a(z)\phi_z(z)\} \\
&= \phi_{z\bar{z}} - a\phi_z - \bar{a}\phi_{\bar{z}} \qquad \text{in} \quad \Omega.
\end{aligned}
\tag{3}
$$

Here we denote the real and imaginary parts of a complex number z by $\operatorname{Re} z$ and $\operatorname{Im} z$, respectively. Now we consider solutions

$$
f_* = f_*(z) \in C^1(\Omega, \mathbb{C}\backslash\{0\})
$$

of the differential equation

$$
\frac{\partial}{\partial\bar{z}}f_*(z) - a(z)f_*(z) = 0 \qquad \text{in} \quad \Omega.
\tag{4}
$$

These appear in the form

$$
f_*(z) = F_*(z)\exp\left\{\frac{-1}{\pi}\iint_\Omega \frac{a(\zeta)}{\zeta - z}\,d\xi\,d\eta\right\}, \qquad z \in \Omega,
\tag{5}
$$

with an arbitrary holomorphic function $F_* : \Omega \to \mathbb{C}\backslash\{0\}$. Furthermore, we have utilized *Hadamard's integral operator*

$$
T_\Omega[a](z) := \frac{-1}{\pi}\iint_\Omega \frac{a(\zeta)}{\zeta - z}\,d\xi\,d\eta, \qquad z \in \Omega \quad (\zeta = \xi + i\eta)
\tag{6}
$$

from Definition 4.11 in Section 4 of Chapter 4. We now consider the *associate gradient function*

$$
f(z) := \frac{2i}{f_*(z)}\phi_z(z), \qquad z \in \Omega.
\tag{7}
$$

With the coefficient function

$$
b(z) := -\frac{1}{f_*(z)}\frac{\partial}{\partial z}\overline{f}_*(z) = -\overline{\left(\frac{1}{\overline{f}_*}\frac{\partial}{\partial\bar{z}}f_*(z)\right)}, \qquad z \in \Omega,
\tag{8}
$$

we calculate

$$\frac{\partial}{\partial \overline{z}} f(z) - b(z)\overline{f}(z) = \frac{\partial}{\partial \overline{z}}\left(\frac{2i}{f_*(z)}\phi_z(z)\right) + \frac{1}{f_*(z)}\left(\frac{\partial}{\partial z}\overline{f}_*(z)\right)\frac{-2i}{\overline{f}_*(z)}\phi_{\overline{z}}(z)$$

$$= \frac{2i}{f_*}\phi_{z\overline{z}} - \frac{2i}{f_*^2}\left(\frac{\partial}{\partial \overline{z}}f_*\right)\phi_z - \frac{2i}{f_*}\left(\frac{1}{\overline{f}_*}\frac{\partial}{\partial z}\overline{f}_*\right)\phi_{\overline{z}}$$

$$= \frac{2i}{f_*}\left\{\phi_{z\overline{z}} - a\phi_z - \overline{a}\phi_{\overline{z}}\right\}$$

$$= \frac{i}{2f_*(z)}\mathcal{L}\phi(x,y), \qquad z = x + iy \in \Omega.$$

$$(9)$$

Theorem 1.1. *a) If the function $\phi = \phi(x,y) \in C^2(\Omega)$ satisfies*

$$\mathcal{L}\phi(x,y) = r(x,y) \quad in \quad \Omega,$$

then its associate gradient function (7) fulfills the following differential equation:

$$\frac{\partial}{\partial \overline{z}} f(z) - b(z)\overline{f}(z) = \frac{i}{2f_*(z)}r(z) =: c(z), \qquad z \in \Omega. \qquad (10)$$

b) On the other hand, if we start with a solution $f \in C^1(\Omega,\mathbb{C})$ of the equation (10) in the simply connected domain $\Omega \subset \mathbb{C}$, then the real contour integral

$$\phi(x,y) := 2\,Re\int_{z_0}^{z}\frac{1}{2i}f_*(\zeta)f(\zeta)\,d\zeta, \quad z \in \Omega \qquad (11)$$

gives us a solution of the differential equation $\mathcal{L}\phi(x,y) = r(x,y)$ in Ω. Here the point $z_0 \in \Omega$ is chosen arbitrarily.

Proof: a) This follows from the identity (9).

b) At first, we infer the following differential equation from (8):

$$\frac{\partial}{\partial \overline{z}}f_*(z) + \overline{b(z)}\,\overline{f}_*(z) = 0, \qquad z \in \Omega.$$

Furthermore, the contour integral from (11) is independent of the path chosen: With $G \subset\subset \Omega$ taking an arbitrary normal domain, the Gaussian integral theorem in the complex form yields

$$Re\int_{\partial G}\frac{1}{2i}f_*(\zeta)f(\zeta)\,d\zeta = Re\iint_{G}\left(f_*(z)f(z)\right)_{\overline{z}}dx\,dy$$

$$= Re\iint_{G}\left\{(\frac{\partial}{\partial \overline{z}}f_*)f + (\frac{\partial}{\partial \overline{z}}f)f_*\right\}dx\,dy$$

$$= Re\iint_{G}\left\{-\overline{b(z)}\overline{f}_*f + b(z)\overline{f}f_* + \frac{i}{2}r(z)\right\}dx\,dy = 0.$$

Furthermore, we have

$$\phi(z) = \frac{1}{2i} \int\limits_{z_0}^{z} \left\{ f_*(\zeta) f(\zeta)\, d\zeta - \overline{f_*(\zeta)}\, \overline{f(\zeta)}\, d\overline{\zeta} \right\}, \qquad z \in \Omega,$$

which implies

$$\phi_z(z) = \frac{1}{2i} f_*(z) f(z), \qquad z \in \Omega. \tag{12}$$

We infer the validity of $\mathcal{L}\phi = r(x,y)$ in Ω from the identity (9).

$$\text{q.e.d.}$$

Theorem 1.2. (P. Hartman, A. Wintner)

Let the nonconstant function $\phi = \phi(x,y) \in C^2(\Omega)$ satisfy the homogeneous elliptic differential equation

$$\mathcal{L}\phi(x,y) = 0, \qquad (x,y) \in \Omega. \tag{13}$$

Then the gradient of ϕ has, at most, isolated zeros in Ω, and at each zero $z_0 \in \Omega$ we have the asymptotic expansion

$$\phi_z(z_0 + \zeta) = c\,\zeta^n + o(|\zeta|^n), \qquad \zeta \to 0. \tag{14}$$

Here we used the numbers $n \in \mathbb{N}$, $c = c_1 + ic_2 \in \mathbb{C} \backslash \{0\}$; the symbol $o(|\zeta|^n)$ denotes a function $\psi = \psi(\zeta) : \mathbb{C} \backslash \{0\} \to \mathbb{C}$ with the property

$$\lim_{\substack{\zeta \to 0 \\ \neq}} \frac{|\psi(\zeta)|}{|\zeta|^n} = 0.$$

Furthermore, the function ϕ reveals the behavior of a saddle point near z_0, namely

$$\phi(z_0 + re^{i\varphi}) = \phi(z_0) + \frac{2}{n+1} r^{n+1} \Big(c_1 \cos(n+1)\varphi - c_2 \sin(n+1)\varphi \Big) + o(r^{n+1})$$
$$\tag{15}$$

with $r \to 0+$. Consequently, the function ϕ does not attain a local minimum nor a local maximum at the point z_0.

Proof: The identity (12) implies

$$\phi_z(z) = \frac{1}{2i} f_*(z) f(z), \qquad z \in \Omega.$$

Now the function f_* is defined by (5), and f satisfies the differential equation

$$\frac{\partial}{\partial z} f(z) = b(z) \overline{f}(z), \qquad z \in \Omega. \tag{16}$$

Consequently, the function f is pseudoholomorphic (compare Chapter 4, Section 6, Definition 6.1), and we obtain the following expansion near a zero z_0 of ϕ_z:

$$\phi_z(z_0 + \zeta) = c\zeta^n + o(|\zeta|^n), \qquad \zeta \to 0.$$

Here we have chosen the numbers $c = c_1 + ic_2 \in \mathbb{C}\setminus\{0\}$ and $n \in \mathbb{N}$. Furthermore, we have the identity

$$\phi(z_0 + re^{i\varphi}) - \phi(z_0)$$

$$= \int_0^r \frac{d}{d\varrho}\phi(z_0 + \varrho e^{i\varphi})\, d\varrho = \int_0^r \frac{d}{d\varrho}\phi(x_0 + \varrho\cos\varphi, y_0 + \varrho\sin\varphi)\, d\varrho$$

$$= \int_0^r \left\{\phi_x(\ldots)\cos\varphi + \phi_y(\ldots)\sin\varphi\right\} d\varrho = 2\int_0^r \mathrm{Re}\left\{\phi_z(z_0 + \varrho e^{i\varphi})e^{i\varphi}\right\} d\varrho.$$

When we insert the asymptotic expansion (14) of ϕ_z, we finally obtain

$$\phi(z_0 + re^{i\varphi}) - \phi(z_0) = 2\int_0^r \mathrm{Re}\left\{c\varrho^n e^{i(n+1)\varphi}\right\} d\varrho + o(r^{n+1})$$

$$= 2\mathrm{Re}\left\{(c_1 + ic_2)(\cos(n+1)\varphi + i\sin(n+1)\varphi)\right\}\frac{r^{n+1}}{n+1} + o(r^{n+1})$$

$$= \frac{2}{n+1}\left\{c_1\cos(n+1)\varphi - c_2\sin(n+1)\varphi\right\}r^{n+1} + o(r^{n+1}) \quad \text{with} \quad r \to 0+ .$$

$$\text{q.e.d.}$$

Let $\Omega \subset \mathbb{C}$ be a simply connected bounded domain, whose boundary consists of a regular C^2-curve in the following sense:

$$\partial\Omega : \; z = \zeta(t) : [0, T] \to \partial\Omega \in C_T^2(\mathbb{R}, \mathbb{C}) \; \text{with} \; |\zeta'(t)| \equiv 1, \; 0 \le t \le T. \quad (17)$$

Here $\zeta'(t)$, $0 \le t \le T$ gives us the tangential vector-field to $\partial\Omega$, and we abbreviate

$$C_T^2(\mathbb{R}, \mathbb{C}) := \left\{g \in C^2(\mathbb{R}, \mathbb{C}) \; : \; g \text{ is periodic with the period } T\right\}.$$

Furthermore, the vector-field $\nu(t) := -i\zeta'(t)$, $0 \le t \le T$ represents the exterior normal to $\partial\Omega$. Now we prescribe the continuous unit vector-field

$$\gamma(t) = \alpha(t) + i\beta(t) \in C_T^0(\mathbb{R}, \mathbb{R}^2) \qquad \text{with} \quad |\gamma(t)| \equiv 1, \; t \in \mathbb{R}$$

on the boundary $\partial\Omega$ and the function $\chi = \chi(t) \in C_T^0(\mathbb{R}, \mathbb{R})$.
Then we consider the following *boundary value problem of Poincaré*

$$\phi = \phi(x,y) \in C^2(\Omega) \cap C^1(\overline{\Omega}),$$

$$\mathcal{L}\phi(x,y) = r(x,y) \quad \text{in} \quad \Omega, \tag{18}$$

$$\phi_x(\zeta(t))\alpha(t) + \phi_y(\zeta(t))\beta(t) = \chi(t), \qquad 0 \le t \le T.$$

Remarks:

1. In the special case $\gamma(t) = \nu(t)$, $0 \le t \le T$, the condition reduces to *Neumann's boundary condition*

$$\frac{\partial}{\partial\nu}\phi(\zeta(t)) = \chi(t), \qquad 0 \le t \le T. \tag{19}$$

2. In the special case $\gamma(t) = \zeta'(t)$, $0 \le t \le T$, the condition reduces to *Dirichlet's boundary condition*

$$\frac{\partial}{\partial t}\phi(\zeta(t)) = \chi(t), \qquad 0 \le t \le T,$$

and consequently

$$\phi(\zeta(t)) = \phi(\zeta(0)) + \int_0^t \chi(\tau)\,d\tau, \qquad 0 \le t \le T.$$

Additionally, we require

$$\int_0^T \chi(\tau)\,d\tau = 0. \tag{20}$$

The associate gradient function $f(z) = \dfrac{2i}{f_*(z)}\phi_z(z)\,, \quad z \in \overline{\Omega}$ satisfies

$$\chi(t) = \phi_x(\zeta(t))\alpha(t) + \phi_y(\zeta(t))\beta(t)$$
$$= 2\mathrm{Re}\big\{\phi_z(\zeta(t))\gamma(t)\big\}$$
$$= \mathrm{Re}\big\{-if_*(z)\gamma(z)f(z)\big\}\big|_{z=\zeta(t)}, \qquad 0 \le t \le T.$$

Introducing the function

$$g(z) := i\overline{f_*(z)}\,\overline{\gamma(z)}, \qquad z \in \partial\Omega, \tag{21}$$

we find the following boundary condition for f:

$$\mathrm{Re}\Big\{\overline{g(\zeta(t))}f(\zeta(t))\Big\} = \chi(t), \qquad 0 \le t \le T. \tag{22}$$

Together with Theorem 1.1 we arrive at the following

Theorem 1.3. *a) If the function ϕ solves the general boundary value problem (18), then the associate gradient function*

$$f = f(z) := \frac{2i}{f_*(z)}\phi_z(z) \in C^1(\Omega) \cap C^0(\overline{\Omega})$$

yields a solution of the Riemann-Hilbert boundary value problem (10), (22).
b) If the complex-valued function $f = f(z) \in C^1(\Omega) \cap C^0(\overline{\Omega})$ solves the Riemann-Hilbert boundary value problem (10), (22), then we obtain a solution of the general boundary value problem (18) by the real contour integral (11).

Now, the Riemann-Hilbert boundary value problem

$$\frac{\partial}{\partial \overline{z}}f(z) - b(z)\overline{f(z)} = c(z), \qquad z \in \Omega,$$

$$\text{Re}\{\overline{g}(z)f(z)\} = \chi(z), \qquad z \in \partial\Omega, \tag{23}$$

is invariant with respect to conformal mappings. Applying the Riemann mapping theorem (compare Chapter 4, Section 7 and Section 8), we shall assume Ω to be the unit disc in the sequel.

2 The Schwarzian Integral Formula

On the unit disc $B := \{z = x + iy \in \mathbb{C} : |z| < 1\}$ with the boundary $\partial B = \{e^{i\varphi} : 0 \leq \varphi \leq 2\pi\}$ and the exterior domain $A := \{z \in \mathbb{C} : |z| > 1\}$ we shall solve boundary value problems for holomorphic functions. We begin with the important

Theorem 2.1. (Plemelj)
Let $F : \partial B \to \mathbb{C}$ be a Hölder continuous function; that means $\varphi \mapsto F(e^{i\varphi})$ defines a 2π-periodic Hölder continuous function. Then the Cauchy principal values

$$H(z) := \lim_{\varepsilon \to 0+} \frac{1}{2\pi i} \oint_{\substack{\zeta \in \partial B \\ |\zeta - z| \geq \varepsilon}} \frac{F(\zeta)}{\zeta - z}d\zeta, \qquad z \in \partial B$$

represent a continuous function. Furthermore, the function

$$G(z) := \frac{1}{2\pi i} \oint_{\zeta \in \partial B} \frac{F(\zeta)}{\zeta - z}d\zeta, \qquad z \in B \cup A$$

has the following boundary behavior at the circle line ∂B:

$$\lim_{\substack{z \to z_0 \\ z \in B}} G(z) = H(z_0) + \frac{1}{2}F(z_0) \qquad \text{for all} \quad z_0 \in \partial B \tag{1}$$

and

$$\lim_{\substack{z \to z_0 \\ z \in A}} G(z) = H(z_0) - \frac{1}{2} F(z_0) \qquad \text{for all} \quad z_0 \in \partial B. \tag{2}$$

Remark: The function G can be continuously extended within the disc B and from the exterior A onto the circle line ∂B. However, there G has a jump of the size $F(z_0)$, $z_0 \in \partial B$.

Proof: We show continuity of the Cauchy principal values by arguments from Proposition 5.9 in Chapter 4, Section 5. For a fixed point $z_0 \in \partial B$ we define $\Gamma_\varepsilon := \{z \in \partial B : |z - z_0| > \varepsilon\}$, $S_\varepsilon^- := \{z \in B : |z - z_0| = \varepsilon\}$, and $S_\varepsilon^+ := \{z \in A : |z - z_0| = \varepsilon\}$. For all points $z \in B \backslash \{0\}$ we deduce

$$G(z) = \frac{1}{2\pi i} \oint_{\partial B} \frac{F(\zeta) - F\left(\frac{z}{|z|}\right)}{\zeta - z} \, d\zeta + F\left(\frac{z}{|z|}\right) \frac{1}{2\pi i} \oint_{\partial B} \frac{1}{\zeta - z} \, d\zeta$$

$$= \frac{1}{2\pi i} \oint_{\partial B} \frac{F(\zeta) - F\left(\frac{z}{|z|}\right)}{\zeta - z} \, d\zeta + F\left(\frac{z}{|z|}\right) \frac{1}{2\pi i} \left\{ \int_{\Gamma_\varepsilon} \frac{1}{\zeta - z} \, d\zeta + \int_{S_\varepsilon^+} \frac{1}{\zeta - z} \, d\zeta \right\}$$

with sufficiently small $\varepsilon > 0$. For $z_0 \in \partial B$ we obtain

$$\lim_{\substack{z \to z_0 \\ z \in B}} G(z) = \frac{1}{2\pi i} \oint_{\partial B} \frac{F(\zeta) - F(z_0)}{\zeta - z_0} \, d\zeta$$

$$+ F(z_0) \frac{1}{2\pi i} \left\{ \int_{\Gamma_\varepsilon} \frac{1}{\zeta - z_0} \, d\zeta + \int_{S_\varepsilon^+} \frac{1}{\zeta - z_0} \, d\zeta \right\}$$

with arbitrary $\varepsilon > 0$. The passage to the limit $\varepsilon \to 0+$ yields

$$\lim_{\substack{z \to z_0 \\ z \in B}} G(z) = \lim_{\varepsilon \to 0+} \frac{1}{2\pi i} \left\{ \int_{\Gamma_\varepsilon} \frac{F(\zeta) - F(z_0)}{\zeta - z_0} \, d\zeta + F(z_0) \int_{\Gamma_\varepsilon} \frac{1}{\zeta - z_0} \, d\zeta \right\} + \frac{1}{2} F(z_0)$$

$$= \lim_{\varepsilon \to 0+} \left\{ \frac{1}{2\pi i} \int_{\Gamma_\varepsilon} \frac{F(\zeta)}{\zeta - z_0} \, d\zeta \right\} + \frac{1}{2} F(z_0)$$

$$= H(z_0) + \frac{1}{2} F(z_0),$$

and we attain (1). By similar calculations we obtain (2), substituting the integrals on S_ε^+ by the corresponding integrals on S_ε^-.   q.e.d.

Theorem 2.2. (Schwarzian integral formula)
Let us consider the Hölder continuous, real-valued function $\phi : \partial B \to \mathbb{R}$, and let the Schwarzian integral be defined as follows:

$$F(z) := \frac{1}{2\pi} \int_0^{2\pi} \frac{e^{i\varphi} + z}{e^{i\varphi} - z} \phi(e^{i\varphi}) \, d\varphi, \qquad |z| < 1. \tag{3}$$

Then the holomorphic function F can be continuously extended onto the closed unit disc $\overline{B}$. Furthermore, its real part

$$Re F(z) : \overline{B} \to \mathbb{R}$$

takes on the boundary values ϕ; more precisely

$$\lim_{\substack{z \to z_0 \\ z \in B}} Re F(z) = \phi(z_0) \qquad \text{for all} \quad z_0 \in \partial B. \tag{4}$$

Proof:

1. We extend F to the function

$$F(z) := \frac{1}{2\pi} \int_0^{2\pi} \frac{e^{i\varphi} + z}{e^{i\varphi} - z} \phi(e^{i\varphi}) \, d\varphi, \qquad z \in B \cup A,$$

and obtain the *reflection condition*

$$\overline{F(z)} = \frac{1}{2\pi} \int_0^{2\pi} \frac{e^{-i\varphi} + \overline{z}}{e^{-i\varphi} - \overline{z}} \phi(e^{i\varphi}) \, d\varphi = \frac{1}{2\pi} \int_0^{2\pi} \frac{\frac{1}{\overline{z}} + e^{i\varphi}}{\frac{1}{\overline{z}} - e^{i\varphi}} \phi(e^{i\varphi}) \, d\varphi$$

$$= -\frac{1}{2\pi} \int_0^{2\pi} \frac{e^{i\varphi} + \frac{1}{\overline{z}}}{e^{i\varphi} - \frac{1}{\overline{z}}} \phi(e^{i\varphi}) \, d\varphi = -F\left(\frac{1}{\overline{z}}\right), \qquad z \in (B \backslash \{0\}) \cup A.$$

2. Furthermore, we have the following identity for all $z \in B \cup A$:

$$F(z) = \frac{1}{2\pi} \int_0^{2\pi} \frac{e^{i\varphi} + z}{e^{i\varphi} - z} \phi(e^{i\varphi}) \, d\varphi = \frac{1}{2\pi i} \oint_{\partial B} \frac{\zeta + z}{\zeta - z} \phi(\zeta) \frac{d\zeta}{\zeta}.$$

We observe

$$\frac{z + \zeta}{\zeta(\zeta - z)} = \frac{z - \zeta + 2\zeta}{\zeta(\zeta - z)} = -\frac{1}{\zeta} + \frac{2}{\zeta - z}$$

and calculate

$$F(z) = \frac{1}{2\pi i} \oint_{\partial B} \left(-\frac{1}{\zeta} + \frac{2}{\zeta - z} \right) \phi(\zeta) \, d\zeta$$

$$= -\frac{1}{2\pi i} \oint_{\partial B} \phi(\zeta) \frac{d\zeta}{\zeta} + \frac{1}{\pi i} \oint_{\partial B} \frac{\phi(\zeta)}{\zeta - z} \, d\zeta$$

$$= -\frac{1}{2\pi} \int_0^{2\pi} \phi(e^{i\varphi}) \, d\varphi + \frac{1}{\pi i} \oint_{\partial B} \frac{\phi(\zeta)}{\zeta - z} \, d\zeta, \qquad z \in B \cup A.$$

Due to Theorem 2.1, the function F can be continuously extended to ∂B, and we see

$$\lim_{\substack{z \to z_0 \\ z \in B}} F(z) = -\frac{1}{2\pi} \int_0^{2\pi} \phi(e^{i\varphi})\, d\varphi + \frac{1}{\pi i} \oint_{\partial B} \frac{\phi(\zeta)}{\zeta - z_0}\, d\zeta + \phi(z_0), \qquad z_0 \in \partial B,$$

$$\lim_{\substack{z \to z_0 \\ z \in A}} F(z) = -\frac{1}{2\pi} \int_0^{2\pi} \phi(e^{i\varphi})\, d\varphi + \frac{1}{\pi i} \oint_{\partial B} \frac{\phi(\zeta)}{\zeta - z_0}\, d\zeta - \phi(z_0), \qquad z_0 \in \partial B.$$

$$(5)$$

By the integrals $\oint\limits_{\partial B} \ldots$ given here, we comprehend the Cauchy principal values according to Theorem 2.1. Finally, we obtain the following identity for all $z_0 \in \partial B$:

$$\lim_{\substack{z \to z_0 \\ z \in B}} \operatorname{Re} F(z) = \frac{1}{2} \lim_{\substack{z \to z_0 \\ z \in B}} \left[F(z) + \overline{F(z)} \right] = \frac{1}{2} \lim_{\substack{z \to z_0 \\ z \in B}} \left[F(z) - F\left(\frac{1}{\bar{z}}\right) \right] = \phi(z_0).$$

q.e.d.

3 The Riemann-Hilbert Boundary Value Problem

We now consider the following *Riemann-Hilbert boundary value problem*: For the Hölder continuous coefficient function $b = b(z) \in C^0(\overline{B}, \mathbb{C})$ being given, let the function

$$f = f(z) = u(x, y) + iv(x, y) \in C^1(B, \mathbb{C}) \cap C^0(\overline{B}, \mathbb{C})$$

satisfy the homogeneous differential equation

$$\frac{\partial}{\partial \bar{z}} f(z) - b(z)\overline{f}(z) = 0, \qquad z \in B. \tag{1}$$

Furthermore, let us take the Hölder continuous directional function

$$a = a(z) = \alpha(x, y) + i\beta(x, y) : \partial B \to \partial B$$

satisfying

$$\alpha^2(z) + \beta^2(z) = 1 \quad \text{for all} \quad z \in \partial B.$$

The *index* $n \in \mathbb{Z}$ *of the Riemann-Hilbert problem* indicates how often the directional vector-field a winds about the origin 0. Therefore, we assume the representation

$$a(z) = z^n\, e^{i\phi(z)}, \qquad z \in \partial B, \tag{2}$$

with a Hölder continuous function $\phi : \partial B \to \mathbb{R}$. Furthermore, we prescribe the Hölder continuous function $\chi : \partial B \to \mathbb{R}$ and require the *Riemann-Hilbert boundary condition*

$$\alpha(z)u(z) + \beta(z)v(z) = Re\left(\overline{a(z)}f(z)\right) = \chi(z), \qquad z \in \partial B. \tag{3}$$

We now solve the Riemann-Hilbert boundary value problem (1), (3) for the indices $n \geq -1$ by the integral equation method of I. N. Vekua. Particularly important is the case $n = -1$: Due to Section 1, Theorem 1.3 we can then solve a mixed boundary value problem for linear elliptic differential equations, especially under Dirichlet and Neumann boundary conditions. We owe to G. Hellwig the fundamental observation: *The solution space for the Riemann-Hilbert problem is interrelated with this index and integral conditions on the right-hand side have to be imposed.*

Based on Theorem 2.2 from Section 2, we consider the following function which is continuous on $\overline{B}$ and holomorphic in B:

$$\phi(z) + i\psi(z) = F(z) := \frac{1}{2\pi} \int\limits_{0}^{2\pi} \frac{e^{i\varphi} + z}{e^{i\varphi} - z} \phi(e^{i\varphi})\, d\varphi, \qquad |z| < 1. \tag{4}$$

We note that

$$\lim_{\substack{z \in B \\ z \to z_0}} \phi(z) = \phi(z_0) \qquad \text{for all} \quad z_0 \in \partial B. \tag{5}$$

We multiply (3) by $e^{\psi(z)}$, $z \in \partial B$ and equivalently obtain

$$\begin{aligned}
\eta(z) &:= e^{\psi(z)}\chi(z) = Re\left(e^{\psi(z)}e^{-i\phi(z)}\frac{f(z)}{z^n}\right) \\
&= Re\left(\frac{e^{-iF(z)}f(z)}{z^n}\right) \qquad \text{for all} \quad z \in \partial B.
\end{aligned} \tag{6}$$

Multiplication of the differential equation (1) by the holomorphic function

$$e^{-iF(z)} = e^{\psi(z)}e^{-i\phi(z)} \neq 0, \qquad z \in \overline{B}$$

yields the equivalent differential equation

$$\frac{\partial}{\partial \overline{z}}\left(e^{-iF(z)}f(z)\right) - b(z)e^{-2i\phi(z)}\overline{\left(e^{-iF(z)}f(z)\right)} = 0, \qquad z \in B. \tag{7}$$

By the transition $f(z) \mapsto e^{-iF(z)}f(z)$ we obtain the *canonical Riemann-Hilbert boundary condition* from (6):

$$Re\left(\frac{f(z)}{z^n}\right) = \chi(z), \qquad z \in \partial B. \tag{8}$$

Therefore, we have to solve the boundary value problem (1) and (8), which we shall transform into an integral equation problem. We obtain the following *Riemann-Hilbert boundary value problem in the normal form:*

$$f = f(z) \in C^1(B, \mathbb{C}) \cap C^0(\overline{B}, \mathbb{C}),$$

$$\frac{\partial}{\partial \overline{z}} f(z) - b(z)\overline{f}(z) = 0 \quad \text{in} \quad B, \tag{9}$$

$$Re\left(\frac{f(z)}{z^n}\right) = \chi(z) \quad \text{on} \quad \partial B.$$

We denote by

$$T_B[g](z) := -\frac{1}{\pi} \iint\limits_B \frac{g(\zeta)}{\zeta - z} \, d\xi \, d\eta, \qquad z \in B \quad (\zeta = \xi + i\eta)$$

Hadamard's integral operator. For $n = 0, 1, 2, \ldots$ we consider the *Riemann-Hilbert operator of order n*

$$\mathbb{V}_n g(z) := -\frac{1}{\pi} \iint\limits_B \frac{g(\zeta)}{\zeta - z} \, d\xi \, d\eta - \frac{z^{2n+1}}{\pi} \iint\limits_B \frac{\overline{g}(\zeta)}{1 - z\overline{\zeta}} \, d\xi \, d\eta$$

$$= T_B[g](z) - z^{2n}\left(-\frac{1}{\pi} \iint\limits_B \frac{g(\zeta)}{\zeta - \frac{1}{\overline{z}}} \, d\xi \, d\eta\right) \tag{10}$$

$$= T_B[g](z) - z^{2n}\left\{\overline{T_B[g]\left(\frac{1}{\overline{z}}\right)}\right\}, \qquad z \in B.$$

The substitution

$$\zeta = \frac{1}{\overline{\gamma}}, \quad \gamma = \alpha + i\beta \in A := \overline{\mathbb{C}} \setminus \overline{B}, \qquad d\xi \, d\eta = \frac{1}{|\gamma|^4} \, d\alpha \, d\beta$$

yields

$$\mathbb{V}_n g(z) = T_B[g](z) - \frac{z^{2n+1}}{\pi} \iint\limits_A \frac{\overline{g\left(\frac{1}{\overline{\gamma}}\right)}}{\left(1 - z\frac{1}{\gamma}\right)\gamma\,\gamma\overline{\gamma}^2} \, d\alpha \, d\beta$$

$$= T_B[g](z) - \frac{z^{2n+1}}{\pi} \iint\limits_A \frac{\overline{g\left(\frac{1}{\overline{\zeta}}\right)}\frac{1}{\zeta\overline{\zeta}^2}}{\zeta - z} \, d\xi \, d\eta \tag{11}$$

$$= T_B[g](z) + z^{2n+1} T_A[\tilde{g}](z), \qquad z \in B,$$

with

$$\tilde{g}(\zeta) := \frac{1}{\zeta\overline{\zeta}^2} \overline{g\left(\frac{1}{\overline{\zeta}}\right)}, \quad \zeta \in A. \tag{12}$$

We note that

$$\frac{\partial}{\partial \overline{z}} \mathbb{V}_n g(z) = g(z) \quad \text{in} \quad B,$$

$$Re\left(\frac{\mathbb{V}_n g(z)}{z^n}\right) = 0 \quad \text{on} \quad \partial B, \tag{13}$$

which follows from (10) immediately. The following Riemann-Hilbert problem

$$f = f(z) \in C^1(B, \mathbb{C}) \cap C^0(\overline{B}, \mathbb{C}),$$

$$\frac{\partial}{\partial \overline{z}} f(z) = 0 \quad \text{in} \quad B, \tag{14}$$

$$\text{Re}\left(\frac{f(z)}{z^n}\right) = \chi(z) \quad \text{on} \quad \partial B$$

can be solved explicitly with the aid of the Schwarzian integral:

$$\Phi(z) = \frac{z^n}{2\pi i} \int_{\partial B} \chi(\zeta) \frac{\zeta + z}{\zeta - z} \frac{d\zeta}{\zeta} + i\gamma z^n + \sum_{k=0}^{n-1} \left\{ \alpha_k(z^k - z^{2n-k}) + i\beta_k(z^k + z^{2n-k}) \right\}.$$

$$\tag{15}$$

Here we have used $2n + 1$ real constants $\alpha_0, \ldots, \alpha_{n-1}, \beta_0, \ldots \beta_{n-1}, \gamma$. We can transfer the boundary value problem (9) equivalently into the integral equation

$$f(z) - \mathbb{V}_n[b\overline{f}](z) = \Phi(z), \qquad z \in B, \tag{16}$$

with the right-hand side $\Phi(z)$ from (15). The linear integral operator $f \mapsto \mathbb{V}_n[b\overline{f}]$ is completely continuous on the Hilbert space $\mathcal{H} = L^2(B, \mathbb{C})$ since the kernel appearing is weakly singular. Applying Theorem 6.19 from Section 6 in Chapter 8 we comprehend: A solution $f \in \mathcal{H}$ of the integral equation (16) belongs to the class $\mathcal{D} := C^0(B, \mathbb{C}) \cap L^\infty(B, \mathbb{C})$.

We need the following

Proposition 3.1. (Vekua)
Let $n \in \{0, 1, 2, \ldots\}$ be given, and $f \in \mathcal{H}$ may solve the integral equation $f - \mathbb{V}_n[b\overline{f}] = 0$. Then we have $f = 0$.

Proof: Let f be a solution of the integral equation $f - \mathbb{V}_n[b\overline{f}] = 0$. This implies

$$f(z) + \frac{1}{\pi} \iint_B \frac{b(\zeta)\overline{f}(\zeta)}{\zeta - z} \, d\xi \, d\eta = -\frac{z^{2n+1}}{\pi} \iint_B \frac{\overline{b(\zeta)f(\zeta)}}{1 - z\overline{\zeta}} \, d\xi \, d\eta, \qquad z \in B. \tag{17}$$

The right-hand side of (17) is holomorphic in B and continuous on $\overline{B}$. The integral on the left-hand side is continuous in the entire Gaussian plane $\mathbb{C}$, vanishes at ∞, and is holomorphic in the exterior domain $A = \overline{\mathbb{C}} \backslash \overline{B}$. We take $z \in \partial B$, multiply both sides of (17) by

$$\frac{1}{2\pi i} \frac{dz}{z - t}, \qquad t \in B,$$

and integrate along ∂B. The Cauchy integral theorem and Cauchy's integral formula yield

$$\frac{1}{2\pi i} \oint_{\partial B} \frac{f(z)}{z - t}\, dz = -\frac{t^{2n+1}}{\pi} \iint_B \frac{\overline{b(\zeta)}f(\zeta)}{1 - t\overline{\zeta}}\, d\xi\, d\eta. \tag{18}$$

We now develop both sides into powers of t about the point 0 and see

$$\oint_{\partial B} f(z)e^{-ik\theta}\, d\theta = 0 \qquad \text{for} \quad k = 0, 1, \dots, 2n \tag{19}$$

with $z = e^{i\theta}$. The similarity principle of Bers and Vekua gives us the following representation

$$f(z) = \psi(z)e^{p(z)}, \qquad z \in B. \tag{20}$$

Here the function ψ is holomorphic in B, and we define

$$p(z) = -\frac{1}{\pi} \iint_B \left\{ \frac{g(\zeta)}{\zeta - z} - \frac{z\overline{g(\zeta)}}{1 - \overline{\zeta}z} \right\} d\xi\, d\eta, \qquad g = b\frac{\overline{f}}{f}. \tag{21}$$

With the aid of the equation $\operatorname{Im} p(z) = 0$ on ∂B and (20), we deduce the boundary condition

$$\operatorname{Re}\!\left(\frac{\psi(z)}{z^n}\right) = 0 \qquad \text{on} \quad \partial B \tag{22}$$

for the holomorphic function ψ. This implies

$$\psi(z) = \sum_{k=0}^{2n} c_k z^k, \tag{23}$$

where the complex constants $c_0, c_1, \dots, c_{2n}$ satisfy the following conditions:

$$c_{2n-k} = -\overline{c}_k, \qquad k = 0, 1, \dots, n. \tag{24}$$

Therefore, we find that

$$f(z) = \left(\sum_{k=0}^{2n} c_k z^k \right) e^{p(z)}, \qquad z \in \overline{B}. \tag{25}$$

Inserting (25) into (19), we infer

$$\sum_{k=0}^{2n} c_k \int_{\partial B} z^k z^{-l} e^{p(z)}\, d\theta = 0, \qquad l = 0, 1, \dots, 2n, \tag{26}$$

and consequently $c_k = 0$ for $k = 0, 1, 2, \dots, 2n$. Here we use that the Gramian determinant for the system of linear independent functions

$$z^k e^{\frac{1}{2}p(z)}, \qquad k = 0, 1, 2, \dots, 2n,$$

satisfying $\operatorname{Im} p(z) = 0$ on ∂B, is different from zero. We consequently obtain $f = 0$ from the representation (25).   q.e.d.

Theorem 3.2. *For the indices $n = 0, 1, 2, 3, \ldots$, the Riemann-Hilbert bound-ary value problem (9) possesses a (1+2n)-dimensional space of solutions.*

Proof: We use the integral equation (16) and Proposition 3.1. With the aid of Theorem 6.19 from Section 6 in Chapter 8, we can solve the integral equation for all right-hand sides Φ in (15) within the class of continuous functions. Therefore, we obtain a $(2n+1)$-dimensional solution space of (9).   q.e.d.

We shall now solve the Riemann-Hilbert problem (9) for the index $n = -1$. Taking a solution f of (9) we make a transition to the continuous function

$$g(z) := zf(z), \qquad z \in \overline{B}. \tag{27}$$

The latter solves the following Riemann-Hilbert problem for the index 0, namely

$$0 = \frac{\partial}{\partial \overline{z}} \left[zf(z) \right] - b(z) \frac{z}{\overline{z}} \overline{\left[zf(z) \right]} = \frac{\partial}{\partial \overline{z}} g(z) - c(z) \overline{g(z)} \qquad \text{in} \quad \dot{B}, \tag{28}$$

$$\chi(z) = \operatorname{Re} g(z) \qquad \text{on} \quad \partial B.$$

Here we abbreviate $\dot{B} := B \backslash \{0\}$ and set

$$c(z) := \frac{z}{\overline{z}} b(z), \qquad z \in \dot{B}.$$

The function $g(z) = zf(z)$, $z \in \overline{B}$ consequently fulfills the integral equation

$$zf(z) - \mathbb{V}_0[c\overline{z}\overline{f}](z) = \frac{1}{2\pi i} \int\limits_{\partial B} \chi(\zeta) \frac{\zeta + z}{\zeta - z} \frac{d\zeta}{\zeta} + i\gamma$$

or equivalently

$$zf(z) - \mathbb{V}_0[zb\overline{f}](z) = \frac{1}{2\pi i} \int\limits_{\partial B} \frac{\chi(\zeta)}{\zeta} d\zeta + \frac{z}{\pi i} \int\limits_{\partial B} \frac{\chi(\zeta)}{\zeta(\zeta - z)} d\zeta + i\gamma, \qquad z \in \overline{B}, \tag{29}$$

with $\gamma \in \mathbb{R}$. We now develop

$$\mathbb{V}_0[zg]\Big|_z = -\frac{1}{\pi} \iint\limits_B g(\zeta)\, d\xi\, d\eta + z \mathbb{W}[g]\Big|_z, \qquad z \in \overline{B}, \tag{30}$$

and define

$$\mathbb{W}[g]\Big|_z := -\frac{1}{\pi} \iint\limits_B \left\{ \frac{g(\zeta)}{\zeta - z} + \frac{\overline{\zeta} \overline{g(\zeta)}}{1 - z\overline{\zeta}} \right\} d\xi\, d\eta, \qquad z \in \overline{B}.$$

In this context we note that

$$-\frac{1}{\pi}\iint_B g(\zeta)\,d\xi\,d\eta - \frac{z}{\pi}\iint_B \left\{\frac{g(\zeta)}{\zeta-z} + \frac{\overline{\zeta}\,\overline{g(\zeta)}}{1-z\overline{\zeta}}\right\}d\xi\,d\eta$$

$$=-\frac{1}{\pi}\iint_B \left\{\frac{\zeta g(\zeta)}{\zeta-z} + \frac{z\overline{\zeta}\,\overline{g(\zeta)}}{1-z\overline{\zeta}}\right\}d\xi\,d\eta.$$

When we insert (30) into (29), the following integral equation is revealed:

$$f(z) - \mathbb{W}[b\overline{f}](z) = \frac{1}{\pi i}\int_{\partial B}\frac{\chi(\zeta)}{\zeta(\zeta-z)}\,d\zeta$$

$$+\frac{1}{z}\left\{i\gamma + \frac{1}{2\pi i}\int_{\partial B}\frac{\chi(\zeta)}{\zeta}\,d\zeta - \frac{1}{\pi}\iint_B b(\zeta)\overline{f(\zeta)}\,d\xi\,d\eta\right\}. \tag{31}$$

In order to obtain a continuous solution of (31), the condition

$$0 = i\gamma + \frac{1}{2\pi i}\int_{\partial B}\frac{\chi(\zeta)}{\zeta}\,d\zeta - \frac{1}{\pi}\iint_B b(\zeta)\overline{f(\zeta)}\,d\xi\,d\eta \tag{32}$$

has to be fulfilled. Then we have to solve the following integral equation

$$f(z) - \mathbb{W}[b\overline{f}](z) = \frac{1}{\pi i}\int_{\partial B}\frac{\chi(\zeta)}{\zeta(\zeta-z)}\,d\zeta, \qquad z\in\overline{B}. \tag{33}$$

We now consider the integral operator

$$\mathbb{W}[g](z) = -\frac{1}{\pi}\iint_B \left\{\frac{g(\zeta)}{\zeta-z} + \frac{\overline{g(\zeta)}}{\frac{1}{\zeta}-z}\right\}d\xi\,d\eta, \qquad z\in\overline{B}.$$

With the aid of the substitution

$$\zeta = \frac{1}{\overline{\gamma}}, \quad \gamma = \alpha + i\beta \in A, \qquad d\xi\,d\eta = \frac{1}{|\gamma|^4}\,d\alpha\,d\beta$$

we obtain

$$\mathbb{W}[g](z) = T_B[g](z) - \frac{1}{\pi}\iint_A \frac{\overline{g\left(\frac{1}{\overline{\gamma}}\right)}}{\gamma-z}\frac{1}{\gamma}\frac{1}{\gamma\overline{\gamma}^2}\,d\alpha\,d\beta$$

$$= T_B[g](z) - \frac{1}{\pi}\iint_A \frac{\overline{g\left(\frac{1}{\overline{\zeta}}\right)}\frac{1}{\overline{\zeta}^2}\frac{1}{\zeta}}{\zeta-z}\,d\xi\,d\eta \tag{34}$$

$$= T_B[g](z) + T_A\left[\frac{\tilde{g}}{z}\right](z), \quad z\in\overline{B},$$

setting

$$\tilde{g}(\zeta) := \frac{1}{\overline{\zeta\zeta}^2}\overline{g\left(\frac{1}{\overline{\zeta}}\right)}, \qquad \zeta\in A. \tag{35}$$

Proposition 3.3. (Vekua) *Let the function $f \in \mathcal{H}$ be a solution of the identity $f - \mathbb{W}[\overline{b}f] = 0$. Then we have $f = 0$.*

Proof: We define the kernel function

$$K(z,\zeta) := \frac{\overline{\zeta}}{1 - z\overline{\zeta}} \qquad \text{for} \quad z, \zeta \in B \tag{36}$$

and calculate

$$K^*(z,\zeta) := \overline{K(\zeta,z)} = \overline{\left(\frac{\overline{z}}{1 - \zeta\overline{z}}\right)} = \frac{z}{1 - z\overline{\zeta}}. \tag{37}$$

We deduce for arbitrary functions $f, g \in C^0(\overline{B}, \mathbb{C})$:

$$\iint_B \left\{ f(z)\mathbb{W}[g](z) + g(z)\mathbb{V}_0[f](z) \right\} dz$$

$$= -\frac{1}{\pi} \iint_B \iint_B \left\{ f(z)\frac{g(\zeta)}{\zeta - z} + g(z)\frac{f(\zeta)}{\zeta - z} \right\} dz\, d\zeta$$

$$-\frac{1}{\pi} \iint_B \iint_B \left\{ f(z)K(z,\zeta)\overline{g(\zeta)} + g(z)\overline{K(\zeta,z)}\,\overline{f(\zeta)} \right\} dz\, d\zeta \tag{38}$$

$$= -\frac{2}{\pi} \operatorname{Re}\left(\iint_B \iint_B f(z)K(z,\zeta)\overline{g(\zeta)}\, dz\, d\zeta \right).$$

Here we naturally comprehend $dz = dx\, dy$ and $d\zeta = d\xi\, d\eta$. By substitution of the functions $f \mapsto b\overline{g}$ and $g \mapsto \overline{b}f$ into the commutator relation (38), we obtain the following identity for arbitrary functions $f, g \in C^0(\overline{B}, \mathbb{C})$:

$$\operatorname{Im} \iint_B \left\{ b(z)\overline{g}(z)\Big(f(z) - \mathbb{W}[\overline{b}f](z) \Big) + \overline{b(z)f(z)}\Big(g(z) - \mathbb{V}_0[b\overline{g}](z) \Big) \right\} dx\, dy = 0.$$

$$\tag{39}$$

If the function $f \in C^0(\overline{B}, \mathbb{C})$ solves the integral equation $f - \mathbb{W}[\overline{b}f] = 0$, we infer

$$\operatorname{Im} \iint_B \overline{b(z)f(z)}\Big(g(z) - \mathbb{V}_0[b\overline{g}](z) \Big) dx\, dy = 0 \qquad \text{for all} \quad g \in C^0(\overline{B}, \mathbb{C}). \tag{40}$$

With the aid of Theorem 3.2, we now determine the solution $g \in C^0(\overline{B}, \mathbb{C})$ of the integral equation for the given right-hand side $ib(z)f(z)$ as follows:

$$g(z) - \mathbb{V}_0[b\overline{g}](z) = ib(z)f(z), \qquad z \in \overline{B}.$$

We insert into (40) and arrive at

$$0 = \mathrm{Im}\left\{i \iint\limits_B |b(z)|^2 |f(z)|^2 \, dx \, dy\right\} = \iint\limits_B |b(z)|^2 |f(z)|^2 \, dx \, dy. \qquad (41)$$

This implies $\overline{b(z)f(z)} \equiv 0$ and consequently

$$f(z) = \mathbb{W}[\overline{bf}](z) \equiv 0 \qquad \text{in} \quad B.$$

q.e.d.

Theorem 3.4. *For the index $n = -1$, the Riemann-Hilbert boundary value problem (9) has a solution if and only if the condition (32) is satisfied.*

Proof: We use Theorem 6.19 from Section 6 in Chapter 8 again together with Proposition 3.3.

q.e.d.

Remark: For the indices $n = -2, -3, \ldots$ as well, we can solve the Riemann-Hilbert problem if and only if $(-n)$ suitable integral conditions are imposed. In this context, we refer the reader to the monograph of I. N. Vekua [V], especially Chapter 4, Section 7, part 3.

4 Potential-theoretic Estimates

We now refer the reader to the results of Chapter 5, Section 1 and Section 2 about Poisson's differential equation. For the unit ball $B := \{x \in \mathbb{R}^n : |x| < 1\}$ we can explicitly give Green's function as follows:

$$\phi(y; x) = \frac{1}{2\pi} \log\left|\frac{y - x}{1 - \overline{x}y}\right|, \qquad y \in \overline{B}, \quad x \in B, \qquad \text{if} \quad \mathbf{n = 2} \qquad (1)$$

and

$$\phi(y; x) = \frac{1}{(2 - n)\omega_n}\left(\frac{1}{|y - x|^{n-2}} - \frac{1}{(1 - 2(x \cdot y) + |x|^2|y|^2)^{\frac{n-2}{2}}}\right), \qquad (2)$$

$$y \in \overline{B}, \quad x \in B, \qquad \text{if} \quad \mathbf{n \geq 3}.$$

Theorem 2.3 on Poisson's integral formula from Section 2 in Chapter 5 is our starting point: A solution u of the problem

$$u = u(x) = u(x_1, \ldots, x_n) \in C^2(B) \cap C^0(\overline{B}),$$
$$\Delta u(x) = f(x), \qquad x \in B, \qquad (3)$$

with the right-hand side

$$f = f(x) \in C^0(\overline{B}) \qquad (4)$$

satisfies Poisson's integral representation

$$u(x) = \frac{1}{\omega_n} \int\limits_{|y|=1} \frac{|y|^2 - |x|^2}{|y - x|^n} u(y) \, d\sigma(y) + \int\limits_{|y|\leq 1} \phi(y; x) f(y) \, dy, \qquad x \in B. \quad (5)$$

Question I: For which right-hand sides $f : \overline{B} \to \mathbb{R}$ and for which boundary values $u : \partial B \to \mathbb{R}$ can we solve the Dirichlet problem of Poisson's equation?

Question II: Under which conditions can the second derivatives $u_{x_i x_j}(x)$, $x \in B$ be extended continuously onto the closure $\overline{B}$ with $i, j = 1, \ldots, n$?

If the function u possesses zero boundary values on ∂B, we only have to consider the singular integral defined on B in (5).

Definition 4.1. *Let $\Omega \subset \mathbb{R}^n$ be a domain, and the parameter $\alpha \in (0,1)$ is regularly used in the following. Then the continuous function $f : \Omega \to \mathbb{R}$ belongs to the regularity class $C^\alpha(\overline{\Omega})$, if we have a Hölder constant $b \in (0, +\infty)$ satisfying*

$$|f(x) - f(y)| \le b|x - y|^\alpha \qquad \text{for all} \quad x, y \in \Omega. \tag{6}$$

Proposition 4.2. (E. Hopf)
For the dimensions $n = 2, 3, \ldots$ let $\Omega \subset \mathbb{R}^n$ be a bounded domain, and we set

$$\Omega \otimes \Omega := \Big\{ (x, y) \in \Omega \times \Omega \ : \ x \ne y \Big\}.$$

Let the symmetric kernel function

$$\phi(y; x) = \phi(x; y) : \Omega \otimes \Omega \to \mathbb{R}$$

be given with the growth conditions

$$|\phi(y; x)| \le \begin{cases} a \log |y - x|, & \text{if } n = 2 \\ a|y - x|^{2-n}, & \text{if } n \ge 3 \end{cases}, \tag{7}$$

and

$$|\phi_{x_i}(y; x)| \le a|y - x|^{1-n},$$
$$|\phi_{x_i x_j}(y; x)| \le a|y - x|^{-n}, \qquad i, j = 1, \ldots, n. \tag{8}$$

Here $a \in (0, +\infty)$ denotes a constant. Furthermore, the functions

$$\Phi_i(x) := \int_\Omega \phi_{x_i}(y; x) dy, \quad x \in \Omega, \qquad \text{with} \quad i = 1, \ldots, n$$

belong to the class $C^1(\Omega)$. Finally, we consider the following parameter integral associated with the function $f \in C^\alpha(\overline{\Omega})$, namely

$$F(x) := \int_\Omega \phi(y; x) f(y) \, dy, \qquad x \in \Omega. \tag{9}$$

Then $F(x) \in C^2(\Omega)$ holds true, and we calculate their derivatives in the form

$$F_{x_i}(x) = \int_\Omega \phi_{x_i}(y;x)f(y)\,dy, \qquad x \in \Omega, \tag{10}$$

and

$$F_{x_ix_j}(x) = \int_\Omega \phi_{x_ix_j}(y;x)\big(f(y) - f(x)\big)\,dy + f(x)\Phi_{ix_j}(x), \qquad x \in \Omega. \tag{11}$$

Proof: The integral (9) converges absolutely due to (7). On account of (8), we can form the difference quotient of $F(x)$ and have a convergent majorizing function. We deduce the identity

$$F_{x_i}(x) = \int_\Omega \phi_{x_i}(y;x)f(y)\,dy, \quad x \in \Omega, \qquad \text{for} \quad i = 1, \dots, n$$

by the convergence theorem for improper Riemannian integrals. We are not allowed to directly differentiate this integral once more, since it does not remain absolutely convergent. Therefore, we consider the rearrangement

$$F_{x_i}(x) = \int_\Omega \phi_{x_i}(y;x)\big(f(y) - f(x_0)\big)\,dy + f(x_0)\Phi_i(x), \qquad x \in \Omega,$$

with the point $x_0 \in \Omega$ being fixed. Now the difference quotient converges again

$$F_{x_ix_j}(x_0) = \int_\Omega \phi_{x_ix_j}(y;x_0)\big(f(y) - f(x_0)\big)\,dy + f(x_0)\Phi_{ix_j}(x_0) \quad \text{for all} \quad x_0 \in \Omega,$$

since the integral possesses the convergent majorant $|y - x_0|^{-n+\alpha}$. \qquad q.e.d.

A very important device in potential theory is the intricate

Proposition 4.3. (Hopf's estimates)
Let $\Omega \subset \mathbb{R}^n$ be a bounded convex domain, on which the singular kernel

$$K(x,y) : \Omega \otimes \Omega \to \mathbb{R} \in C^1(\Omega \otimes \Omega)$$

is defined with the growth conditions

$$|K(x,y)| \le \frac{a}{|x-y|^n},$$

$$\sum_{i=1}^n |K_{x_i}(x,y)| \le \frac{a}{|x-y|^{n+1}} \qquad \text{for} \quad (x,y) \in \Omega \otimes \Omega. \tag{12}$$

Furthermore, let the function $f = f(x) \in C^{\alpha}(\overline{\Omega})$ be given satisfying

$$|f(x'') - f(x')| \leq b|x'' - x'|^{\alpha} \qquad \text{for all} \quad x', x'' \in \Omega. \tag{13}$$

Here the quantities $a, b \in (0, +\infty)$ and $\alpha \in (0, 1)$ are fixed constants.
Then the parameter integral

$$F(x) := \int_{\Omega} K(x, y)\big(f(y) - f(x)\big)\, dy, \qquad x \in \Omega,$$

fulfills the following estimates

$$|F(x)| \leq M_0(\alpha, n, \operatorname{diam}(\Omega))ab, \qquad x \in \Omega, \tag{14}$$

and

$$\left|(F(x'') - F(x')) + (f(x'') - f(x')) \cdot \int_{\substack{y \in \Omega \\ |y-x'| \geq 3|x''-x'|}} K(x', y)\, dy\right| \leq M_1(\alpha, n)ab|x'' - x'|^{\alpha} \tag{15}$$

for all $x', x'' \in \Omega$.

Proof:

1. We have the inequality

$$|F(x)| \leq \int_{\Omega} |K(x, y)|\, |f(y) - f(x)|\, dy \leq ab \int_{\Omega} |y - x|^{-n+\alpha}\, dy$$
$$\leq M_0(\alpha, n, \operatorname{diam}(\Omega))ab \quad \text{for all} \quad x \in \Omega.$$

2. We choose arbitrary points $x', x'' \in \Omega$, set $\delta := |x'' - x'|$, and calculate

$$F(x'') - F(x')$$

$$= \int_{\Omega} K(x'',y)\big(f(y) - f(x'')\big)\,dy - \int_{\Omega} K(x',y)\big(f(y) - f(x')\big)\,dy$$

$$= \int_{|y-x'|\leq 3\delta} K(x'',y)\big(f(y) - f(x'')\big)\,dy - \int_{|y-x'|\leq 3\delta} K(x',y)\big(f(y) - f(x')\big)\,dy$$

$$+ \int_{|y-x'|\geq 3\delta} K(x'',y)\big(f(y) - f(x'')\big)\,dy - \int_{|y-x'|\geq 3\delta} K(x',y)\big(f(y) - f(x')\big)\,dy$$

$$= \int_{|y-x'|\leq 3\delta} K(x'',y)\big(f(y) - f(x'')\big)\,dy - \int_{|y-x'|\leq 3\delta} K(x',y)\big(f(y) - f(x')\big)\,dy$$

$$+ \int_{|y-x'|\geq 3\delta} \big(K(x'',y) - K(x',y)\big)\big(f(y) - f(x'')\big)\,dy$$

$$+ \big(f(x') - f(x'')\big) \int_{|y-x'|\geq 3\delta} K(x',y)\,dy$$

$$=: I_1 + I_2 + I_3 + \big(f(x') - f(x'')\big) \int_{|y-x'|\geq 3\delta} K(x',y)\,dy. \tag{16}$$

This implies

$$\left| \big(F(x'') - F(x')\big) + \big(f(x'') - f(x')\big) \int_{|y-x'|\geq 3\delta} K(x',y)\,dy \right| \leq |I_1| + |I_2| + |I_3|. \tag{17}$$

3. On account of (12), we can estimate I_1 as follows:

$$|I_1| \leq \int_{|y-x'|\leq 3\delta} \frac{a}{|y - x''|^n} b|y - x''|^\alpha\,dy \leq ab \int_{|y-x''|\leq 4\delta} |y - x''|^{\alpha-n}\,dy$$

$$= ab \int_0^{4\delta} r^{\alpha-n} r^{n-1}\omega_n\,dr = ab\,\omega_n \int_0^{4\delta} r^{\alpha-1}\,dr = \omega_n \frac{ab}{\alpha}\big[r^\alpha\big]_0^{4\delta} \tag{18}$$

$$= \frac{ab\,\omega_n}{\alpha}(4\delta)^\alpha = \frac{4^\alpha ab\,\omega_n}{\alpha}|x'' - x'|^\alpha.$$

Correspondingly, we deduce

$$|I_2| \leq \frac{3^\alpha ab\,\omega_n}{\alpha}|x'' - x'|^\alpha. \tag{19}$$

4. The mean value theorem of differential calculus implies

$$K(x'', y) - K(x', y) = \sum_{i=1}^{n} K_{x_i}(\zeta, y)(x_i'' - x_i'),$$

with an intermediate point $\zeta = x' + t(x'' - x') \in \Omega$ and a parameter $t \in (0, 1)$. From $|y - x'| \geq 3\delta$ we infer $|y - x''| \geq 2\delta$ and therefore

$$|y - \zeta| \geq |y - x''| - |x'' - \zeta| \geq |y - x''| - |x'' - x'| \geq \frac{1}{2}|y - x''|.$$

Noting (12), we obtain for all $y \in \Omega$ with $|y - x'| \geq 3\delta$ an inequality as follows:

$$|K(x'', y) - K(x', y)| \leq |x'' - x'| \sum_{i=1}^{n} |K_{x_i}(\zeta, y)|$$

$$\leq a\delta \frac{1}{|y - \zeta|^{n+1}} \leq a\delta 2^{n+1} \frac{1}{|y - x''|^{n+1}}. \tag{20}$$

Inserting into I_3 we get

$$|I_3| \leq \int_{|y-x'| \geq 3\delta} |K(x'', y) - K(x', y)| \, |f(y) - f(x'')| \, dy$$

$$\leq 2^{n+1} ab \, \delta \int_{|y-x'| \geq 3\delta} |y - x''|^{-n-1+\alpha} \, dy$$

$$\leq 2^{n+1} ab \, \delta \int_{|y-x''| \geq 2\delta} |y - x''|^{-n-1+\alpha} \, dy$$

$$\leq 2^{n+1} ab \, \delta \int_{2\delta}^{+\infty} r^{-n-1+\alpha} \omega_n r^{n-1} \, dr$$

$$= 2^{n+1} \omega_n \, ab \, \delta \int_{2\delta}^{+\infty} r^{\alpha-2} \, dr$$

$$= 2^{n+1} \omega_n \, ab \, \delta \left[r^{\alpha-1} \right]_{2\delta}^{+\infty} \frac{1}{\alpha - 1}$$

$$= \frac{2^{n+1}}{1 - \alpha} \omega_n \, ab \, \delta (2\delta)^{\alpha-1} = \frac{2^{n+\alpha}}{1 - \alpha} \omega_n \, ab \, \delta^{\alpha}$$

and consequently

$$|I_3| \leq \frac{2^{n+\alpha}}{1 - \alpha} \omega_n \, ab |x'' - x'|^{\alpha}. \tag{21}$$

5. From (17)-(19) and (21) we now obtain a constant $M_1 = M_1(\alpha, n)$, such that the estimate (15) is valid.   q.e.d.

For the function $f \in C^\alpha(\overline{\Omega})$ given in the domain $\Omega \subset \mathbb{R}^n$ we define the quantities

$$\|f\|_0^\Omega := \sup_{x \in \Omega} |f(x)|,$$

$$\|f\|_{0,\alpha}^\Omega := \sup_{\substack{x',x'' \in \Omega \\ x' \neq x''}} \frac{|f(x') - f(x'')|}{|x' - x''|^\alpha}, \tag{22}$$

$$\|f\|_\alpha^\Omega := \|f\|_0^\Omega + \|f\|_{0,\alpha}^\Omega.$$

By the norm (22) the set $C^\alpha(\overline{\Omega})$ becomes a Banach space. Furthermore, we easily show the following inequality for two functions $f, g \in C^\alpha(\overline{\Omega})$, namely

$$\|fg\|_\alpha^\Omega \leq \|f\|_\alpha^\Omega \|g\|_\alpha^\Omega.$$

In the function space

$$C^{2+\alpha}(\overline{\Omega}) := \left\{ u \in C^2(\overline{\Omega}) \ : \ u_{x_i x_j} \in C^\alpha(\overline{\Omega}) \text{ für } i, j = 1, \ldots, n \right\}$$

we define the following quantities

$$\|u\|_0^\Omega := \sup_{x \in \Omega} |u(x)|,$$

$$\|u\|_1^\Omega := \sup_{x \in \Omega} \sum_{i=1}^n |u_{x_i}(x)|,$$

$$\|u\|_2^\Omega := \sup_{x \in \Omega} \sum_{i,j=1}^n |u_{x_i x_j}(x)|, \tag{23}$$

$$\|u\|_{2+\alpha}^\Omega := \|u\|_0^\Omega + \|u\|_1^\Omega + \|u\|_2^\Omega + \|u\|_{2,\alpha}^\Omega.$$

Here we have abbreviated

$$\|u\|_{2,\alpha}^\Omega := \sup_{\substack{x',x'' \in \Omega \\ x' \neq x''}} \sum_{i,j=1}^n \frac{|u_{x_i x_j}(x') - u_{x_i x_j}(x'')|}{|x' - x''|^\alpha}.$$

With the aid of the norm (23) the set $C^{2+\alpha}(\overline{\Omega})$ becomes a Banach space. By the symbol

$$C_*^{2+\alpha}(\overline{\Omega}) := \left\{ u \in C^{2+\alpha}(\overline{\Omega}) \ : \ u|_{\partial\Omega} = 0 \right\}$$

we denote the closed subspace of $C^{2+\alpha}(\overline{\Omega})$, consisting of those functions with zero boundary values.

We now prove the following

Theorem 4.4. *Let the function* $f \in C^\alpha(\overline{B})$ *be given. Then the parameter integral*

$$u(x) := \int\limits_{|y|<1} \phi(y;x)f(y)\,dy, \qquad x \in B,$$

belongs to the class $C_*^{2+\alpha}(\overline{B})$ and satisfies Poisson's differential equation $\Delta u = f$ in $\overline{B}$. Furthermore, we have the estimate

$$\|u\|_{2+\alpha}^B \le C(\alpha,n)\|f\|_\alpha^B \tag{24}$$

with a constant $C(\alpha,n) \in (0,+\infty)$.

Proof:

1. From the representation

$$\phi(y;x) = \frac{-1}{(n-2)\omega_n}\left\{\frac{1}{|x-y|^{n-2}} - \frac{1}{|y|^{n-2}}\frac{1}{\left|x-\frac{y}{|y|^2}\right|^{n-2}}\right\}, \quad x,y \in B$$

of Green's function, one easily derives the following estimates

$$\sum_{i=1}^n |\phi_{x_i}(y;x)| \le \frac{a(n)}{|x-y|^{n-1}} \quad \text{for all} \quad x,y \in B \quad \text{with} \quad x \ne y,$$

$$\sum_{i,j=1}^n |\phi_{x_i x_j}(y;x)| \le \frac{a(n)}{|x-y|^n} \quad \text{for all} \quad x,y \in B \quad \text{with} \quad x \ne y,$$

$$\sum_{i,j,k=1}^n |\phi_{x_i x_j x_k}(y;x)| \le \frac{a(n)}{|x-y|^{n+1}} \quad \text{for all} \quad x,y \in B \quad \text{with} \quad x \ne y,$$

$$\tag{25}$$

with a universal constant $a = a(n) \in (0,+\infty)$.

2. We consider the function $w(x) := \dfrac{|x|^2-1}{2n}$, $x \in \overline{B}$ of the class $C_*^{2+\alpha}(\overline{B})$, satisfying the differential equation

$$\Delta w(x) = \sum_{i=1}^n w_{x_i x_i}(x) = \frac{1}{2n}\sum_{i=1}^n 2 = 1, \qquad x \in \overline{B}.$$

The Poisson integral representation yields

$$\int\limits_B \phi(y;x)\,dy = \frac{|x|^2-1}{2n}, \qquad x \in \overline{B}, \tag{26}$$

with the nonpositive Green's function $\phi(y;x)$. For all $x \in B$ we estimate as follows:

$$|u(x)| = \left|\int\limits_{|y|<1} \phi(y;x)f(y)\,dy\right| \le \|f\|_0^B \left|\int\limits_B \phi(y;x)\,dy\right| \le \frac{1-|x|^2}{2n}\|f\|_0^B.$$

This implies

$$u(x) = 0 \qquad \text{for all} \quad x \in \partial B \qquad (27)$$

and

$$\|u\|_0^B \le \frac{1}{2n}\|f\|_0^B. \qquad (28)$$

Noting (25) we can differentiate (26) and obtain the functions

$$\Phi_i(x) := \int_B \phi_{x_i}(y;x)\,dy = \frac{1}{n}x_i, \quad x \in \overline{B}, \quad \text{for} \quad i = 1,\dots,n \qquad (29)$$

of class $C^1(\overline{B})$ with

$$\Phi_{ix_j}(x) = \frac{1}{n}\delta_{ij}, \quad x \in \overline{B}, \qquad \text{for} \quad i,j = 1,\dots,n. \qquad (30)$$

3. On account of (25) we have

$$u_{x_i}(x) = \int_B \phi_{x_i}(y;x)f(y)\,dy, \quad x \in B, \qquad \text{for} \quad i = 1,\dots,n. \qquad (31)$$

This implies the estimate

$$\sum_{i=1}^n |u_{x_i}(x)| \le \|f\|_0^B \int_B \sum_{i=1}^n |\phi_{x_i}(y;x)|\,dy$$

$$\le \|f\|_0^B \int_B \frac{a(n)}{|y-x|^{n-1}}\,dy, \qquad x \in B,$$

and consequently

$$\|u\|_1^B \le c_1(n)\|f\|_0^B. \qquad (32)$$

Proposition 4.2 yields the representation

$$u_{x_i x_j}(x) = \int_B \phi_{x_i x_j}(y;x)\big(f(y) - f(x)\big)\,dy + f(x)\Phi_{ix_j}(x), \quad x \in B \qquad (33)$$

for $i,j = 1,\dots,n$. With the aid of (30) we deduce the differential equation

$$\Delta u(x) = \int_B \Delta_x \phi(y;x)\big(f(y) - f(x)\big)\,dy + f(x) \cdot \sum_{i=1}^n \phi_{ix_i}(x)$$

$$= f(x) \cdot \sum_{i=1}^n \frac{1}{n} = f(x), \qquad x \in B. \qquad (34)$$

Here we have utilized the property $\Delta_x \phi(y;x) = 0$. For all $x \in B$ we infer the following estimate from (25):

$$\sum_{i,j=1}^{n} |u_{x_i x_j}(x)| \leq \int_B \sum_{i,j=1}^{n} |\phi_{x_i x_j}(y;x)| \, |f(y) - f(x)| \, dy + \|f\|_0^B$$

$$\leq \|f\|_{0,\alpha}^B \int_B \frac{a}{|y-x|^n} |y-x|^\alpha \, dy + \|f\|_0^B \leq c_2(n,\alpha) \|f\|_\alpha^B \qquad (35)$$

and consequently

$$\|u\|_2^B \leq c_2(n,\alpha) \|f\|_\alpha^B. \qquad (36)$$

4. We still have to estimate $\|u\|_{2,\alpha}^B$. The indices $i,j \in \{1,\ldots,n\}$ being fixed, we consider the kernel

$$K(x,y) := \phi_{x_i x_j}(y;x) : B \otimes B \to \mathbb{R}$$

and utilize Hopf's estimate for the function

$$F(x) := \int_B K(x,y) \big(f(y) - f(x)\big) dy, \qquad x \in B.$$

With the aid of the Gaussian integral theorem, we show the uniform boundedness of the Cauchy principal values

$$\left| \int_{\substack{y \in B : |y-x| \geq \delta}} K(x,y) \, dy \right| = \left| \int_{\substack{y \in B \\ |y-x| \geq \delta}} \phi_{x_i x_j}(y;x) \, dy \right| \leq c_3(n), \, x \in B \text{ for all } \delta > 0. \qquad (37)$$

We obtain the following estimate

$$\left| |F(x'') - F(x')| - |f(x'') - f(x')| \middle| \int_{\substack{y \in B \\ |y-x'| \geq 3|x''-x'|}} K(x',y) \, dy \right|$$

$$\leq \left| \big(F(x'') - F(x')\big) + \big(f(x'') - f(x')\big) \int_{\substack{y \in B \\ |y-x'| \geq 3|x''-x'|}} K(x',y) \, dy \right| \qquad (38)$$

$$\leq M_1(\alpha,n) \, a \, \|f\|_{0,\alpha}^B |x''-x'|^\alpha \quad \text{for all} \quad x', x'' \in B.$$

This implies

$$|F(x'') - F(x')| \leq \Big\{ c_3(n) + a \, M_1(\alpha,n) \Big\} \|f\|_{0,\alpha}^B |x''-x'|^\alpha,$$

and we see

$$\frac{|F(x'') - F(x')|}{|x''-x'|^\alpha} \leq \tilde{c}_3(n,\alpha) \|f\|_{0,\alpha}^B \quad \text{for all } x', x'' \in B \text{ with } x' \neq x''. \qquad (39)$$

Taking (33) and (30) into account, we deduce

$$\|u\|_{2,\alpha}^B \leq c_4(n,\alpha) \|f\|_{0,\alpha}^B. \qquad (40)$$

5. Finally, we infer from (28), (32), (36) and (40) a constant $C(n, \alpha)$, such that

$$\|u\|_{2+\alpha}^B \leq C(n, \alpha)\|f\|_\alpha^B$$

is valid.   q.e.d.

For later use in Chapter 12, we derive a potential-theoretic estimate of the solutions for the inhomogeneous Cauchy-Riemann equation. The Hölder norms defined in (22) and (23) are naturally transferred to complex-valued functions

$$w = f(z) : B \to \mathbb{C} \tag{41}$$

on the unit disc $B := \{z = x + iy \in \mathbb{C} : |z| < 1\}$. We consider functions $f \in C^\alpha(\overline{B}, \mathbb{C})$ and define the *Riemann-Hilbert operator* $(\zeta = \xi + i\eta \in B)$:

$$\mathbb{V}f(z) := -\frac{1}{\pi} \left\{ \iint_B \frac{f(\zeta)}{\zeta - z} \, d\xi \, d\eta + \iint_B \frac{z\overline{f(\zeta)}}{1 - z\overline{\zeta}} \, d\xi \, d\eta \right.$$
$$\left. - \iint_B \frac{1}{2|\zeta|^2} \left(\overline{\zeta} f(\zeta) - \zeta \overline{f(\zeta)} \right) d\xi \, d\eta \right\}, \qquad z \in B. \tag{42}$$

Theorem 4.5. *Assuming $f = f(z) \in C^\alpha(\overline{B}, \mathbb{C})$ the associate function*

$$g(z) := \mathbb{V}f(z), \quad z \in B$$

solves the uniquely determined Riemann-Hilbert boundary value problem

$$g = g(z) \in C^1(B, \mathbb{C}) \cap C^0(\overline{B}, \mathbb{C}),$$
$$\frac{\partial}{\partial \overline{z}} g(z) = f(z), \qquad z \in B,$$
$$\operatorname{Re} g(z) = 0, \qquad z \in \partial B, \tag{43}$$
$$\operatorname{Im} g(0) = 0.$$

Furthermore, we have $g \in C^{1+\alpha}(\overline{B}, \mathbb{C})$, and there exists a constant $C(\alpha) \in (0, +\infty)$ such that

$$\|g\|_{C^{1+\alpha}(\overline{B}, \mathbb{C})} \leq C(\alpha)\|f\|_{C^\alpha(\overline{B}, \mathbb{C})}. \tag{44}$$

Proof: By the identity (10) from Section 3 one easily realizes that the function

$$g(z) = \mathbb{V}f(z), \quad z \in B$$

solves the boundary value problem (43). Applying the maximum principle for harmonic functions on the real part of the difference of two solutions, we directly see the unique solvability of the problem (43). With the right-hand side $f(z) \equiv 1$, $z \in B$ in particular, this boundary value problem yields the solution $g(z) = \bar{z} - z$, $z \in B$. Corresponding to the formula (26), we obtain the identity

$$-\frac{1}{\pi}\left\{ \iint\limits_B \frac{1}{\zeta - z}\, d\xi\, d\eta + \iint\limits_B \frac{z}{1 - z\bar{\zeta}}\, d\xi\, d\eta - \frac{1}{2}\iint\limits_B \left(\frac{1}{\zeta} - \frac{1}{\bar{\zeta}}\right) d\xi\, d\eta \right\} = \bar{z} - z \quad (45)$$

for all $z \in B$. Using Proposition 5.9 and Proposition 5.11 from Section 5 in Chapter 4, we can differentiate the function $g(z) = \mathbb{V}f(z)$ with respect to z and $\bar{z}$. Parallel to the proof of Theorem 4.4 we attain the a priori-inequality (44) via the Hopf estimate.

q.e.d.

As a corollary we obtain

Theorem 4.6. (Privalov)
To the boundary function $\phi(z) : \partial B \to \mathbb{R} \in C^{1+\alpha}(\partial B)$ we consider the Schwarzian integral

$$F(z) := \frac{1}{2\pi}\int\limits_0^{2\pi} \frac{e^{i\varphi} + z}{e^{i\varphi} - z}\phi(e^{i\varphi})\, d\varphi, \quad |z| < 1.$$

Then we have a constant $\widehat{C}(\alpha) \in (0, +\infty)$ satisfying

$$\|F\|_{C^{1+\alpha}(\overline{B})} \leq \widehat{C}(\alpha)\|\phi\|_{C^{1+\alpha}(\partial B)}. \quad (46)$$

Proof: We compare $F(z)$ with the function

$$G(z)\Big|_{z=re^{i\varphi}} := r^{1+\alpha}\psi(\varphi), \quad 0 \leq r \leq 1, \quad 0 \leq \varphi \leq 2\pi,$$

setting $\psi(\varphi) := \phi(e^{i\varphi})$, $0 \leq \varphi \leq 2\pi$. For all $z \in B \setminus \{0\}$ we calculate

$$\frac{\partial}{\partial \bar{z}}G(z)\Big|_{z=re^{i\varphi}} = \frac{1}{2}\left\{\left(\frac{\partial}{\partial x} + i\frac{\partial}{\partial y}\right)G(z)\right\}\Big|_{z=re^{i\varphi}}$$

$$= \frac{e^{i\varphi}}{2}\left(\frac{\partial}{\partial r} + \frac{i}{r}\frac{\partial}{\partial \varphi}\right)G(re^{i\varphi})$$

$$= \frac{e^{i\varphi}}{2}\left((1+\alpha)r^{\alpha}\psi(\varphi) + ir^{\alpha}\psi'(\varphi)\right) =: f(z)\Big|_{z=re^{i\varphi}}.$$

We note that

$$\|f\|_{C^{\alpha}(\overline{B})} \leq \widetilde{C}(\alpha)\|\phi\|_{C^{1+\alpha}(\partial B)}. \quad (47)$$

Using Theorem 2.2 from Section 2, the function $g(z) := G(z) - F(z)$, $z \in B$ solves the boundary value problem (43), and Theorem 4.5 yields

$$\|G - F\|_{C^{1+\alpha}(\overline{B})} \leq C(\alpha)\|f\|_{C^{\alpha}(\overline{B})} \leq C(\alpha)\widetilde{C}(\alpha)\|\phi\|_{C^{1+\alpha}(\partial B)}. \qquad (48)$$

We finally obtain (46).   q.e.d.

Without requiring boundary conditions,we supplement the following

Theorem 4.7. *Considering the Hadamard integral operator*

$$T_B[f](z) := -\frac{1}{\pi} \iint\limits_{B} \frac{f(\zeta)}{\zeta - z}\, d\xi\, d\eta, \qquad z \in B,$$

we have the estimate

$$\|T_B[f]\|_{C^{1+\alpha}(\overline{B})} \leq C(\alpha)\|f\|_{C^{\alpha}(\overline{B})}$$

for all $f \in C^{\alpha}(\overline{B})$ *with a constant* $C(\alpha) \in (0, \infty)$.

Proof: We remark

$$\frac{\partial}{\partial z} T_B[f](z) = \Pi_B[f](z) := \lim_{\varepsilon \to 0+} \left\{ -\frac{1}{\pi} \iint\limits_{\substack{\zeta \in B \\ |\zeta - z| > \varepsilon}} \frac{f(\zeta)}{(\zeta - z)^2}\, d\xi\, d\eta \right\}, \qquad z \in B.$$

Then we apply the Hopf estimates from Proposition 4.3 to the Vekua integral operator $\Pi_B[f]$.   q.e.d.

5 Schauder's Continuity Method

We now follow the arguments in Chapter 6, Section 1 and define the differential operator

$$\mathcal{L}(u) := \sum_{i,j=1}^{n} a_{ij}(x)\frac{\partial^2 u}{\partial x_i \partial x_j} + \sum_{i=1}^{n} b_i(x)\frac{\partial u}{\partial x_i} + c(x)u(x) = f(x), \qquad x \in \Omega.$$

Assumption C_1: The solution $u(x)$ of $\mathcal{L}(u) = f$ belongs to the class $C^{2+\alpha}(\overline{\Omega})$. Furthermore, we have $u(x) = 0$ on $\partial\Omega$.

Assumption C_2: The coefficients $a_{ij}(x)$, $b_i(x)$, $c(x)$ with $i, j = 1, \ldots, n$ belong to the regularity class $C^{\alpha}(\overline{\Omega})$. Furthermore, the matrix $(a_{ij}(x))_{i,j=1,\ldots,n}$ is real, symmetric and positive-definite for all points $x \in \overline{\Omega}$.

Assumption C_3: For each point $\xi \in \partial\Omega$ there exists a positive number $\varrho = \varrho(\xi)$ and a function

$$G(x) \in C^{2+\alpha}(\{x \in \mathbb{R}^n \, : \, |x - \xi| < \varrho\}, \mathbb{R})$$

satisfying

$$\sum_{i=1}^{n} G_{x_i}(x)^2 > 0 \quad \text{for } |x - \xi| < \varrho, \qquad G(\xi) = 0,$$

such that

$$\Omega \cap \left\{ x \in \mathbb{R}^n \ : \ |x - \xi| < \varrho \right\} = \left\{ x \in \mathbb{R}^n \ : \ |x - \xi| < \varrho, \ G(x) < 0 \right\}$$

is valid; this means $\partial\Omega \in C^{2+\alpha}$. Furthermore, $\Omega \subset \mathbb{R}^n$ is a bounded domain.

Assumption C_4: We have $c(x) \leq 0$ for all $x \in \Omega$.

We need the following profound result, which we shall prove in Section 7.

Theorem 5.1. (Schauder's estimates)
Let the assumptions C_1, C_2, C_3 be satisfied. Furthermore, we require the bound

$$\sum_{i,j=1}^{n} \|a_{ij}\|_\alpha^\Omega + \sum_{i=1}^{n} \|b_i\|_\alpha^\Omega + \|c\|_\alpha^\Omega \leq H$$

and the ellipticity condition

$$m^2 \sum_{i=1}^{n} \lambda_i^2 \leq \sum_{i,j=1}^{n} a_{ij}(x)\lambda_i\lambda_j \leq M^2 \sum_{i=1}^{n} \lambda_i^2$$
for all $\lambda = (\lambda_1, \ldots, \lambda_n) \in \mathbb{R}^n$ *and all* $x \in \overline{\Omega}$

with the constants $H > 0$ and $0 < m \leq M < +\infty$.
Then we can determine a number $\theta = \theta(\alpha, n, m, M, H, \Omega)$ such that

$$\|u\|_1^\Omega + \|u\|_2^\Omega + \|u\|_{2,\alpha}^\Omega \leq \theta \left(\|u\|_0^\Omega + \|f\|_0^\Omega + \|f\|_{0,\alpha}^\Omega \right) \tag{1}$$

holds true.

Generalizing Theorem 4.4 from Section 4, we obtain as a corollary the following

Theorem 5.2. *In addition to the assumptions of Theorem 5.1 let the condition C_4 be required.*
Then we have a fixed positive number $\theta = \theta(\alpha, n, m, M, H, \Omega)$ such that the following a priori estimate

$$\|u\|_{2+\alpha}^\Omega \leq \theta \|f\|_\alpha^\Omega \tag{2}$$

for all solutions $u \in C^{2+\alpha}(\overline{\Omega})$ of the Dirichlet problem

$$\begin{aligned}
\mathcal{L}(u) &= f \quad \text{in} \quad \Omega, \\
u &= 0 \quad \text{on} \quad \partial\Omega
\end{aligned} \tag{3}$$

holds true.

Proof: Theorem 1.3 from Section 1 in Chapter 6 yields

$$\|u\|_0^\Omega \leq \gamma \|f\|_0^\Omega$$

with a constant $\gamma = \gamma(\Omega, m, M)$. We combine this inequality with the Schauder estimate and obtain (2).

q.e.d.

We additionally need the following

Assumption C_0: For all $f \in C^\alpha(\overline{\Omega})$ the partial differential equation $\Delta u = f$ possesses a solution in the regularity class $u \in C_*^{2+\alpha}(\overline{\Omega})$.

Remark: Because of Theorem 4.4 from Section 4, the condition (C_0) is satisfied for the unit ball $\Omega = B$. Later we shall show the implication $(C_3) \Rightarrow (C_0)$ and eliminate this assumption.

Theorem 5.3. (Continuity method)
We require the assumptions C_0, C_2, C_3, C_4 and consider the differential operator $\mathcal{L}$ on the domain Ω.
Then the boundary value problem

$$\begin{aligned} \mathcal{L}(u) &= f \quad in \quad \Omega, \\ u &= 0 \quad on \quad \partial\Omega \end{aligned} \tag{4}$$

has exactly one solution $u \in C_*^{2+\alpha}(\overline{\Omega})$ for each right-hand side $f \in C^\alpha(\overline{\Omega})$.

Proof: We take $0 \leq \tau \leq 1$ and define the family of differential operators

$$\mathcal{L}_\tau(u) := \sum_{i,j=1}^n a_{ij}(x,\tau)\frac{\partial^2 u}{\partial x_i \partial x_j} + \sum_{i=1}^n b_i(x,\tau)\frac{\partial u}{\partial x_i} + c(x,\tau)u$$

with the coefficients

$$\begin{aligned} a_{ij}(x,\tau) &:= \tau a_{ij}(x) + (1-\tau)\delta_{ij}, \quad i,j = 1,\ldots,n, \\ b_i(x,\tau) &:= \tau b_i(x), \quad i = 1,\ldots,n, \\ c(x,\tau) &:= \tau c(x). \end{aligned}$$

This means $\mathcal{L}_\tau = (1-\tau)\Delta + \tau\mathcal{L}$ briefly. Due to Theorem 5.2, we have the following a priori estimate

$$\|u\|_{2+\alpha} \leq \theta\|f\|_\alpha, \quad \tau \in [0,1], \tag{5}$$

for all solutions of the Dirichlet problem $\mathcal{L}_\tau(u) = f$ in B and $u \in C_*^{2+\alpha}(\overline{\Omega})$. Here we abbreviate $\|u\|_{2+\alpha} := \|u\|_{2+\alpha}^\Omega$ and $\|f\|_\alpha := \|f\|_\alpha^\Omega$ for the domain Ω being fixed.

We now start with a solution $u = u_{\tau_0} \in C_*^{2+\alpha}(\overline{\Omega})$ of the problem $\mathcal{L}_{\tau_0}(u) = f$ for an arbitrary $\tau_0 \in [0,1]$. On account of (C_0), this is possible for $\tau_0 = 0$, and we consider

$$\mathcal{L}_\tau(u) = f \quad \Longleftrightarrow \quad \mathcal{L}_{\tau_0}(u) = \mathcal{M}_\tau(u) + f$$

$$\text{with} \quad \mathcal{M}_\tau = (\mathcal{L}_{\tau_0} - \mathcal{L}_\tau)(u) = (\tau - \tau_0)(\Delta - \mathcal{L})(u). \tag{6}$$

We set $u_0 \equiv 0$ and successively define the approximating sequence $\{u_k\}_{k=0,1,\dots}$ by the prescription

$$\mathcal{L}_{\tau_0}(u_k) = \mathcal{M}_\tau(u_{k-1}) + f, \qquad k = 1, 2, \dots \tag{7}$$

We start with the statement

$$A_{\tau_0} : \begin{cases} \text{For each } f \in C^\alpha(\overline{\Omega}) \text{ the differential equation} \\ \mathcal{L}_{\tau_0}(u) = f \text{ has a solution } u_{\tau_0} \in C_*^{2+\alpha}(\overline{\Omega}) \end{cases}. \tag{8}$$

Then we shall investigate the convergence of the sequence

$$\{u_k\}_{k=0,1,\dots} \subset C_*^{2+\alpha}(\overline{\Omega})$$

with respect to the $\|\cdot\|_{2+\alpha}$-norm. Taking an arbitrary $u \in C_*^{2+\alpha}(\overline{\Omega})$ we infer

$$\|\mathcal{M}_\tau(u)\|_\alpha = |\tau - \tau_0| \, \|(\Delta - \mathcal{L})u\|_\alpha \le |\tau - \tau_0|\eta(H)\|u\|_{2+\alpha} \tag{9}$$

with a constant $\eta = \eta(H)$. From (7) we deduce

$$\mathcal{L}_{\tau_0}(u_k - u_{k-1}) = \mathcal{M}_\tau(u_{k-1} - u_{k-2}), \qquad k = 2, 3, \dots \tag{10}$$

The Schauder estimate (5) together with (9) yields the inequality

$$\|u_k - u_{k-1}\|_{2+\alpha} \le \theta \|\mathcal{M}_\tau(u_{k-1} - u_{k-2})\|_\alpha$$

$$\le |\tau - \tau_0|\theta \, \eta(H)\|u_{k-1} - u_{k-2}\|_{2+\alpha}, \quad k = 2, 3, \dots \tag{11}$$

Choosing $|\tau - \tau_0| \le \frac{1}{2\theta \, \eta(H)}$, we deduce

$$\|u_k - u_{k-1}\|_{2+\alpha} \le \frac{1}{2}\|u_{k-1} - u_{k-2}\|_{2+\alpha}, \qquad k = 2, 3, \dots,$$

and

$$\|u_k - u_{k-1}\|_{2+\alpha} \le \frac{\|u_1\|_{2+\alpha}}{2^{k-1}}, \qquad k = 2, 3, \dots.$$

This implies

$$\sum_{k=1}^{+\infty} \|u_k - u_{k-1}\|_{2+\alpha} < +\infty.$$

Therefore, the series $\sum\limits_{k=1}^{+\infty}(u_k - u_{k-1})$ converges in the Banach space $C_*^{2+\alpha}(\overline{\Omega})$.
For all τ satisfying $|\tau - \tau_0| \leq \frac{1}{2\theta\,\eta(H)}$ we have a function $u_\tau \in C_*^{2+\alpha}(\overline{\Omega})$ such that $\mathcal{L}_{\tau_0}(u_\tau) = \mathcal{M}_\tau(u_\tau) + f$ holds true, and finally

$$\mathcal{L}_\tau(u_\tau) = f.$$

Consequently, the statement (A_τ) is valid for all $|\tau - \tau_0| \leq \frac{1}{2\vartheta\,\eta(H)}$. By the usual continuation process we attain the statement (A_1) after finitely many steps.   q.e.d.

The following profound result will be proved in Section 7 as well:

Theorem 5.4. (Interior Schauder estimates)
The coefficients of the differential operator $\mathcal{L}$ defined on the bounded domain $\Omega \subset \mathbb{R}^n$ satisfy the assumption C_2. Furthermore, we require the inequalities

$$\sum_{i,j=1}^{n}\|a_{ij}\|_\alpha^\Omega + \sum_{i=1}^{n}\|b_i\|_\alpha^\Omega + \|c\|_\alpha^\Omega \leq H$$

and

$$m^2\sum_{i=1}^{n}\lambda_i^2 \leq \sum_{i,j=1}^{n} a_{ij}(x)\lambda_i\lambda_j \leq M^2\sum_{i=1}^{n}\lambda_i^2$$

for all $\lambda \in \mathbb{R}^n$ and all $x \in \overline{\Omega}$; with the constants $H > 0$ and $0 < m \leq M < +\infty$ given. The function $u = u(x) \in C^{2+\alpha}(\Omega) \cap C^0(\overline{\Omega})$ solves the differential equation

$$\mathcal{L}(u) = f \quad in \quad \Omega$$

with the right-hand side $f \in C^\alpha(\overline{\Omega})$. Finally, we consider the set

$$\Omega_d := \Big\{x \in \Omega \,:\, dist(x, \partial\Omega) > d\Big\},$$

where we choose $d > 0$ sufficiently small.
Then we have an a-priori-bound $\kappa = \kappa(\alpha, n, m, M, H, d) > 0$, such that

$$\|u\|_1^{\Omega_d} + \|u\|_2^{\Omega_d} + \|u\|_{2,\alpha}^{\Omega_d} \leq \kappa\big(\|u\|_0^\Omega + \|f\|_0^\Omega + \|f\|_{0,\alpha}^\Omega\big) \tag{12}$$

is satisfied.

Remark: The abbreviation $u \in C^{2+\alpha}(\Omega)$ means that $u \in C^{2+\alpha}(\Theta)$ is fulfilled for each compact subset $\Theta \subset \Omega$.

Theorem 5.5. *With the assumptions C_0, C_2, C_3, C_4 we consider the differential operator $\mathcal{L}$ on the domain Ω. Consequently, for all $f \in C^\alpha(\overline{\Omega})$ and all continuous functions $g : \partial\Omega \to \mathbb{R}$ the Dirichlet problem*

$$\begin{aligned} \mathcal{L}(u) &= f \quad in \quad \Omega, \\ u &= g \quad on \quad \partial\Omega \end{aligned} \tag{13}$$

has exactly one solution in the regularity class $C^{2+\alpha}(\Omega) \cap C^0(\overline{\Omega})$.

Proof: We construct a sequence of polynomials $\{g_n\}_{n=1,2,\ldots}$, which converge on the boundary $\partial\Omega$ uniformly towards $g(x)$. For each index $n = 1, 2, \ldots$ we solve the problem

$$u_n \in C^{2+\alpha}(\overline{\Omega}),$$

$$\mathcal{L}(u_n) = f \quad \text{in} \quad \Omega, \tag{14}$$

$$u_n = g_n \quad \text{on} \quad \partial\Omega.$$

With the aid of Theorem 5.3 we construct a sequence

$$\{v_n\}_{n=1,2,\ldots} \subset C_*^{2+\alpha}(\overline{\Omega})$$

satisfying

$$\mathcal{L}(v_n) = f - \mathcal{L}(g_n) =: f_n \in C^{\alpha}(\overline{\Omega}) \quad \text{in} \quad \Omega,$$

$$v_n = 0 \quad \text{on} \quad \partial\Omega. \tag{15}$$

Evidently, the functions $u_n := v_n + g_n$ solve the boundary value problems (14) for $n = 1, 2, \ldots$. Because of $\mathcal{L}(u_m - u_n) = 0$ in Ω and (C$_4$) the maximum principle yields

$$\|u_n - u_m\|_0^{\Omega} \leq \max_{x \in \partial\Omega} |g_n(x) - g_m(x)| \to 0 \ (m, n \to \infty)$$

$$\|u_n\|_0^{\Omega} \leq \text{const}, \qquad n = 1, 2, \ldots \tag{16}$$

Choosing $d > 0$ sufficiently small, we obtain the following inequality by the interior Schauder estimate:

$$\|u_n - u_m\|_1^{\Omega_d} + \|u_n - u_m\|_2^{\Omega_d} + \|u_n - u_m\|_{2,\alpha}^{\Omega_d}$$

$$\leq \kappa(d)\|u_n - u_m\|_0^{\Omega} \to 0 \ (m, n \to \infty).$$

Setting

$$u(x) := \lim_{n \to \infty} u_n(x), \qquad x \in \overline{\Omega},$$

we deduce $u_n \to u \, (n \to \infty)$ in $C^{2+\alpha}(\Theta)$ for each compact subset $\Theta \subset \Omega$. Therefore, the function u belongs to the class $C^0(\overline{\Omega}) \cap C^{2+\alpha}(\Omega)$ and represents the unique solution of (13).

q.e.d.

6 Existence and Regularity Theorems

At first, we shall eliminate the assumption C$_0$.

Definition 6.1. *Two bounded domains $\Omega_1, \Omega_2 \subset \mathbb{R}^n$ are $C^{2+\alpha}$-diffeomorphic, if we have a one-to-one mapping*

$$y = y(x) : \overline{\Omega}_1 \to \overline{\Omega}_2 \in C^{2+\alpha}(\overline{\Omega}_1)$$

with the inverse mapping

$$x = x(y) : \overline{\Omega}_2 \to \overline{\Omega}_1 \in C^{2+\alpha}(\overline{\Omega}_2)$$

satisfying

$$\frac{\partial(y_1, \ldots, y_n)}{\partial(x_1, \ldots, x_n)} \neq 0 \quad in \quad \overline{\Omega}_1.$$

When the set Ω is $C^{2+\alpha}$-diffeomorphic to the unit ball $B \subset \mathbb{R}^n$, we speak of a $C^{2+\alpha}$-ball.

We need the following

Theorem 6.2. (Reconstruction)

In the $C^{2+\alpha}$-ball $\Omega \subset \mathbb{R}^n$ the coefficients of the differential operator $\mathcal{L}$ fulfill the assumptions C_2 and C_4.
For all right-hand sides $f \in C^{\alpha}(\overline{\Omega})$ and all boundary values $g \in C^0(\partial\Omega)$, then there exists a solution $u = u(x)$ of the regularity class $C^{2+\alpha}(\Omega) \cap C^0(\overline{\Omega})$ for the Dirichlet problem

$$\begin{aligned} \mathcal{L}(u) &= f \quad in \quad \Omega, \\ u &= g \quad on \quad \partial\Omega. \end{aligned} \tag{1}$$

At a boundary point $\xi \in \partial\Omega$ we define the set

$$\Omega(\xi, \varrho) := \{x \in \Omega : |x - \xi| < \varrho\} \quad with \quad \varrho > 0$$

and additionally require the boundary condition

$$g(x) = 0 \quad for \ all \quad x \in \partial\Omega \cap \partial\Omega(\xi, \varrho). \tag{2}$$

In this situation we have

$$u \in C^{2+\alpha}(\overline{\Omega(\xi, r)}) \tag{3}$$

for all sufficiently small $0 < r < \varrho$.

Proof:

1. Since the set Ω is $C^{2+\alpha}$-diffeomorphic to B, there exists a $C^{2+\alpha}$-diffeomorphism

$$y = (y_1(x), \ldots, y_n(x)) \in C^{2+\alpha}(\overline{\Omega})$$

of $\overline{\Omega}$ onto $\overline{B}$ with the inverse mapping

$$x = (x_1(y), \ldots, x_n(y)) \in C^{2+\alpha}(\overline{B}).$$

We define the function $u(x) = \tilde{u}(y(x))$, $x \in \overline{\Omega}$ and deduce

$$u_{x_i} = \sum_{k=1}^n \tilde{u}_{y_k} \frac{\partial y_k}{\partial x_i},$$

$$u_{x_i x_j} = \sum_{k,l=1}^n \tilde{u}_{y_k y_l} \frac{\partial y_k}{\partial x_i} \frac{\partial y_l}{\partial x_j} + \sum_{k=1}^n \tilde{u}_{y_k} \frac{\partial^2 y_k}{\partial x_i \partial x_j}.$$

For all $y \in \overline{B}$ we obtain

$$
\begin{aligned}
\mathcal{L}(u)|_{x=x(y)} &= \Big\{ \sum_{i,j=1}^{n} a_{ij} u_{x_i x_j} + \sum_{i=1}^{n} b_i u_{x_i} + cu \Big\}\Big|_{x=x(y)} \\
&= \sum_{k,l=1}^{n} \Big(\sum_{i,j=1}^{n} a_{ij} \frac{\partial y_k}{\partial x_i} \frac{\partial y_l}{\partial x_j} \Big)\Big|_{x=x(y)} \tilde{u}_{y_k y_l} \\
&\quad + \sum_{k=1}^{n} \Big(\sum_{i,j=1}^{n} a_{ij} \frac{\partial^2 y_k}{\partial x_i \partial x_j} + \sum_{i=1}^{n} b_i \frac{\partial y_k}{\partial x_i} \Big)\Big|_{x=x(y)} \tilde{u}_{y_k} \\
&\quad + c|_{x=x(y)} \tilde{u}(y) =: \sum_{k,l=1}^{n} \tilde{a}_{kl}(y) \tilde{u}_{y_k y_l} + \sum_{k=1}^{n} \tilde{b}_k(y) \tilde{u}_{y_k} + \tilde{c}(y) \tilde{u}.
\end{aligned}
\tag{4}
$$

Because of Theorem 4.4 from Section 4 the set B satisfies the assumption C_0, and we can solve the Dirichlet problem (1) in B with the aid of Theorem 5.5 from Section 5. On account of the behavior for the coefficients in (4) with respect to the given transformations, namely

$$
\begin{aligned}
&\tilde{a}_{kl}(y), \tilde{b}_k(y) \in C^\alpha(\overline{B}), \qquad k, l = 1, \ldots, n, \\
&0 \geq \tilde{c}(y) \in C^\alpha(\overline{B}),
\end{aligned}
\tag{5}
$$

we can solve the Dirichlet problem (1) on the domain Ω as well.

2. We control the construction in the proof of Theorem 5.5 from Section 5 as follows: With the additional assumption (2) we approximate the function g uniformly on $\partial\Omega$ by a sequence $\{g_k\}_{k=1,2,\ldots}$ of polynomials. We use the mollifier

$$
\Theta(t) := \begin{cases} 0, & 0 \leq t \leq \frac{1}{2} \\ 1, & 1 \leq t < +\infty \end{cases} \in C^\infty(\mathbb{R})
$$

and consider the functions

$$
\tilde{g}_k(x) := g_k(x) \Theta\Big(\frac{|x - \xi|}{\varrho} \Big), \qquad x \in \mathbb{R}^n, \qquad k = 1, 2, \ldots
$$

We observe the uniform convergence $\tilde{g}_k(x) \to g(x)$ $(k \to \infty)$ on $\partial\Omega$ and the fact that

$$
\tilde{g}_k(x) = 0 \qquad \text{for all} \quad x \in \mathbb{R}^n \quad \text{with} \quad |x - \xi| \leq \frac{1}{2}\varrho.
\tag{6}
$$

Following the proof of Theorem 5.5 in Section 5 we obtain the solutions

$$
\begin{aligned}
u_k &\in C^{2+\alpha}(\overline{\Omega}), \\
\mathcal{L}(u_k) &= f \qquad \text{in} \quad \Omega, \\
u_k &= \tilde{g}_k \qquad \text{on} \quad \partial\Omega.
\end{aligned}
\tag{7}
$$

We utilize an interpolation of the Schauder estimates from the Theorems 5.1 and 5.4 in Section 5 (compare Section 7) and obtain the following estimate for arbitrary radii $0 < r < \frac{1}{2}\varrho$ in

$$\|u_k - u_l\|_1^{\Omega(\xi,r)} + \|u_k - u_l\|_2^{\Omega(\xi,r)} + \|u_k - u_l\|_{2,\alpha}^{\Omega(\xi,r)}$$

$$\leq \vartheta\|u_k - u_l\|_0^{\Omega(\xi,\frac{1}{2}\varrho)} \leq \vartheta\|u_k - u_l\|_0^{\Omega} \tag{8}$$

$$\leq \vartheta \sup_{x \in \partial\Omega} |\tilde{g}_k(x) - \tilde{g}_l(x)| \to 0 \ (k,l \to \infty).$$

This implies

$$u(x) := \lim_{k\to\infty} u_k(x) \in C^{2+\alpha}(\Omega) \cap C^0(\overline{\Omega}) \cap C^{2+\alpha}(\overline{\Omega(\xi,r)})$$

for all $0 < r < \frac{1}{2}\varrho$, and the function u evidently solves the boundary value problem (1).   q.e.d.

Proposition 6.3. *Let the function* $G = G(x) \in C^{2+\alpha}(\{x \in \mathbb{R}^n : |x - \xi| < \varrho\})$ *with* $\varrho > 0$ *be given, which satisfies* $G(\xi) = 0$ *and*

$$\nabla G(x) \neq 0 \quad \text{for all} \quad x \quad \text{with} \quad |x - \xi| < \varrho.$$

Then we have a $C^{2+\alpha}$*-ball*

$$D \subset \left\{x \in \mathbb{R}^n : |x - \xi| < \varrho, \ G(x) < 0\right\},$$

whose boundary fulfills

$$\partial D \cap \{x \in \mathbb{R}^n : |x - \xi| < \varrho'\} = \left\{x \in \mathbb{R}^n : |x - \xi| < \varrho', \ G(x) = 0\right\} \tag{9}$$

for a number $0 < \varrho' < \varrho$.

Proof: We leave the proof as an exercise for our readers.

Theorem 6.4. (Existence theorem for linear elliptic equations)
With the assumptions C_2, C_3, C_4 *let the differential operator* $\mathcal{L}$ *be defined on the domain* Ω. *Furthermore, we take the functions* $f \in C^{\alpha}(\overline{\Omega})$ *and* $g : \partial\Omega \to \mathbb{R} \in C^0(\partial\Omega)$ *arbitrarily. Then the Dirichlet problem*

$$\begin{aligned} \mathcal{L}(u) = f \quad &\text{in} \quad \Omega, \\ u = g \quad &\text{on} \quad \partial\Omega \end{aligned} \tag{10}$$

possesses exactly one solution in the regularity class $C^{2+\alpha}(\Omega) \cap C^0(\overline{\Omega})$.

Proof: We only have to eliminate the assumption C_0 in Theorem 5.5 from Section 5. With $f \in C^\alpha(\overline{\Omega})$, we consider the following function in the dimensions $n \geq 3$

$$v(x) := \frac{1}{(2-n)\omega_n} \int_\Omega \frac{f(y)}{|y-x|^{n-2}} \, dy, \qquad x \in \Omega. \tag{11}$$

We derive

$$v \in C^2(\Omega) \cap C^0(\overline{\Omega}),$$
$$\Delta v(x) = f(x) \qquad \text{in} \quad \Omega. \tag{12}$$

Following Section 3 in Chapter 5, we solve the boundary value problem by Perron's method

$$w \in C^2(\Omega) \cap C^0(\overline{\Omega}),$$
$$\Delta w = 0 \qquad \text{in} \quad \Omega, \tag{13}$$
$$w = -v \qquad \text{on} \quad \partial\Omega.$$

Then the function $u(x) := v(x) + w(x)$, $x \in \overline{\Omega}$ represents a solution of the boundary value problem

$$\Delta u(x) = f(x) \qquad \text{in} \quad \Omega,$$
$$u = 0 \qquad \text{on} \quad \partial\Omega. \tag{14}$$

With the aid of Theorem 6.2 we locally reconstruct the solution u in the interior and at the boundary via Proposition 6.3 as well. Then we obtain $u \in C^{2+\alpha}(\overline{\Omega})$. In this context we refer the reader to the subsequent proofs of Theorem 6.5 and Theorem 6.7. \hfill q.e.d.

Theorem 6.5. (Inner regularity)
Let the differential operator $\mathcal{L}$ be defined on the domain $\Omega \subset \mathbb{R}^n$ with the assumption C_2, and let the right-hand side $f \in C^\alpha(\Omega)$ be given.
Then a solution $u \in C^2(\Omega)$ of the differential equation

$$\mathcal{L}(u) = f \qquad \text{in} \quad \Omega \tag{15}$$

belongs to the regularity class $C^{2+\alpha}(\Omega)$.

Proof: On account of (15) the function $u \in C^2(\Omega)$ satisfies the differential equation

$$\tilde{\mathcal{L}}(u) = \tilde{f} \qquad \text{in} \quad \Omega$$

abbreviating

$$\tilde{\mathcal{L}}(u) := \sum_{i,j=1}^n a_{ij}(x) \frac{\partial^2 u}{\partial x_i \partial x_j} + \sum_{i=1}^n b_i(x) \frac{\partial u}{\partial x_i},$$
$$\tilde{f} := f - cu \in C^\alpha(\Omega).$$

Since the operator $\tilde{\mathcal{L}}$ satisfies the assumption C_4, we can reconstruct the solution with the aid of Theorem 6.2 as follows: We take $\xi \in \Omega$, choose $\varrho > 0$ sufficiently small, and consider the set

$$D := \{x \in \mathbb{R}^n : |x - \xi| < \varrho\} \subset\subset \Omega.$$

There exists a solution $v \in C^{2+\alpha}(D) \cap C^0(\overline{D})$ of the problem

$$\tilde{\mathcal{L}}(v) = \tilde{f} \quad \text{in} \quad D,$$
$$v = u \quad \text{on} \quad \partial D. \tag{16}$$

The maximum principle implies $u(x) \equiv v(x)$ in D, and consequently $u \in C^{2+\alpha}(D)$.   q.e.d.

With the assumption C_3 we comprehend the set $\partial\Omega$ as an $(n-1)$-dimensional manifold of regularity class $C^{2+\alpha}$. Therefore, we naturally define boundary functions

$$g : \partial\Omega \to \mathbb{R} \in C^{2+\alpha}(\partial\Omega).$$

We easily show the following

Proposition 6.6. *Let the function* $g : \partial\Omega \to \mathbb{R} \in C^{2+\alpha}(\partial\Omega)$ *be prescribed. For each* $\xi \in \partial\Omega$ *and each sufficiently small* $\varepsilon > 0$, *then there exists a function*

$$h = h(x_1, \ldots, x_n) \in C^{2+\alpha}(\{x \in \mathbb{R}^n : |x - \xi| \le \varepsilon\})$$

satisfying $h = g$ *on* $\partial\Omega \cap \{x \in \mathbb{R}^n : |x - \xi| \le \varepsilon\}$.

Theorem 6.7. (Boundary regularity)
With the assumptions C_2 *and* C_3, *let the differential operator* $\mathcal{L}$ *be defined on the domain* $\Omega \subset \mathbb{R}^n$. *For the boundary distribution* $g \in C^{2+\alpha}(\partial\Omega)$ *and the right-hand side* $f \in C^\alpha(\overline{\Omega})$ *let the solution* u *of the following Dirichlet problem be given:*

$$u \in C^2(\Omega) \cap C^0(\overline{\Omega}),$$
$$\mathcal{L}(u) = f \quad \text{in} \quad \Omega, \tag{17}$$
$$u = g \quad \text{on} \quad \partial\Omega.$$

Then we have $u = u(x) \in C^{2+\alpha}(\overline{\Omega})$.

Proof:

1. We choose the point $\xi = (\xi_1, \ldots, \xi_n) \in \partial\Omega$ arbitrarily. Furthermore, we consider the function $w(x) := e^{-\mu(x_1-\xi_1)^2} > 0$ in $\mathbb{R}^n$ with the parameter $\mu > 0$ to be fixed later. On account of

$$w_{x_1} = -2\mu(x_1 - \xi_1)e^{-\mu(x_1-\xi_1)^2},$$
$$w_{x_1x_1} = \{4\mu^2(x_1 - \xi_1)^2 - 2\mu\}e^{-\mu(x_1-\xi_1)^2}$$

we obtain

$$\mathcal{L}w|_{x=\xi} = -2\mu a_{11}(\xi) + c(\xi) < 0$$

taking $\mu > 0$ sufficiently large. Now we choose $\varrho > 0$ sufficiently small and observe

$$\mathcal{L}w(x) \leq 0, \quad w(x) > 0 \qquad \text{for all} \quad x \in \overline{\Omega} \text{ with } |x - \xi| \leq \varrho. \qquad (18)$$

By the product device $u(x) = w(x)v(x)$ presented in Section 1 from Chapter 6, we find that the differential operator relevant for v satisfies the assumption C_4. Therefore, we additionally require the assumption (C_4) in the sequel.

2. Due to Proposition 6.6 we can locally extend $g : \partial\Omega \to \mathbb{R}$ about the point ξ to a function

$$h \in C^{2+\alpha}(\{x \in \mathbb{R}^n : |x - \xi| \leq \varrho\}).$$

Now we choose a $C^{2+\alpha}$-ball D described in Proposition 6.3, such that

$$D \subset \left\{x \in \mathbb{R}^n : |x - \xi| \leq \varrho\right\} \cap \Omega$$

holds true. The function $v(x) := u(x) - h(x) \in C^{2+\alpha}(D) \cap C^0(\overline{D})$ satisfies

$$\mathcal{L}v(x) = \mathcal{L}u(x) - \mathcal{L}h(x) = f(x) - \mathcal{L}h(x), \qquad x \in D. \qquad (19)$$

Here the right-hand side of (19) belongs to the class $C^\alpha(\overline{D})$. Furthermore, $v(x) = 0$ for all $x \in \partial\Omega(\xi, \varrho') \cap \partial\Omega$ holds true with a sufficiently small $\varrho' > 0$. In this context we defined

$$\Omega(\xi, \varrho') := \left\{x \in \Omega : |x - \xi| < \varrho'\right\}.$$

Reconstructing the solution v on $\overline{D}$ with the aid of Theorem 6.2 as in the proof of Theorem 6.5, we obtain

$$v \in C^{2+\alpha}(\overline{\Omega(\xi, \varrho'')}) \qquad (20)$$

for a parameter $0 < \varrho'' < \varrho' < \varrho$. The point $\xi \in \partial\Omega$ being chosen arbitrarily, we finally see

$$u \in C^{2+\alpha}(\overline{\Omega}).$$

$$\text{q.e.d.}$$

Remark: Since the proof of Theorem 6.7 is of a local nature as described in Theorem 6.2, we could prove a local regularity result as well.

We now attain the goal of our theory, namely

Theorem 6.8. (Fundamental theorem for elliptic differential operators)

With the assumptions C_2 and C_3 let the differential operator $\mathcal{L}$ be defined on the domain Ω, and we require the condition:

> *The homogeneous problem $\mathcal{L}(u) = 0$ in Ω, $u = 0$ on $\partial\Omega$, $u \in C^2(\overline{\Omega})$ admits only the trivial solution $u \equiv 0$.* $\quad$ (21)

For all functions $f \in C^\alpha(\overline{\Omega})$ and $g \in C^{2+\alpha}(\partial\Omega)$ given, then the boundary value problem

$$u \in C^{2+\alpha}(\overline{\Omega}),$$
$$\mathcal{L}(u) = f \quad in \quad \Omega, \tag{22}$$
$$u = g \quad on \quad \partial\Omega$$

possesses exactly one solution.

Proof: We consider the reduced differential operator

$$\mathcal{L}_0(u) := \sum_{i,j=1}^{n} a_{ij}(x) \frac{\partial^2 u}{\partial x_i \partial x_j} + \sum_{i=1}^{n} b_i(x) \frac{\partial u}{\partial x_i}$$

and solve the following Dirichlet problem with the aid of Theorems 6.4 and 6.7:

$$u_0 \in C^{2+\alpha}(\overline{\Omega}),$$
$$\mathcal{L}_0(u_0) = 0 \quad in \quad \Omega, \tag{23}$$
$$u_0 = g \quad on \quad \partial\Omega.$$

The right-hand side $f \in C^\alpha(\overline{\Omega})$ given, we solve the problem (22) by the ansatz

$$u = u_0 + u_1, \qquad u_1 \in C_*^{2+\alpha}(\overline{\Omega}). \tag{24}$$

For the function u_1 we find the condition

$$f = \mathcal{L}(u) = \mathcal{L}(u_0 + u_1) = \mathcal{L}_0(u_0 + u_1) + c(u_0 + u_1)$$
$$= \mathcal{L}_0(u_0) + \mathcal{L}_0(u_1) + cu_0 + cu_1 = \mathcal{L}_0(u_1) + cu_0 + cu_1$$

or equivalently

$$u_1 + \mathcal{L}_0^{-1}(cu_1) = \mathcal{L}_0^{-1}(f) - \mathcal{L}_0^{-1}(cu_0) = \tilde{f} \in C_*^{2+\alpha}(\overline{\Omega}). \tag{25}$$

We consider the Banach space

$$\mathcal{B} := \left\{ u : \overline{\Omega} \to \mathbb{R} \in C^2(\overline{\Omega}) \ : \ u = 0 \text{ on } \partial\Omega \right\}$$

with the norm

$$\|u\| := \|u\|_0^\Omega + \|u\|_1^\Omega + \|u\|_2^\Omega, \qquad u \in \mathcal{B}.$$

Now we introduce the linear operator

$$K(u) := -\mathcal{L}_0^{-1}\big[c(x)u(x)\big], \qquad u \in \mathcal{B}. \tag{26}$$

The Schauder estimate yields

$$\|K(u)\| \le \|K(u)\|_{2+\alpha}^{\Omega} = \big\|\mathcal{L}_0^{-1}\big[c(x)u(x)\big]\big\|_{2+\alpha}^{\Omega} \le \vartheta\|cu\|_{\alpha}^{\Omega}$$

$$\le \vartheta\|c\|_{\alpha}^{\Omega}\|u\| = \tilde{\vartheta}\|u\|, \qquad u \in \mathcal{B}. \tag{27}$$

Consequently, the linear operator K is bounded on the Banach space $\mathcal{B}$ and even is completely continuous, due to the theorem of Arzelà-Ascoli. On account of (21) the homogeneous equation

$$u + \mathcal{L}_0^{-1}(cu) = 0, \qquad u \in \mathcal{B}, \tag{28}$$

admits only the trivial solution $u \equiv 0$. We apply the Theorem of F. Riesz from Chapter 7, Section 4 and obtain exactly one solution of the operator equation

$$u - K(u) = \tilde{f}, \qquad u \in \mathcal{B}, \tag{29}$$

for each right-hand side $\tilde{f} \in \mathcal{B}$. With the aid of Theorem 6.7 we obtain the desired function

$$u_1 \in C_*^{2+\alpha}(\overline{\Omega})$$

satisfying (25). Therefore, the function $u = u_0 + u_1$ solves the Dirichlet problem (22).   q.e.d.

7 The Schauder Estimates

For $\xi \in \mathbb{R}^n$ and $R > 0$ given, we consider the set

$$B = B(\xi, R) := \Big\{x = (x_1, \ldots, x_n) \in \mathbb{R}^n \; : \; |x - \xi| < R, \; x_n > 0\Big\}.$$

With $x = (x_1, \ldots, x_n)$ we set $x^* = (x_1, \ldots, x_{n-1}, -x_n)$. We abbreviate

$$E := \Big\{x \in \mathbb{R}^n \; : \; x_n = 0\Big\},$$

and for $n \ge 3$ we define Green's function on the half-space $x_n > 0$

$$\phi(x, y) := \frac{1}{(n-2)\omega_n}\Big(\frac{1}{|y - x|^{n-2}} - \frac{1}{|y - x^*|^{n-2}}\Big).$$

Evidently, we have $\phi(x, y) = 0$ for all $y \in E$. We now define the class of functions

$$C_*^{2+\alpha}(\overline{B}) := \Big\{u \in C^{2+\alpha}(\overline{B(\xi, R)}) \; : \; u(x) = 0 \text{ for all } x \in \partial B(\xi, R) \cap E\Big\}.$$

Proposition 7.1. *Let the function* $u \in C_*^{2+\alpha}(\overline{B})$ *be given, then we have the following identity for all* $x \in B$:

$$u(x) = \int\limits_{\substack{|y-\xi|=R \\ y_n \geq 0}} \left(\phi(x,y) \frac{\partial u}{\partial \nu}(y) - u(y) \frac{\partial \phi(x,y)}{\partial \nu} \right) d\sigma(y) - \int\limits_{\substack{|y-\xi|\leq R \\ y_n \geq 0}} \phi(x,y) \Delta u(y)\, dy.$$

Proof: At first, the Gaussian integral theorem yields

$$\int\limits_{\substack{B(\xi,R) \\ |y-x|>\varepsilon}} \Big(\phi(x,y)\Delta u(y) - u(y)\Delta_y \phi(x,y) \Big)\, d\sigma(y)$$

$$= \int\limits_{\partial B} \left(\phi(x,y) \frac{\partial u}{\partial \nu} - u(y) \frac{\partial \phi}{\partial \nu} \right) d\sigma(y)$$

$$+ \int\limits_{|y-x|=\varepsilon} \left(\phi(x,y) \frac{\partial u}{\partial \nu} - u(y) \frac{\partial \phi}{\partial \nu} \right) d\sigma(y)$$

for all $\varepsilon > 0$ and $x \in B$. We note that $u \in C_*^{2+\alpha}(\overline{B})$ and consider the limit process $\varepsilon \downarrow 0$

$$u(x) = \int\limits_{\substack{|y-\xi|=R \\ y_n \geq 0}} \left(\phi \frac{\partial u}{\partial \nu} - u \frac{\partial \phi}{\partial \nu} \right) d\sigma(y) - \int\limits_{\substack{|y-\xi|<R \\ y_n>0}} \phi(x,y)\Delta u(y)\, dy.$$

q.e.d.

Remarks:

1. It is important that u vanishes on a plane portion of the boundary. A contraction with $\xi \in E$ and $R \downarrow 0$ transfers this part into itself. Very important in the sequel are half-balls.
2. With Green's function for the half-ball $B = B(\xi, R)$ and $\xi \in E$ at our disposal, we could directly derive potential-theoretic estimates extending to the boundary $\partial B \cap E$ as in Section 4.
3. In order to construct Green's function for the half-ball, we have to solve the Dirichlet problem for the ball by Poisson's integral formula. However, we do not yet know of the latter solution whether their derivatives are continuous in $\overline{B}$.

Proposition 7.2. *The function* $u \in C_*^{2+\alpha}(\overline{B})$ *may satisfy Poisson's differential equation* $\Delta u = f$ *in* B *with* $f \in C^\alpha(\overline{B})$. *Then, we have the following equations for* $x \in B$ *and* $i,j = 1,\ldots,n$:

$$u_{x_i x_j}(x) = \int\limits_{\substack{|y-\xi|=R \\ y_n \geq 0}} \left(\phi_{x_i x_j}(x,y) \frac{\partial u}{\partial \nu}(y) - u(y) \frac{\partial \phi_{x_i x_j}(x,y)}{\partial \nu} \right) d\sigma(y)$$

$$-f(x)\psi_{x_i x_j}(x,R) - \int\limits_{B} \phi_{x_i x_j}(x,y)\big(f(y) - f(x)\big)\, dy.$$

Here, we define

$$\psi(x,R) := \int\limits_{\substack{|y-\xi|=R \\ y_n \geq 0}} \left(\phi(x,y)\frac{\partial(\frac{1}{2}y_n^2)}{\partial \nu} - \frac{1}{2}y_n^2\frac{\partial \phi}{\partial \nu}(x,y) \right) d\sigma(y) - \frac{1}{2}x_n^2.$$

Proof: From the integral representation of Proposition 7.1 we can easily differentiate the surface integral twice. The questionable volume integral

$$F(x) = \int\limits_{B} \phi(x,y)f(y)\,dy, \qquad x \in B,$$

can be directly differentiated only once:

$$F_{x_i}(x) = \int\limits_{B} \phi_{x_i}(x,y)f(y)\,dy, \quad x \in B \quad \text{for} \quad i = 1,\ldots,n.$$

We apply Proposition 7.1 and insert the function $u(x) := \frac{1}{2}x_n^2$ with $u(x) = 0$ for $x \in E$ and $\Delta u(x) = 1$ in $\mathbb{R}^n$. This implies

$$\int\limits_{B} \phi(x,y)dy = \int\limits_{\substack{|y-\xi|=R \\ y_n \geq 0}} \left(\phi(x,y)\frac{\partial(\frac{1}{2}y_n^2)}{\partial \nu} - \frac{1}{2}y_n^2\frac{\partial \phi}{\partial \nu}(x,y) \right) d\sigma(y) - \frac{1}{2}x_n^2$$

$$= \psi(x,R), \qquad x \in B.$$

Therefore, the function

$$\Phi_i(x) := \int\limits_{B} \phi_{x_i}(x,y)\,dy = \psi_{x_i}(x,R), \qquad x \in B,$$

belongs to the class $C^1(B)$, and Proposition 4.2 from Section 4 by E. Hopf yields

$$F_{x_ix_j}(x) = \int\limits_{B} \phi_{x_ix_j}(x,y)\big(f(y) - f(x)\big)\,dy + f(x)\psi_{x_ix_j}(x,R), \qquad x \in B$$

for $i,j = 1,\ldots,n$.   q.e.d.

Proposition 7.3. *The function* $u = u(x) \in C_*^{2+\alpha}(\overline{B(\xi,1)})$ *may satisfy the partial differential equation* $\Delta u(x) = f(x)$ *in* $B(\xi,1)$ *with the right-hand side* $f \in C^\alpha(\overline{B})$. *Then, we have a constant* $C = C(n,\alpha)$, *such that the inequality*

$$\|u\|_2^{B(\xi,\frac{1}{2})} + \|u\|_{2,\alpha}^{B(\xi,\frac{1}{2})} \leq C(\alpha,n)\left(\|u\|_0^{B(\xi,1)} + \|u\|_1^{B(\xi,1)} + \|f\|_0^{B(\xi,1)} + \|f\|_{0,\alpha}^{B(\xi,1)} \right)$$

holds true.

Proof: The indices $i, j \in \{1, \ldots, n\}$ being fixed, we consider the functions

$$g(x) := \int\limits_{\substack{|y-\xi|=1 \\ y_n \geq 0}} \left(\phi_{x_i x_j}(x,y) \frac{\partial u}{\partial \nu}(y) - u(y) \frac{\partial \phi_{x_i x_j}}{\partial \nu}(x,y) \right) d\sigma(y),$$

$$h(x) := f(x)\psi_{x_i x_j}(x,1),$$

$$F(x) := \int\limits_{B(\xi,1)} \phi_{x_i x_j}(x,y)\big(f(y) - f(x)\big)\, dy, \qquad x \in B(\xi,1).$$

Proposition 7.2 yields

$$u_{x_i x_j}(x) = g(x) - h(x) - F(x), \qquad x \in B(\xi,1). \tag{1}$$

At first, we deduce

$$\|g\|_0^{B(\xi,\frac{1}{2})} + \|g\|_{0,\alpha}^{B(\xi,\frac{1}{2})} \leq C_1(\alpha,n)\left(\|u\|_0^{B(\xi,1)} + \|u\|_1^{B(\xi,1)} \right). \tag{2}$$

Furthermore, we have

$$\|h\|_0^{B(\xi,\frac{1}{2})} + \|h\|_{0,\alpha}^{B(\xi,\frac{1}{2})} = \|h\|_\alpha^{B(\xi,\frac{1}{2})} = \|f \cdot \psi_{x_i x_j}(\cdot,1)\|_\alpha^{B(\xi,\frac{1}{2})}$$

$$\leq \|\psi_{x_i x_j}(\cdot,1)\|_\alpha^{B(\xi,\frac{1}{2})} \|f\|_\alpha^{B(\xi,\frac{1}{2})} \leq C_2(\alpha,n)\left(\|f\|_0^{B(\xi,1)} + \|f\|_{0,\alpha}^{B(\xi,1)} \right). \tag{3}$$

We utilize Proposition 4.3 from Section 4 in order to estimate $F(x)$ and obtain

$$|F(x)| \leq C_3(\alpha,n)\|f\|_{0,\alpha}^{B(\xi,1)}, \qquad x \in B(\xi,1). \tag{4}$$

Furthermore, the estimate of E. Hopf yields

$$\left| F(x'') - F(x') + \big(f(x'') - f(x')\big) \int\limits_{\substack{y \in B(\xi,1) \\ |y-x'| \geq 3|x''-x'|}} \phi_{x_i x_j}(x',y)\, dy \right|$$

$$\leq C_4(\alpha,n)\|f\|_{0,\alpha}^{B(\xi,1)}|x''-x'|^\alpha \qquad \text{for all} \quad x', x'' \in B(\xi,1). \tag{5}$$

With the aid of the Gaussian integral theorem we show the uniform boundedness of the Cauchy principal values:

$$\left| \int\limits_{\substack{y \in B(\xi,1) \\ |y-x'| \geq \delta}} \phi_{x_i x_j}(x',y)\, dy \right| \leq C_5 \qquad \text{for all} \quad x' \in B\!\left(\xi, \frac{1}{2}\right) \quad \text{and} \quad \delta > 0. \tag{6}$$

Together with (5) we see

$$|F(x'') - F(x')| \leq \{C_4(\alpha,n) + C_5\}\|f\|_{0,\alpha}^{B(\xi,1)}|x''-x'|^\alpha$$

$$\text{for all} \qquad x \in B\!\left(\xi, \frac{1}{2}\right) \quad \text{and} \quad x'' \in B(\xi,1). \tag{7}$$

The inequalities (4) and (7) now imply

$$\|F\|_0^{B(\xi,\frac{1}{2})} + \|F\|_{0,\alpha}^{B(\xi,\frac{1}{2})} \leq \left\{ C_3(\alpha,n) + C_4(\alpha,n) + C_5 \right\} \|f\|_{0,\alpha}^{B(\xi,1)}. \qquad (8)$$

From (1)-(3) and (8) we obtain the inequality

$$\|u_{x_i x_j}\|_0^{B(\xi,\frac{1}{2})} + \|u_{x_i x_j}\|_{0,\alpha}^{B(\xi,\frac{1}{2})}$$

$$\leq \tilde{C}(\alpha,n) \left\{ \|u\|_0^{B(\xi,1)} + \|u\|_1^{B(\xi,1)} + \|f\|_0^{B(\xi,1)} + \|f\|_{0,\alpha}^{B(\xi,1)} \right\}$$

for $i,j = 1,\ldots,n$, which gives us the desired estimate.   q.e.d.

By means of a scaling argument we now show the following

Theorem 7.4. *Let $u = u(x) \in C_*^{2+\alpha}(\overline{B(\xi,R)})$ with $\xi \in \mathbb{R}^n$ and $R > 0$ be a solution of Poisson's equation $\Delta u(x) = f(x)$, $x \in B(\xi,R)$.*
Then, we have the estimates

$$\|u\|_2^{B(\xi,\frac{1}{2}R)} \leq C(\alpha,n) \left(\frac{\|u\|_0^{B(\xi,R)}}{R^2} + \frac{\|u\|_1^{B(\xi,R)}}{R} + \frac{\|f\|_0^{B(\xi,R)}}{1} + R^\alpha \|f\|_{0,\alpha}^{B(\xi,R)} \right)$$

and

$$\|u\|_{2,\alpha}^{B(\xi,\frac{1}{2}R)} \leq C(\alpha,n) \left(\frac{\|u\|_0^{B(\xi,R)}}{R^{2+\alpha}} + \frac{\|u\|_1^{B(\xi,R)}}{R^{1+\alpha}} + \frac{\|f\|_0^{B(\xi,R)}}{R^\alpha} + \|f\|_{0,\alpha}^{B(\xi,R)} \right).$$

Proof: We apply Proposition 7.3 to the function $v(y) := u(Ry)$, $y \in B(\frac{\xi}{R},1)$ of the class $C_*^{2+\alpha}\left(\overline{B(\frac{\xi}{R},1)}\right)$ and obtain

$$\|v\|_2^{B(\frac{\xi}{R},\frac{1}{2})} + \|v\|_{2,\alpha}^{B(\frac{\xi}{R},\frac{1}{2})}$$

$$\leq C(\alpha,n) \left\{ \|v\|_0^{B(\frac{\xi}{R},1)} + \|v\|_1^{B(\frac{\xi}{R},1)} + \|g\|_0^{B(\frac{\xi}{R},1)} + \|g\|_{0,\alpha}^{B(\frac{\xi}{R},1)} \right\}. \qquad (9)$$

Furthermore, we calculate

$$v_{y_i}(y) = R\, u_{x_i}(Ry), \qquad i = 1,\ldots,n,$$

$$v_{y_i y_k}(y) = R^2\, u_{x_i x_k}(Ry), \qquad i,k = 1,\ldots,n,$$

$$\Delta v(y) = R^2\, f(Ry) =: g(y), \qquad y \in B\left(\frac{\xi}{R},1\right)$$

and note that

$$\|v\|_l^{B(\frac{\xi}{R},1)} = R^l \|u\|_l^{B(\xi,R)}, \qquad \|v\|_l^{B(\frac{\xi}{R},\frac{1}{2})} = R^l \|u\|_l^{B(\xi,\frac{R}{2})}, \qquad l = 0,1,2.$$

Finally, we use the identities

$$\|v\|_{2,\alpha}^{B(\frac{\xi}{R},\frac{1}{2})} = R^{2+\alpha}\|u\|_{2,\alpha}^{B(\xi,\frac{R}{2})},$$

$$\|g\|_{0}^{B(\frac{\xi}{R},1)} = R^{2}\|f\|_{0}^{B(\xi,R)},$$

$$\|g\|_{0,\alpha}^{B(\frac{\xi}{R},1)} = R^{2+\alpha}\|f\|_{0,\alpha}^{B(\xi,R)}.$$

From (9) we infer the inequality

$$R^2\|u\|_2^{B(\xi,\frac{R}{2})} + R^{2+\alpha}\|u\|_{2,\alpha}^{B(\xi,\frac{R}{2})}$$

$$\leq C(\alpha,n)\Big\{\|u\|_0^{B(\xi,R)} + R\|u\|_1^{B(\xi,R)} + R^2\|f\|_0^{B(\xi,R)} + R^{2+\alpha}\|f\|_{0,\alpha}^{B(\xi,R)}\Big\}.$$

This relation implies the estimates stated above.   q.e.d.

We now present the transition to elliptic differential operators with constant coefficients and prove the preparatory

Proposition 7.5. *Let* $A = (a_{ij})_{i,j=1,\ldots,n}$ *be a real, symmetric, positive-definite matrix satisfying*

$$m^2\sum_{i=1}^n \xi_i^2 \leq \sum_{i,j=1}^n a_{ij}\xi_i\xi_j \leq M^2\sum_{i=1}^n \xi_i^2 \qquad \text{for all} \quad \xi = (\xi_1,\ldots,\xi_n) \in \mathbb{R}^n$$

with the constants $0 < m \leq M < +\infty$.
Then we have a real matrix $T = (t_{ij})_{i,j=1,\ldots,n}$ *with the entries* $t_{nj} = 0$ *for* $j = 1,\ldots,n-1$ *and* $t_{nn} > 0$, *such that*

$$T \circ A \circ T^* = E$$

holds true. Furthermore, the following dilation estimates are valid:

$$M^{-1}|x| \leq |Tx| \leq m^{-1}|x|, \qquad x \in \mathbb{R}^n;$$

$$m|y| \leq |T^{-1}y| \leq M|y|, \qquad y \in \mathbb{R}^n.$$

Proof: Since A is a real, symmetric matrix, there exists an orthogonal matrix B with $B \circ B^* = E = B^* \circ B$, such that

$$B \circ A \circ B^* = \Lambda =: \mathrm{diag}(\lambda_1,\ldots,\lambda_n)$$

is transformed into a diagonal matrix with the eigenvalues $\lambda_i \in [m^2, M^2]$ for $i = 1,\ldots,n$. We multiply this equation from the left- and the right-hand side by the matrix

$$\Lambda^{-1/2} := \mathrm{diag}(\lambda_1^{-1/2},\ldots,\lambda_n^{-1/2}).$$

With $C := \Lambda^{-1/2} \circ B$ we obtain the identity

$$E = \Lambda^{-1/2} \circ B \circ A \circ B^* \circ (\Lambda^{-1/2})^* = C \circ A \circ C^*.$$

The multiplication of this equation by an arbitrary orthogonal matrix D yields the following identity with $T := D \circ C$:

$$E = D \circ C \circ A \circ C^* \circ D^* = T \circ A \circ T^*.$$

We now choose D in such a way that the conditions $t_{nj} = 0$ $(j = 1, \ldots, n-1)$ and $t_{nn} > 0$ are fulfilled. We remark the relation

$$T = D \circ \Lambda^{-1/2} \circ B$$

with the orthogonal matrices B and D, and the diagonal matrix $\Lambda^{-1/2}$ with the elements $\lambda_i^{-1/2} \in [M^{-1}, m^{-1}]$ for $i = 1, \ldots, n$. This representation gives us the estimate

$$M^{-1}|x| \le |Tx| \le m^{-1}|x|, \qquad x \in \mathbb{R}^n.$$

Setting $x = T^{-1}y$ we obtain the second dilation estimate

$$M^{-1}|T^{-1}y| \le |y| \le m^{-1}|T^{-1}y|, \qquad y \in \mathbb{R}^n.$$

q.e.d.

Theorem 7.6. *The real, symmetric matrix $A = (a_{ij})_{i,j=1,\ldots,n}$ may satisfy the condition*

$$m^2 \sum_{i=1}^{n} \xi_i^2 \le \sum_{i,j=1}^{n} a_{ij}\xi_i\xi_j \le M^2 \sum_{i=1}^{n} \xi_i^2, \qquad \xi \in \mathbb{R}^n,$$

with the constants $0 < m \le M < +\infty$. Let $u = u(x) \in C_*^{2+\alpha}(\overline{B(\xi, R)})$ solve the partial differential equation

$$L(u)\Big|_x := \sum_{i,j=1}^{n} a_{ij} \frac{\partial^2 u}{\partial x_i \partial x_j}(x) = f(x), \qquad x \in B(\xi, R).$$

Then we have a constant $C = C(\alpha, n, m, M) \in (0, +\infty)$, such that the relations

$$\|u\|_2^{B(\xi, \frac{m}{M}\frac{R}{2})} \le C\left(\|f\|_0^{B(\xi, R)} + R^\alpha \|f\|_{0,\alpha}^{B(\xi, R)} + \frac{\|u\|_0^{B(\xi, R)}}{R^2} + \frac{\|u\|_1^{B(\xi, R)}}{R} \right)$$

and

$$\|u\|_{2,\alpha}^{B(\xi, \frac{m}{M}\frac{R}{2})} \le C\left(\frac{\|f\|_0^{B(\xi, R)}}{R^\alpha} + \|f\|_{0,\alpha}^{B(\xi, R)} + \frac{\|u\|_0^{B(\xi, R)}}{R^{2+\alpha}} + \frac{\|u\|_1^{B(\xi, R)}}{R^{1+\alpha}} \right)$$

hold true.

Proof: The matrix A being given, we use the transformation $y = Tx$ according to Proposition 7.5. On account of $y_n = t_{nn}x_n$ we have

$$T : \begin{array}{l} \{x \in \mathbb{R}^n : x_n = 0\} \leftrightarrow \{y \in \mathbb{R}^n : y_n = 0\}, \\ \{x \in \mathbb{R}^n : x_n > 0\} \leftrightarrow \{y \in \mathbb{R}^n : y_n > 0\}. \end{array}$$

Furthermore, the following inclusions are valid:

$$T\left(B\left(\xi, \frac{m}{2M}R\right)\right) \subset B\left(T\xi, \frac{R}{2M}\right), \tag{10}$$

$$B\left(T\xi, \frac{R}{M}\right) \subset T(B(\xi, R)). \tag{11}$$

Starting with $y \in T(B(\xi, \frac{m}{2M}R))$, we see

$$|T^{-1}(y - T\xi)| = |T^{-1}y - \xi| < \frac{m}{2M}R.$$

Proposition 7.5 then implies

$$m|y - T\xi| \leq |T^{-1}(y - T\xi)| < \frac{m}{2M}R \quad \text{or equivalently} \quad |y - T\xi| < \frac{R}{2M}.$$

This means $y \in B(T\xi, \frac{R}{2M})$ and (10) is proved.
Starting with $y \in B(T\xi, \frac{R}{M})$, which means $|y - T\xi| < \frac{R}{M}$, our Proposition 7.5 yields the inequality

$$|T^{-1}y - \xi| = |T^{-1}(y - T\xi)| \leq M|y - T\xi| < R.$$

Therefore, $T^{-1}y \in B(\xi, R)$ and $y \in T(B(\xi, R))$ hold true. Now (11) is proved as well.

We consider the function $v(y) := u(T^{-1}y)$ of the class $C_*^{2+\alpha}(T(\overline{B(\xi, R)}))$ and consequently $u(x) = v(Tx)$ of the class $C_*^{2+\alpha}(\overline{B(\xi, R)})$. Noting that

$$u_{x_i} = \sum_{k=1}^{n} v_{y_k} t_{ki}, \quad u_{x_i x_j} = \sum_{k,l=1}^{n} v_{y_k y_l} t_{ki} t_{lj} \quad \text{for} \quad i, j = 1, \ldots, n$$

holds true, we deduce

$$L(u)\Big|_x = \sum_{i,j=1}^{n} a_{ij} \frac{\partial^2 u}{\partial x_i \partial x_j}(x) = \sum_{k,l=1}^{n} \left(\sum_{i,j=1}^{n} a_{ij} t_{ki} t_{lj} \right) v_{y_k y_l}\Big|_{Tx}$$

$$= \sum_{k,l=1}^{n} \delta_{kl} v_{y_k y_l}\Big|_{Tx} = \sum_{k=1}^{n} v_{y_k y_k}\Big|_{Tx} = \Delta v(Tx), \quad x \in B(\xi, R).$$

Consequently, we obtain

$$\Delta v(y) = g(y), \quad y \in T(B(\xi, R)) \qquad \text{with} \quad g(y) := f(T^{-1}y). \tag{12}$$

Due to the formula (11), we can apply Theorem 7.4 to the function v in the ball $B(T\xi, \frac{R}{M})$: There exists a constant $\tilde{C} = \tilde{C}(\alpha, n)$, such that

$$\|v\|_2^{B(T\xi, \frac{R}{2M})} \le \tilde{C}\Big(\frac{M^2}{R^2}\|v\|_0^{B(T\xi, \frac{R}{M})} + \frac{M}{R}\|v\|_1^{B(T\xi, \frac{R}{M})}$$

$$+\|g\|_0^{B(T\xi, \frac{R}{M})} + \frac{R^\alpha}{M^\alpha}\|g\|_{0,\alpha}^{B(T\xi, \frac{R}{M})}\Big)$$

holds true. We note (10) and deduce

$$\|u\|_2^{B(\xi, \frac{m}{2M}R)} \le \mu(m, M)\|v\|_2^{B(T\xi, \frac{R}{2M})}.$$

Finally, we obtain

$$\|u\|_2^{B(\xi, \frac{m}{2M}R)} \le \mu(m, M)\tilde{C}\Big\{\frac{M^2}{R^2}\|u\|_0^{B(\xi, R)} + \frac{M}{R}\mu_1(m, M)\|u\|_1^{B(\xi, R)}$$

$$+\|f\|_0^{B(\xi, R)} + \frac{R^\alpha}{M^\alpha}\mu_2(m, M)\|f\|_{0,\alpha}^{B(\xi, R)}\Big\}$$

$$\le C(\alpha, n, m, M)\Big\{\frac{\|u\|_0^{B(\xi, R)}}{R^2} + \frac{\|u\|_1^{B(\xi, R)}}{R}$$

$$+\|f\|_0^{B(\xi, R)} + R^\alpha\|f\|_{0,\alpha}^{B(\xi, R)}\Big\},$$

taking (11) into account. Analogously, we estimate the quantity $\|u\|_{2,\alpha}^{B(\xi, \frac{m}{2M}R)}$. q.e.d.

For the functions $u \in C_*^{2+\alpha}(\overline{B(\xi, R)})$ we introduce the following *weighted norms*, abbreviating $d(x) := R - |x - \xi|$:

$A_0 := \sup\limits_{x \in B} |u(x)|,$

$A_1 := \sup\limits_{x \in B} \Big\{ d(x) \sum\limits_{i=1}^n |u_{x_i}(x)| \Big\},$

$A_2 := \sup\limits_{x \in B} \Big\{ d(x)^2 \sum\limits_{i,j=1}^n |u_{x_i x_j}(x)| \Big\},$

$A_{2,\alpha} := \sup\limits_{\substack{x', x'' \in B \\ x' \ne x''}} \Big\{ \big(\min[d(x'), d(x'')] \big)^{2+\alpha} \sum\limits_{i,j=1}^n \frac{|u_{x_i x_j}(x') - u_{x_i x_j}(x'')|}{|x' - x''|^\alpha} \Big\}.$

$$\tag{13}$$

Proposition 7.7. (Norm-interpolation)
Given the functions $u = u(x) \in C_*^{2+\alpha}(\overline{B(\xi, R)})$, we have the following estimate:

$$A_1 \leq \frac{2n}{\kappa} A_0 + \frac{n\kappa}{(1-\kappa)^2} A_2 \qquad \text{for all} \quad \kappa \in (0, 1). \tag{14}$$

Proof: We assume $A_1 > 0$ and choose a point $x' = (x_1', \ldots, x_n') \in \overline{B}$ satisfying

$$A_1 = d(x') \sum_{i=1}^{n} |u_{x_i}(x')| \qquad \text{and} \qquad d(x') > 0.$$

For an arbitrary index $j \in \{1, \ldots, n\}$ we define $x'' = (x_1'', \ldots, x_n'')$ by $x_i'' := x_i'$ for $i \neq j$ and $x_j'' := x_j' + \kappa d(x')$, with an arbitrary $\kappa \in (0, 1)$. We remark that $x'' \in \overline{B}$ holds true. The mean value theorem of differential calculus gives us a value $\tilde{\kappa} \in (0, \kappa)$ with the adjoint point $\tilde{x} = (\tilde{x}_1, \ldots, \tilde{x}_n)$ satisfying $\tilde{x}_i = x_i'$, $i \neq j$ and $\tilde{x}_j = x_j' + \tilde{\kappa} d(x')$, such that the relation

$$u_{x_j}(\tilde{x}) = \frac{u(x'') - u(x')}{\kappa d(x')}$$

holds true. This implies

$$|u_{x_j}(\tilde{x})| \leq \frac{2A_0}{\kappa d(x')}. \tag{15}$$

Furthermore, we calculate

$$u_{x_j}(\tilde{x}) - u_{x_j}(x') = \int_{x_j'}^{\tilde{x}_j} u_{x_j x_j}(x_1', \ldots, x_{j-1}', t, x_{j+1}', \ldots, x_n') \, dt.$$

For $x = (x_1', \ldots, x_{j-1}', t, x_{j+1}', \ldots, x_n')$ and $x_j' \leq t \leq \tilde{x}_j$ we infer

$$d(x) = R - |x - \xi| \geq R - |x' - \xi| - |x - x'| \geq d(x')(1 - \kappa)$$

and consequently

$$|u_{x_j x_j}(x)| \leq \frac{A_2}{d(x)^2} \leq \frac{A_2}{(1-\kappa)^2 d(x')^2}.$$

We then obtain

$$|u_{x_j}(\tilde{x}) - u_{x_j}(x')| \leq \frac{A_2 \kappa d(x')}{(1-\kappa)^2 (d(x'))^2} = \frac{\kappa A_2}{(1-\kappa)^2 d(x')}. \tag{16}$$

The relations (15) and (16) imply

$$|u_{x_j}(x')| \leq |u_{x_j}(\tilde{x}) - u_{x_j}(x')| + |u_{x_j}(\tilde{x})| \leq \frac{\kappa A_2}{(1-\kappa)^2 d(x')} + \frac{2A_0}{\kappa d(x')}$$

and
$$d(x')|u_{x_j}(x')| \le \frac{\kappa A_2}{(1-\kappa)^2} + \frac{2A_0}{\kappa} \quad \text{for} \quad j = 1, \ldots, n.$$

We summarize and get
$$A_1 = d(x') \sum_{i=1}^{n} |u_{x_j}(x')| \le \frac{2nA_0}{\kappa} + \frac{n\kappa A_2}{(1-\kappa)^2},$$

with the arbitrary parameter $\kappa \in (0,1)$.   q.e.d.

We now consider general elliptic differential operators

$$\mathcal{L}(u) := \sum_{i,j=1}^{n} a_{ij}(x)\frac{\partial^2 u}{\partial x_i \partial x_j} + \sum_{i=1}^{n} b_i(x)\frac{\partial u}{\partial x_i} + c(x)u, \quad x \in \overline{B(\xi,R)}, \quad (17)$$

and require the following conditions on the coefficients:

Assumption D: For $i, j = 1, \ldots, n$ let the coefficients
$$a_{ij}(x), b_i(x), c(x) \in C^\alpha(\overline{B(\xi,R)})$$

with the bound P be given:

$$\sum_{i,j=1}^{n} \|a_{ij}\|_\alpha^{B(\xi,R)} + \sum_{i=1}^{n} \|b_i\|_\alpha^{B(\xi,R)} + \|c\|_\alpha^{B(\xi,R)} + R \le P.$$

With the ellipticity constants $0 < m \le M < +\infty$, we have the following inequalities for all $x \in B(\xi,R)$ and $\lambda \in \mathbb{R}^n$:

$$m^2 \sum_{i=1}^{n} \lambda_i^2 \le \sum_{i,j=1}^{n} a_{ij}(x)\lambda_i\lambda_j \le M^2 \sum_{i=1}^{n} \lambda_i^2.$$

Proposition 7.8. *Under the assumption D, let the function $u \in C_*^{2+\alpha}(\overline{B})$ satisfy the differential equation*

$$\mathcal{L}(u) = f \quad \text{in } B \quad \text{with} \quad f \in C^\alpha(\overline{B}).$$

Then we have the following estimates for each point $\tilde{x} \in B$ and each number $\kappa \in (0, \frac{1}{2})$:

$$\|u\|_2^{B(\tilde{x}, \frac{m}{2M}\kappa d(\tilde{x}))} \le \frac{C}{d(\tilde{x})^2}\left\{ \|f\|_0^B + \kappa^\alpha\|f\|_{0,\alpha}^B + \frac{A_0}{\kappa^2} + \frac{A_1}{\kappa} + \kappa^\alpha A_2 + \kappa^{2\alpha} A_{2,\alpha} \right\}$$

and

$$\|u\|_{2,\alpha}^{B(\tilde{x}, \frac{m}{2M}\kappa d(\tilde{x}))} \le \frac{C}{d(\tilde{x})^{2+\alpha}}\left\{ \frac{\|f\|_0^B}{\kappa^\alpha} + \|f\|_{0,\alpha}^B + \frac{A_0}{\kappa^{2+\alpha}} + \frac{A_1}{\kappa^{1+\alpha}} + A_2 + \kappa^\alpha A_{2,\alpha} \right\}$$

with a constant $C = C(\alpha, n, m, M, P) \in (0, +\infty)$.

Proof: We show this proposition by a method usually called *freezing of coefficients.* For each $x \in B(\tilde{x}, \kappa d(\tilde{x}))$ we have

$$d(x) = R - |x - \xi| \geq R - |\tilde{x} - \xi| - |x - \tilde{x}| \geq (1 - \kappa)d(\tilde{x}) \geq \frac{1}{2}d(\tilde{x}).$$

Then we obtain

$$A_0 = \sup_{x \in B} |u(x)| \geq \sup_{x \in B(\tilde{x}, \kappa d(\tilde{x}))} |u(x)| = \|u\|_0^{B(\tilde{x}, \kappa d(\tilde{x}))} \tag{18}$$

and

$$\frac{2A_1}{d(\tilde{x})} = \sup_{x \in B} \left\{ \frac{2d(x)}{d(\tilde{x})} \sum_{i=1}^{n} |u_{x_i}(x)| \right\}$$

$$\geq \sup_{x \in B(\tilde{x}, \kappa d(\tilde{x}))} \left\{ \sum_{i=1}^{n} |u_{x_i}(x)| \right\} = \|u\|_1^{B(\tilde{x}, \kappa d(\tilde{x}))}. \tag{19}$$

Furthermore, we note that

$$\frac{4A_2}{d(\tilde{x})^2} \geq \|u\|_2^{B(\tilde{x}, \kappa d(\tilde{x}))} \tag{20}$$

and

$$\frac{2^{2+\alpha} A_{2,\alpha}}{d(\tilde{x})^{2+\alpha}} \geq \|u\|_{2,\alpha}^{B(\tilde{x}, \kappa d(\tilde{x}))}. \tag{21}$$

Since u satifies the differential equation $\mathcal{L}(u) = f$, we infer

$$\tilde{\mathcal{L}}(u) := \sum_{i,j=1}^{n} a_{ij}(\tilde{x}) \frac{\partial^2 u}{\partial x_j \partial x_j}(x) = g(x), \quad x \in B(\tilde{x}, \kappa d(\tilde{x})), \tag{22}$$

with the right-hand side

$$g(x) := f(x) - \left\{ \sum_{i,j=1}^{n} \left(a_{ij}(x) - a_{ij}(\tilde{x}) \right) \frac{\partial^2 u}{\partial x_i \partial x_j} + \sum_{i=1}^{n} b_i(x) \frac{\partial u}{\partial x_i} + c(x)u \right\}. \tag{23}$$

To the p.d.e. (22) we now apply Theorem 2 with $\xi = \tilde{x}$, $R = \kappa d(\tilde{x})$. Then we obtain

$$\|u\|_2^{B(\tilde{x}, \frac{m}{2M} \kappa d(\tilde{x}))} \leq \tilde{C} \left(\|g\|_0^{B(\tilde{x}, \kappa d(\tilde{x}))} + \|g\|_{0,\alpha}^{B(\tilde{x}, \kappa d(\tilde{x}))} \kappa^{\alpha} d(\tilde{x})^{\alpha} \right.$$

$$\left. + \frac{A_0}{\kappa^2 d(\tilde{x})^2} + \frac{2A_1}{\kappa d(\tilde{x})^2} \right) \tag{24}$$

and

$$\|u\|_{2,\alpha}^{B(\tilde{x}, \frac{m}{2M} \kappa d(\tilde{x}))} \leq \tilde{C} \left(\frac{\|g\|_0^{B(\tilde{x}, \kappa d(\tilde{x}))}}{\kappa^{\alpha} d(\tilde{x})^{\alpha}} + \|g\|_{0,\alpha}^{B(\tilde{x}, \kappa d(\tilde{x}))} \right.$$

$$\left. + \frac{A_0}{\kappa^{2+\alpha} d(\tilde{x})^{2+\alpha}} + \frac{2A_1}{\kappa^{1+\alpha} d(\tilde{x})^{2+\alpha}} \right) \tag{25}$$

with the constant $\tilde{C} = \tilde{C}(\alpha, n, m, M) \in (0, +\infty)$.

The quantity $\|g\|_0^{B(\tilde{x}, \kappa d(\tilde{x}))}$ is estimated as follows: Taking $x \in B(\tilde{x}, \kappa d(\tilde{x}))$ we have

$$|g(x)| \le \|f\|_0^{B(\tilde{x},\kappa d(\tilde{x}))} + \sum_{i,j=1}^{n} |a_{ij}(x) - a_{ij}(\tilde{x})| \left| \frac{\partial^2 u}{\partial x_i \partial x_j}(x) \right|$$

$$+ \sum_{i=1}^{n} |b_i(x)| \left| \frac{\partial u}{\partial x_i}(x) \right| + |c(x)| \, |u(x)|$$

and consequently

$$|g(x)| \le \|f\|_0^{B(\tilde{x},\kappa d(\tilde{x}))} + \|u\|_2^{B(\tilde{x},\kappa d(\tilde{x}))} \kappa^\alpha d(\tilde{x})^\alpha \sum_{i,j=1}^{n} \|a_{ij}\|_{0,\alpha}^{B(\tilde{x},\kappa d(\tilde{x}))}$$

$$+ \|u\|_1^{B(\tilde{x},\kappa d(\tilde{x}))} \sum_{i=1}^{n} \|b_i\|_0^{B(\tilde{x},\kappa d(\tilde{x}))} + \|u\|_0^{B(\tilde{x},\kappa d(\tilde{x}))} \|c\|_0^{B(\tilde{x},\kappa d(\tilde{x}))}.$$

Therefore, we find

$$\|g\|_0^{B(\tilde{x},\kappa d(\tilde{x}))} \le \|f\|_0^B + P\left(A_0 + \frac{2A_1}{d(\tilde{x})} + \frac{4\kappa^\alpha d(\tilde{x})^\alpha}{d(\tilde{x})^2} A_2 \right)$$

$$\le \|f\|_0^B + \frac{k_0(P)}{d(\tilde{x})^2} (A_0 + A_1 + \kappa^\alpha A_2), \tag{26}$$

with a constant $k_0 = k_0(P)$.

In order to estimate the quantity $\|g\|_{0,\alpha}^{B(\tilde{x},\kappa d(\tilde{x}))}$, we calculate for each two points $x', x'' \in B(\tilde{x}, \kappa d(\tilde{x}))$ as follows:

$$|g(x') - g(x'')| \le |f(x') - f(x'')| + \sum_{i,j=1}^{n} \Big\{ |a_{ij}(x') - a_{ij}(x'')| \, |u_{x_i x_j}(x')|$$

$$+ |a_{ij}(x'') - a_{ij}(\tilde{x})| \, |u_{x_i x_j}(x') - u_{x_i x_j}(x'')| \Big\}$$

$$+ \sum_{i=1}^{n} \Big\{ |b_i(x') - b_i(x'')| \, |u_{x_i}(x')| + |b_i(x'')| \, |u_{x_i}(x') - u_{x_i}(x'')| \Big\}$$

$$+ \Big\{ |c(x') - c(x'')| \, |u(x')| + |c(x'')| \, |u(x') - u(x'')| \Big\}.$$

This implies

$$|g(x') - g(x'')| \leq |x' - x''|^\alpha \left\{ \|f\|_{0,\alpha}^{B(\tilde{x},\kappa d(\tilde{x}))} + \|u\|_2^{B(\tilde{x},\kappa d(\tilde{x}))} \sum_{i,j=1}^n \|a_{ij}\|_{0,\alpha}^{B(\tilde{x},\kappa d(\tilde{x}))} \right.$$

$$+ \|u\|_{2,\alpha}^{B(\tilde{x},\kappa d(\tilde{x}))} \kappa^\alpha d(\tilde{x})^\alpha \sum_{i,j=1}^n \|a_{ij}\|_{0,\alpha}^{B(\tilde{x},\kappa d(\tilde{x}))} + \|u\|_1^{B(\tilde{x},\kappa d(\tilde{x}))} \sum_{i=1}^n \|b_i\|_{0,\alpha}^{B(\tilde{x},\kappa d(\tilde{x}))}$$

$$+ \|u\|_2^{B(\tilde{x},\kappa d(\tilde{x}))} (2\kappa d(\tilde{x}))^{1-\alpha} \sum_{i=1}^n \|b_i\|_0^{B(\tilde{x},\kappa d(\tilde{x}))}$$

$$\left. + \|c\|_{0,\alpha}^{B(\tilde{x},\kappa d(\tilde{x}))} \|u\|_0^{B(\tilde{x},\kappa d(\tilde{x}))} + \|c\|_0^{B(\tilde{x},\kappa d(\tilde{x}))} \|u\|_1^{B(\tilde{x},\kappa d(\tilde{x}))} (2\kappa d(\tilde{x}))^{1-\alpha} \right\}$$

$$\leq |x' - x''|^\alpha \left\{ \|f\|_{0,\alpha}^B + \frac{4A_2 P}{d(\tilde{x})^2} + \kappa^\alpha P \frac{2^{2+\alpha} A_{2,\alpha}}{d(\tilde{x})^2} + \frac{2A_1 P}{d(\tilde{x})} \right.$$

$$\left. + \frac{4A_2 P}{d(\tilde{x})^2} (2\kappa)^{1-\alpha} (d(\tilde{x}))^{1-\alpha} + P A_0 + \frac{2P A_1}{d(\tilde{x})} (2\kappa)^{1-\alpha} (d(\tilde{x}))^{1-\alpha} \right\}.$$

Then we obtain

$$\|g\|_{0,\alpha}^{B(\tilde{x},\kappa d(\tilde{x}))} \leq \|f\|_{0,\alpha}^B + \frac{k_1(P)}{d(\tilde{x})^2} \{ A_0 + A_1 + A_2 + \kappa^\alpha A_{2,\alpha} \} \tag{27}$$

with a constant $k_1 = k_1(P)$.
Combining the estimates (24), (26), and (27) we deduce

$$\|u\|_2^{B(\tilde{x},\frac{m}{2M}\kappa d(\tilde{x}))} \leq \tilde{C} \left\{ \|f\|_0^B + \frac{k_0(P)}{d(\tilde{x})^2} (A_0 + A_1 + \kappa^\alpha A_2) + \kappa^\alpha d(\tilde{x})^\alpha \|f\|_{0,\alpha}^B \right.$$

$$\left. + \kappa^\alpha \frac{k_1(P)}{d(\tilde{x})^{2-\alpha}} (A_0 + A_1 + A_2 + \kappa^\alpha A_{2,\alpha}) + \frac{A_0}{\kappa^2 d(\tilde{x})^2} + \frac{2A_1}{\kappa d(\tilde{x})^2} \right\}$$

$$\leq \frac{C(\alpha,n,m,M,P)}{d(\tilde{x})^2} \left\{ \|f\|_0^B + \kappa^\alpha \|f\|_{0,\alpha}^B + \frac{A_0}{\kappa^2} + \frac{A_1}{\kappa} + \kappa^\alpha A_2 + \kappa^{2\alpha} A_{2,\alpha} \right\}.$$

Furthermore, we estimate with the aid of (25), (26), and (27) as follows:

$$\|u\|_{2,\alpha}^{B(\tilde{x},\frac{m}{2M}\kappa d(\tilde{x}))} \leq \tilde{C} \left\{ \frac{\|f\|_0^B}{\kappa^\alpha d(\tilde{x})^\alpha} + \frac{k_0(P)}{\kappa^\alpha d(\tilde{x})^{2+\alpha}} (A_0 + A_1 + \kappa^\alpha A_2) + \|f\|_{0,\alpha}^B \right.$$

$$\left. + \frac{k_1(P)}{d(\tilde{x})^2} (A_0 + A_1 + A_2 + \kappa^\alpha A_{2,\alpha}) + \frac{A_0}{\kappa^{2+\alpha} d(\tilde{x})^{2+\alpha}} + \frac{2A_1}{\kappa^{1+\alpha} d(\tilde{x})^{2+\alpha}} \right\}$$

$$\leq \frac{C(\alpha,n,m,M,P)}{d(\tilde{x})^{2+\alpha}} \left\{ \frac{\|f\|_0^B}{\kappa^\alpha} + \|f\|_{0,\alpha}^B + \frac{A_0}{\kappa^{2+\alpha}} + \frac{A_1}{\kappa^{1+\alpha}} + A_2 + \kappa^\alpha A_{2,\alpha} \right\}.$$

This completes the proof.   q.e.d.

Theorem 7.9. *Let the assumption D be fulfilled; and $u \in C_*^{2+\alpha}(\overline{B(\xi,R)})$ may satisfy the differential equation*

$$\mathcal{L}(u) = f \quad in \quad B = B(\xi, R)$$

with the right-hand side $f \in C^\alpha(\overline{B})$.
Then we have a constant $C = C(\alpha, n, m, M, P) \in (0, +\infty)$, such that

$$A_1 + A_2 + A_{2,\alpha} \le C(A_0 + \|f\|_0^B + \|f\|_{0,\alpha}^B). \tag{28}$$

Proof: We choose $\kappa \in (0, \frac{1}{2})$ and infer the following inequality from Proposition 7.7:

$$A_1 \le \frac{2n}{\kappa^{1+\alpha}} A_0 + \frac{n\kappa^{1+\alpha}}{(1 - \kappa^{1+\alpha})^2} A_2. \tag{29}$$

Together with Proposition 7.8 we obtain the estimates

$$A_2 \le C\left\{ \|f\|_0^B + \kappa^\alpha \|f\|_{0,\alpha}^B + \left(\frac{1}{\kappa^2} + \frac{2n}{\kappa^{2+\alpha}} \right) A_0 \right.$$
$$\left. + \left(\frac{n}{(1 - \kappa^{1+\alpha})^2} + 1 \right) \kappa^\alpha A_2 + \kappa^{2\alpha} A_{2,\alpha} \right\}$$

and

$$\kappa^\alpha A_{2,\alpha} \le C\left\{ \|f\|_0^B + \kappa^\alpha \|f\|_{0,\alpha}^B + \left(\frac{1}{\kappa^2} + \frac{2n}{\kappa^{2+\alpha}} \right) A_0 \right.$$
$$\left. + \left(\frac{n}{(1 - \kappa^{1+\alpha})^2} + 1 \right) \kappa^\alpha A_2 + \kappa^{2\alpha} A_{2,\alpha} \right\}.$$

Their addition yields

$$A_2 + \kappa^\alpha A_{2,\alpha} \le 2C\left\{ \|f\|_0^B + \kappa^\alpha \|f\|_{0,\alpha}^B + \left(\frac{1}{\kappa^2} + \frac{2n}{\kappa^{2+\alpha}} \right) A_0 \right.$$
$$\left. + \kappa^\alpha \left(1 + \frac{n}{(1 - \kappa^{1+\alpha})^2} \right) (A_2 + \kappa^\alpha A_{2,\alpha}) \right\}.$$

Choosing $0 < \kappa_0$ so small that the condition

$$2C\kappa_0^\alpha \left(1 + \frac{n}{(1 - \kappa_0^{1+\alpha})^2} \right) \le \frac{1}{2}$$

is fulfilled, we deduce

$$A_2 + \kappa_0^\alpha A_{2,\alpha} \le 4C\left\{ \|f\|_0^B + \kappa_0^\alpha \|f\|_{0,\alpha}^B + \left(\frac{1}{\kappa_0^2} + \frac{2n}{\kappa_0^{2+\alpha}} \right) A_0 \right\}.$$

Consequently, the quantities A_2 and $A_{2,\alpha}$ are estimated in the desired form. Utilizing Proposition 7.7 once more, we attain the stated inequality (28). q.e.d.

Proof of Theorem 5.4 from Section 5: The quantity $d > 0$ being chosen arbitrarily small, we consider the set

$$\Omega_d := \{x \in \Omega \ : \ \mathrm{dist}(x, \partial\Omega) > d\}.$$

In the ball $B = B(x_0, d) \subset \Omega$ we apply Theorem 7.9 with $x_0 \in \Omega_d$ and $R = d$. Then we obtain

$$\tilde{C}\Big(\sup_{x \in \Omega} |u(x)| + \|f\|_0^\Omega + \|f\|_{0,\alpha}^\Omega\Big) \geq A_1 + A_2 + A_{2,\alpha}$$

$$\geq d \sum_{i=1}^n |u_{x_i}(x_0)| + d^2 \sum_{i,j=1}^n |u_{x_i x_j}(x_0)|$$

$$+ \Big(\frac{d}{2}\Big)^{2+\alpha} \sum_{i,j=1}^n \sup_{\substack{x', x'' \in B(x_0, d/2) \\ x' \neq x''}} \frac{|u_{x_i x_j}(x') - u_{x_i x_j}(x'')|}{|x' - x''|^\alpha}$$

for all $x_0 \in \Omega_d$. This implies

$$\|u\|_1^{\Omega_d} + \|u\|_2^{\Omega_d} + \|u\|_{2,\alpha}^{\Omega_d} \leq C(\ldots, d)\big(\|u\|_0^\Omega + \|f\|_0^\Omega + \|f\|_{0,\alpha}^\Omega\big).$$

q.e.d.

Let the domain Ω satisfy the assumption C_3. For each boundary point $x_0 \in \partial\Omega$ we then have a half-neighborhood Ω_0, which can be mapped onto a half-ball $B(\xi, R)$ with $\xi \in E = \{x \in \mathbb{R}^n \ : \ x_n = 0\}$ in such away that $\overline{B(\xi, R)} \cap E$ is related to $\partial\Omega \cap \partial\Omega_0$. This mapping represents a diffeomorphism of $B(\xi, R)$ onto $\overline{\Omega_0}$ in the class $C^{2+\alpha}$. The differential equation $\mathcal{L}(u) = f$ is transformed - similar to the proof of Theorem 6.2 from Section 6 - into an elliptic differential equation on the half-ball with zero boundary values on E. The Schauder estimates utilized in the proof of Theorem 6.2 from Section 6 can be directly inferred from Theorem 7.9 with the aid of the transformation above.

Proof of Theorem 5.1 from Section 5: Following the arguments above, to each point $x_0 \in \partial\Omega$ there exists a neighborhood $\Omega_0 := \overline{\Omega} \cap B(x_0, \varepsilon_0)$ with $\varepsilon_0 > 0$, such that

$$\|u\|_1^{\Omega_0} + \|u\|_2^{\Omega_0} + \|u\|_{2,\alpha}^{\Omega_0} \leq \tilde{C}\big(\|u\|_0^\Omega + \|f\|_0^\Omega + \|f\|_{0,\alpha}^\Omega\big)$$

holds true. The boundary $\partial\Omega$ being compact, finitely many such neighborhoods Ω_j, $j = 1, \ldots, N$ suffice in order to cover this set. Then we obtain

$$\|u\|_1^{\Omega_j} + \|u\|_2^{\Omega_j} + \|u\|_{2,\alpha}^{\Omega_j} \leq \tilde{C}\big(\|u\|_0^\Omega + \|f\|_0^\Omega + \|f\|_{0,\alpha}^\Omega\big) \qquad \text{for} \quad j = 1, \ldots, N.$$

We choose $d > 0$ sufficiently small and attain, with a constant $C = C(\alpha, n, m, M, P, \Omega) \in (0, +\infty)$, the global Schauder estimate:

$$\|u\|_1^{\Omega} + \|u\|_2^{\Omega} + \|u\|_{2,\alpha}^{\Omega}$$

$$\leq \|u\|_1^{\Omega_d} + \|u\|_2^{\Omega_d} + \|u\|_{2,\alpha}^{\Omega_d} + \sum_{j=1}^{N} \left(\|u\|_1^{\Omega_j} + \|u\|_2^{\Omega_j} + \|u\|_{2,\alpha}^{\Omega_j} \right)$$

$$\leq \left(\tilde{C}N + \tilde{C}(d) \right) \left(\|u\|_0^{\Omega} + \|f\|_0^{\Omega} + \|f\|_{0,\alpha}^{\Omega} \right) \leq C \left(\|u\|_0^{\Omega} + \|f\|_0^{\Omega} + \|f\|_{0,\alpha}^{\Omega} \right).$$

Now we have completely proved all the Schauder estimates, which we have already applied in Section 5 and Section 6.

8 Some Historical Notices to Chapter 9

Boundary value problems for holomorphic functions have already been considered by B. Riemann. The just established theory of integral equations enabled D. Hilbert in 1904, to obtain new results for this problem. G. Hellwig observed in 1952 the nonuniqueness of the Riemann-Hilbert problem and the intricate structure for the set of their solutions. For a thorough treatment of these questions we refer the reader to the profound monograph by I.N. Vekua.

The boundary value problem for elliptic differential equations was solved by J. Schauder from 1932–1934 via functional analytic methods. At about the same time, G. Giraud and E. Hopf treated similar problems with alternative methods. For a detailed account we refer the reader to the book by D. Gilbarg and N. Trudinger.

J. Schauder worked, as a brilliant student of S. Banach, in the intellectually excellent atmosphere of the University at Lwów, now in the Ukraine. However, his life ended already in 1943 – within the tragical times of World War II.

Figure 2.4   PORTRAIT OF JULIUSZ PAWEL SCHAUDER (1899–1943)

This photo can be found under the electronic address
http://www-history.mcs.st-andrews.ac.uk/PictDisplay/Schauder.html
in the *Mac Tutor History of Mathematics archive* at the School of Mathematics and Statistics in the University of St Andrews, Scotland. The picture has been taken from a photo of the participants in the *First International Topology Conference* 1935 at Moscow. In this context, we refer to the article by R.S. Ingarden: *Juliusz Schauder – Personal Reminiscences* within *Topological Methods in Nonlinear Analysis* – Journal of the Juliusz Schauder Center, Volume 2 (1993).

Chapter 10

Weak Solutions of Elliptic Differential Equations

In this chapter we consider Sobolev spaces in Section 1 and prove the Sobolev embedding theorem and the Rellich selection theorem in Section 2. Then we establish the existence of weak solutions in Section 3. With the aid of Moser's iteration method we show the boundedness of weak solutions in Section 4. In the subsequent Section 5 we deduce Hölder continuity of weak solutions with the aid of the weak Harnack inequality by J. Moser. The necessary regularity theorem of John and Nirenberg will be derived in Section 6. Finally, we investigate the boundary regularity of weak solutions in Section 7. Then we apply our results to equations in divergence form (compare Section 8). At the end of this chapter we present Green's function for elliptic operators with the aid of capacity methods, and we treat the eigenvalue problem for the Laplace-Beltrami operator.

1 Sobolev Spaces

Let $\Omega \subset \mathbb{R}^n$ be a bounded open set. Then the space $C_0^\infty(\Omega)$ is dense in the Lebesgue space $L^p(\Omega)$ for all $1 \leq p < +\infty$. We shall now construct a space $W^{k,p}(\Omega)$ of the k-times weakly differentiable functions - with weak derivatives in the space $L^p(\Omega)$.

To the element $f \in L^p(\Omega)$ we attribute the following functional in a natural way:

$$A_f(\varphi) := \int_\Omega f(x)\varphi(x)\,dx, \qquad \varphi \in C_0^\infty(\Omega). \tag{1}$$

Taking the multi-index $\alpha = (\alpha_1, \ldots, \alpha_n) \in \mathbb{N}_0^n$, $\mathbb{N}_0 := \mathbb{N} \cup \{0\}$, with $|\alpha| := \alpha_1 + \ldots + \alpha_n \in \mathbb{N}_0$, we consider the functionals

$$A_{f,\alpha}(\varphi) := (-1)^{|\alpha|} \int_\Omega f(x)\partial^\alpha \varphi(x)\,dx, \qquad \varphi \in C_0^\infty(\Omega). \tag{2}$$

F. Sauvigny, *Partial Differential Equations 2*, Universitext,
DOI 10.1007/978-1-4471-2984-4_4, © Springer-Verlag London 2012

Here the symbol

$$\partial^\alpha \varphi(x) := \frac{\partial^{|\alpha|}}{\partial x_1^{\alpha_1} \dots \partial x_n^{\alpha_n}} \varphi(x), \qquad x \in \Omega,$$

denotes the corresponding partial derivative of φ. We note that

$$A_{f,0} = A_f.$$

With the function $f \in C^{|\alpha|}(\Omega)$, an $|\alpha|$-times partial integration yields

$$A_{f,\alpha}(\varphi) = \int_\Omega \left(\partial^\alpha f(x) \right) \varphi(x) \, dx, \qquad \varphi \in C_0^\infty(\Omega). \tag{3}$$

On account of $\varphi \in C_0^\infty(\Omega)$, the boundary integrals vanish during the partial integration.

In case the linear functional $A_{f,\alpha}$ – defined in (2) above – is bounded with respect to the $L^q(\Omega)$-norm on the set $C_0^\infty(\Omega)$, we can extend this functional to the Lebesgue space $L^q(\Omega)$ for $1 \leq q < +\infty$. Via the Riesz representation theorem we have an element $g \in L^p(\Omega)$ with $p = \frac{q}{q-1} \in (1, +\infty]$, such that

$$A_{f,\alpha}(\varphi) = \int_\Omega g(x)\varphi(x) \, dx = A_g(\varphi) \qquad \text{for all} \quad \varphi \in C_0^\infty(\Omega) \tag{4}$$

holds true.

Due to (3) the following definition is justified:

Definition 1.1. *Take the multi-index* $\alpha = (\alpha_1, \dots, \alpha_n) \in \mathbb{N}_0^n$, *the exponent* $1 \leq p \leq +\infty$, *and the element* $f \in L^p(\Omega)$. *Then the element*

$$g(x) := D^\alpha f(x) \in L^p(\Omega)$$

is called the weak partial derivative of order α for f *if*

$$\int_\Omega g(x)\varphi(x) \, dx = (-1)^{|\alpha|} \int_\Omega f(x)\partial^\alpha \varphi(x) \, dx \qquad \text{for all} \quad \varphi \in C_0^\infty(\Omega) \tag{5}$$

holds true.

Remarks:

1. Let the elements $g_1, g_2 \in L^p(\Omega)$ with $p > 1$ satisfy the identity (5). We then obtain

$$\int_\Omega g_1(x)\varphi(x) \, dx = A_{f,\alpha}(\varphi) = \int_\Omega g_2(x)\varphi(x) \, dx \qquad \text{for all} \quad \varphi \in C_0^\infty(\Omega).$$

For the conjugate exponent $q \in [1, +\infty)$ with $p^{-1} + q^{-1} = 1$ we infer

$$\int\limits_{\Omega} (g_1 - g_2)(x)\varphi(x)\,dx = 0 \qquad \text{for all} \quad \varphi \in L^q(\Omega).$$

Utilizing Theorem 8.2 from Chapter 2, Section 8 we obtain

$$0 = \|A_{(g_1 - g_2)}\| = \|g_1 - g_2\|_p$$

and the identities $g_1 = g_2$ in $L^p(\Omega)$ respectively $g_1 = g_2$ almost everywhere (briefly a.e.) in Ω follow. Even if the weak derivative exists only in $L^1(\Omega)$, it is uniquely determined in this space.

2. In the classical situation $f \in C^{|\alpha|}(\Omega)$ we have the coincidence

$$\partial^\alpha f(x) = D^\alpha f(x) \quad \text{in} \quad \Omega$$

according to the relation (3).

Definition 1.2. *Let the numbers $k \in \mathbb{N}_0$ and $1 \le p \le +\infty$ be given. Then we define the Sobolev space*

$$W^{k,p}(\Omega) := \left\{ f \in L^p(\Omega) : D^\alpha f \in L^p(\Omega), \ |\alpha| \le k \right\}$$

with the Sobolev norm

$$\|f\|_{W^{k,p}(\Omega)} := \|f\|_{k,p,\Omega} := \left(\sum_{|\alpha| \le k} \int\limits_{\Omega} |D^\alpha f(x)|^p\,dx \right)^{\frac{1}{p}}. \tag{6}$$

Remarks:

1. An equivalent norm is given by

$$\|f\|'_{k,p} := \sum_{|\alpha| \le k} \|D^\alpha f\|_p.$$

Therefore, there exist constants $0 < c_1 \le c_2 < +\infty$ satisfying

$$c_1 \|f\|_{k,p} \le \|f\|'_{k,p} \le c_2 \|f\|_{k,p} \qquad \text{for all} \quad f \in W^{k,p}(\Omega).$$

2. Endowed with the norm from Definition 1.2, the space $W^{k,p}(\Omega)$ becomes a Banach space.

3. With $1 < p \le +\infty$ and $q = \frac{p}{p-1}$, the preliminary considerations yield

$$W^{k,p}(\Omega) = \left\{ f \in L^p(\Omega) : A_{f,\alpha} \in (L^q(\Omega))^*, \ |\alpha| \le k \right\}. \tag{7}$$

Here the symbol $(L^q(\Omega))^*$ denotes the continuous dual space of $L^q(\Omega)$.

4. In the special case $p = 2$ we obtain the Hilbert spaces

$$H^k(\Omega) := W^{k,2}(\Omega)$$

with the inner product

$$(f, g)_{H^k(\Omega)} := \sum_{|\alpha| \le k} \int_{\Omega} D^\alpha f(x) D^\alpha g(x)\, dx, \qquad f, g \in H^k(\Omega).$$

5. One immediately shows the linearity of the weak derivative: Let the numbers $c, d \in \mathbb{R}$, the multi-index α from the set $\mathbb{N}_0^n$, and the elements f, g in $W^{k,p}(\Omega)$ be given. Then we have

$$D^\alpha(cf + dg) = cD^\alpha f + dD^\alpha g.$$

We shall now present a mollifying process which we owe to K. Friedrichs. By $\varrho \in C^\infty(\mathbb{R}^n)$ we denote the mollifier

$$\varrho(x) = \begin{cases} c\exp\left(\frac{1}{|x|^2 - 1}\right), & |x| < 1 \\ 0, & |x| \ge 1 \end{cases}$$

satisfying

$$\int_{\mathbb{R}^n} \varrho(x)\, dx = 1.$$

Here we have to choose $c > 0$ suitably. A function $u(x) \in L^p(\Omega)$ with $1 \le p \le +\infty$ is extended onto the whole Euclidean space $\mathbb{R}^n$ as follows:

$$u(x) = \begin{cases} u(x), & x \in \Omega \\ 0, & x \in \mathbb{R}^n \setminus \Omega \end{cases}.$$

Theorem 1.3. (Friedrichs)
Taking the exponent $1 \le p \le +\infty$ and the function $u(x) \in L^p(\Omega)$, we attribute the regularized function

$$u_h(x) := h^{-n} \int_{\mathbb{R}^n} \varrho\left(\frac{x - y}{h}\right) u(y)\, dy, \qquad x \in \mathbb{R}^n,$$

for each $h > 0$. Then the mapping $u \mapsto u_h$ is linear from $L^p(\mathbb{R}^n)$ into $L^p(\mathbb{R}^n)$, and we have the estimate

$$\|u_h\|_p \le \|u\|_p \qquad \text{for all} \quad h > 0, \quad u \in L^p(\mathbb{R}^n).$$

Proof: The linearity of the map $u \mapsto u_h$ is evident, and we only have to show the norm estimate. The transformation formula for multiple integrals remains applicable to L^1-functions via approximation, and we deduce

$$u_h(x) = h^{-n} \int_{\mathbb{R}^n} \varrho\left(\frac{x-y}{h}\right) u(y)\, dy$$

$$= \int_{\mathbb{R}^n} \varrho(z) u(x - hz)\, dz \tag{8}$$

$$= \int_{|z| \leq 1} \varrho(z) u(x - hz)\, dz$$

for all $h > 0$. With the aid of Hölder's inequality for $1 < p < +\infty$ and the identity $p^{-1} + q^{-1} = 1$ we arrive at

$$|u_h(x)| \leq \int_{|z| \leq 1} \varrho^{\frac{1}{p}}(z) |u(x - hz)| \varrho^{\frac{1}{q}}(z)\, dz$$

$$\leq \left(\int_{|z| \leq 1} \varrho(z) |u(x - hz)|^p\, dz \right)^{\frac{1}{p}} \left(\int_{|z| \leq 1} \varrho(z)\, dz \right)^{\frac{1}{q}}$$

and

$$|u_h(x)|^p \leq \int_{|z| \leq 1} \varrho(z) |u(x - hz)|^p\, dz \qquad \text{for all} \quad x \in \mathbb{R}^n.$$

Integration via Fubini's theorem yields (for $p = 1$ as well)

$$\int_{\mathbb{R}^n} |u_h(x)|^p\, dx \leq \int_{x \in \mathbb{R}^n} \left(\int_{|z| \leq 1} \varrho(z) |u(x - hz)|^p\, dz \right) dx$$

$$= \int_{|z| \leq 1} \varrho(z) \left(\int_{x \in \mathbb{R}^n} |u(x - hz)|^p\, dx \right) dz$$

$$= \left(\int_{\mathbb{R}^n} |u(x)|^p\, dx \right) \left(\int_{|z| \leq 1} \varrho(z)\, dz \right).$$

This implies

$$\|u_h\|_p \leq \|u\|_p \qquad \text{for all} \quad u \in L^p(\mathbb{R}^n), \quad h > 0, \quad 1 \leq p < +\infty.$$

In the case $p = \infty$ we obtain $|u_h(x)| \leq \|u\|_\infty$ a. e. in $\mathbb{R}^n$ and consequently $\|u_h\|_\infty \leq \|u\|_\infty$.   q.e.d.

Theorem 1.4. (Friedrichs)
We have the following statements:

1. *Taking $u(x) \in C_0^0(\Omega)$ we observe*

$$\sup_{x \in \mathbb{R}^n} |u(x) - u_h(x)| \longrightarrow 0 \qquad for \quad h \to 0+.$$

2. *For $u \in L^p(\Omega)$ with $1 \le p < +\infty$ we infer*

$$\|u - u_h\|_p \longrightarrow 0 \qquad for \quad h \to 0+.$$

Proof:

1. We depart from $u \in C_0^0(\Omega)$: For each $\varepsilon > 0$ we find a number $\delta > 0$, such that the estimate
$$|u(x) - u(y)| \le \varepsilon$$
is valid for all $x, y \in \mathbb{R}^n$ with $|x - y| \le \delta$. Via (8) we obtain

$$|u_h(x) - u(x)| \le \int_{|z| \le 1} \varrho(z)|u(x - hz) - u(x)|\, dz$$

$$\le \varepsilon \qquad \text{for all} \quad 0 < h \le \delta(\varepsilon), \quad x \in \mathbb{R}^n.$$

Consequently, we observe

$$\sup_{x \in \mathbb{R}^n} |u_h(x) - u(x)| \longrightarrow 0 \qquad \text{for} \quad h \to 0+.$$

2. Now we consider $u \in L^p(\Omega)$ with the exponent $1 \le p < +\infty$. Due to Theorem 7.13 in Chapter 2, Section 7, each given $\varepsilon > 0$ admits a function

$$v \in C_0^0(\Omega) \quad \text{satisfying} \quad \|u - v\|_p \le \varepsilon.$$

Utilizing part 1 of our proof, we choose a number $h_0(\varepsilon) > 0$ so small that

$$\|v - v_h\|_p \le \varepsilon \qquad \text{for all} \quad 0 < h \le h_0(\varepsilon)$$

is correct. For all $0 < h \le h_0(\varepsilon)$ we obtain the following inequality

$$\|u - u_h\|_p \le \|u - v\|_p + \|v - v_h\|_p + \|v_h - u_h\|_p$$

$$\le 2\|u - v\|_p + \|v - v_h\|_p \le 3\varepsilon,$$

taking Theorem 1.3 into account. This implies $\|u - u_h\|_p \to 0$ for $h \to 0+$.
q.e.d.

One can interchange weak differentiation and mollification via the subsequent

Theorem 1.5. (Friedrichs)
Let us extend the function $f \in W^{k,p}(\Omega)$ onto the whole Euclidean space setting
$f \equiv 0$ *on* $\mathbb{R}^n \setminus \Omega$. *With the number* $\varepsilon > 0$ *given, we define the regularized*
function of class $C^\infty(\Omega)$ *by*

$$f_\varepsilon(x) := \frac{1}{\varepsilon^n} \int\limits_{\mathbb{R}^n} \varrho\Big(\frac{x-y}{\varepsilon}\Big) f(y)\, dy, \qquad x \in \mathbb{R}^n.$$

For all multi-indices $\alpha \in \mathbb{N}_0^n$ *with* $|\alpha| \le k$ *and all numbers* ε *which satisfy*
$0 < \varepsilon < \operatorname{dist}(x, \mathbb{R}^n \setminus \Omega)$, *we have the identity*

$$\partial^\alpha f_\varepsilon(x) = (D^\alpha f)_\varepsilon(x), \qquad x \in \Omega.$$

Proof: We calculate

$$\partial^\alpha f_\varepsilon(x) = \frac{1}{\varepsilon^n} \int\limits_{\mathbb{R}^n} \partial_x^\alpha \varrho\Big(\frac{x-y}{\varepsilon}\Big) f(y)\, dy$$

$$= (-1)^{|\alpha|} \frac{1}{\varepsilon^n} \int\limits_{\mathbb{R}^n} \partial_y^\alpha \varrho\Big(\frac{x-y}{\varepsilon}\Big) f(y)\, dy$$

$$= \frac{1}{\varepsilon^n} \int\limits_{\mathbb{R}^n} \varrho\Big(\frac{x-y}{\varepsilon}\Big) D^\alpha f(y)\, dy = (D^\alpha f)_\varepsilon(x).$$

q.e.d.

Theorem 1.6. (Meyers, Serrin)
With the exponent $1 \le p < +\infty$ *given, the linear subspace* $C^\infty(\Omega) \cap W^{k,p}(\Omega)$
is dense in the Sobolev space $W^{k,p}(\Omega)$.

Proof: We choose the open sets $\Omega_j \subset \mathbb{R}^n$ for $j \in \mathbb{N}_0$ satisfying

$$\emptyset = \Omega_0 \subset \Omega_1 \subset \Omega_2 \subset \ldots \subset \Omega \qquad \text{and} \qquad \overline{\Omega_j} \subset \Omega_{j+1}, \quad j \in \mathbb{N}_0,$$

such that

$$\bigcup_{j=1}^{\infty} \Omega_j = \Omega.$$

Furthermore, let $\Psi_j \in C_0^\infty(\Omega)$ denote a partition of unity subordinate to the set system $\{\Omega_{j+1} \setminus \overline{\Omega_{j-1}}\}_{j=1,2,\ldots}$. This means

$$\operatorname{supp} \Psi_j \subset \Omega_{j+1} \setminus \overline{\Omega_{j-1}} \qquad \text{and} \qquad \sum_{j=1}^{\infty} \Psi_j(x) = 1, \quad x \in \Omega.$$

For the given $\varepsilon > 0$, let us choose $\varepsilon_j > 0$ such that the condition
$\varepsilon_j < \operatorname{dist}(\Omega_{j+1}, \partial\Omega)$ as well as the inequality

$$\|(\Psi_j f)_{\varepsilon_j} - (\Psi_j f)\|_{W^{k,p}(\Omega)} \leq \varepsilon\, 2^{-j}$$

are valid. This can be achieved with the aid of Theorem 1.4 and Theorem 1.3. We now observe

$$g(x) := \sum_{j=1}^{\infty} (\Psi_j f)_{\varepsilon_j}(x) \in C^\infty(\Omega)$$

and furthermore

$$\|g - f\|_{W^{k,p}(\Omega)} = \left\| \sum_{j=1}^{\infty} (\Psi_j f)_{\varepsilon_j} - \sum_{j=1}^{\infty} (\Psi_j f) \right\|_{W^{k,p}(\Omega)}$$

$$\leq \sum_{j=1}^{\infty} \|(\Psi_j f)_{\varepsilon_j} - (\Psi_j f)\|_{W^{k,p}(\Omega)} \leq \sum_{j=1}^{\infty} \frac{\varepsilon}{2^j} = \varepsilon.$$

From the property $f \in W^{k,p}(\Omega)$ we infer $g \in W^{k,p}(\Omega)$.   q.e.d.

Because of this theorem we comprehend the Sobolev space $W^{k,p}(\Omega)$ as the *completion* of the linear set of functions $C^\infty(\Omega)$ with respect to the Sobolev norm $\|\cdot\|_{W^{k,p}(\Omega)}$. If the boundary $\partial\Omega$ represents a smooth C^1-hypersurface in the Euclidean space $\mathbb{R}^n$, one can prove that even the linear space $C^\infty(\overline{\Omega})$ lies densely in the Sobolev space $W^{k,p}(\Omega)$. However, only in the case $k = 0$ and $p < +\infty$ is the set $C_0^\infty(\Omega)$ dense in the space $W^{k,p}(\Omega) = L^p(\Omega)$. For $k > 0$ we obtain the Sobolev space $W_0^{k,p}(\Omega)$ with *weak zero boundary values*.

Definition 1.7. *Let the numbers $k \in \mathbb{N}$ and $1 \leq p \leq +\infty$ be prescribed, then we define the* Sobolev space

$$W_0^{k,p}(\Omega) := \left\{ f \in W^{k,p}(\Omega) : \begin{array}{l} \text{There is a sequence } \{f_l\}_{l=1,2,\ldots} \subset C_0^\infty(\Omega) \\ \text{with } \|f - f_l\|_{W^{k,p}(\Omega)} \to 0 \text{ for } l \to \infty \end{array} \right\}.$$

In the sequel we concentrate on the Sobolev spaces $W^{1,p}(\Omega)$ and $W_0^{1,p}(\Omega)$. Let the symbol $e_i := (\delta_{1i}, \ldots, \delta_{ni}) \in \mathbb{R}^n$ with $i \in \{1, \ldots, n\}$ denote a unit vector. Taking the point $x \in \Omega$ and the number ε with $0 < |\varepsilon| < \mathrm{dist}\,(x, \mathbb{R}^n \setminus \Omega)$ arbitrarily, we define the *difference quotient in the direction* e_i by

$$\triangle_{i,\varepsilon} f(x) := \frac{f(x + \varepsilon e_i) - f(x)}{\varepsilon}.$$

We can characterize the Sobolev functions as follows:

Theorem 1.8. *Let the exponent $1 < p < +\infty$ and the element $f \in L^p(\Omega)$ be given, then the following two statements are equivalent:*

  i) We have the property $f \in W^{1,p}(\Omega)$.

ii) There exists a constant $C \in [0, +\infty)$, such that we have the uniform estimate

$$\|\triangle_{i,\varepsilon} f\|_{L^p(\Theta)} \leq C$$

for all open sets $\Theta \subset \Omega$ with $\overline{\Theta} \subset \Omega$, and all indices $i \in \{1, \ldots, n\}$, and all numbers ε with $0 < |\varepsilon| < dist(\Theta, \mathbb{R}^n \setminus \Omega)$.

Proof:

1. At first, we prove the direction: i) $\Rightarrow$ ii).
   Choosing $i \in \{1, \ldots, n\}$, $f \in C^\infty(\Omega) \cap W^{1,p}(\Omega)$, and $\overline{\Theta} \subset \Omega$ we calculate

$$\triangle_{i,\varepsilon} f(x) = \frac{f(x + \varepsilon e_i) - f(x)}{\varepsilon} = \frac{1}{\varepsilon} \int_0^\varepsilon \frac{\partial}{\partial x_i} f(x + t e_i) \, dt$$

for all $0 < |\varepsilon| < dist\,(\Theta, \mathbb{R}^n \setminus \Omega)$. With the aid of Hölder's inequality we deduce the following estimate for all $x \in \Theta$:

$$|\triangle_{i,\varepsilon} f(x)|^p \leq \frac{1}{\varepsilon^p} \left(\int_0^\varepsilon 1 \left| \frac{\partial}{\partial x_i} f(x + t e_i) \right| dt \right)^p$$

$$\leq \frac{\varepsilon^{\frac{p}{q}}}{\varepsilon^p} \int_0^\varepsilon \left| \frac{\partial}{\partial x_i} f(x + t e_i) \right|^p dt$$

$$= \frac{1}{\varepsilon} \int_0^\varepsilon \left| \frac{\partial}{\partial x_i} f(x + t e_i) \right|^p dt.$$

Here we observe $1 \in L^q(\Omega)$ and $p^{-1} + q^{-1} = 1$. Then Fubini's theorem yields

$$\int_\Theta |\triangle_{i,\varepsilon} f(x)|^p \, dx \leq \frac{1}{\varepsilon} \int_0^\varepsilon \left(\int_{\mathbb{R}^n} |D^{e_i} f(x + t e_i)|^p \, dx \right) dt$$

$$= \int_{\mathbb{R}^n} |D^{e_i} f(x)|^p \, dx.$$

Thus we obtain the subsequent estimate via the Meyers-Serrin theorem, namely

$$\|\triangle_{i,\varepsilon} f\|_{L^p(\Theta)} \leq \|f\|_{W^{1,p}(\Omega)} =: C.$$

2. It remains to show the inverse direction: ii) $\Rightarrow$ i).
   For $i \in \{1, \ldots, n\}$, $\overline{\Theta} \subset \Omega$, and an ε satisfying $0 < |\varepsilon| < dist\,(\Theta, \mathbb{R}^n \setminus \Omega)$, we have the inequality

$$\|\triangle_{i,\varepsilon} f\|_{L^p(\Theta)} \leq C.$$

Due to Theorem 8.9 in Chapter 2, Section 8, we have a sequence $\varepsilon_k \downarrow 0$ and an element $g_i \in L^p(\Omega)$ such that

$$\int\limits_{\mathbb{R}^n} \varphi(x)\triangle_{i,\varepsilon_k} f(x)\, dx \longrightarrow \int\limits_{\mathbb{R}^n} \varphi(x)g_i(x)\, dx \qquad \text{for all} \quad \varphi \in C_0^\infty(\Omega)$$

is valid. Now we see that

$$\int\limits_{\mathbb{R}^n} \varphi(x)\triangle_{i,\varepsilon_k} f(x)\, dx = -\int\limits_{\mathbb{R}^n} \Big(\triangle_{i,-\varepsilon_k}\varphi(x)\Big) f(x)\, dx$$
$$\longrightarrow -\int\limits_{\mathbb{R}^n} f(x)\frac{\partial}{\partial x_i}\,\varphi(x)\, dx \tag{9}$$

for $k \to \infty$, and consequently

$$-\int\limits_{\mathbb{R}^n} f(x)\frac{\partial}{\partial x_i}\varphi(x)\, dx = \int\limits_{\mathbb{R}^n} \varphi(x)g_i(x)\, dx \quad \text{for all} \quad \varphi \in C_0^\infty(\Omega).$$

This implies

$$\triangle_{i,\varepsilon_k} f \overset{L^p(\Omega)}{\longrightarrow} D^{e_i} f = g_i \in L^p(\Omega),$$

and therefore $f \in W^{1,p}(\Omega)$ holds true. In order to comprehend the identity (9), we integrate as follows:

$$\triangle_{i,\varepsilon_k}(\varphi(x)f(x))$$
$$= \frac{1}{\varepsilon_k}\left\{\Big(\varphi(x+\varepsilon_k e_i) - \varphi(x)\Big)f(x+\varepsilon_k e_i) + \varphi(x)\Big(f(x+\varepsilon_k e_i) - f(x)\Big)\right\}$$
$$= \left\{\frac{1}{\varepsilon_k}\Big(\varphi(y) - \varphi(y-\varepsilon_k e_i)\Big)f(y)\right\}\Bigg|_{y=x+\varepsilon_k e_i} + \varphi(x)\triangle_{i,\varepsilon_k} f(x)$$
$$= \left\{f(y)\triangle_{i,-\varepsilon_k}\varphi(y)\right\}\Big|_{y=x+\varepsilon_k e_i} + \varphi(x)\triangle_{i,\varepsilon_k} f(x). \tag{10}$$

This completes the proof.   q.e.d.

Remark: With the Sobolev function $f \in W^{1,p}(\Omega)$ we consider the weakly convergent sequence of difference quotients $\{\triangle_{i,\varepsilon_k} f\}_{k=1,2,\ldots}$, where $\varepsilon_k \downarrow 0$ holds true. Then the proof above yields

$$\triangle_{i,\varepsilon_k} f \to D^{e_i} f \qquad \text{in} \quad L^p(\Omega), \quad i = 1,\ldots,n.$$

This fact explains the notion *weak derivative*.

Theorem 1.9. (Weak product rule)
Let the Sobolev functions $f, g \in W^{1,p}(\Omega) \cap L^{\infty}(\Omega)$ with exponents $1 < p < +\infty$ be given. Then we have the property $h := fg \in W^{1,p}(\Omega) \cap L^{\infty}(\Omega)$ and the formula

$$D^{\alpha} h = f D^{\alpha} g + g D^{\alpha} f, \qquad \text{for all} \quad \alpha \in \mathbb{N}_0^n \quad \text{with} \quad |\alpha| = 1.$$

Proof: Choose the function $\varphi \in C_0^{\infty}(\Omega)$ and a sufficiently small number $\varepsilon > 0$. When we apply the identity (10) twice, we obtain the following equation:

$$\triangle_{i,\varepsilon}\Big(\varphi(x)h(x)\Big) = \Big\{ h(y)\triangle_{i,-\varepsilon}\varphi(y) \Big\}_{y=x+\varepsilon e_i} + \varphi(x)\triangle_{i,\varepsilon} h(x)$$

$$= \Big\{ h(y)\triangle_{i,-\varepsilon}\varphi(y) \Big\}_{y=x+\varepsilon e_i}$$

$$+ \varphi(x)\Big(\Big\{ f(y)\triangle_{i,-\varepsilon} g(y) \Big\}_{y=x+\varepsilon e_i} + g(x)\triangle_{i,\varepsilon} f(x) \Big)$$

$$= \Big\{ h(y)\triangle_{i,-\varepsilon}\varphi(y) \Big\}_{y=x+\varepsilon e_i} \tag{11}$$

$$+ \varphi(x)g(x)\triangle_{i,\varepsilon} f(x) + \Big\{ \varphi(y)f(y)\triangle_{i,-\varepsilon} g(y) \Big\}_{y=x+\varepsilon e_i}$$

$$+ \Big\{ \Big(\varphi(y - \varepsilon e_i) - \varphi(y) \Big) f(y)\triangle_{i,-\varepsilon} g(y) \Big\}_{y=x+\varepsilon e_i}.$$

Noting $f, g \in W^{1,p}(\Omega)$, Theorem 1.8 and Theorem 8.9 from Chapter 2, Section 8 allow us to choose a zero sequence $\varepsilon_k \downarrow 0$, such that

$$\triangle_{i,\varepsilon_k} f(x) \rightarrow D^{e_i} f(x) \qquad \text{and} \qquad \triangle_{i,-\varepsilon_k} g(x) \rightarrow D^{e_i} g(x) \qquad \text{in} \quad L^p(\Omega).$$

From (11) we obtain

$$0 = \int_{\Omega} h(x)\triangle_{i,-\varepsilon_k}\varphi(x)\,dx + \int_{\Omega} \varphi(x)g(x)\triangle_{i,\varepsilon_k} f(x)\,dx$$

$$+ \int_{\Omega} \varphi(x)f(x)\triangle_{i,-\varepsilon_k} g(x)\,dx + \int_{\Omega} \Big(\varphi(x - \varepsilon_k e_i) - \varphi(x) \Big) f(x)\triangle_{i,-\varepsilon_k} g(x)\,dx$$

by integration via the transformation formula. The passage to the limit $k \rightarrow \infty$ yields

$$0 = \int_{\Omega} h(x)\frac{\partial}{\partial x_i}\varphi(x)\,dx + \int_{\Omega} \varphi(x)g(x)D^{e_i} f(x)\,dx + \int_{\Omega} \varphi(x)f(x)D^{e_i} g(x)\,dx$$

for all $\varphi \in C_0^{\infty}(\Omega)$. This implies $D^{\alpha} h = f D^{\alpha} g + g D^{\alpha} f$ for all $\alpha \in \mathbb{N}_0^n$ with $|\alpha| = 1$.

q.e.d.

Theorem 1.10. (Weak chain rule)
On the bounded open set $\Omega \subset \mathbb{R}^n$ let the function $f \in W^{1,p}(\Omega) \cap L^\infty(\Omega)$ be defined with the exponent $1 < p < +\infty$. Furthermore, we have the scalar function $g : \mathbb{R} \to \mathbb{R} \in C^1$. Then the composition $h := g \circ f$ belongs to the class $W^{1,p}(\Omega) \cap L^\infty(\Omega)$ as well, and we have the chain rule

$$D^\alpha h(x) = g'\big(f(x)\big) D^\alpha f(x), \qquad x \in \Omega, \tag{12}$$

for all multi-indices $\alpha \in \mathbb{N}_0^n$ with $|\alpha| = 1$.

Proof:

1. For monomials $g(y) = y^m$, $m \in \mathbb{N}$, we show the chain rule by induction. We start the induction with the evident case $m = 1$. From the validity of the statement for m we infer the correctness of the statement for $m + 1$ with the aid of Theorem 1.9:

$$D^\alpha\Big\{(f(x))^{m+1}\Big\} = D^\alpha\Big\{f(x)(f(x))^m\Big\}$$

$$= (D^\alpha f(x))(f(x))^m + f(x) D^\alpha\Big\{(f(x))^m\Big\}$$

$$= (D^\alpha f(x))(f(x))^m + f(x) m (f(x))^{m-1} D^\alpha f(x)$$

$$= (m+1)(f(x))^m D^\alpha f(x) = g'(f(x)) D^\alpha f(x).$$

2. When

$$g(y) = \sum_{k=0}^m a_k y^k, \qquad a_k \in \mathbb{R}, \quad k = 0, \dots, m$$

is an arbitrary polynomial, we deduce

$$D^\alpha\Big\{g(f(x))\Big\} = \sum_{k=0}^m a_k D^\alpha\Big\{(f(x))^k\Big\} = \sum_{k=0}^m k a_k (f(x))^{k-1} D^\alpha f(x)$$

$$= g'(f(x)) D^\alpha f(x).$$

3. In the general case $g : \mathbb{R} \to \mathbb{R} \in C^1$ we invoke the Weierstraß approximation theorem: We obtain a sequence of polynomials g_k, $k = 1, 2, \dots$, which converge together with their first derivatives g_k' locally uniformly on $\mathbb{R}$. Following part 2, the functions $h_k := g_k(f) \in W^{1,p}(\Omega) \cap L^\infty(\Omega)$ satisfy

$$D^\alpha h_k(x) = g_k'(f(x)) D^\alpha f(x) \qquad \text{for all} \quad \alpha \in \mathbb{N}_0^n \quad \text{with} \quad |\alpha| = 1.$$

For all $\varphi \in C_0^\infty(\Omega)$ this implies

$$\int\limits_\Omega g_k'(f(x))(D^\alpha f(x))\varphi(x)\, dx = \int\limits_\Omega (D^\alpha h_k(x))\varphi(x)\, dx$$

$$= (-1)^{|\alpha|} \int\limits_\Omega h_k(x) D^\alpha \varphi(x)\, dx$$

with arbitrary $|\alpha| = 1$. The passage to the limit $k \to \infty$ yields

$$\int_{\Omega} g'(f(x))(D^{\alpha}f(x))\varphi(x)\,dx = (-1)^{|\alpha|}\int_{\Omega} h(x)\partial^{\alpha}\varphi(x)\,dx$$

for all $\varphi \in C_0^{\infty}(\Omega)$. Finally, we obtain $D^{\alpha}h = g'(f)D^{\alpha}f \in L^p(\Omega)$, $|\alpha| = 1$.
$$\text{q.e.d.}$$

If the function $g : \mathbb{R} \to \mathbb{R}$ satisfies a Lipschitz condition

$$|g(y_1) - g(y_2)| \le C\,|y_1 - y_2| \qquad \text{for all} \quad y_1, y_2 \in \mathbb{R}$$

and the property $f \in W^{1,p}(\Omega)$ holds true, then the composition $h := g \circ f$ belongs to the class $W^{1,p}(\Omega)$ as well. This is shown with the aid of Theorem 1.8, since we have

$$|\triangle_{i,\varepsilon}h(x)| = \left|\frac{g(f(x + \varepsilon e_i)) - g(f(x))}{\varepsilon}\right| \le C\frac{|f(x + \varepsilon e_i) - f(x)|}{|\varepsilon|}$$
$$= C|\triangle_{i,\varepsilon}f(x)| \qquad \text{for all} \quad x \in \Omega \text{ and } 0 < |\varepsilon| < \text{dist}\{x, \mathbb{R}^n \setminus \Omega\}.$$

In order to establish the chain rule in this situation, one needs the a.e.-differentiability for the absolutely continuous function g.
Now we shall prove an important special case of this statement directly.

Theorem 1.11. (Lattice property)
Taking the Sobolev function $f \in W^{1,p}(\Omega)$ with the exponent $1 < p < +\infty$, then the following functions

$$f^+(x) := max\{f(x), 0\}, \quad f^-(x) := -min\{f(x), 0\}, \quad |f|(x) := |f(x)|,$$

$$f_{-c,+c}(x) := \begin{cases} -c, & f(x) \le -c \\ f(x), & -c < f(x) < +c \\ +c, & +c \le f(x) \end{cases}$$

belong to the Sobolev space $W^{1,p}(\Omega)$, and we have

$$Df^+ = \begin{cases} Df, & \text{if } f > 0 \\ 0, & \text{if } f \le 0 \end{cases}, \qquad Df^- = \begin{cases} 0, & \text{if } f \ge 0 \\ -Df, & \text{if } f < 0 \end{cases},$$

$$D|f| = \begin{cases} Df, & \text{if } f > 0 \\ 0, & \text{if } f = 0 \\ -Df, & \text{if } f < 0 \end{cases}, \qquad Df_{-c,+c} = \begin{cases} Df, & \text{if } -c < f < +c \\ 0, & \text{else} \end{cases}.$$

$$(13)$$

Here the symbol $Df = (D^{e_1}f, \ldots, D^{e_n}f)$ denotes the weak gradient of f.

Proof:

1. On account of the identities $f^- = (-f)^+$, $|f| = f^+ + f^-$, and $f_{-c,+c} = (2c - (f - c)^-)^+ - c$ it suffices to investigate the function f^+.

2. We consider $f \in W^{1,p}(\Omega)$ with the exponent $1 < p < +\infty$. Then the element f^+ belongs to $L^p(\Omega)$ as well, and its difference quotient satisfies

$$|\triangle_{i,\varepsilon} f^+(x)| = \left| \frac{f^+(x + \varepsilon e_i) - f^+(x)}{\varepsilon} \right| \leq \left| \frac{f(x + \varepsilon e_i) - f(x)}{\varepsilon} \right| = |\triangle_{i,\varepsilon} f(x)|$$

for all $x \in \Omega$ and all $0 < |\varepsilon| < \mathrm{dist}\{x, \mathbb{R}^n \setminus \Omega\}$. According to Theorem 1.8 the function f^+ belongs to the class $W^{1,p}(\Omega)$ as well.

3. Let the function

$$g(y) := \begin{cases} y, & y > 0 \\ 0, & y \leq 0 \end{cases}$$

be given. For all $\delta > 0$ we approximate this function by the C^1-functions

$$g_\delta(y) := \begin{cases} \sqrt{y^2 + \delta^2} - \delta, & y > 0 \\ 0, & y \leq 0 \end{cases}$$

with their derivatives

$$g_\delta'(y) = \begin{cases} \dfrac{y}{\sqrt{y^2 + \delta^2}}, & y > 0 \\ 0, & y \leq 0 \end{cases}.$$

Evidently, the inequalities

$$0 \leq g_\delta(y) \leq g(y), \quad 0 \leq g_\delta'(y) \leq \begin{cases} 1, & y > 0 \\ 0, & y \leq 0 \end{cases}$$

hold true for all $\delta > 0$, and we observe $g_\delta'(y) \uparrow 1$ ($\delta \downarrow 0$) for all $y > 0$.

4. Assuming $f \in W^{1,p}(\Omega)$, we consider the regularized function $f_\varepsilon \in C^\infty(\Omega)$ for all $\varepsilon > 0$. We differentiate the $C^1(\Omega)$-function

$$h_{\varepsilon,\delta}(x) := g_\delta(f_\varepsilon(x)), \qquad x \in \Omega,$$

and obtain

$$\partial^\alpha h_{\varepsilon,\delta}(x) = g_\delta'(f_\varepsilon(x)) \partial^\alpha f_\varepsilon(x) = g_\delta'(f_\varepsilon(x))(D^\alpha f)_\varepsilon(x)$$

for all $\alpha \in \mathbb{N}_0^n$ with $|\alpha| = 1$. Taking an arbitrary test function $\varphi \in C_0^\infty(\Omega)$, we infer

$$\int_\Omega g_\delta'(f_\varepsilon(x))(D^\alpha f)_\varepsilon(x)\varphi(x)\,dx = \int_\Omega (\partial^\alpha h_{\varepsilon,\delta}(x))\varphi(x)\,dx$$

$$= (-1)^{|\alpha|} \int_\Omega h_{\varepsilon,\delta}(x)\partial^\alpha \varphi(x)\,dx = (-1)^{|\alpha|} \int_\Omega g_\delta(f_\varepsilon(x))\partial^\alpha \varphi(x)\,dx$$

for all $\alpha \in \mathbb{N}_0^n$ with $|\alpha| = 1$.

5. Due to the convergence $f_\varepsilon \to f$ and $(D^\alpha f)_\varepsilon \to D^\alpha f$ for $\varepsilon \to 0$ in $L^p(\Omega)$, Theorem 4.12 from Section 4 in Chapter 2 gives us a subsequence $\varepsilon_k \downarrow 0$, such that $f_{\varepsilon_k} \to f$ and $(D^\alpha f)_{\varepsilon_k} \to D^\alpha f$ a.e. in Ω are correct. Via the Lebesgue convergence theorem we obtain the following identity observing $\varepsilon_k \downarrow 0$:

$$\int_\Omega g'_\delta(f(x))(D^\alpha f(x))\varphi(x)\,dx = (-1)^{|\alpha|} \int_\Omega g_\delta(f(x))\partial^\alpha \varphi(x)\,dx$$

for all $\varphi \in C_0^\infty(\Omega)$. The transition to the limit $\delta \to 0+$ yields

$$\int_{x\in\Omega:f(x)>0} (D^\alpha f(x))\varphi(x)\,dx = (-1)^{|\alpha|} \int_{x\in\Omega:f(x)>0} f(x)\partial^\alpha \varphi(x)\,dx$$

$$= (-1)^{|\alpha|} \int_\Omega f^+(x)\partial^\alpha \varphi(x)\,dx.$$

Finally, we obtain

$$D^\alpha f^+(x) = \begin{cases} D^\alpha f(x), & f > 0 \\ 0, & f \le 0 \end{cases}.$$

q.e.d.

2 Embedding and Compactness

We begin with the fundamental

Theorem 2.1. (Sobolev's embedding theorem)
Let the open bounded set $\Omega \subset \mathbb{R}^n$ with $n \ge 3$ and the exponent $1 \le p < n$ be given. Then the Sobolev space $W_0^{1,p}(\Omega) \subset L^{\frac{np}{n-p}}(\Omega)$ is continuously embedded into the specified Lebesgue space: This means that the following estimate

$$\|f\|_{L^{\frac{np}{n-p}}(\Omega)} \le C\|Df\|_{L^p(\Omega)} \qquad \text{for all} \quad f \in W_0^{1,p}(\Omega) \tag{1}$$

holds true with a constant $C = C(n,p) \in (0, +\infty)$. Here we denote the weak gradient by $Df := (D^{e_1}f, \ldots, D^{e_n}f) \in L^p(\Omega) \times \ldots \times L^p(\Omega)$.

Proof: (L. Nirenberg)

1. Because of the Definition 1.7 from Section 1 it suffices to prove the inequality (1) for all $f \in C_0^\infty(\Omega)$. In this context we need the *generalized Hölder inequality*, which can easily be deduced from Hölder's inequality by induction. For the integer $m \in \mathbb{N}$ with $m \ge 2$ we choose the exponents $p_1, \ldots, p_m \in (1, \infty)$ satisfying $p_1^{-1} + \ldots + p_m^{-1} = 1$. For all $f_j \in L^{p_j}(\Omega)$ with $j = 1, \ldots, m$ then the following inequality holds true:

$$\int\limits_{\Omega} f_1(x) \dots f_m(x)\, dx \le \|f_1\|_{L^{p_1}(\Omega)} \dots \|f_m\|_{L^{p_m}(\Omega)}. \tag{2}$$

2. At first, we deduce the estimate (1) in the case $p = 1$. Noting $f \in C_0^\infty(\Omega)$, we have the following representation for all $x \in \mathbb{R}^n$:

$$f(x) = \int\limits_{-\infty}^{x_i} D^{e_i} f(x_1, \dots, x_{i-1}, t, x_{i+1}, \dots, x_n)\, dt.$$

This implies

$$|f(x)| \le \int\limits_{-\infty}^{x_i} |D^{e_i} f|\, dt \le \int\limits_{-\infty}^{+\infty} |D^{e_i} f|\, dx_i,$$

and consequently

$$|f(x)|^{\frac{n}{n-1}} \le \left(\prod_{i=1}^{n} \int\limits_{-\infty}^{+\infty} |D^{e_i} f|\, dx_i \right)^{\frac{1}{n-1}}.$$

We integrate this inequality successively with respect to the variables $x_1, \dots, x_n$, using each time the generalized Hölder inequality with $p_1 = \dots = p_m = n - 1$ and $m = n - 1$. We then obtain

$$\int\limits_{-\infty}^{+\infty} |f(x)|^{\frac{n}{n-1}}\, dx_1$$

$$\le \left(\int\limits_{-\infty}^{+\infty} |D^{e_1} f|\, dx_1 \right)^{\frac{1}{n-1}} \int\limits_{-\infty}^{+\infty} \prod_{i=2}^{n} \left(\int\limits_{-\infty}^{+\infty} |D^{e_i} f|\, dx_i \right)^{\frac{1}{n-1}} dx_1$$

$$\le \left(\int\limits_{-\infty}^{+\infty} |D^{e_1} f|\, dx_1 \right)^{\frac{1}{n-1}} \prod_{i=2}^{n} \left(\int\limits_{-\infty}^{+\infty}\int\limits_{-\infty}^{+\infty} |D^{e_i} f|\, dx_i dx_1 \right)^{\frac{1}{n-1}}.$$

A similar integration over the variables $x_2, \dots, x_n$ yields

$$\int\limits_{\mathbb{R}^n} |f(x)|^{\frac{n}{n-1}} dx \le \left(\prod_{i=1}^{n} \int\limits_{\mathbb{R}^n} |D^{e_i} f|\, dx \right)^{\frac{1}{n-1}},$$

and finally

$$\|f\|_{\frac{n}{n-1}} \le \left(\prod_{i=1}^{n} \int\limits_{\mathbb{R}^n} |D^{e_i} f|\, dx \right)^{\frac{1}{n}} \le \frac{1}{n} \int\limits_{\Omega} \left(\sum_{i=1}^{n} |D^{e_i} f| \right) dx$$

$$\le \frac{1}{\sqrt{n}} \int\limits_{\Omega} |Df|\, dx = \frac{1}{\sqrt{n}} \|Df\|_1 \tag{3}$$

for all $f \in C_0^\infty(\Omega)$.

3. We now consider the case $1 < p < n$. Here we insert $|f|^\gamma$ with $\gamma > 1$ into (3) and obtain the following relation with the aid of Hölder's inequality and the condition $p^{-1} + q^{-1} = 1$:

$$\| |f|^\gamma \|_{\frac{n}{n-1}} \le \frac{1}{\sqrt{n}} \int_\Omega \left| D|f|^\gamma \right| dx = \frac{\gamma}{\sqrt{n}} \int_\Omega |f|^{\gamma-1} |Df| \, dx$$

$$\le \frac{\gamma}{\sqrt{n}} \| |f|^{\gamma-1} \|_q \|Df\|_p, \tag{4}$$

and consequently

$$\|f\|_{\frac{\gamma n}{n-1}}^\gamma \le \frac{\gamma}{\sqrt{n}} \|f\|_{(\gamma-1)q}^{\gamma-1} \|Df\|_p.$$

Choosing

$$\gamma := \frac{(n-1)p}{n-p} = \frac{np-p}{n-p},$$

we infer

$$\frac{\gamma n}{n-1} = (\gamma-1)q = \frac{np}{n-p}.$$

Finally, we arrive at

$$\|f\|_{\frac{np}{n-p}} \le \frac{\gamma}{\sqrt{n}} \|Df\|_p \qquad \text{for all} \quad f \in C_0^\infty(\Omega).$$

With the constant

$$C := \frac{np-p}{\sqrt{n}(n-p)}$$

the statement above follows.   q.e.d.

Theorem 2.2. (Continuous embedding)
Let the assumptions of Theorem 2.1 with $p > n$ be satisfied. Then we have a constant $C = C(n, p, |\Omega|) \in (0, +\infty)$, such that

$$\|f\|_{C^0(\overline{\Omega})} := \sup_{x \in \overline{\Omega}} |f(x)| \le C\|Df\|_{L^p(\Omega)} \qquad \text{for all} \quad f \in C_0^\infty(\Omega) \tag{5}$$

holds true. This implies $W_0^{1,p}(\Omega) \hookrightarrow C^0(\overline{\Omega})$, which means this Sobolev space is continuously embedded into the space $C^0(\overline{\Omega})$.

Proof:

1. When we have proved this inequality for open bounded sets $\Omega \subset \mathbb{R}^n$ whose measure fulfills $|\Omega| = 1$, we then obtain the inequality stated above by the transformation

$$y = x|\Omega|^{-\frac{1}{n}}, \qquad x \in \Omega.$$

Therefore, we can assume $|\Omega| = 1$ in the sequel.

2. We utilize the inequality (4) and set

$$n' := \frac{n}{n-1} > p' := \frac{p}{p-1}, \quad \delta := \frac{n'}{p'} \in (1, \infty).$$

For all $\gamma \in (1, \infty)$ we therefore obtain the estimate

$$\left\| \frac{\sqrt{n}|f|^{\gamma}}{\|Df\|_p} \right\|_{n'} \leq \gamma \| |f|^{\gamma-1} \|_{p'}.$$

By multiplication with $\frac{\sqrt{n}}{\|Df\|_p}^{\gamma-1}$ this implies

$$\left\| \left(\frac{\sqrt{n}|f|}{\|Df\|_p} \right)^{\gamma} \right\|_{n'} \leq \gamma \left\| \left(\frac{\sqrt{n}|f|}{\|Df\|_p} \right)^{\gamma-1} \right\|_{p'}.$$

Setting $g := \frac{\sqrt{n}}{\|Df\|_p} |f|$ we find

$$\|g^{\gamma}\|_{n'} \leq \gamma \|g^{\gamma-1}\|_{p'} \quad \text{for all} \quad \gamma > 1$$

and consequently

$$\|g\|_{n'\gamma}^{\gamma} \leq \gamma \|g\|_{p'(\gamma-1)}^{\gamma-1} \leq \gamma \|g\|_{p'\gamma}^{\gamma-1}.$$

Finally, we obtain

$$\|g\|_{n'\gamma} \leq \gamma^{\frac{1}{\gamma}} \|g\|_{p'\gamma}^{1-\frac{1}{\gamma}} \quad \text{for all} \quad \gamma > 1.$$

3. We now insert $\gamma := \delta^{\nu}$ with $\nu = 1, 2, \ldots$ into the inequality above and get

$$\|g\|_{n'\delta^{\nu}} \leq \delta^{\nu\delta^{-\nu}} \|g\|_{n'\delta^{\nu-1}}^{1-\delta^{-\nu}}. \tag{6}$$

From (3), the fact that $|\Omega| = 1$, and Hölder's inequality we deduce

$$\|g\|_{n'} = \frac{\sqrt{n}}{\|Df\|_p} \|f\|_{n'} \leq \frac{\|Df\|_1}{\|Df\|_p} \leq 1.$$

With the aid of $|\Omega| = 1$ and Hölder's inequality we see that the sequence $\|g\|_{n'\delta^{\nu}}$, $\nu = 0, 1, 2, \ldots$ increases weakly monotonically. Therefore, we have the following alternative: $\|g\|_{n'\delta^{\nu}} \leq 1$ for all ν – or there exists an index $\lambda > 0$ satisfying $\|g\|_{n'\delta^{\nu}} \leq 1$ for all $\nu \leq \lambda$ and $\|g\|_{n'\delta^{\nu}} > 1$ for all $\nu > \lambda$. In the second case, we obtain the following estimate from the iteration formula (6) for $\mu > \lambda$:

$$\|g\|_{n'\delta^{\mu}} \leq \delta^{\sum_{\nu=\lambda+1}^{\mu} \nu\delta^{-\nu}} \|g\|_{n'\delta^{\lambda}}^{1-\delta^{-(\lambda+1)}} \leq \delta^{\sum_{\nu=1}^{\infty} \nu\delta^{-\nu}} =: c \in \mathbb{R}.$$

In each case we have $\|g\|_{L^{\infty}(\Omega)} \leq c$ and therefore

$$\|f\|_{L^{\infty}(\Omega)} \leq \frac{c}{\sqrt{n}} \|Df\|_p \quad \text{for all} \quad f \in C_0^{\infty}(\Omega).$$

This implies the statement above.   q.e.d.

If one intends to treat eigenvalue problems for partial differential equations with the aid of direct variational methods, we need the subsequent

Theorem 2.3. (Selection theorem of Rellich and Kondrachov)
Let $\Omega \subset \mathbb{R}^n$ with $n \geq 3$ denote a convex open bounded set and let $1 \leq p < n$ be an exponent. For all $1 \leq q < \frac{np}{n-p}$ and all $s \in [0, +\infty)$ then the set

$$\mathcal{K} := \left\{ f \in W_0^{1,p}(\Omega) \cap L^q(\Omega) : \|f\|_{W^{1,p}(\Omega)} \leq s \right\} \subset L^q(\Omega)$$

is compact: This means for each sequence $\{f_k\}_{k=1,2,\ldots} \subset \mathcal{K}$ we can select a subsequence $\{f_{k_l}\}_{l=1,2,\ldots}$ and an element $f \in L^q(\Omega)$ satisfying

$$\lim_{l \to \infty} \|f_{k_l} - f\|_{L^q(\Omega)} = 0.$$

Remarks:

1. F. Rellich discovered this result for the Hilbert spaces $W_0^{1,2}(\Omega) \hookrightarrow L^2(\Omega)$ in the year 1930. The general case was investigated later by Kondrachov.
2. The Banach space $\{\mathcal{B}_1, \|\cdot\|_1\}$ may be continuously embedded into the Banach space $\{\mathcal{B}_2, \|\cdot\|_2\}$. We call $\mathcal{B}_1$ *compactly embedded* into $\mathcal{B}_2$ if the injective mapping $I_1 : \mathcal{B}_1 \to \mathcal{B}_2$ is *compact*; this means that bounded sets in $\mathcal{B}_1$ are mapped on precompact sets in $\mathcal{B}_2$. Here a set $A \subset \mathcal{B}_2$ is called *precompact* if each sequence $\{f_k\}_{k=1,2,\ldots} \subset A$ contains a subsequence converging in $\mathcal{B}_2$ with respect to the norm. Therefore, the theorem above indicates that $W_0^{1,p}(\Omega)$ is compactly embedded into the Lebesgue space $L^q(\Omega)$.

Proof of Theorem 2.3:

1. We start with an arbitrary sequence $\{f_k\}_{k=1,2,\ldots} \subset \mathcal{K}$, and make the transition to a sequence $\{g_k\}_{k=1,2,\ldots} \subset C_0^\infty(\Omega)$ with the property

$$\|g_k - f_k\|_{W^{1,p}(\Omega)} \leq \frac{1}{k}.$$

The last sequence satisfies the restriction

$$\|g_k\|_{W^{1,p}(\Omega)} \leq 1 + s \tag{7}$$

for all $k \in \mathbb{N}$. If we manage to select a subsequence $\{g_{k_l}\}_{l=1,2,\ldots}$ convergent in $L^1(\Omega)$ from the sequence $\{g_k\}_{k=1,2,\ldots}$, then the associate sequence $\{f_{k_l}\}_{l=1,2,\ldots}$ is convergent in $L^1(\Omega)$ as well. Here we observe the inequality

$$\|g_k - f_k\|_{L^1(\Omega)} \leq c\|g_k - f_k\|_{W^{1,p}(\Omega)} \leq \frac{c}{k}.$$

2. In order to show then that the sequence $\{f_{k_l}\}_{l=1,2,\ldots}$ converges even in the space $L^q(\Omega)$ with $1 < q < \frac{np}{n-p}$, we apply the following *interpolation inequality:*

If the exponents $1 \le p \le q \le r$ fulfill $\frac{1}{q} = \frac{\lambda}{p} + \frac{(1-\lambda)}{r}$ with $\lambda \in [0,1]$, we conclude:

$$\|f\|_q \le \|f\|_p^\lambda \|f\|_r^{1-\lambda} \qquad \text{for all} \quad f \in L^r(\Omega). \tag{8}$$

The proof of this interpolation estimate is established via Hölder's inequality. Noting

$$1 = \frac{\lambda q}{p} + \frac{(1-\lambda)q}{r} = \left(\frac{p}{\lambda q}\right)^{-1} + \left(\frac{r}{(1-\lambda)q}\right)^{-1}$$

we obtain

$$\|f\|_q = \left(\int\limits_\Omega |f|^{\lambda q} |f|^{(1-\lambda)q} dx \right)^{\frac{1}{q}}$$

$$\le \left(\int\limits_\Omega |f|^p dx \right)^{\frac{\lambda}{p}} \left(\int\limits_\Omega |f|^r dx \right)^{\frac{1-\lambda}{r}}$$

$$= \|f\|_p^\lambda \|f\|_r^{1-\lambda}.$$

We now choose a number $\lambda \in (0,1)$ with the property $\frac{1}{q} = \lambda + (1-\lambda)\frac{n-p}{np}$, and Theorem 2.1 yields the estimate

$$\|f\|_q \le \|f\|_1^\lambda \|f\|_{\frac{np}{n-p}}^{1-\lambda} \le \|f\|_1^\lambda (C\|Df\|_p)^{1-\lambda}$$

for all $f \in W_0^{1,p}(\Omega)$. Therefore, we have

$$\|f_{k_l} - f_{k_m}\|_q \le \widetilde{C}\|f_{k_l} - f_{k_m}\|_1^\lambda \longrightarrow 0 \qquad \text{for} \quad l,m \to \infty.$$

Consequently, the sequence $\{f_{k_l}\}_{l=1,2,\dots}$ converges in $L^q(\Omega)$ if $\{g_{k_l}\}_{l=1,2,\dots}$ is convergent in $L^1(\Omega)$.

3. It still remains to select a subsequence convergent in $L^1(\Omega)$ from the sequence

$$\{g_k\}_{k=1,2,\dots} \subset C_0^\infty(\Omega).$$

Therefore, we take an arbitrary $\varepsilon \in (0,1]$ and consider the sequence of functions

$$g_{k,\varepsilon}(x) := \frac{1}{\varepsilon^n} \int\limits_{\mathbb{R}^n} \varrho\left(\frac{x-y}{\varepsilon}\right) g_k(y)\, dy = \int\limits_{\mathbb{R}^n} \varrho(z) g_k(x-\varepsilon z)\, dz \in C_0^\infty(\Theta)$$

with

$$\Theta := \Big\{ x \in \mathbb{R}^n : \operatorname{dist}(x,\Omega) < 1 \Big\}.$$

For each fixed $\varepsilon \in (0,1]$ the sequence of functions $\{g_{k,\varepsilon}\}_{k=1,2,\dots}$ is uniformly bounded and equicontinuous: We namely have the following estimates for all $x \in \Theta$:

$$|g_{k,\varepsilon}(x)| \leq \frac{1}{\varepsilon^n} \int\limits_{\mathbb{R}^n} \varrho\left(\frac{x-y}{\varepsilon}\right) |g_k(y)|\, dy \leq \frac{C_0}{\varepsilon^n} \sup_{|z|\leq 1} \varrho(z)$$

and

$$|Dg_{k,\varepsilon}(x)| \leq \frac{1}{\varepsilon^{n+1}} \int\limits_{\mathbb{R}^n} \left| D\varrho\left(\frac{x-y}{\varepsilon}\right) \right| |g_k(y)|\, dy$$

$$\leq \varepsilon^{-(n+1)} \sup_{|z|\leq 1} |D\varrho(z)| \int\limits_{\mathbb{R}^n} |g_k(y)|\, dy$$

$$\leq \frac{C_0}{\varepsilon^{n+1}} \sup_{|z|\leq 1} |D\varrho(z)|.$$

4. We apply the Arzelà-Ascoli theorem as follows: For each $\varepsilon > 0$ we have a subsequence $\{g_{k_l,\varepsilon}\}_{l=1,2,\dots}$ of the sequence $\{g_{k,\varepsilon}\}_{k=1,2,\dots}$ converging uniformly in the set $\overline{\Omega}$. We now set $\varepsilon_m = \frac{1}{m}$ with $m = 1,2,\dots$; with the aid of Cantor's diagonal procedure we select a subsequence $\{g_{k_l}\}_{l=1,2,\dots}$ of the sequence $\{g_k\}_{k=1,2,\dots}$ with the following property: For each fixed $m \in \mathbb{N}$ the sequence $\{g_{k_l,\varepsilon_m}\}_{l=1,2,\dots}$ converges uniformly in the set $\overline{\Omega}$.
5. We have the inequality

$$|g_k(x) - g_{k,\varepsilon}(x)| \leq \int\limits_{|z|\leq 1} \varrho(z)|g_k(x) - g_k(x-\varepsilon z)|\, dz$$

$$\leq \int\limits_{|z|\leq 1} \varrho(z) \int\limits_0^\varepsilon |Dg_k(x-tz)|\, dt\, dz,$$

for all $x \in \Omega$, which implies the estimate

$$\int\limits_{\Omega} |g_k(x) - g_{k,\varepsilon}(x)|\, dx \leq \varepsilon \int\limits_{\Omega} |Dg_k(x)|\, dx \leq C_1\, \varepsilon$$

for all $k \in \mathbb{N}$. Choosing an arbitrary number $\varepsilon > 0$, we obtain the relation

$$\|g_{k_{l_1}} - g_{k_{l_2}}\|_{L^1(\Omega)} \leq \|g_{k_{l_1}} - g_{k_{l_1},\varepsilon_m}\|_{L^1(\Omega)} + \|g_{k_{l_1},\varepsilon_m} - g_{k_{l_2},\varepsilon_m}\|_{L^1(\Omega)}$$

$$+ \|g_{k_{l_2},\varepsilon_m} - g_{k_{l_2}}\|_{L^1(\Omega)}$$

$$\leq (2C_1 + |\Omega|)\varepsilon \qquad \text{for all} \quad l_1, l_2 \geq l_0(\varepsilon).$$

In this context, we determine $m = m(\varepsilon) \in \mathbb{N}$ sufficiently large and afterwards we choose $l_1, l_2 \geq l_0(\varepsilon, m(\varepsilon)) =: l_0(\varepsilon)$. Consequently, $\{g_{k_l}\}_{l=1,2,\dots}$ represents a Cauchy sequence in the space $L^1(\Omega)$ possessing a limit in $L^1(\Omega)$ – according to Theorem 7.6 (Fischer, Riesz) from Section 7 in Chapter 2. $\hfill$ q.e.d.

3 Existence of Weak Solutions

From now on, we require $n \geq 3$ for the space dimension in this chapter. With adequate regularity assumptions, we consider a solution $v = v(x) : \Omega \to \mathbb{R}$ on the open bounded set $\Omega \subset \mathbb{R}^n$ of the following elliptic differential equation in divergence form

$$\mathcal{L}v(x) := \sum_{i,j=1}^{n} \frac{\partial}{\partial x_j}\left(a_{ij}(x)\frac{\partial}{\partial x_i}v(x)\right) + c(x)v(x) = f(x), \qquad x \in \Omega, \qquad (1)$$

under Dirichlet's boundary conditions

$$v(x) = g(x), \qquad x \in \partial\Omega. \qquad (2)$$

Extending the boundary values $g = g(x)$ onto $\overline{\Omega}$, then the function

$$u(x) := v(x) - g(x), \qquad x \in \overline{\Omega}$$

solves the Dirichlet problem

$$-\sum_{i,j=1}^{n} \frac{\partial}{\partial x_j}\left(a_{ij}(x)\frac{\partial}{\partial x_i}u(x)\right) - c(x)u(x)$$
$$= -f(x) + c(x)g(x) + \sum_{i,j=1}^{n} \frac{\partial}{\partial x_j}\left(a_{ij}(x)\frac{\partial}{\partial x_i}g(x)\right), \qquad x \in \Omega, \qquad (3)$$

under zero boundary conditions

$$u(x) = 0, \qquad x \in \partial\Omega. \qquad (4)$$

We now define the bilinear form

$$B(u,v) := \int_{\Omega} \left\{ \sum_{i,j=1}^{n} a_{ij}(x)D^{e_i}u(x)D^{e_j}v(x) - c(x)u(x)v(x) \right\} dx \qquad (5)$$

and the linear form

$$F(v) := \int_{\Omega} \left\{ \Big(-f(x)+c(x)g(x)\Big)v(x) - \sum_{i,j=1}^{n} a_{ij}(x)D^{e_i}g(x)D^{e_j}v(x) \right\} dx. \qquad (6)$$

Here the symbol D^{e_i} again denotes the weak derivative in the direction $e_i = (\delta_{1i}, \ldots, \delta_{ni})$ with $i = 1, \ldots, n$. Multiplying (3) by a test function $\varphi = \varphi(x) \in C_0^{\infty}(\Omega)$, the Gaussian integral theorem gives us the differential equation (3) in the weak form

$$B(u,\varphi) = F(\varphi) \qquad \text{for all} \quad \varphi \in C_0^{\infty}(\Omega) \qquad (7)$$

under zero boundary conditions (4).

Now we fix the assumptions for the coefficients of the differential equation:

$$a_{ij}(x) \in L^\infty(\Omega) \qquad \text{for} \quad i,j = 1,\ldots,n,$$

$$a_{ij}(x) = a_{ji}(x) \qquad \text{a.e. in } \Omega \qquad \text{for} \quad i,j = 1,\ldots,n, \tag{8}$$

$$\frac{1}{M}|\xi|^2 \le \sum_{i,j=1}^{n} a_{ij}(x)\xi_i\xi_j \le M|\xi|^2 \qquad \text{a.e. in } \Omega \qquad \text{for all} \quad \xi \in \mathbb{R}^n$$

and

$$0 \le -c(x) \quad \text{a.e. in } \Omega, \qquad \|c\|_{L^\infty(\Omega)} \le N \tag{9}$$

with the constants $M \in [1, +\infty)$ and $N \in [0, +\infty)$. We work in the Hilbert space $\mathcal{H} := W_0^{1,2}(\Omega)$ with the inner product

$$(u,v)_{\mathcal{H}} := \int_\Omega \Big\{ Du(x) \cdot Dv(x) \Big\}\, dx = \int_\Omega \Big\{ \sum_{i=1}^{n} D^{e_i} u(x) D^{e_i} v(x) \Big\}\, dx, \qquad u, v \in \mathcal{H}. \tag{10}$$

Because of the Sobolev embedding theorem, the induced norm

$$\|u\|_{\mathcal{H}} := \left(\int_\Omega |Du(x)|^2\, dx \right)^{\frac{1}{2}}, \qquad u \in \mathcal{H},$$

is equivalent to the norm of the space $W^{1,2}(\Omega)$ specified in Section 1, Definition 1.2. For the right-hand side and the boundary condition we now assume

$$f(x) \in L^2(\Omega) \qquad \text{and} \qquad g(x) \in W^{1,2}(\Omega). \tag{11}$$

Then $F(v)$ defined in (6) becomes a bounded linear functional on $\mathcal{H}$. More precisely, we have a constant

$$b = b(\|f\|_{L^2(\Omega)}, \|g\|_{W^{1,2}(\Omega)}, M, N) \in [0, +\infty)$$

with the property

$$|F(v)| \le b\|v\|_{\mathcal{H}} \qquad \text{for all} \quad v \in \mathcal{H}. \tag{12}$$

The representation theorem of Fréchet-Riesz in the Hilbert space $\mathcal{H}$ implies the existence of an element $w \in \mathcal{H}$ satisfying

$$(w,v)_{\mathcal{H}} = F(v) \qquad \text{for all} \quad v \in \mathcal{H}. \tag{13}$$

In the special situation $a_{ij}(x) = \delta_{ij}$ for $i,j = 1,\ldots,n$ and $c(x) = 0$ a.e. in Ω, we have already found a solution $u = w$ of the weak differential equation (7). We emphasize that the representation theorem used above has been proved in Chapter 2, Section 6 by direct variational methods.

In the general situation the coefficients satisfy the conditions (8) and (9), and we consider the symmetric bilinear form $B(u,v)$ for $u, v \in \mathcal{H}$ defined in (5). The latter is bounded and coercive, and therefore we have constants $c^{\pm} = c^{\pm}(M, N)$ with $0 < c^- \leq c^+ < +\infty$, such that the inequalities

$$|B(u,v)| \leq c^+ \|u\|_{\mathcal{H}} \|v\|_{\mathcal{H}} \qquad \text{for all} \quad u, v \in \mathcal{H} \qquad (14)$$

and

$$B(u,u) \geq c^- \|u\|_{\mathcal{H}}^2 \qquad \text{for all} \quad u \in \mathcal{H} \qquad (15)$$

are satisfied. Based on the Lax-Milgram theorem (compare Chapter 8, Section 4 Theorem 4.20), we find a bounded symmetric operator $T : \mathcal{H} \to \mathcal{H}$ with $\|T\| \leq c^+$ possessing a bounded inverse $T^{-1} : \mathcal{H} \to \mathcal{H}$ with $\|T^{-1}\| \leq \frac{1}{c^-}$, such that

$$B(u,v) = (Tu,v)_{\mathcal{H}} \qquad \text{for all} \quad u, v \in \mathcal{H}. \qquad (16)$$

This existence result is established by direct variational methods as well. The weak differential equation (7) therefore is transformed as follows:

$$(Tu,v)_{\mathcal{H}} = F(v) = (w,v)_{\mathcal{H}} \qquad \text{for all} \quad v \in \mathcal{H}. \qquad (17)$$

With the element $u := T^{-1}w \in \mathcal{H}$ we obtain a solution of the weak differential equation (7).

Theorem 3.1. *With the assumptions (8) and (9) for the coefficients, the weak differential equation (7) has exactly one solution $u \in \mathcal{H}$ for all data (11).*

Proof: If we have two solutions u_1 and u_2 of (7), then the function $u = u_1 - u_2 \in \mathcal{H}$ satisfies the weak differential equation

$$B(u,\varphi) = 0 \qquad \text{for all} \quad \varphi \in \mathcal{H}. \qquad (18)$$

We especially insert $\varphi = u$ and obtain

$$0 = B(u,u) \geq \frac{1}{M} \int\limits_{\Omega} |Du(x)|^2 \, dx$$

and consequently $u(x) \equiv \text{const}$ in Ω. On account of $u \in W_0^{1,2}(\Omega)$ we conclude $u \equiv 0$ a.e. in Ω and finally $u_1 = u_2$.

$$\text{q.e.d.}$$

We now eliminate the sign condition in (9) and substitute this by the weaker assumption

$$c(x) \in L^{\infty}(\Omega), \qquad \|c\|_{L^{\infty}(\Omega)} \leq N. \qquad (19)$$

In order to solve the equation (7), we consider the transferred bilinear form to a given $\sigma \in \mathbb{R}$, namely

$$B_{\sigma}(u,v) := \int\limits_{\Omega} \left\{ \sum_{i,j=1}^{n} a_{ij}(x) D^{e_i} u(x) D^{e_j} v(x) + \Big(\sigma - c(x) \Big) u(x) v(x) \right\} dx \qquad (20)$$

for $u, v \in \mathcal{H}$. Furthermore, we need the identical bilinear form

$$I(u,v) := \int_{\Omega} u(x)v(x)\,dx, \qquad u, v \in \mathcal{H}. \tag{21}$$

The equation (7) then appears in the equivalent form

$$B_\sigma(u,\varphi) - \sigma I(u,\varphi) = F(\varphi) \qquad \text{for all} \quad \varphi \in \mathcal{H}. \tag{22}$$

We now choose $\sigma \in \mathbb{R}$ so large that

$$\sigma - c(x) \geq 0 \qquad \text{a.e. in} \quad \Omega \tag{23}$$

is satisfied and the bilinear form $B_\sigma(u,v)$ becomes coercive. We additionally need the following

Proposition 3.2. *The mapping $K : \mathcal{H} \to \mathcal{H}$ satisfying*

$$(Ku,v)_{\mathcal{H}} = I(u,v) \qquad \text{for all} \quad u, v \in \mathcal{H} \tag{24}$$

is completely continuous.

Proof: Let $\{u_k\}_{k=1,2,\ldots} \subset \mathcal{H}$ denote a sequence with $\|u_k\|_{\mathcal{H}} \leq$ const for all $k \in \mathbb{N}$. We then consider the continuous linear functionals

$$T_k := I(u_k, \cdot) : \mathcal{H} \to \mathbb{R} \in \mathcal{H}^*, \qquad k = 1, 2, \ldots$$

We apply the representation theorem of Fréchet-Riesz, and for each $k \in \mathbb{N}$ we have exactly one element $v_k =: Ku_k \in \mathcal{H}$, such that

$$I(u_k, \cdot) = T_k(\cdot) = (v_k, \cdot)_{\mathcal{H}} = (Ku_k, \cdot)_{\mathcal{H}}$$

is valid. The selection theorem of Rellich-Kondrachov allows the transition to a subsequence $\{u_{k_l}\}_{l=1,2,\ldots}$ of $\{u_k\}_{k=1,2,\ldots}$ satisfying

$$\|u_{k_l} - u_{k_m}\|_{L^2(\Omega)} \to 0 \ (l, m \to \infty).$$

We obtain

$$\|Ku_{k_l} - Ku_{k_m}\|_{\mathcal{H}} = \|T_{k_l} - T_{k_m}\| \leq c\|u_{k_l} - u_{k_m}\|_{L^2(\Omega)} \to 0 \ (l, m \to \infty).$$

Therefore, the operator $K : \mathcal{H} \to \mathcal{H}$ is completely continuous.   q.e.d.

With the aid of the representations (13), (16), and (24) we transform (22) equivalently for σ from (23):

$$(T_\sigma u, \varphi)_{\mathcal{H}} - \sigma(Ku, \varphi)_{\mathcal{H}} = (w, \varphi)_{\mathcal{H}} \qquad \text{for all} \quad \varphi \in \mathcal{H}. \tag{25}$$

When we insert $\varphi = T_\sigma^{-1} v$ into this equation, we obtain

$$(u,v)_{\mathcal{H}} - \sigma(T_\sigma^{-1} \circ Ku, v)_{\mathcal{H}} = (T_\sigma^{-1}w, v) \qquad \text{for all} \quad v \in \mathcal{H} \qquad (26)$$

and consequently

$$\left(\mathrm{Id}_{\mathcal{H}} - \sigma T_\sigma^{-1} \circ K\right)u = T_\sigma^{-1}w \qquad (27)$$

with the completely continuous operator $T_\sigma^{-1} \circ K : \mathcal{H} \to \mathcal{H}$. According to Fredholm's theorem (compare Chapter 8, Section 6 Theorem 6.15) the null space

$$\mathcal{N} := \left\{ u \in \mathcal{H} : B(u,v) = 0 \text{ for all } v \in \mathcal{H} \right\} \qquad (28)$$

is finite-dimensional with the orthogonal space

$$\mathcal{N}^\perp := \left\{ u \in \mathcal{H} : (u,v)_{\mathcal{H}} = 0 \text{ for all } v \in \mathcal{N} \right\}. \qquad (29)$$

Choosing the right-hand side f and the boundary condition g from (11) such that its representation w from (13) satisfies the condition

$$T_\sigma^{-1}w \in \mathcal{N}^\perp, \qquad (30)$$

then the weak differential equation (7) possesses a solution $u \in \mathcal{H}$. We finally obtain the following

Theorem 3.3. *With the assumptions (8) and (19) the solution space $\mathcal{N}$ of the homogeneous equation from (28) is finite-dimensional. To those data (11), whose linear form (6) allows a representation w from (13) such that $T_\sigma^{-1}w \in \mathcal{N}^\perp$ with $\sigma \in \mathbb{R}$ from (23) is satisfied, the weak differential equation (7) has a solution $u \in \mathcal{H}$.*

4 Boundedness of Weak Solutions

We continue our considerations from Section 3 and quote those results by the added symbol *. We refer the reader to the bilinear form $B(u,v)$ from (5*) with the coefficients (8*) and (19*). With the aid of Moser's iteration method we prove the following

Theorem 4.1. (Stampacchia)
There exists a constant $C = C(M, N, n, |\Omega|) \in (0, +\infty)$, such that each weak solution $u \in \mathcal{H} := W_0^{1,2}(\Omega)$ of the elliptic differential equation

$$B(u,v) = 0 \qquad \text{for all} \quad v \in \mathcal{H} \qquad (1)$$

satisfies the following estimate

$$\|u\|_{L^\infty(\Omega)} \le C\|u\|_{L^2(\Omega)}. \qquad (2)$$

Proof:

1. We refer the reader to the proof of Theorem 2.2 from Section 2 for an orientation. Having already proved the inequality (2) for open bounded sets $\Omega \subset \mathbb{R}^n$ with the measure $|\Omega| = 1$, we obtain the general case by the following transformation

$$y = |\Omega|^{-\frac{1}{n}} x, \qquad x \in \Omega. \tag{3}$$

The coefficients of the weak differential equation then additionally depend on $|\Omega|$. Therefore, we assume $|\Omega| = 1$ in the sequel, and the norm $\|u\|_p := \|u\|_{L^p(\Omega)}$ becomes weakly monotonically increasing with respect to $1 \leq p \leq \infty$ via Hölder's inequality.

2. We choose $K \in (0, +\infty)$ arbitrarily, and consider the function

$$\bar{u}(x) := \begin{cases} K, & u(x) \geq K \\ u(x), & -K < u(x) < K \\ -K, & u(x) \leq -K \end{cases} \tag{4}$$

of the class $W_0^{1,2}(\Omega) \cap L^\infty(\Omega)$. With the exponent

$$\beta \in [+1, +\infty) \tag{5}$$

we insert the test functions

$$v(x) := \bar{u}(x)^\beta, \qquad x \in \Omega, \tag{6}$$

into the weak differential equation (1). Together with the Sobolev embedding theorem, we obtain

$$\int_\Omega c(x)u(x)\bar{u}(x)^\beta \, dx$$

$$= \beta \int_\Omega \left\{ \sum_{i,j=1}^n a_{ij}(x) D^{e_i}\bar{u}(x) D^{e_j}\bar{u}(x) \right\} \bar{u}(x)^{\beta-1} \, dx$$

$$= \frac{4\beta}{(\beta+1)^2} \int_\Omega \left\{ \sum_{i,j=1}^n a_{ij}(x) D^{e_i}\left(\bar{u}(x)^{\frac{1}{2}(\beta+1)}\right) D^{e_j}\left(\bar{u}(x)^{\frac{1}{2}(\beta+1)}\right) \right\} dx$$

$$\geq \frac{4\beta}{M(\beta+1)^2} \left\| D\left(\bar{u}^{\frac{1}{2}(\beta+1)}\right) \right\|_2^2$$

$$\geq \frac{4\beta}{M(\beta+1)^2 C(n,2)^2} \left\| \bar{u}^{\frac{1}{2}(\beta+1)} \right\|_{2\frac{n}{n-2}}^2. \tag{7}$$

3. For all $\beta \in [+1, +\infty)$ and $K \in (0, +\infty)$ we infer

$$\|\bar{u}\|_{\frac{n}{n-2}(\beta+1)}^{\beta+1} \leq \beta M N C(n,2)^2 \|u\|_{\beta+1}^{\beta+1}, \tag{8}$$

if $u \in L^{\beta+1}(\Omega)$ is satisfied. In (8) we pass to the limit $K \to +\infty$ and set

$$\delta := \frac{n}{n-2} \in (+1, +\infty) \qquad \text{and} \qquad \Gamma := MNC(n,2)^2 \in [0, +\infty).$$

Then we find an *iteration inequality*

$$\|u\|_{\delta(\beta+1)} \leq {}^{\beta+1}\!\!\sqrt{\beta+1}\;{}^{\beta+1}\!\!\sqrt{\Gamma}\;\|u\|_{\beta+1} \tag{9}$$

for all $\beta \in [+1, +\infty)$ if $u \in L^{\beta+1}(\Omega)$.

4. Noticing $u \in L^2(\Omega)$ we start the iteration with $\beta = 1$ and obtain

$$\|u\|_{2\delta} \leq \sqrt[2]{2}\sqrt[2]{\Gamma}\,\|u\|_2. \tag{10}$$

We then choose $\beta \in (1, +\infty)$ such that $\beta + 1 = 2\delta$, and from (9) we infer the inequality

$$\|u\|_{2\delta^2} \leq \sqrt[2\delta]{2\delta}\sqrt[2]{2}\sqrt[2\delta]{\Gamma}\sqrt[2]{\Gamma}\,\|u\|_2. \tag{11}$$

Continuation of this procedure yields for all $k \in \mathbb{N}$:

$$\|u\|_{2\delta^k} \leq \left(\prod_{j=0}^{k-1} \sqrt[2\delta^j]{2\delta^j}\right)\left(\prod_{j=0}^{k-1} \sqrt[2\delta^j]{\Gamma}\right)\|u\|_2$$

$$= (\sqrt{2})^{\sum_{j=0}^{k-1}(\frac{1}{\delta})^j}\left\{\prod_{j=0}^{k-1}\left(\sqrt{\delta^j}\right)^{\delta^{-j}}\right\}(\sqrt{\Gamma})^{\sum_{j=0}^{k-1}(\frac{1}{\delta})^j}\|u\|_2 \tag{12}$$

$$\leq (\sqrt{2})^{\sum_{j=0}^{\infty}(\frac{1}{\delta})^j}(\sqrt{\delta})^{\sum_{j=0}^{\infty}j(\frac{1}{\delta})^j}(\sqrt{\Gamma})^{\sum_{j=0}^{\infty}(\frac{1}{\delta})^j}\|u\|_2.$$

Observing $k \to +\infty$, we finally obtain the desired estimate

$$\|u\|_{L^\infty(\Omega)} \leq C(M, N, n, |\Omega|)\|u\|_2. \tag{13}$$

q.e.d.

Now we shall estimate weak solutions of the Dirichlet problem by their boundary values. In the bilinear form (5*) we require

$$c(x) = 0 \qquad \text{a.e. in} \quad \Omega, \tag{14}$$

and we obtain the *Dirichlet-Riemann bilinear form*

$$R(u,v) := \int_\Omega \left\{ \sum_{i,j=1}^n a_{ij}(x)D^{e_i}u(x)D^{e_j}v(x)\right\} dx \tag{15}$$

with the coefficients from (8*). The boundary function is prescribed as follows:

$$g = g(x) \in W^{1,2}(\Omega) \cap C^0(\overline{\Omega}). \tag{16}$$

Theorem 4.2. (L^∞-**boundary-estimate**)
Let $u = u(x) \in W^{1,2}(\Omega)$ denote a weak solution of the differential equation

$$R(u, v) = 0 \qquad \text{for all} \quad v \in \mathcal{H} \tag{17}$$

with the weak boundary values

$$u - g \in W_0^{1,2}(\Omega). \tag{18}$$

Then we have

$$\mu := \inf_{y \in \partial\Omega} g(y) \leq u(x) \leq \sup_{y \in \partial\Omega} g(y) =: \nu \qquad \text{for almost all} \quad x \in \Omega. \tag{19}$$

Proof: Since the problem is invariant with respect to translations, we can always assume $\mu = 0$ by the transition $u(x) \mapsto u(x) - \mu$. We now show

$$u(x) \geq 0 \qquad \text{a.e. in} \quad \Omega. \tag{20}$$

If (20) were violated, we then would consider the nonvanishing function

$$u^-(x) := \begin{cases} u(x), & u(x) < 0 \\ 0, & u(x) \geq 0 \end{cases} \tag{21}$$

of the class $W_0^{1,2}(\Omega)$. When we insert this function into (17), we attain a contradiction with the relation

$$0 = R(u, u^-) > 0. \tag{22}$$

Therefore, the inequality (20) is valid. Due to the invariance with respect to translations, we can additionally achieve $\nu = 0$. Then we can reduce the second part of the inequality (19) to the statement (20) by the reflection $u(x) \mapsto -u(x)$.

q.e.d.

Remark: Further L^∞-estimates for weak solutions are contained in [GT] 8.5.

5 Hölder Continuity of Weak Solutions

We quote the results from Section 3 by the added symbol * and those from Section 4 by the superscript **. With

$$K_r(y) := \left\{ x \in \mathbb{R}^n \ : \ |x - y| \leq r \right\}$$

we denote the closed balls of radius $r \in (0, +\infty)$ about the center $y \in \mathbb{R}^n$. We consider the bilinear form $B(u, v)$ from (5*) again, with the coefficients (8*) and (19*). With the aid of Moser's iteration method we now show the profound

Theorem 5.1. (Moser's inequality)
Let $u = u(x) \in W^{1,2}(\Omega) \cap L^\infty(\Omega)$ with the property $u(x) \geq 0$ a.e. in Ω be a solution of the weak differential inequality

$$B(u,v) \geq 0 \quad \text{for all} \quad v \in \mathcal{H} \quad \text{satisfying} \quad v(x) \geq 0 \quad a.e. \quad in \quad \Omega. \tag{1}$$

Then we have a constant $C = C(M, Nr^2, n) \in (0, +\infty)$, such that the integral means over all balls $K_{4r}(y) \subset \overline{\Omega}$ satisfy the following inequality:

$$\fint_{K_{2r}(y)} u(x)\, dx := \frac{1}{|K_{2r}(y)|} \int_{K_{2r}(y)} u(x)\, dx \leq C \inf_{x \in K_r(y)} u(x). \tag{2}$$

Remarks:

1. With a function $u(x) \in W^{1,2}(\Omega) \cap L^\infty(\Omega)$ we naturally define

$$\inf_{x \in \Omega} u(x) := \inf\Big\{c \in \mathbb{R} : \{x \in \Omega : u(x) \leq c\} \text{ is not a null-set}\Big\}. \tag{3}$$

2. If $\Omega \subset \mathbb{R}^n$ is a domain, Theorem 5.1 implies the *principle of unique continuation*: A nonnegative solution of (1) vanishes on the set Ω, if we have a point $y \in \Omega$ and a radius $r_0 > 0$ such that

$$\inf_{x \in K_r(y)} u(x) = 0 \quad \text{for all balls} \quad K_r(y) \subset \Omega \quad \text{with} \quad 0 < r < r_0.$$

Proof of Theorem 5.1:

1. We choose $r_0 = r_0(n) > 0$ such that

$$|K_{3r_0}(y)| = 1 \tag{4}$$

is valid. For all $0 < r \leq 3r_0$ then the $\|.\|_{L^p(K_r(y))}$-norm becomes monotonically increasing with respect to $1 \leq p \leq +\infty$. Let the point $y \in \Omega$ be chosen to be fixed such that $K_{4r_0}(y) \subset \overline{\Omega}$ is correct. At first, we show the estimate (2) with $r = r_0$ and afterwards we prove the general case by a scaling argument. For measurable functions $v = v(x) : \Omega \to \mathbb{R}$ satisfying

$$0 < \varepsilon \leq v(x) \leq \frac{1}{\varepsilon} \quad \text{a.e. in} \quad \Omega \tag{5}$$

with fixed $\varepsilon > 0$, we define the positive-homogeneous function

$$\|v\|_{p, K_r(y)} := \left(\int_{K_r(y)} v(x)^p\, dx \right)^{\frac{1}{p}} \tag{6}$$

for all $p \in \mathbb{R}$ and all $0 < r \leq 3r_0$. In the interval $p \geq 1$ we obtain the familiar L^p-norm, and we see that

$$\lim_{p \to -\infty} \|v\|_{p,K_r(y)} = \cfrac{1}{\lim\limits_{p \to -\infty} \left(\int\limits_{K_r(y)} \left(\frac{1}{v(x)}\right)^{-p} dx \right)^{\frac{1}{-p}}}$$

$$= \frac{1}{\left\|\frac{1}{v}\right\|_{L^\infty(K_r(y))}} = \frac{1}{\sup\limits_{K_r(y)} \frac{1}{v}} = \inf_{K_r(y)} v.$$

(7)

By the symbol $\eta = \eta_{r,\varrho}(x) : \Omega \to \mathbb{R} \in W^{1,\infty}(\Omega)$ for $0 < r + \varrho \leq 3r_0$ we denote the piecewise linear, radially symmetric, annihilating function with the properties

$$\eta(x) \begin{cases} = 1, & x \in K_r(y) \\ \in [0,1], & x \in K_{r+\varrho}(y) \setminus K_r(y) \\ = 0, & x \in \Omega \setminus K_{r+\varrho}(y) \end{cases}$$

(8)

and

$$|D\eta(x)| \leq \frac{1}{\varrho} \qquad \text{a.e. in} \quad \Omega.$$

(9)

2. Into the weak differential inequality (1) we now insert the following test function

$$v(x) := \eta(x)^2 \bar{u}(x)^\beta, \qquad x \in \Omega,$$

(10)

with

$$\bar{u}(x) := u(x) + \varepsilon, \qquad x \in \Omega,$$

(11)

and the exponents

$$-\infty < \beta < -1 \qquad \text{and} \qquad -1 < \beta < 0.$$

(12)

Here we have chosen $\varepsilon > 0$ in (11) to be fixed. We observe $v \in W_0^{1,2}(\Omega)$ and calculate

$$\int_\Omega c(x)u(x)\bar{u}(x)^\beta \eta(x)^2 \, dx$$

$$\leq \beta \int_\Omega \left\{ \sum_{i,j=1}^n a_{ij}(x) D^{e_i}\bar{u}(x) D^{e_j}\bar{u}(x) \right\} \bar{u}(x)^{\beta-1}\eta(x)^2 \, dx$$

$$+2 \int_\Omega \left\{ \sum_{i,j=1}^n a_{ij}(x) D^{e_i}\bar{u}(x) D^{e_j}\eta(x) \right\} \bar{u}(x)^\beta \eta(x) \, dx$$

$$= \frac{4\beta}{(\beta+1)^2} \int_\Omega \left\{ \sum_{i,j=1}^n a_{ij}(x) D^{e_i}\left(\bar{u}(x)^{\frac{1}{2}(\beta+1)}\eta(x)\right) D^{e_j}\left(\bar{u}(x)^{\frac{1}{2}(\beta+1)}\eta(x)\right) \right\} dx$$

$$-\frac{4\beta}{(\beta+1)^2} \int_\Omega \left\{ \sum_{i,j=1}^n a_{ij}(x) D^{e_i}\eta(x) D^{e_j}\eta(x) \right\} \bar{u}(x)^{\beta+1} \, dx$$

$$+\left(2 - \frac{4\beta}{\beta+1}\right) \int_\Omega \left\{ \sum_{i,j=1}^n a_{ij}(x) D^{e_i}\bar{u}(x) D^{e_j}\eta(x) \right\} \bar{u}(x)^\beta \eta(x) \, dx$$

$$= \frac{4\beta}{(\beta+1)^2} \int_\Omega \left\{ \sum_{i,j=1}^n a_{ij}(x) D^{e_i}\left(\bar{u}(x)^{\frac{1}{2}(\beta+1)}\eta(x)\right) D^{e_j}\left(\bar{u}(x)^{\frac{1}{2}(\beta+1)}\eta(x)\right) \right\} dx$$

$$-\frac{4}{(\beta+1)^2} \int_\Omega \left\{ \sum_{i,j=1}^n a_{ij}(x) D^{e_i}\eta(x) D^{e_j}\eta(x) \right\} \bar{u}(x)^{\beta+1} \, dx$$

$$-4\frac{\beta-1}{(\beta+1)^2} \int_\Omega \left\{ \sum_{i,j=1}^n a_{ij}(x) D^{e_i}\left(\bar{u}(x)^{\frac{1}{2}(\beta+1)}\eta(x)\right) D^{e_j}\eta(x) \right\} \bar{u}(x)^{\frac{1}{2}(\beta+1)} \, dx.$$

$$(13)$$

Then we obtain the following inequality for all $\beta \in (-\infty, -1) \cup (-1, 0)$:

$$\int_\Omega \left\{ \sum_{i,j=1}^n a_{ij}(x) D^{e_i}\left(\bar u(x)^{\frac12(\beta+1)}\eta(x) \right) D^{e_j}\left(\bar u(x)^{\frac12(\beta+1)}\eta(x) \right) \right\} dx =$$

$$\le \frac14 (1+\beta)\left(1+\frac1\beta\right) \int_\Omega c(x) u(x) \bar u(x)^\beta \eta(x)^2\, dx$$

$$+\left(1-\frac1\beta\right) \int_\Omega \left\{ \sum_{i,j=1}^n a_{ij}(x) D^{e_i}\left(\bar u(x)^{\frac12(\beta+1)}\eta(x) \right) D^{e_j}\eta(x) \right\} \bar u(x)^{\frac12(\beta+1)}\, dx$$

$$+\frac1\beta \int_\Omega \left\{ \sum_{i,j=1}^n a_{ij}(x) D^{e_i}\eta(x) D^{e_j}\eta(x) \right\} \bar u(x)^{\beta+1}\, dx$$

$$\le \frac14 (1+\beta)\left(1+\frac1\beta\right) \int_\Omega c(x) u(x) \bar u(x)^\beta \eta(x)^2\, dx$$

$$+\frac12 \int_\Omega \left\{ \sum_{i,j=1}^n a_{ij}(x) D^{e_i}\left(\bar u(x)^{\frac12(\beta+1)}\eta(x) \right) D^{e_j}\left(\bar u(x)^{\frac12(\beta+1)}\eta(x) \right) \right\} dx$$

$$+\left\{ \frac1\beta + \frac12\left(1-\frac1\beta\right)^2 \right\} \int_\Omega \left\{ \sum_{i,j=1}^n a_{ij}(x) D^{e_i}\eta(x) D^{e_j}\eta(x) \right\} \bar u(x)^{\beta+1}\, dx.$$

$$(14)$$

Finally, we deduce the estimate

$$\int_\Omega \left\{ \sum_{i,j=1}^n a_{ij}(x) D^{e_i}\left(\bar u(x)^{\frac12(\beta+1)}\eta(x) \right) D^{e_j}\left(\bar u(x)^{\frac12(\beta+1)}\eta(x) \right) \right\} dx$$

$$\le \frac12 (1+\beta)\left(1+\frac1\beta\right) \int_\Omega c(x) u(x) \bar u(x)^\beta \eta(x)^2\, dx \qquad (15)$$

$$+\left(1+\frac{1}{\beta^2}\right) \int_\Omega \left\{ \sum_{i,j=1}^n a_{ij}(x) D^{e_i}\eta(x) D^{e_j}\eta(x) \right\} \bar u(x)^{\beta+1}\, dx$$

for all $\beta \in (-\infty, -1) \cup (-1, 0)$.

3. We now apply Sobolev's embedding theorem with $p = 2$. Let $\delta \in (1, \frac{n}{n-2}]$ be chosen, and furthermore we assume $0 < r + \varrho \le 3r_0$. We take the definition of η into account and obtain the following estimate for all $\beta \in (-\infty, -1) \cup (-1, 0)$:

$$\|\bar{u}\|_{\delta(\beta+1),K_r}^{\beta+1} = \left(\int_{K_r} |\bar{u}(x)|^{\delta(\beta+1)}\,dx\right)^{\frac{1}{\delta}} = \left\|\bar{u}^{\frac{1}{2}(\beta+1)}\eta\right\|_{L^{2\delta}(K_r)}^2$$

$$\leq \left\|\bar{u}^{\frac{1}{2}(\beta+1)}\eta\right\|_{L^{\frac{2n}{n-2}}(\Omega)}^2 \leq C(n,2)^2\left\|D\big(\bar{u}^{\frac{1}{2}(\beta+1)}\eta\big)\right\|_{L^2(\Omega)}^2$$

$$\leq MC(n,2)^2\int_\Omega\left\{\sum_{i,j=1}^n a_{ij}D^{e_i}\big(\bar{u}^{\frac{1}{2}(\beta+1)}\eta\big)D^{e_j}\big(\bar{u}^{\frac{1}{2}(\beta+1)}\eta\big)\right\}dx$$

$$\leq MC(n,2)^2\left\{\frac{1}{2}|1+\beta|\left|1+\frac{1}{\beta}\right|N + \left(1+\frac{1}{\beta^2}\right)\frac{M}{\varrho^2}\right\}\|\bar{u}\|_{\beta+1,K_{r+\varrho}}^{\beta+1}.\tag{16}$$

4. In part 8 of the proof, we determine a number $p_0 = p_0(M,N,n) > 0$ and a constant $C_0 = C_0(M,N,n) > 0$, such that

$$\|\bar{u}\|_{p_0,K_{3r_0}} \leq C_0\|\bar{u}\|_{-p_0,K_{3r_0}}.\tag{17}$$

We now choose $\delta \in (1,\frac{n}{n-2}]$ and $\nu \in \mathbb{N}_0$ satisfying

$$\begin{aligned}\delta^j p_0 &\in (0,1) \qquad \text{for} \quad j = 0,\dots,\nu-1,\\ \delta^\nu p_0 &\in (1,+\infty).\end{aligned}\tag{18}$$

Taking $j = 0,\dots,\nu$ we consider the balls $K_{\varrho_j} \subset \Omega$ with the radii $\varrho_j := 3r_0 - j\frac{r_0}{\nu}$. Formula (16) then yields a constant $\widetilde{C}_+ = \widetilde{C}_+(M,N,n) > 0$, such that

$$\|\bar{u}\|_{\delta^j p_0,K_{\varrho_j}} \leq \widetilde{C}_+\|\bar{u}\|_{\delta^{j-1}p_0,K_{\varrho_{j-1}}} \qquad \text{for} \quad j = 1,\dots,\nu\tag{19}$$

holds true. An iteration ν-times finally reveals the following estimate:

$$\|\bar{u}\|_{L^1(K_{2r_0})} \leq \|\bar{u}\|_{\delta^\nu p_0,K_{\varrho_\nu}} \leq C_+(M,N,n)\|\bar{u}\|_{p_0,K_{3r_0}}.\tag{20}$$

5. From (16) we obtain the following inequalities for all $\beta \leq -1 - p_0$ abbreviating $\delta := \frac{n}{n-2}$:

$$\|\bar{u}\|_{\delta(\beta+1),K_r}^{\beta+1} \leq MC(n,2)^2\left\{\frac{1}{2}|\beta+1|N + \frac{2M}{\varrho^2}\right\}\|\bar{u}\|_{\beta+1,K_{r+\varrho}}^{\beta+1}$$

$$\leq \frac{\widetilde{C}_-(M,N,n)|\beta+1|}{\varrho^2}\|\bar{u}\|_{\beta+1,K_{r+\varrho}}^{\beta+1}$$

with a constant $\widetilde{C}_- = \widetilde{C}_-(M,N,n) > 0$ and

$$\|\bar{u}\|_{\beta+1,K_{r+\varrho}} \leq \left(\frac{\widetilde{C}_-(M,N,n)|\beta+1|}{\varrho^2}\right)^{\frac{1}{|\beta+1|}}\|\bar{u}\|_{\delta(\beta+1),K_r}\tag{21}$$

assuming $0 < r + \varrho \leq 3r_0$. When we choose

$$\varrho_j := 3r_0 - 2r_0 \sum_{l=1}^{j} \frac{1}{2^l}, \qquad j = 0, 1, 2, \ldots,$$

the relation (21) yields the iteration inequality

$$\|\bar{u}\|_{-\delta^j p_0, K_{\varrho_j}} \leq \widetilde{C}_-^{\frac{\delta-j}{p_0}} (\delta^j p_0)^{\frac{\delta-j}{p_0}} \left(\frac{2^{2j}}{r_0^2}\right)^{\frac{\delta-j}{p_0}} \|\bar{u}\|_{-\delta^{j+1} p_0, K_{\varrho_{j+1}}} \tag{22}$$

for $j = 0, 1, 2, \ldots$ This implies the estimate

$$\|\bar{u}\|_{-p_0, K_{3r_0}} \leq \left\{ \left(\sqrt[p_0]{\widetilde{C}_-}\right)^{\sum_{j=0}^{k} (\frac{1}{\delta})^j} \left(\sqrt[p_0]{\delta}\right)^{\sum_{j=0}^{k} j(\frac{1}{\delta})^j} \left(\sqrt[p_0]{p_0}\right)^{\sum_{j=0}^{k} (\frac{1}{\delta})^j} \right.$$
$$\left. \cdot \left(\sqrt[p_0]{4}\right)^{\sum_{j=0}^{k} j(\frac{1}{\delta})^j} \left(\sqrt[p_0]{r_0^{-2}}\right)^{\sum_{j=0}^{k} (\frac{1}{\delta})^j} \right\} \|\bar{u}\|_{-\delta^{k+1} p_0, K_{\varrho_{k+1}}} \tag{23}$$

$$\leq C_-(M, N, n) \|\bar{u}\|_{-\delta^{k+1} p_0, K_{\varrho_{k+1}}}, \qquad k = 0, 1, 2, \ldots$$

The passage to the limit $k \to \infty$ finally gives

$$\|\bar{u}\|_{-p_0, K_{3r_0}} \leq C_-(M, N, n) \inf_{x \in K_{r_0}} \bar{u}(x). \tag{24}$$

6. From (20), (17), and (24) we obtain

$$\|\bar{u}\|_{L^1(K_{2r_0})} \leq C_+ \|\bar{u}\|_{p_0, K_{3r_0}} \leq C_+ C_0 \|\bar{u}\|_{-p_0, K_{3r_0}} \leq C_+ C_0 C_- \inf_{x \in K_{r_0}} \bar{u}(x).$$

Setting $C = C(M, N, n) := C_+ C_0 C_-$ and observing the independence of this constant from $\varepsilon > 0$, the transition to the limit $\varepsilon \to 0+$ yields the following inequality:

$$\|u\|_{L^1(K_{2r_0})} \leq C(M, N, n) \inf_{x \in K_{r_0}} u(x). \tag{25}$$

This implies Moser's inequality (2) for the case $r = r_0$ with $r_0 = r_0(n) > 0$ from (4).

Having chosen $y \in \Omega$ and $r > 0$ with $K_{4r}(y) \subset \overline{\Omega}$, we then observe the transition from u to

$$u^*(x) := u\left(\frac{r}{r_0} x\right), \qquad x \in K_{4r_0}\left(\frac{r_0}{r} y\right). \tag{26}$$

The function $u^* = u^*(x)$ satisfies a weak differential inequality (1) in $\overset{\circ}{K}_{4r_0}\left(\frac{r_0}{r} y\right)$ with the coefficients $a_{ij} \in L^\infty(K_{4r_0}(\frac{r_0}{r} y))$ defined in (8*) and

$$c \in L^\infty\left(K_{4r_0}\left(\frac{r_0}{r} y\right)\right), \qquad \|c\|_{L^\infty(K_{4r_0}(\frac{r_0}{r} y))} \leq \frac{Nr^2}{r_0^2}.$$

The arguments above therefore yield the inequalities

$$\|u^\star\|_{L^1\left(K_{2r_0}\left(\frac{r_0}{r}y\right)\right)} \leq C(M, Nr^2, n) \inf_{x \in K_{r_0}\left(\frac{r_0}{r}y\right)} u^\star(x)$$

and

$$\frac{r_0^n}{r^n} \int\limits_{K_{2r}(y)} u(x)\,dx \leq C(M, Nr^2, n) \inf_{x \in K_r(y)} u(x).$$

Now the proof of our theorem is complete, if we still show (17).

7. To this aim we deduce a growth condition for Dirichlet's integral from the weak differential inequality. With the annihilating function $\eta(x)$ from (8) for $\varrho = r$ and with $\bar{u}(x)$ from (11), we insert the following test function into the inequality (1):

$$v(x) := \eta(x)^2 \bar{u}(x)^{-1}, \qquad x \in \Omega. \tag{27}$$

We remark that this function coincides with v from (10) for $\beta = -1$! We then obtain

$$\int\limits_{\Omega} c(x)u(x)\bar{u}(x)^{-1}\eta(x)^2\,dx$$

$$\leq -\int\limits_{\Omega} \left\{ \sum_{i,j=1}^{n} a_{ij}(x)D^{e_i}\bar{u}(x)D^{e_j}\bar{u}(x) \right\} \bar{u}(x)^{-2}\eta(x)^2\,dx$$

$$+2\int\limits_{\Omega} \left\{ \sum_{i,j=1}^{n} a_{ij}(x)D^{e_i}\bar{u}(x)D^{e_j}\eta(x) \right\} \bar{u}(x)^{-1}\eta(x)\,dx$$

$$= -\int\limits_{\Omega} \left\{ \sum_{i,j=1}^{n} a_{ij}(x)D^{e_i}\big(\log\bar{u}(x)\big)D^{e_j}\big(\log\bar{u}(x)\big) \right\} \eta(x)^2\,dx$$

$$+2\int\limits_{\Omega} \left\{ \sum_{i,j=1}^{n} a_{ij}(x)\left[\frac{1}{\sqrt{2}}\eta(x)D^{e_i}\big(\log\bar{u}(x)\big)\right]\left[\sqrt{2}D^{e_j}\eta(x)\right] \right\} dx$$

$$\leq -\frac{1}{2}\int\limits_{\Omega} \left\{ \sum_{i,j=1}^{n} a_{ij}(x)D^{e_i}\big(\log\bar{u}(x)\big)D^{e_j}\big(\log\bar{u}(x)\big) \right\} \eta(x)^2\,dx$$

$$+2\int\limits_{\Omega} \left\{ \sum_{i,j=1}^{n} a_{ij}(x)D^{e_i}\eta(x)D^{e_j}\eta(x) \right\} dx.$$

$$\tag{28}$$

We now define the function

$$w(x) := \log\bar{u}(x), \qquad x \in \Omega,$$

and from (28) we infer the estimate

$$\int\limits_{K_r(y)} |Dw(x)|^2 \, dx \leq M \int\limits_{\Omega} \left\{ \sum_{i,j=1}^{n} a_{ij}(x) D^{e_i} w(x) D^{e_j} w(x) \right\} \eta(x)^2 \, dx$$

$$\leq 2M \int\limits_{\Omega} |c(x)| \bar{u}(x) \bar{u}(x)^{-1} \eta(x)^2 \, dx$$

$$+ 4M \int\limits_{\Omega} \left\{ \sum_{i,j=1}^{n} a_{ij}(x) D^{e_i} \eta(x) D^{e_j} \eta(x) \right\} dx$$

$$\leq 2M \left\{ N + \frac{2M}{r^2} \right\} |K_{2r}(y)| \leq C_1(M,N,n) r^{n-2}$$

for $r \leq r_0(n)$. Therefore, the growth condition

$$\int\limits_{K_r(y)} |Dw(x)| \, dx \leq \sqrt{\kappa_n} \, r^{\frac{n}{2}} \left(\int\limits_{K_r(y)} |Dw(x)|^2 \, dx \right)^{\frac{1}{2}} \leq C_2(M,N,n) r^{n-1}$$

(29)

follows for all balls $K_{2r}(y) \subset \Omega$ with $r \leq r_0(n)$. Here κ_n denotes the volume of the n-dimensional unit ball.

8. We now apply the regularity theorem of John and Nirenberg to the function $w(x)$ (see Theorem 6.7 in Section 6): Taking $y \in \Omega$ with $K_{4r_0}(y) \subset \Omega$ we define

$$w_0 := |K_{3r_0}(y)|^{-1} \int\limits_{K_{3r_0}(y)} w(x) \, dx.$$

Then there exists a constant $p_0 = p_0(M,N,n) > 0$, such that

$$\int\limits_{K_{3r_0}(y)} \exp \left\{ p_0 |w(x) - w_0| \right\} dx \leq C_3(M,N,n).$$

(30)

This implies

$$\int\limits_{K_{3r_0}(y)} \exp \left\{ p_0 \left(\pm w(x) \mp w_0 \right) \right\} dx \leq C_3(M,N,n)$$

and consequently

$$\int\limits_{K_{3r_0}(y)} \exp \left\{ \pm p_0 w(x) \right\} \leq e^{\pm p_0 w_0} C_3(M,N,n).$$

We then obtain by multiplication

$$\int\limits_{K_{3r_0}(y)} \exp\left\{p_0 w(x)\right\} dx \cdot \int\limits_{K_{3r_0}(y)} \exp\{-p_0 w(x)\} dx \leq C_3(M, N, n)^2$$

and finally

$$\|\bar{u}\|_{p_0, K_{3r_0}(y)} \leq C_4(M, N, n) \|\bar{u}\|_{-p_0, K_{3r_0}(y)}.$$

This is the desired estimate (17).   q.e.d.

We now prove the important

Theorem 5.2. (de Giorgi, Nash)

Let $u = u(x) \in W^{1,2}(\Omega) \cap L^\infty(\Omega)$ denote a solution of the weak differential equation

$$R(u, v) = 0 \qquad \text{for all} \quad v \in \mathcal{H} = W_0^{1,2}(\Omega) \tag{31}$$

with the Dirichlet-Riemann bilinear form $R(u, v)$ from (15**). Then we have constants $C = C(M, n) \in (0, +\infty)$ and $\alpha = \alpha(M, n) \in (0, 1)$, such that the oscillation estimate

$$\operatorname*{osc}_{K_r(y)} u \leq C \left(\frac{r}{r_0}\right)^\alpha \operatorname*{osc}_{K_{r_0}(y)} u, \qquad 0 < r \leq r_0, \tag{32}$$

holds true for all balls $K_{r_0}(y) \subset \Omega$. Here the oscillation is defined by

$$\operatorname*{osc}_{K_r(y)} u := \sup_{K_r(y)} u - \inf_{K_r(y)} u. \tag{33}$$

Proof:

1. We abbreviate $K_r = K_r(y)$, and for $0 < r \leq \frac{1}{4}r_0$ we introduce the quantities

$$M_4 = \sup_{K_{4r}} u, \quad m_4 = \inf_{K_{4r}} u, \quad M_1 = \sup_{K_r} u, \quad m_1 = \inf_{K_r} u.$$

The functions $M_4 - u$ and $u - m_4$ are nonnegative in $K_{4r} \subset K_{r_0} \subset \Omega$, and they satify the following weak differential equation (31) there. Moser's inequality now yields

$$|K_{2r}|^{-1} \int\limits_{K_{2r}} \left\{M_4 - u(x)\right\} dx \leq C(M, n)(M_4 - M_1),$$

$$|K_{2r}|^{-1} \int\limits_{K_{2r}} \left\{u(x) - m_4\right\} dx \leq C(M, n)(m_1 - m_4). \tag{34}$$

Addition gives us

$$M_4 - m_4 \leq C\left\{(M_4 - m_4) - (M_1 - m_1)\right\}$$

and

$$M_1 - m_1 \leq \left(1 - \frac{1}{C}\right)(M_4 - m_4),$$

respectively. Therefore, we obtain the oscillation estimate

$$\underset{K_r}{\mathrm{osc}}\, u \leq \gamma \underset{K_{4r}}{\mathrm{osc}}\, u \quad \text{with} \quad \gamma := 1 - \frac{1}{C} \in (0,1). \tag{35}$$

2. We then consider the monotonically increasing function

$$\omega(r) := \underset{K_r}{\mathrm{osc}}\, u, \quad 0 < r \leq r_0, \tag{36}$$

with the growth property

$$\omega(r) \leq \gamma \omega(4r) \quad \text{for} \quad 0 < r \leq \frac{1}{4}r_0. \tag{37}$$

To each number $r \in (0, \frac{1}{4}r_0]$ we have an integer $k \in \mathbb{N}$ satisfying

$$\left(\frac{1}{4}\right)^{k+1} r_0 < r \leq \left(\frac{1}{4}\right)^k r_0, \tag{38}$$

and we choose $\alpha = \alpha(M, n) \in (0, 1)$, such that

$$\gamma \leq \left(\frac{1}{4}\right)^\alpha. \tag{39}$$

From (37)-(39) we infer the estimate

$$\omega(r) \leq \omega\left(\left(\frac{1}{4}\right)^k r_0\right) \leq \gamma^k \omega(r_0)$$

$$\leq \left(\frac{1}{4^k}\right)^\alpha \omega(r_0) \leq \left(\frac{4r}{r_0}\right)^\alpha \omega(r_0), \quad 0 < r \leq \frac{1}{4}r_0,$$

or equivalently

$$\underset{K_r}{\mathrm{osc}}\, u \leq 4^\alpha \left(\frac{r}{r_0}\right)^\alpha \underset{K_{r_0}}{\mathrm{osc}}\, u, \quad 0 < r \leq r_0. \tag{40}$$

The monotonicity of $\omega(r)$ namely implies (40) for $\frac{1}{4}r_0 \leq r \leq r_0$.   q.e.d.

Remarks:

1. Requiring an exterior cone condition for the domain Ω, one can even prove Hölder continuity of the solution up to the boundary. We refer the reader to Theorem 7.2 from Section 7 in this context.
2. With suitable assumptions on the coefficient matrix $(a_{ij}(x))_{i,j=1,\ldots,n}$, one obtains higher regularity of the solution by the Schauder theory from Chapter 9. Here one should locally reconstruct the weak solution by the classical $C^{2+\alpha}$-solution.

Theorem 5.3. (J. Moser)

Let $u = u(x) \in W^{1,2}(\mathbb{R}^n) \cap L^\infty(\mathbb{R}^n)$ denote an entire solution of the weak differential equation

$$R(u, v) = 0 \qquad \text{for all} \quad v \in C_0^\infty(\mathbb{R}^n). \tag{41}$$

Then we have

$$u(x) \equiv \text{const} \qquad \text{in} \quad \mathbb{R}^n.$$

Proof: We infer the following estimate from Theorem 5.2:

$$\operatorname*{osc}_{K_r(0)} u \le C\left(\frac{r}{r_0}\right)^\alpha \operatorname*{osc}_{\mathbb{R}^n} u, \qquad 0 < r \le r_0 < +\infty. \tag{42}$$

The passage to the limit $r_0 \to +\infty$ yields

$$\operatorname*{osc}_{K_r(0)} u = 0 \qquad \text{for all} \quad 0 < r < +\infty.$$

Therefore, the solution u is constant. $\hfill$ q.e.d.

For later use in Section 9 we still provide the following

Theorem 5.4. (Harnack-Moser inequality)

Let $u = u(x) \in W^{1,2}(\Omega) \cap C^0(\overline{\Omega})$ with the property $u(x) > 0$, $x \in \overline{\Omega}$ be a positive solution of the weak differential equation

$$R(u, v) = 0 \qquad \text{for all} \quad v \in \mathcal{H}.$$

Then we have a constant $C = C(M, n) \in (0, +\infty)$, such that the following inequality

$$\sup_{x \in K_r(y)} u(x) \le C \inf_{x \in K_r(y)} u(x)$$

is fulfilled for all balls $K_{4r}(y) \subset \overline{\Omega}$.

Proof: We have to supplement the proof of Moser's inequality in Theorem 5.1 as follows: We insert the test functions (10) with arbitrary positive powers $\beta \in (0, +\infty)$ into the weak differential equation. Now the constant $N = 0$ vanishes, the inequality (13) turns into an equation $0 = \ldots$, and via the estimates (14) and (15) we arrive at the decisive inequality

$$\|\bar{u}\|_{\delta(\beta+1), K_r}^{\beta+1} \le M C(n, 2)^2 \left(1 + \frac{1}{\beta^2}\right) \frac{M}{\varrho^2} \|\bar{u}\|_{\beta+1, K_{r+\varrho}}^{\beta+1} \qquad \text{for all} \quad \beta \in (0, +\infty) \tag{43}$$

parallel to (16). When we define $p_1 := \delta^\nu p_0 \in (1 + \infty)$ from (18), we easily obtain a constant $\widetilde{C}_{++}(M, n)$ such that the estimate

$$\|\bar{u}\|_{\delta(\beta+1), K_r} \le \left(\frac{\widetilde{C}_{++}(M, n)}{\varrho^2}\right)^{\frac{1}{\beta+1}} \|\bar{u}\|_{\beta+1, K_{r+\varrho}} \qquad \text{for all} \quad \beta \ge p_1 - 1 \tag{44}$$

holds true - parallel to (21) - with $\delta := \frac{n}{n-2}$. Now we introduce the radii

$$\varrho_j := 2r_0 - r_0 \sum_{l=1}^{j} \frac{1}{2^l}, \qquad j = 0, 1, 2, \ldots$$

and employ Moser's iteration technique - as described in (22) and (23). Thus we obtain a constant $C_{++}(M, n)$ such that

$$\|\bar{u}\|_{\delta^{k+1}p_1, K_{\varrho_{k+1}}} \leq$$

$$\left\{ \left(\sqrt[p_1]{\widetilde{C}_{++}} \right)^{\sum_{j=0}^{k} (\frac{1}{\delta})^j} \left(\sqrt[p_1]{4} \right)^{\sum_{j=0}^{k} (j+1)(\frac{1}{\delta})^j} \left(\sqrt[p_1]{r_0^{-2}} \right)^{\sum_{j=0}^{k} (\frac{1}{\delta})^j} \right\} \|\bar{u}\|_{p_1, K_{\varrho_0}} \qquad (45)$$

$$\leq C_{++}(M, n) \|\bar{u}\|_{p_1, K_{2r_0}}, \quad k = 0, 1, 2, \ldots$$

is valid. Then we evaluate the limit

$$\lim_{p \to +\infty} \|\bar{u}\|_{p, K_{r_0}(y)} = \sup_{x \in K_{r_0}(y)} \bar{u}(x) \qquad (46)$$

in formula (45) and arrive at the estimate

$$\sup_{x \in K_{r_0}(y)} \bar{u}(x) \leq C_{++}(M, n) \|\bar{u}\|_{p_1, K_{2r_0}}. \qquad (47)$$

Combining this inequality with the estimates (20), (17), and (24) we finally obtain the Harnack-Moser inequality.   q.e.d.

6 Weak Potential-theoretic Estimates

Let $\Omega \subset \mathbb{R}^n$ denote an open ball with radius $R > 0$ about the center $x_0 \in \mathbb{R}^n$. By the symbol ω_n we denote the area of the unit sphere in $\mathbb{R}^n$.

Definition 6.1. *For the numbers $\mu \in (0, 1]$ we define the* Riesz operator

$$\mathbb{V}_\mu f(x) := \int_\Omega |x - y|^{n(\mu-1)} f(y) \, dy \qquad \text{for all} \quad f \in C_0^\infty(\Omega). \qquad (1)$$

Proposition 6.2. *The linear operator $\mathbb{V}_\mu : L^1(\Omega) \to L^1(\Omega)$ is continuous for all $\mu \in (0, 1]$ and satisfies*

$$\|\mathbb{V}_\mu f\|_{L^1(\Omega)} \leq \frac{\omega_n}{n\mu} (2R)^{n\mu} \|f\|_{L^1(\Omega)} \qquad \text{for all} \quad f \in L^1(\Omega). \qquad (2)$$

Proof: We choose $f \in C_0^\infty(\Omega)$ and estimate as follows:

$$\|\nabla_\mu f\|_{L^1(\Omega)} = \int_\Omega |\nabla_\mu f(x)|\, dx$$

$$\leq \int_\Omega \left\{ \int_\Omega |x - y|^{n(\mu-1)} |f(y)|\, dy \right\} dx \tag{3}$$

$$\leq \int_\Omega |f(y)| \left\{ \int_{x:|x-y|\leq 2R} |x - y|^{n(\mu-1)}\, dx \right\} dy.$$

In polar coordinates we deduce

$$\int_{x:|x-y|\leq 2R} |x - y|^{n(\mu-1)}\, dx = \int_0^{2R} \varrho^{n\mu-n} \omega_n \varrho^{n-1}\, d\varrho = \frac{\omega_n}{n\mu}(2R)^{n\mu}. \tag{4}$$

From (3) and (4) we infer the statement (2).   q.e.d.

Definition 6.3. *For $1 \leq p \leq +\infty$ the measurable function f belongs to Morrey's class of functions $M^p(\Omega)$ if and only if*

$$\int_{\Omega \cap K_r(x)} |f(y)|\, dy \leq L r^{n\left(1-\frac{1}{p}\right)} \qquad \text{for all} \quad x \in \Omega, \quad r > 0 \tag{5}$$

is satisfied, with a constant $L \in [0, +\infty)$.

Remark: Evidently $L^p(\Omega) \subset M^p(\Omega)$ holds true for all $1 \leq p \leq +\infty$.

When we remember part 7 in the proof of Theorem 5.1 from Section 5, we should concentrate on the class $M^n(\Omega)$.

Proposition 6.4. *Let $f \in M^n(\Omega)$ and $\frac{1}{n} < \mu \leq 1$ be satisfied. Then we have*

$$|\nabla_\mu f(x)| \leq (2R)^{n\mu-1} \frac{n-1}{n\mu-1} L \qquad \text{a.e. in} \quad \Omega. \tag{6}$$

Proof: We fix the point $x \in \Omega$ and consider the function

$$\Phi(r) := \int_{\Omega \cap K_r(x)} |f(y)|\, dy, \qquad 0 < r < 2R, \tag{7}$$

with the derivative

$$\Phi'(r) = \int_{\Omega \cap \partial K_r(x)} |f(y)|\, d\sigma(y). \tag{8}$$

Then we obtain the following estimate for almost all $x \in \Omega$

$$|\mathbb{V}_\mu f(x)| \le \int_\Omega |y - x|^{n\mu - n} |f(y)|\, dy$$

$$= \int_0^{2R} r^{n\mu - n} \left\{ \int_{\Omega \cap \partial K_r(x)} |f(y)|\, d\sigma(y) \right\} dr = \int_0^{2R} r^{n\mu - n} \Phi'(r)\, dr$$

$$= \left[r^{n\mu - n} \Phi(r) \right]_{0+}^{2R} - (n\mu - n) \int_0^{2R} r^{n\mu - n - 1} \Phi(r)\, dr$$

$$\le (2R)^{n\mu - n} L(2R)^{n-1} + n(1 - \mu) \int_0^{2R} r^{n\mu - n - 1} L r^{n-1}\, dr$$

$$= L \left\{ (2R)^{n\mu - 1} + n(1 - \mu) \frac{1}{n\mu - 1} \left[r^{n\mu - 1} \right]_0^{2R} \right\}$$

$$= L(2R)^{n\mu - 1} \frac{n - 1}{n\mu - 1},$$

and consequently (6). $\hspace{6cm}$ q.e.d.

Proposition 6.5. *The functions $f \in M^n(\Omega)$ are subject to the following estimate*

$$\int_\Omega \exp \left\{ \frac{\gamma}{(n-1)L} |\mathbb{V}_{\frac{1}{n}} f(x)| \right\} dx \le C(n, \gamma) R^n \tag{9}$$

for each $\gamma \in (0, \frac{1}{e})$, with a constant $C = C(n, \gamma) > 0$.

Proof: For $k = 1, 2, \ldots$ we note that

$$|x - y|^{1-n} = |x - y|^{n \left(\frac{1}{nk} - 1 \right) \frac{1}{k}} |x - y|^{n \left(\frac{1}{nk} + \frac{1}{n} - 1 \right) \left(1 - \frac{1}{k} \right)}.$$

With the aid of Hölder's inequality we deduce

$$|\mathbb{V}_{\frac{1}{n}} f(x)|$$

$$\le \int_\Omega \left\{ |x - y|^{n \left(\frac{1}{nk} - 1 \right) \frac{1}{k}} |f(y)|^{\frac{1}{k}} \right\} \left\{ |x - y|^{n \left(\frac{1}{nk} + \frac{1}{n} - 1 \right) \left(1 - \frac{1}{k} \right)} |f(y)|^{1 - \frac{1}{k}} \right\} dy$$

$$\le \left(\int_\Omega |x - y|^{n \left(\frac{1}{nk} - 1 \right)} |f(y)|\, dy \right)^{\frac{1}{k}} \left(\int_\Omega |x - y|^{n \left(\frac{1}{nk} + \frac{1}{n} - 1 \right)} |f(y)|\, dy \right)^{1 - \frac{1}{k}}$$

and therefore

$$|\mathbb{V}_{\frac{1}{n}} f(x)|^k \le \left(\mathbb{V}_{\frac{1}{nk}} |f|(x) \right) \left(\mathbb{V}_{\left(\frac{1}{n} + \frac{1}{nk} \right)} |f|(x) \right)^{k-1} \tag{10}$$

for all $x \in \Omega$ and $k = 1, 2, \ldots$

Via Proposition 6.2 and 6.4 we estimate for $k = 1, 2, \ldots$ as follows:

$$\int_\Omega |\mathbb{V}_{\frac{1}{n}} f(x)|^k \, dx \leq \left\{ \sup_{x \in \Omega} \mathbb{V}_{(\frac{1}{n} + \frac{1}{nk})} |f|(x) \right\}^{k-1} \int_\Omega \mathbb{V}_{\frac{1}{nk}} |f|(x) \, dx$$

$$\leq (2R)^{\frac{k-1}{k}} \{k(n-1)\}^{k-1} L^{k-1} k \, \omega_n (2R)^{\frac{1}{k}} \|f\|_{L^1(\Omega)}$$

$$\leq 2R \, k^k (n-1)^{k-1} L^{k-1} \omega_n L R^{n-1}$$

$$= 2 \frac{\omega_n}{n-1} R^n \{(n-1)L\}^k k^k.$$

Consequently, we arrive at

$$\int_\Omega \frac{1}{k!} \left\{ \frac{\gamma}{(n-1)L} |\mathbb{V}_{\frac{1}{n}} f(x)| \right\}^k dx \leq 2 \frac{\omega_n}{n-1} R^n \frac{(\gamma k)^k}{k!} \qquad \text{for} \quad k = 0, 1, 2, \ldots$$

$$(11)$$

The summation over $k = 0, 1, 2, \ldots$ yields

$$\int_\Omega \exp \left\{ \frac{\gamma}{(n-1)L} |\mathbb{V}_{\frac{1}{n}} f(x)| \right\} dx \leq 2 \frac{\omega_n}{n-1} R^n \sum_{k=0}^\infty \frac{(\gamma k)^k}{k!}.$$

We investigate the convergence of the series $\sum_{k=0}^\infty a_k$ by the quotient test with $a_k := \dfrac{(\gamma k)^k}{k!}$:

$$\frac{a_{k+1}}{a_k} = \frac{\{\gamma(k+1)\}^{k+1} k!}{(k+1)!(\gamma k)^k} = \gamma \left(1 + \frac{1}{k}\right)^k \stackrel{k \to \infty}{\longrightarrow} \gamma e < 1.$$

Therefore, we find a constant $C = C(n, \gamma) \in (0, +\infty)$ satisfying

$$\int_\Omega \exp \left\{ \frac{\gamma}{(n-1)L} |\mathbb{V}_{\frac{1}{n}} f(x)| \right\} dx \leq C(n, \gamma) R^n.$$

q.e.d.

Proposition 6.6. *We take $u = u(x) \in W^{1,1}(\Omega)$ and set*

$$u_0 := \frac{1}{|\Omega|} \int_\Omega u(x) \, dx.$$

Then we have the inequality

$$|u(x) - u_0| \leq \frac{2^n}{n \kappa_n} \int_\Omega |x - y|^{1-n} |Du(y)| \, dy, \qquad (12)$$

where κ_n denotes the volume of the n-dimensional unit ball.

Proof: Because of the Meyers-Serrin theorem it suffices to prove the inequality (12) in the class of functions $u = u(x) \in C^1(\Omega) \cap W^{1,1}(\Omega)$. We choose $x, y \in \Omega$ arbitrarily and note that

$$u(x) - u(y) = -\int_0^{|x-y|} \frac{d}{dr} u(x + r\zeta) \, dr \quad \text{with} \quad \zeta := \frac{y - x}{|y - x|}.$$

We integrate over Ω with respect to y and obtain

$$|\Omega|(u(x) - u_0) = -\int_\Omega \left\{ \int_0^{|x-y|} \frac{d}{dr} u(x + r\zeta) \, dr \right\} dy.$$

Now we define the $L^1(\mathbb{R}^n)$-function

$$v(x) := \begin{cases} |Du(x)|, & x \in \Omega \\ 0, & x \notin \Omega \end{cases}.$$

On account of $|\frac{d}{dr} u(x + r\zeta)| \leq |Du(x + r\zeta)|$ we get the estimate

$$|u(x) - u_0| \leq \frac{1}{|\Omega|} \int_\Omega \left\{ \int_0^{|x-y|} |Du(x + r\zeta)| \, dr \right\} dy$$

$$\leq \frac{1}{|\Omega|} \int_{K_{2R}(x)} \left\{ \int_0^\infty v(x + r\zeta) \, dr \right\} dy$$

$$= \frac{1}{|\Omega|} \int_0^\infty \left\{ \int_{K_{2R}(x)} v(x + r\zeta) \, dy \right\} dr.$$

We introduce polar coordinates $y = x + \varrho\zeta$, and for fixed $r \in (0, +\infty)$ we obtain

$$\int_{K_{2R}(x)} v(x + r\zeta) \, dy = \int_{|\zeta|=1} \left\{ \int_0^{2R} v(x + r\zeta) \varrho^{n-1} \, d\varrho \right\} d\sigma(\zeta)$$

$$= \frac{(2R)^n}{n} \int_{|\zeta|=1} v(x + r\zeta) \, d\sigma(\zeta)$$

and consequently

$$|u(x) - u_0| \leq \frac{(2R)^n}{n|\Omega|} \int_0^\infty \left\{ \int_{|\zeta|=1} v(x + r\zeta) \, d\sigma(\zeta) \right\} dr. \qquad (13)$$

With the notation $z = x + r\zeta$, $dz = |x - z|^{n-1} dr\, d\sigma(\zeta)$ and the definition of v we infer the following inequality from (13):

$$|u(x) - u_0| \leq \frac{2^n}{n\kappa_n} \int_\Omega |x - z|^{1-n} |Du(z)|\, dz.$$

q.e.d.

We summarize our results to the subsequent

Theorem 6.7. (John, Nirenberg)
Let the function $u = u(x) \in W^{1,1}(\Omega)$ satisfy the growth condition

$$\int_{\Omega \cap K_r(y)} |Du(x)|\, dx \leq L r^{n-1} \qquad \text{for all } y \in \Omega, \quad r > 0 \tag{14}$$

with a constant $L > 0$. Then we have a constant $C = C(n, \gamma) > 0$ for each $\gamma \in (0, \frac{1}{e})$, such that

$$\int_\Omega \exp\left\{ \frac{n\kappa_n\gamma}{2^n(n-1)L} |u(x) - u_0| \right\} dx \leq C(n, \gamma) R^n \tag{15}$$

holds true.

Proof: Due to (14) the function $f(x) := |Du(x)|$, $x \in \Omega$, belongs to Morrey's class $M^n(\Omega)$. From Proposition 6.6 we infer

$$\frac{n\kappa_n}{2^n} |u(x) - u_0| \leq \mathbb{V}_{\frac{1}{n}} f(x), \qquad x \in \Omega.$$

Then Proposition 6.5 yields the desired estimate (15).

q.e.d.

Now we require a higher growth condition in (14) and deduce the Hölder continuity directly. In this context we modify Proposition 6.4 to the following

Proposition 6.8. *Let $f \in M^p(\Omega)$ with $n < p < +\infty$ be given. Then we have the estimate*

$$|\mathbb{V}_{\frac{1}{n}} f(x)| \leq L \cdot C(n, p) \cdot R^\alpha \qquad \text{a.e. in } \Omega \tag{16}$$

with the Hölder constant $C(n, p) \in (0, +\infty)$ and the Hölder exponent $\alpha = 1 - \frac{n}{p} \in (0, 1)$.

Proof: We follow the arguments in the proof of Proposition 6.4 utilizing (7) and (8). Then we obtain the subsequent estimate for almost all $x \in \Omega$, namely

$$|\mathbb{V}_{\frac{1}{n}} f(x)| \leq \int_0^{2R} r^{1-n} \Phi'(r)\, dr = \left[r^{1-n}\Phi(r) \right]_{0+}^{2R} + (n-1) \int_0^{2R} r^{-n} \Phi(r)\, dr$$

$$\leq (2R)^{1-n}\Phi(2R) + (n-1)L \int_0^{2R} r^{-\frac{n}{p}}\, dr \leq L\left\{ (2R)^{1-\frac{n}{p}} + \frac{n-1}{1-\frac{n}{p}} \left[r^{1-\frac{n}{p}} \right]_0^{2R} \right\}$$

$$= L\left\{ 2^\alpha + \frac{n-1}{1-\frac{n}{p}} 2^\alpha \right\} \cdot R^\alpha =: LC(n, p) \cdot R^\alpha.$$

q.e.d.

In order to establish regularity for solutions of variational problems, we prove the fundamental

Theorem 6.9. (C.B. Morrey)

Let $\Theta \subset \mathbb{R}^n$ denote a bounded domain where the function $u = u(x) \in W^{1,1}(\Theta)$ may satisfy Morrey's growth condition

$$\int\limits_{\Theta \cap K_r(x)} |Du(y)|\, dy \leq Lr^{n-\frac{n}{p}} \quad \text{for almost all } x \in \Theta \quad \text{and all } r > 0; \quad (17)$$

with $n < p < +\infty$ and $L \in (0, +\infty)$.
Then we find a constant $C = C(n, p, \Theta_0) > 0$ for each open set $\Theta_0 \subset\subset \Theta$, such that the Hölder estimate

$$|u(y) - u(z)| \leq L \cdot C \cdot |y - z|^\alpha \quad \text{for all} \quad y, z \in \Theta_0 \quad (18)$$

holds true; with the Hölder exponent $\alpha = 1 - \frac{n}{p} \in (0, 1)$.

Proof: We take $y, z \in \Theta_0$ with $R := |y - z| \leq \text{dist}(\Theta_0, \partial\Theta)$ and define

$$\Omega := \{x \in \mathbb{R}^n : |x - x_0| \leq R\},$$

where $x_0 := y$ is chosen. Due to (17), the function

$$f(x) := |Du(x)|, \quad x \in \Omega$$

belongs to Morrey's class $M^p(\Omega)$. Proposition 6.8 in combination with Proposition 6.6 implies

$$|u(x) - u_0| \leq \frac{2^n}{n\kappa_n} \cdot L \cdot C(n, p) \cdot R^\alpha \quad \text{a.e. in } \Omega. \quad (19)$$

Now we arrive at the inequality

$$|u(y) - u(z)| \leq |u(y) - u_0| + |u(z) - u_0| \leq L \cdot C(n, p, \Theta_0) \cdot |y - z|^\alpha$$

for all $y, z \in \Theta_0 \subset\subset \Theta$.   q.e.d.

Remark: When we require the following growth condition for Dirichlet's integral, namely

$$\int\limits_{\Theta \cap K_r(x)} |Du(y)|^2\, dy \leq Lr^{n-\frac{2n}{p}} \quad \text{for almost all } x \in \Theta \quad \text{and all } r > 0 \quad (20)$$

with $2 \leq n < p < +\infty$ and $0 < L < +\infty$, the Morrey growth condition (17) is obviously satisfied. This regularity criterion has been originally invented by C.B. Morrey in the case $n = 2$.

Finally we note the

Theorem 6.10. (Morrey's embedding theorem)
Let $\Theta \subset \mathbb{R}^n$ denote a bounded domain and $u \in W_0^{1,p}(\Theta)$ a Sobolev function with the exponent $n < p < +\infty$. Then u belongs to the class $C^\alpha(\overline{\Theta})$ with the Hölder exponent $\alpha = 1 - \frac{n}{p}$.

Proof: We continue u trivially beyond $\partial\Theta$ and preserve the $W^{1,1}$-regularity. On account of

$$Du \in L^p(\Theta) \subset M^p(\Theta) \quad \text{with} \quad n < p < +\infty,$$

we have Morrey's growth condition globally on Θ. Then Theorem 6.9 implies our corollary.   q.e.d.

7 Boundary Behavior of Weak Solutions

We continue the considerations from Section 5 and need the following variant of Moser's inequality in this section.

Theorem 7.1. (Trudinger)
Let $u = u(x) \in W^{1,2}(\Omega) \cap L^\infty(\Omega)$ with $u(x) \geq 0$ a.e. in Ω denote a weak solution of the differential equation

$$R(u,v) = 0 \qquad \text{for all} \quad v \in \mathcal{H} := W_0^{1,2}(\Omega) \tag{1}$$

with the Dirichlet-Riemann bilinear form $R(u,v)$ given in formula (15) from Section 4. With $y \in \partial\Omega$ and $r > 0$ we furthermore assume $u \in C^0(\partial\Omega \cap K_{4r}(y))$, and we set

$$m \in \left[0, \inf_{\partial\Omega \cap K_{4r}(y)} u(x)\right].$$

Then we have a constant $C = C(M,n) \in (0, +\infty)$, such that the extended function

$$w(x) = [u]^m(x) := \begin{cases} m, & x \in K_{4r}(y) \setminus \Omega \\ \inf\{u(x), m\}, & x \in \Omega \end{cases} \tag{2}$$

satisfies the following estimate

$$\fint_{K_{2r}(y)} w(x)\, dx := \frac{1}{|K_{2r}(y)|} \int_{K_{2r}(y)} w(x)\, dx \leq C \inf_{x \in K_r(y)} w(x). \tag{3}$$

Proof: We have only to consider the case $m > 0$ and transfer the proof of Theorem 5.1 in Section 5 to this situation. Here we define the set

$$\Omega^m := \left\{x \in \Omega : u(x) < m\right\}.$$

The function u is continuous in Ω according to Section 5, Theorem 5.2 and therefore Ω^m represents an open set. In the case $\Omega^m \cap K_{4r}(y) = \emptyset$ we have nothing to show. Otherwise we define the positive function

$$\bar{w}(x) := \frac{1}{m+\varepsilon}(w(x) + \varepsilon), \qquad x \in K_{4r}(y) \cup \Omega \tag{4}$$

with $\varepsilon > 0$ fixed. In Ω^m this function $\bar{w}$ satisfies the weak equation

$$R(\bar{w}, v) = 0 \qquad \text{for all} \quad v \in W_0^{1,2}(\Omega^m). \tag{5}$$

Furthermore, we have

$$\bar{w}(x) = 1 \qquad \text{for all} \quad x \in K_{4r}(y) \setminus \Omega^m. \tag{6}$$

We choose the powers

$$\beta \in (-\infty, 0) \tag{7}$$

and insert the following test functions into the weak differential equation (5):

$$v(x) := (\bar{w}(x)^\beta - 1)\eta(x)^2 \in W_0^{1,2}(\Omega^m). \tag{8}$$

Here the function $\eta = \eta(x)$ is defined as in the proof of Theorem 5.1 from Section 5. We now obtain

$$-\beta \int_{\Omega^m} \left\{ \sum_{i,j=1}^n a_{ij}(x) D^{e_i} \bar{w}(x) D^{e_j} \bar{w}(x) \right\} \bar{w}(x)^{\beta-1} \eta(x)^2 \, dx$$

$$= 2 \int_{\Omega^m} \left\{ \sum_{i,j=1}^n a_{ij}(x) D^{e_i} \bar{w}(x) D^{e_j} \eta(x) \right\} (\bar{w}(x)^\beta - 1)\eta(x) \, dx$$

$$= 2 \int_{\Omega^m} \left\{ \left[\sum_{i,j=1}^n a_{ij}(x) D^{e_i} \bar{w}(x) D^{e_j} \eta(x) \right] \sqrt{\frac{-\beta}{2}} (\bar{w}(x)^{\frac{\beta}{2}} - 1) \frac{1}{\sqrt{\bar{w}(x)}} \eta(x) \right.$$

$$\left. \cdot \sqrt{\frac{2}{-\beta}} (\bar{w}(x)^{\frac{\beta}{2}} + 1) \sqrt{\bar{w}(x)} \right\} dx$$

$$\leq \frac{-\beta}{2} \int_{\Omega^m} \left\{ \sum_{i,j=1}^n a_{ij}(x) D^{e_i} \bar{w}(x) D^{e_j} \bar{w}(x) \right\} \bar{w}(x)^{\beta-1} \eta(x)^2 \, dx$$

$$+ \frac{8}{-\beta} \int_{\Omega^m} \left\{ \sum_{i,j=1}^n a_{ij}(x) D^{e_i} \eta(x) D^{e_j} \eta(x) \right\} \bar{w}(x)^{\beta+1} \, dx$$

$$\tag{9}$$

and therefore

$$\int_{\Omega^m} \left\{ \sum_{i,j=1}^n a_{ij}(x) D^{e_i} \bar{w}(x) D^{e_j} \bar{w}(x) \right\} \bar{w}(x)^{\beta-1} \eta(x)^2 \, dx$$

$$\leq \frac{16}{\beta^2} \int_{\Omega^m} \left\{ \sum_{i,j=1}^n a_{ij}(x) D^{e_i} \eta(x) D^{e_j} \eta(x) \right\} \bar{w}(x)^{\beta+1} \, dx. \tag{10}$$

Since (10) in $K_{4r}(y) \setminus \Omega^m$ is trivially satisfied, we therefore can deduce estimates analogous to (15) and (28), respectively, from the proof of Theorem 5.1 in Section 5. Here we substitute $\bar{u}$ by $\bar{w}$ and Ω by $\Omega^m \cup K_{4r}(y)$. With the considerations given there, one derives the inequality (3) stated above.

q.e.d.

Theorem 7.2. (Boundary behavior)
In the bounded domain $\Omega \subset \mathbb{R}^n$ the boundary point $y \in \partial\Omega$ is assumed to satisfy the Wiener *condition*

$$\beta \le \frac{|K_r(y) \setminus \Omega|}{|K_r(y)|} \qquad \text{for} \quad 0 < r \le r_0 \tag{11}$$

with the constants $\beta \in (0,1)$ and $r_0 > 0$. Let

$$u = u(x) \in W^{1,2}(\Omega) \cap L^\infty(\Omega) \cap C^0(\partial\Omega \cap K_{r_0}(y))$$

denote a solution of the weak differential equation (1), and we define its boundary oscillation

$$\sigma(r) := \operatorname*{osc}_{x \in \partial\Omega \cap K_r(y)} u(x), \qquad 0 < r \le r_0. \tag{12}$$

Then we have constants $C = C(M, n, \beta) \in (0, +\infty)$ and $\alpha = \alpha(M, n, \beta) \in (0,1)$, such that the following estimate

$$\operatorname*{osc}_{\Omega \cap K_r(y)} u \le C \left(\frac{r}{r_0}\right)^\alpha \operatorname*{osc}_{\Omega \cap K_{r_0}(y)} u + \sigma(r_0), \qquad 0 < r \le r_0 \tag{13}$$

holds true.

Proof:

1. We designate the sets $K_r = K_r(y)$, $\Omega_r := \Omega \cap K_r(y)$, $(\partial\Omega)_r := \partial\Omega \cap K_r(y)$ and use the quantities

$$M_4 = \sup_{\Omega_{4r}} u, \quad m_4 = \inf_{\Omega_{4r}} u, \quad M_1 = \sup_{\Omega_r} u, \quad m_1 = \inf_{\Omega_r} u.$$

In the ball K_{4r} we apply Theorem 7.1 to the functions $M_4 - u(x)$ and $u(x) - m_4$ which are nonnegative in Ω_{4r}, and we set

$$M := \sup_{(\partial\Omega)_{4r}} u, \quad m := \inf_{(\partial\Omega)_{4r}} u.$$

For all $0 < r \le \frac{1}{4} r_0$ we obtain the estimates

$$\beta(M_4 - M) \le (M_4 - M) \frac{|K_{2r} \setminus \Omega|}{|K_{2r}|}$$

$$\le \int_{K_{2r}} [M_4 - u(x)]^{M_4 - M} \, dx \tag{14}$$

$$\le C(M_4 - M_1)$$

and

$$\beta(m - m_4) \le (m - m_4) \frac{|K_{2r} \setminus \Omega|}{|K_{2r}|}$$

$$\le \fint_{K_{2r}} [u(x) - m_4]^{m - m_4} \, dx \tag{15}$$

$$\le C(m_1 - m_4).$$

Addition of (14) and (15) yields

$$\beta(M_4 - m_4) - \beta(M - m) \le C(M_4 - m_4) - C(M_1 - m_1)$$

and

$$M_1 - m_1 \le \left(1 - \frac{\beta}{C}\right)(M_4 - m_4) + \frac{\beta}{C}(M - m),$$

and therefore

$$\underset{\Omega_r}{\operatorname{osc}} \, u \le \gamma \underset{\Omega_{4r}}{\operatorname{osc}} \, u + (1 - \gamma)\sigma(4r), \qquad 0 < r \le \frac{1}{4}r_0, \tag{16}$$

with $\gamma := 1 - \frac{\beta}{C} \in (0, 1)$.

2. Analogously to the proof of Theorem 5.2 in Section 5, we consider the monotonically increasing function

$$\omega(r) := \underset{\Omega_r}{\operatorname{osc}} \, u, \qquad 0 < r \le r_0. \tag{17}$$

The latter satisfies the growth condition

$$\omega(r) \le \gamma\omega(4r) + (1 - \gamma)\sigma(4r), \qquad 0 < r \le \frac{1}{4}r_0. \tag{18}$$

For each $r \in (0, \frac{1}{4}r_0]$ we now have an integer $k \in \mathbb{N}$, such that

$$\left(\frac{1}{4}\right)^{k+1} r_0 < r \le \left(\frac{1}{4}\right)^{k} r_0 \tag{19}$$

is satisfied. Additionally choosing $\alpha \in (0, 1)$ with

$$\gamma \le \left(\frac{1}{4}\right)^{\alpha}, \tag{20}$$

we can calculate

$$\omega(r) \leq \omega\left(\left(\frac{1}{4}\right)^k r_0\right) \leq \gamma\omega\left(\left(\frac{1}{4}\right)^{k-1} r_0\right) + (1-\gamma)\sigma(r_0)$$

$$\leq \gamma\left\{\gamma\omega\left(\left(\frac{1}{4}\right)^{k-2} r_0\right) + (1-\gamma)\sigma(r_0)\right\} + (1-\gamma)\sigma(r_0)$$

$$\vdots$$

$$\leq \gamma^k \omega(r_0) + \left\{1 + \gamma + \ldots + \gamma^{k-1}\right\}(1-\gamma)\sigma(r_0) \tag{21}$$

$$\leq \left(\frac{1}{4^k}\right)^\alpha \omega(r_0) + \left(\sum_{l=0}^\infty \gamma^l\right)(1-\gamma)\sigma(r_0)$$

$$\leq 4^\alpha \left(\frac{r}{r_0}\right)^\alpha \omega(r_0) + \sigma(r_0), \qquad 0 < r \leq \frac{1}{4}r_0.$$

Since (21) for $\frac{1}{4}r_0 \leq r \leq r_0$ is trivially satisfied, we obtain the desired estimate (13).

q.e.d.

Remark: Because of $\sigma(r) \to 0 \, (r \to 0+)$, we prescribe $\varepsilon > 0$ in (13) and choose $r_0 > 0$ sufficiently small and afterwards $\delta(\varepsilon) > 0$ such that the estimate

$$\underset{\Omega \cap K_r(y)}{\mathrm{osc}} \, u \leq \varepsilon \qquad \text{for all} \quad 0 < r \leq \delta(\varepsilon) \tag{22}$$

is realized.

8 Equations in Divergence Form

When we construct minima of energy functionals in the Sobolev space by direct variational methods, we obtain weak solutions of differential equations in divergence form. More precisely, we take a vector-field

$$A(p) = \left(A^1(p), \ldots, A^n(p)\right)^* : \mathbb{R}^n \to \mathbb{R}^n \in C^{1+\alpha}(\mathbb{R}^n) \tag{1}$$

with $\alpha \in (0,1)$, whose Jacobi matrix

$$\partial A(p) := \left(\frac{\partial A^j}{\partial p_k}(p)\right)_{j,k=1,\ldots,n}, \qquad p \in \mathbb{R}^n, \tag{2}$$

is symmetric and satisfies the ellipticity condition

$$\frac{1}{M}|\xi|^2 \leq \sum_{j,k=1}^n \frac{\partial A^j}{\partial p_k}(p)\xi_j\xi_k \leq M|\xi|^2 \qquad \text{for all} \quad \xi, p \in \mathbb{R}^n \tag{3}$$

with a constant $M \in [1, +\infty)$. We now consider bounded weak solutions

$$u = u(x) \in W^{1,2}(\Omega) \cap L^\infty(\Omega) \tag{4}$$

of the differential equation

$$\operatorname{div} A\big(Du(x)\big) = 0 \qquad \text{in} \quad \Omega, \tag{5}$$

and therefore we start with the integral relation

$$\int_\Omega \big\{ \nabla\varphi(x) \cdot A(Du(x)) \big\}\, dx = 0 \qquad \text{for all} \quad \varphi \in C_0^\infty(\Omega). \tag{6}$$

We utilize the difference quotient

$$\Delta_{i,\varepsilon}\varphi(x) := \frac{\varphi(x + \varepsilon e_i) - \varphi(x)}{\varepsilon} \tag{7}$$

in the direction e_i with sufficiently small $\varepsilon \neq 0$. This notion has been introduced in Section 1, and we calculate similar to the proofs of Theorem 1.8 and 1.9 there. When we insert (7) into (6), we obtain

$$0 = \int_\Omega \big\{ \nabla(\Delta_{i,\varepsilon}\varphi(x)) \cdot A(Du(x)) \big\}\, dx = - \int_\Omega \big\{ \nabla\varphi(x) \cdot \Delta_{i,\varepsilon} A(Du(x)) \big\}\, dx. \tag{8}$$

We deduce

$$\Delta_{i,\varepsilon} A(Du(x)) = \frac{1}{\varepsilon}\Big\{ A(Du(x + \varepsilon e_i)) - A(Du(x)) \Big\}$$

$$= \Big\{ \int_0^1 \partial A\big(Du(x) + t[Du(x + \varepsilon e_i) - Du(x)]\big)\, dt \Big\} \Delta_{i,\varepsilon} Du(x) \tag{9}$$

and define the symmetric matrix

$$B_\varepsilon(x) := \int_0^1 \partial A\big(Du(x) + t[Du(x + \varepsilon e_i) - Du(x)]\big)\, dt, \qquad x \in \Omega, \tag{10}$$

satisfying the uniform ellipticity condition

$$\frac{1}{M}|\xi|^2 \leq \xi \circ B_\varepsilon(x) \circ \xi^* \leq M|\xi|^2 \qquad \text{for all} \quad \xi \in \mathbb{R}^n, x \in \Omega, |\varepsilon| \leq \varepsilon_0. \tag{11}$$

The combination of (8), (9), (10) yields the following weak uniformly elliptic differential equation for the difference quotient $\Delta_{i,\varepsilon} u(x)$:

$$0 = \int_\Omega \big\{ \nabla\varphi(x) \circ B_\varepsilon(x) \circ D(\Delta_{i,\varepsilon} u(x)) \big\}\, dx \qquad \text{for all} \quad \varphi \in C_0^\infty(\Omega). \tag{12}$$

This difference quotient satisfies a Hölder condition independent of ε, according to Theorem 5.2 from Section 5 of de Giorgi - Nash. The passage to the limit $\varepsilon \to 0+$ yields

$$u \in C^{1+\mu}(\Omega) \tag{13}$$

for a sufficiently small $\mu \in (0,1)$. We then consider the coefficient matrix

$$B(x) := \partial A\big(Du(x)\big), \qquad x \in \Omega, \tag{14}$$

of the class $C^\mu(\Omega)$. The transition to the limit $\varepsilon \to 0+$ in (12) reveals the following weak differential equation in divergence form for the partial derivatives $u_{x_i}(x)$, $i = 1, \ldots, n$, namely

$$0 = \int\limits_\Omega \big\{ \nabla\varphi(x) \circ B(x) \circ Du_{x_i}(x) \big\}\, dx \qquad \text{for all} \quad \varphi \in C_0^\infty(\Omega) \tag{15}$$

with Hölder continuous coefficients. The higher regularity of u is shown by local reconstruction.

Theorem 8.1. *We prescribe the boundary values* $\psi : \partial K \to \mathbb{R} \in C^{1+\mu}(\partial K)$ *on the boundary of the open ball* $K \subset\subset \Omega$. *Then the following Dirichlet problem adjoint to the vector field (1)-(3) above possesses a solution*

$$\begin{aligned}
&v = v(x) \in C^{2+\mu}(K) \cap C^0(\overline{K}) \cap W^{1,2}(K), \\
&div \ \ A\big(Dv(x)\big) = 0 \quad in \ \ K, \\
&v(x) = \psi(x) \quad on \ \ \partial K.
\end{aligned} \tag{16}$$

Proof:

1. In the first part of our proof, we utilize a method proposed by A. Haar for variational problems. At each point $x_0 \in \partial K$ we have the linear support functions $\eta_-^+(x) : \mathbb{R}^n \to \mathbb{R}$ with

$$\eta_-^+(x_0) = \psi(x_0) \quad \text{and}$$

$$\eta_-(x) \leq \psi(x) \leq \eta^+(x) \quad \text{for all } x \in \partial K, \quad \text{where} \tag{17}$$

$$|D\eta_-^+(x_0)| \leq C\big(\|\psi\|_{C^{1+\mu}(\partial K)}\big) \quad \text{for all } x_0 \in \partial K$$

is satisfied. For the solution $v \in C^1(\overline{K})$ of (16) we then deduce the inequality

$$|Dv(x_0)| \leq C \qquad \text{for all} \quad x_0 \in \partial K \tag{18}$$

from the inclusion

$$\eta_-(x) \leq v(x) \leq \eta^+(x), \qquad x \in \overline{K}. \tag{19}$$

The latter is inferred from (17) by the maximum principle applied to the quasilinear elliptic equation

$$\sum_{j,k=1}^n \frac{\partial A^j}{\partial p_k}\big(Dv(x)\big) v_{x_j x_k}(x) = 0, \qquad x \in K. \tag{20}$$

Now the derivatives v_{x_i} in K are subject to the weak elliptic differential equation (15) as well and therefore satisfy the maximum principle:

$$|Dv(x)| \leq C\big(\|\psi\|_{C^{1+\mu}(\partial K)}\big), \qquad x \in \overline{K}. \tag{21}$$

2. When we have solved our boundary value problem (16) for the boundary values

$$\overline{\psi} : \partial K \to \mathbb{R} \in C^{2+\mu}(\partial K),$$

we then approximate the given function ψ by a sequence

$$\psi_k \to \psi \quad \text{in} \quad C^{1+\mu}(\partial K) \quad (k \to \infty).$$

The adjoint solutions v_k of (16) are equicontinuous because of (21). Therefore, we make the transition to a subsequence which is uniformly convergent in $\overline{K}$ with a limit function v satisfying the inequality

$$|Dv(x)| \leq C, \qquad x \in K. \tag{22}$$

From the inner Hölder estimate for Dv given above, we can achieve via the differential equation (20) by the inner Schauder estimates that the sequence converges in $C^{2+\mu}(\overline{\Theta})$ for each open set $\Theta \subset\subset K$. Consequently, the limit function belongs to the class

$$C^{2+\mu}(K) \cap C^0(\overline{K}) \cap W^{1,2}(K).$$

3. It remains to solve the Dirichlet problem (16) for $C^{2+\mu}$-boundary-values ψ. Here we have to establish a global Hölder estimate for the gradient of the solution with the aid of Theorem 7.2 from Section 7. A result of O. Ladyzhenskaya and N. Uraltseva (see [GT] Theorem 13.2) yields

$$\|Dv\|_{C^\mu(\overline{K})} \leq C(\|\psi\|_{C^2(\partial K)}) \tag{23}$$

for an exponent $\mu \in (0,1)$. This estimate is inferred from the Hölder continuous boundary values of v_{x_i} and the weak differential equation (15) for the derivatives. We insert this inequality into the quasilinear differential equation (20). Applied to the sequence of boundary values

$$\psi_k \to \psi \quad \text{in} \quad C^{2+\mu}(\partial K) \quad (k \to \infty)$$

the global Schauder estimates imply the following statement

$$v_k \to v \quad \text{in} \quad C^{2+\mu}(\overline{K}) \quad (k \to \infty)$$

for the adjoint solutions of the boundary value problem (16).

4. By a nonlinear continuity method deforming the boundary values, we can solve the boundary value problem (16) for all $\psi \in C^{2+\mu}(\partial K)$. This procedure will be presented in Section 1 from Chapter 13 for the nonparametric

equation of prescribed mean curvature. Similar to Proposition 1.8 there, we start with a solution v of (20) and solve the following nonlinear differential equation for small boundary values with the aid of Banach's fixed point theorem:

$$
\begin{aligned}
0 &= \sum_{j,k=1}^{n} \frac{\partial A^j}{\partial p_k}\big(Dv(x) + Dw(x)\big)[v_{x_j x_k} + w_{x_j x_k}] \\
&= \sum_{j,k=1}^{n} \Big[\frac{\partial A^j}{\partial p_k}\big(Dv(x)\big) + \sum_{l=1}^{n} \frac{\partial A^j}{\partial p_k \partial p_l}\big(Dv(x)\big)w_{x_l} + \dots\Big][v_{x_j x_k} + w_{x_j x_k}] \\
&= \sum_{j,k=1}^{n} \frac{\partial A^j}{\partial p_k}\big(Dv(x)\big)w_{x_j x_k} \\
&\quad + \sum_{l=1}^{n}\Big(\sum_{j,k=1}^{n} v_{x_j x_k}\frac{\partial A^j}{\partial p_k \partial p_l}\big(Dv(x)\big)\Big)w_{x_l} + \dots, \qquad x \in K.
\end{aligned}
$$

$$(24)$$

Here we assume polynomial coefficients in the differential equation (20) at first, and we denote by $\dots$ the superlinear terms in the partial derivatives of w. As in Theorem 1.7 of Section 1 from Chapter 13 we then deform the trivial solution $v = 0$ into the solution of the Dirichlet problem posed. By an adequate approximation, we finally solve the differential equation with the given coefficients.

q.e.d.

We obtain the fundamental

Theorem 8.2. (Regularity theorem of de Giorgi)
A bounded weak solution u of (4) and (6) with the vector field (1)-(3) belongs to the regularity class $C^{2+\alpha}(\Omega)$.

Proof: In each ball $K \subset\subset \Omega$ we reconstruct the solution u for the boundary values $\psi := u|_{\partial K}$ by a solution of (16) from Theorem 8.1. With the aid of the Gaussian energy method one easily shows that the boundary value problem (16) for weak solutions is uniquely determined. This implies

$$u(x) = v(x) \qquad \text{in} \quad \overline{K},$$

and consequently $u \in C^{2+\mu}(K)$. By a renewed reconstruction within the C^2-solutions we obtain

$$u \in C^{2+\alpha}(\Omega).$$

q.e.d.

Remarks:

1. The regularity questions are situated in the center of the modern calculus of variations, especially in the monograph [Gi] by M. Giaquinta. In this context we recommend the beautiful presentation in [Jo] 11.3 by J. Jost.

2. By the methods of this chapter a general theory for quasilinear elliptic differential equations in n variables can be developed as in the pioneering book [GT] Part II of D. Gilbarg and N. Trudinger.
3. We want to address the theory of two-dimensional partial differential equations in the next chapters. Here one can transform the equations into a normal form in the hyperbolic and in the elliptic situation as well, and both cases are interrelated via the complex space. For intuitive geometry the two-dimensional theory is of central importance.

We finally treat the *regularity question for the minimal surface equation:*

In the bounded domain $\Omega \subset \mathbb{R}^n$ let $u = u(x) \in W^{1,\infty}(\Omega)$ denote a weak solution of the nonparametric minimal surface equation in divergence form

$$\operatorname{div}\left\{\left(1 + |Du(x)|^2\right)^{-\frac{1}{2}} Du(x)\right\} = 0 \quad \text{in} \quad \Omega. \tag{25}$$

This equation will be derived differential-geometrically in the second part of Section 1 from Chapter 11. Because of $Du(x) \in L^\infty(\Omega)$, the differential equation (25) is uniformly elliptic and Theorem 2.2 from Section 2 reveals $u \in C^0(\overline{\Omega})$. Now the regularity result of de Georgi, Theorem 8.2 implies $u \in C^{2+\alpha}(\Omega)$. Since one can easily construct solutions in the class $W^{1,2}(\Omega)$ within the calculus of variations, the central task remains to estimate $\|Du\|_{L^\infty(\Omega)}$. Therefore, gradient estimates have to be established!

With $\mu \in (0,1)$ given, we prescribe

> a bounded convex domain $\quad \Omega \subset \mathbb{R}^n \quad$ with $\quad C^{2+\mu}$-boundary $\quad \partial\Omega$
> and the boundary values $\quad \psi : \partial\Omega \to \mathbb{R} \in C^{2+\mu}(\partial\Omega)$. $\tag{26}$

We consider a solution of the Dirichlet problem

> $u = u(x) \in C^2(\Omega) \cap C^1(\overline{\Omega}) \quad$ satisfies (25)
> and the boundary condition $\quad u(x) = \psi(x) \quad$ on $\quad \partial\Omega$, $\tag{27}$

and deduce the following boundary-gradient-estimate:

$$|Du(x)| \leq C\big(\partial\Omega, \|\psi\|_{C^{2+\mu}(\partial\Omega)}\big), \qquad x \in \partial\Omega. \tag{28}$$

In this context we show: At each boundary point $(x, u(x))$, $x \in \partial\Omega$ the tangential plane for the surface

$$(x, u(x)), \qquad x \in \overline{\Omega}$$

has an angle with the support plane of the boundary manifold, whose modulus can be estimated from below by a number $\omega > 0$ independent of the point $x \in \partial\Omega$. Here one considers the minimal surface in its height representation

$$v : \overline{\Theta} \to [0, \infty)$$

above the support plane, satisfying the now differentiated minimal surface equation (compare Section 1 in Chapter 11):

$$\big(1 + |Dv(x)|^2\big)\Delta v(x) - \sum_{i,j=1}^{n} v_{x_i} v_{x_j} v_{x_i x_j}(x) = 0 \quad \text{in} \quad \Theta. \tag{29}$$

With the aid of E. Hopf's boundary point lemma from Section 1 in Chapter 6, the statement (28) follows. The weak maximum principle applied to the derivatives u_{x_i} then implies

$$\|u\|_{C^1(\overline{\Omega})} \le C\big(\partial\Omega, \|\psi\|_{C^{2+\mu}(\partial\Omega)}\big). \tag{30}$$

With the aid of methods presented in part 3 and 4 of the proof for Theorem 8.1, one finally shows the following statement whose complete derivation, however, is left to the reader.

Theorem 8.3. (Jenkins, Serrin)
With the data (26) there exists exactly one solution $u \in C^{2+\mu}(\overline{\Omega})$ of the Dirichlet problem (27) for the minimal surface equation (25).

9 Green's Function for Elliptic Operators

In the present section, we shall construct Green's function for elliptic differential operators in divergence form with the aid of Schauder's theory. For instance, this enables us to transform the eigenvalue problem of elliptic operators into an integral equation and then to proceed similar to the considerations for the Laplace operator in Chapter 8.

We take a bounded domain $\Omega \subset \mathbb{R}^n$ such that the regular boundary $\partial\Omega$ is of the class $C^{2+\mu}$ with $\mu \in (0,1)$ and an integer $n \in \mathbb{N}$ satisfying $n \ge 3$. Furthermore, we denote the exterior normal by $\nu = \nu(x)$ and the diameter of Ω by $R > 0$. For the differential operator

$$\mathcal{L}(u) := \sum_{i=1}^{n} \frac{\partial}{\partial x_i}\Big(\sum_{j=1}^{n} a_{ij}(x) u_{x_j}(x) \Big), \quad x \in \Omega \tag{1}$$

in divergence form, we require the coefficient matrix

$$\big(a_{ij}(x)\big)_{i,j=1,\dots,n}, \quad x \in \Omega \quad \text{of the class} \quad C^{1+\mu}(\overline{\Omega}) \tag{2}$$

to be real and symmetric satisfying the following ellipticity condition

$$\frac{1}{M}|\xi|^2 \le \sum_{i,j=1}^{n} a_{ij}(x)\xi_i\xi_j \le M|\xi|^2 \quad \text{in} \ \Omega \quad \text{for all} \ \ \xi \in \mathbb{R}^n, \tag{3}$$

with the ellipticity constant $M \in [1, +\infty)$. For a fixed point $y \in \Omega$ we define the neighborhood

$$\mathcal{U} := \{x \in \mathbb{R}^n : |x - y| < r_0\} \subset \Omega \tag{4}$$

with the fixed radius $0 < r_0 < \rho = \rho(y) := dist(y, \partial\Omega) \in (0, R)$. At first, we assume our coefficients to fulfill

$$a_{ij}(x) = \delta_{ij} \quad \text{for} \quad x \in \mathcal{U} \quad \text{and} \quad i, j = 1, \ldots, n \tag{5}$$

such that the differential operator $\mathcal{L}$ coincides with the Laplacian near the point y. With the aid of Schauder's theory from Chapter 9, we determine the unique solution of the following boundary value problem

$$\phi = \phi(x, y) \in C^{2+\mu}(\overline{\Omega}), \quad \mathcal{L}(\phi) = -\mathcal{L}\left(\frac{1}{(n-2)\omega_n}|x - y|^{2-n}\right) \quad \text{in} \quad \Omega$$

$$\phi(x, y) = -\frac{1}{(n-2)\omega_n}|x - y|^{2-n}, \quad x \in \partial\Omega. \tag{6}$$

Naturally, the quantity ω_n denotes the area of the unit sphere in $\mathbb{R}^n$. We now obtain the *approximate Green's function*

$$g = g(x, y) := \frac{1}{(n-2)\omega_n}|x - y|^{2-n} + \phi(x, y) \in C^{2+\mu}(\overline{\Omega} \setminus \{y\}) \tag{7}$$

satisfying

$$\mathcal{L}(g) = 0, \quad x \in \Omega \setminus \{y\} \quad \text{and} \quad g(x, y) = 0, \quad x \in \partial\Omega. \tag{8}$$

Setting $g(x, y) = 0$ for $x \in \mathbb{R}^n \setminus \Omega$, $y \in \Omega$ we trivially continue the function onto the whole space.

For all functions $u \in C^1(\overline{\Omega})$ with zero boundary values $u(x) = 0$ on $\partial\Omega$ we now calculate

$$\sum_{i=1}^n \frac{\partial}{\partial x_i}\left(u \sum_{j=1}^n a_{ij}(x)g_{x_j}(x, y)\right) = u\mathcal{L}(g) + \sum_{i,j=1}^n a_{ij}(x)u_{x_i}(x)g_{x_j}(x, y) =$$

$$\sum_{i,j=1}^n a_{ij}(x)u_{x_i}(x)g_{x_j}(x, y) \quad \text{for} \quad x \in \Omega \setminus \{y\}. \tag{9}$$

We apply the Gaussian integral theorem on the domain

$$\Omega_\varepsilon := \{x \in \Omega : |x - y| > \varepsilon\}$$

with the exterior normal $\nu = \nu(x)$ and $0 < \varepsilon < r_0$. Then we obtain the following relation:

$$\int_{\Omega_\varepsilon} \left(\sum_{i,j=1}^n a_{ij}(x)u_{x_i}(x)g_{x_j}(x, y)\right)dx =$$

$$\int_{|x-y|=\varepsilon} \left(u(x)\sum_{i=1}^n \{\nu_i(x)\sum_{j=1}^n a_{ij}(x)g_{x_j}(x, y)\}\right)d\sigma(x) = \tag{10}$$

$$\int_{|x-y|=\varepsilon} u(x)\frac{\partial g(x, y)}{\partial \nu(x)}d\sigma(x) \rightarrow u(y) \quad \text{for} \quad \varepsilon \rightarrow 0+ .$$

Consequently, we receive the fundamental

Proposition 9.1. *For all functions* $u \in C^1(\overline{\Omega})$ *with zero boundary values* $u(x) = 0$ *on* $\partial\Omega$ *we have the following identity:*

$$R(u,g) := \int_{\Omega} \Big(\sum_{i,j=1}^{n} a_{ij}(x)u_{x_i}(x)g_{x_j}(x,y) \Big) dx = u(y).$$

Here, the symbol $R(.,.)$ *denotes the Dirichlet-Riemann bilinear form.*

Now we observe $g(x,y) > 0$ for $x \in \Omega \setminus \{y\}$ due to E. Hopf's maximum principle and $g(x,y) \to +\infty$ when $x \to y$ and $x \neq y$ holds true. For all $t \in (0,+\infty)$ we consider the level sets

$$\Omega(t) := \{x \in \Omega : g(x,y) < t\}   \text{ and }   \Theta(t) := \{x \in \Omega : g(x,y) \geq t\},$$

where y is an interior point of the closed set $\Theta(t)$. We define the *truncated approximate Green's function*

$$g^t(x,y) = \begin{cases} g(x,y), & x \in \Omega(t) \\ t, & x \in \Theta(t) \end{cases}. \tag{11}$$

Parallel to Theorem 1.11 in Section 1 we prove by approximation that this function belongs to the class $W_0^{1,2}(\Omega) \cap C^0(\overline{\Omega})$, and $u(x) = g^t(x,y)$ can be inserted as test function in Proposition 9.1. Thus we obtain the *evaluation formula*

$$\int_{\Omega(t)} \Big(\sum_{i,j=1}^{n} a_{ij}(x)g_{x_i}(x,y)g_{x_j}(x,y) \Big) dx = R(g^t,g) = g^t(y,y) = t, \, 0 < t < +\infty. \tag{12}$$

The subsequent concept is of central importance:

Definition 9.2. *For a measurable subset* $E \subset\subset \Omega$ *we define by*

$$cap_{\Omega,\mathcal{L}}(E) := \inf\{R(v,v) : v \in W_0^{1,2}(\Omega), \quad v(x) = 1 \quad a.e. \quad in \quad E\}$$

the capacity *of the set* E *in* Ω.

With the function $v(x) := \frac{1}{t}g^t(x,y), x \in \Omega$ we obviously have the unique minimizer in the variational problem above for the set $E := \Theta(t)$ with the energy $R(v,v) = \frac{1}{t}$ because of the evaluation formula above. Thus we have shown

Proposition 9.3. *The relation* $cap_{\Omega,\mathcal{L}}(\Theta(t)) = \dfrac{1}{t}$, $0 < t < +\infty$ *for the capacities of the level sets is correct.*

We need the following elementary comparison properties of the capacity:

$$E_1 \subset E_2 \subset \Omega \quad \text{implies} \quad cap_{\Omega,\mathcal{L}}(E_1) \leq cap_{\Omega,\mathcal{L}}(E_2), \tag{13}$$

$$E \subset \Omega_1 \subset \Omega \quad \text{implies} \quad cap_{\Omega_1,\mathcal{L}}(E) \geq cap_{\Omega,\mathcal{L}}(E), \tag{14}$$

$$\frac{1}{M} cap_{\Omega}(E) \leq cap_{\Omega,\mathcal{L}}(E) \leq M cap_{\Omega}(E) \quad \text{for} \quad E \subset \Omega, \tag{15}$$

abbreviating $cap_{\Omega}(E) := cap_{\Omega,\Delta}(E)$ for the *standard capacity*. With the radii $r > 0$ we consider the balls

$$U_r := \{x \in \mathbb{R}^n : |x - y| < r\}$$

and define the quantities

$$a(r) := \inf\{g(x,y) : x \in \partial U_r\} \quad \text{and} \quad b(r) := \sup\{g(x,y) : x \in \partial U_r\}$$

for $0 < r < \rho$. We note that $0 < a(r) \leq b(r) < +\infty$, and Hopf's maximum principle implies

$$g(x,y) > a(r), x \in U_r \quad \text{and} \quad g(x,y) < b(r), x \in \overline{\Omega} \setminus \overline{U_r}.$$

Therefore, we obtain the inclusions

$$U_r \subset \Theta(a(r)) \quad \text{and} \quad \Theta(b(r)) \subset U_r.$$

The comparison properties for capacities together with Proposition 9.3 yield

$$cap_{\Omega,\mathcal{L}}(U_r) \leq cap_{\Omega,\mathcal{L}}(\Theta(a(r))) = \frac{1}{a(r)} \tag{16}$$

and

$$cap_{\Omega,\mathcal{L}}(U_r) \geq cap_{\Omega,\mathcal{L}}(\Theta(b(r))) = \frac{1}{b(r)}. \tag{17}$$

To the function $v(x) := g(y + rx, y), \frac{1}{2} < |x| < 2$ we apply the Harnack-Moser inequality from Theorem 5.4 in Section 5. Using the homogeneity in r for the differential equation of v, we see that the ellipticity constant is independent of the radius r. Therefore, we obtain a constant $c = c(M) \in [1, +\infty)$ such that

$$b(r) \leq \sup\{v(x) : \tfrac{1}{2} < |x| < 2\} \leq c(M) \inf\{v(x) : \tfrac{1}{2} < |x| < 2\}$$
$$\leq c(M)a(r), \quad 0 < r < \tfrac{1}{2}\rho \tag{18}$$

holds true. In combination with (16) and (17) we deduce

$$b(r) \leq c(M)\big(cap_{\Omega,\mathcal{L}}(U_r)\big)^{-1} \tag{19}$$

and

$$a(r) \geq \big(c(M)cap_{\Omega,\mathcal{L}}(U_r)\big)^{-1}. \tag{20}$$

From the inclusion $U_\rho \subset \Omega \subset U_R$ we infer the inequalities

$$\frac{1}{M} cap_{U_R}(U_r) \le \frac{1}{M} cap_\Omega(U_r) \le cap_{\Omega,\mathcal{L}}(U_r) \le M cap_\Omega(U_r) \le M cap_{U_\rho}(U_r)$$

via the comparison properties and therefore

$$\frac{1}{M}\big(cap_{U_\rho}(U_r)\big)^{-1} \le \big(cap_{\Omega,\mathcal{L}}(U_r)\big)^{-1} \le M\big(cap_{U_R}(U_r)\big)^{-1} \text{ for } 0 < r < \frac{1}{2}\rho.$$
(21)

The standard capacity of concentric balls can be determined as follows: On the domain U_R we have the standard Green's function

$$g^*(x,y) := |x-y|^{2-n} - R^{2-n}, x \in U_R.$$

For $0 < r < R$ we have the quantities

$$a^*(r) = b^*(r) = r^{2-n} - R^{2-n},$$

and we observe $\overline{U_r} = \Theta(a^*(r))$. Then Proposition 9.3 implies

$$\big(cap_{U_R}(U_r)\big)^{-1} = a^*(r) = r^{2-n} - R^{2-n}.$$
(22)

The combination of (19), (20), (21), and (22) gives us the constants $0 < c_1 = c_1(M,n) \le c_2 = c_2(M,n) < +\infty$ such that

$$c_1 r^{2-n} \le a(r) \le b(r) \le c_2 r^{2-n}, \quad 0 < r < \frac{1}{2}\rho$$
(23)

holds true. Now we easily show

Proposition 9.4. *For the approximate Green's function we have the following estimates:*

$$0 < g(x,y) \le c_2(M,n)|x-y|^{2-n}, \quad x \in \Omega, \quad x \ne y$$

and

$$g(x,y) \ge c_1(M,n)|x-y|^{2-n}, \quad x \in \Omega, \quad |x-y| \le \frac{1}{2}\rho(y).$$

Proof: The second estimate can be directly inferred from (23). However, we still have to show the validity of the first inequality on the whole set Ω. To this aim we introduce the exhausting set $\Omega^\delta := \{x \in \Omega : \rho(x) > \delta\}$ for sufficiently small $\delta > 0$ and take a test function $\chi = \chi_\delta(x) \in C_0^\infty(\mathbb{R}^n,[0,1])$ with *supp* $\chi_\delta \subset \Omega^{\frac{1}{2}\delta}$ satisfying $\chi(x) = 1, \quad x \in \Omega^\delta$. Now we continue the coefficients of our operator differentiably onto the ball U_{2R} as follows:

$$a_{ij}^\delta(x) = \begin{cases} \chi_\delta(x)a_{ij}(x) + (1-\chi_\delta(x))\delta_{ij}, & x \in \Omega \\ \delta_{ij}, & x \in U_{2R} \setminus \Omega \end{cases}.$$
(24)

The operator $\mathcal{L}_\delta$ in divergence form (1) with the coefficients

$$a_{ij}^\delta(x) \quad \text{for} \quad i, j = 1, \ldots, n$$

possesses the approximate Green's function $g_\delta(x, y)$ on the domain U_{2R}. The auxiliary function

$$w(x) := g_\delta(x, y) - g(x, y) + \sup\{g(z, y) : z \in \partial\Omega^\delta\}, \quad x \in \Omega^\delta$$

satisfies

$$\mathcal{L}(w) = 0, \quad x \in \Omega^\delta; \quad w(x) > 0, \quad x \in \partial\Omega^\delta.$$

The maximum principle of E. Hopf yields

$$g(x, y) \leq g_\delta(x, y) + \sup\{g(z, y) : z \in \partial\Omega^\delta\}$$

$$\leq c_2(M, n)|x - y|^{2-n} + \sup\{g(z, y) : z \in \partial\Omega^\delta\}, \quad x \in \Omega^\delta.$$

The transition to the limit $\delta \to 0+$ implies

$$g(x, y) \leq c_2(M, n)|x - y|^{2-n}, \quad x \in \Omega.$$

$$\text{q.e.d.}$$

With the aid of the hole-filling technique, we estimate $|\nabla g|$ in the $L^p(\Omega)$-norm.

Proposition 9.5. *The approximate Green's function satisfies the estimate*

$$\int_\Omega |\nabla g(x, y)|^p dx \quad \leq C(M, n, p)$$

for all exponents $p \in [1, \frac{n}{n-1})$, *with* $0 < C(M, n, p) < +\infty$ *as a priori constant.*

Proof: We utilize the auxiliary function $\chi = \chi(x) \in C^1(\mathbb{R}^n, [0, 1])$ satisfying

$$\chi(x) = 0 \quad \text{for all} \quad x \in \mathbb{R}^n \quad \text{with} \quad |x - y| \leq \frac{1}{4}R \quad \text{or} \quad |x - y| \geq 2R,$$

$$\chi(x) = 1 \quad \text{for all} \quad x \in \mathbb{R}^n \quad \text{with} \quad \frac{1}{2}R \leq |x - y| \leq R,$$

$$\text{and} \quad |\nabla\chi(x)| \leq \frac{c}{R} \quad \text{for all} \quad x \in \mathbb{R}^n.$$

Then we insert the test function $u(x) = \chi(x)^2 g(x, y)$ into the integral equation of Proposition 9.1 and obtain

$$0 = \int_\Omega \chi^2 \sum_{i,j=1}^n (a_{ij}(x) g_{x_i} g_{x_j}) dx + \int_\Omega 2\chi g \sum_{i,j=1}^n (a_{ij}(x) \chi_{x_i} g_{x_j}) dx.$$

Standard estimates for quadratic forms as in Section 5 imply

$$\int\limits_{x\in\Omega:\frac{1}{2}R\leq|x-y|\leq R} |\nabla g(x,y)|^2 \quad dx$$

$$\leq \int\limits_{\Omega} \chi^2 |\nabla g(x,y)|^2 dx$$

$$\leq c(M,n) \int\limits_{\Omega} g^2 |\nabla \chi(x)|^2 dx \qquad (25)$$

$$\leq c(M,n) \int\limits_{x\in\Omega:\frac{1}{4}R\leq|x-y|\leq 2R} g(x,y)^2 |\nabla\chi(x)|^2 \quad dx$$

$$\leq c(M,n)R^{4-2n}\cdot R^{-2}\cdot R^n = c(M,n)R^{2-n},$$

using the growth condition for Green's function from Proposition 9.4.
We take a quantity $1 \leq p < 2$ fixed and apply Hölder's inequality with the
conjugate exponents $q = \frac{2}{p},\quad q' = \frac{2}{2-p}$ as follows:

$$\int\limits_{x\in\Omega:\frac{1}{2}R\leq|x-y|\leq R} |\nabla g(x,y)|^p \quad dx$$

$$= \int\limits_{x\in\Omega:\frac{1}{2}R\leq|x-y|\leq R} |\nabla g(x,y)|^p \cdot 1 \quad dx$$

$$\leq \left(\int\limits_{x\in\Omega:\frac{1}{2}R\leq|x-y|\leq R} |\nabla g(x,y)|^2 \quad dx\right)^{\frac{p}{2}} \cdot \left(\int\limits_{x\in\Omega:\frac{1}{2}R\leq|x-y|\leq R} 1 \quad dx\right)^{\frac{2-p}{2}} \qquad (26)$$

$$\leq c(M,n,p)R^{(2-n)\frac{p}{2}}\cdot R^{n\cdot\frac{2-p}{2}} = c(M,n,p)R^{p-\frac{np}{2}+n-\frac{np}{2}}$$

$$= c(M,n,p)R^{n-(n-1)p}.$$

Now we replace R by $\frac{R}{2^k}$ for $k = 0,1,2,\ldots$ and obtain

$$\int\limits_{x\in\Omega:2^{-k-1}R\leq|x-y|\leq 2^{-k}R} |\nabla g(x,y)|^p \quad dx \leq c(M,n,p)R^{n-(n-1)p}\cdot 2^{-k(n-(n-1)p)}.$$

With $1 \leq p < \frac{n}{n-1}$ we observe that $(n - (n-1)p) > 0$ holds true and the
summation over all k yields the desired estimate

$$\int\limits_{\Omega} |\nabla g(x,y)|^p dx \leq c(M,n,p)R^{n-(n-1)p}\left(\sum_{k=0}^{\infty} 2^{-k(n-(n-1)p)}\right) := C(M,n,p).$$

Here we observe that the series above converges.   q.e.d.

We are now prepared to prove the central

Theorem 9.6. (Generalized Green's function)
For the elliptic differential operator $\mathcal{L}$ with the properties (1), (2), (3) from above we have a function

$$G = G(x,y) : \overline{\Omega} \times \Omega \to \mathbb{R}$$

such that $G(.,y)$ belongs to the class $W_0^{1,p}(\Omega) \cap C^{2+\mu}(\overline{\Omega} \setminus \{y\})$ for all $y \in \Omega$ with the exponent $p \in [1, \frac{n}{n-1})$ satisfying the growth conditions

$$0 < G(x,y) \leq c_2(M,n)|x-y|^{2-n}, \quad x \in \Omega, \quad x \neq y$$

and

$$G(x,y) \geq c_1(M,n)|x-y|^{2-n}, \quad x \in \Omega, \quad |x-y| \leq \frac{1}{2}\rho(y).$$

The function G implies the following representation formula

$$u(y) = \int_\Omega \left(\sum_{i,j=1}^n a_{ij}(x)D^{e_i}u(x)G_{x_j}(x,y) \right) dx, \quad y \in \Omega$$

for all functions $u \in W_0^{1,q}(\Omega)$ with the exponent $q \in (n, +\infty]$. Here we denote by D^{e_i} the weak derivatives and remark that the Sobolev space above is continuously embedded into $C^0(\overline{\Omega})$ due to Theorem 2.2 in Section 2.

Proof: At first, we consider test functions $u = u(x) \in C_0^\infty(\Omega)$ and choose a fixed $y \in \Omega$. Then we take a sequence of auxiliary functions $\chi_k = \chi_k(x) \in C^\infty(\mathbb{R}^n, [0,1])$ satisfying

$$\chi_k(x) = 0, \quad x \in U_{\frac{1}{2k}\rho} \quad \text{and} \quad \chi_k(x) = 1, \quad x \in \Omega \setminus U_{\frac{1}{k}\rho} \quad \text{for} \quad k = 1,2,3,\ldots$$

For our operator (1) we define the coefficients

$$a_{ij}^k(x) = \chi_k(x)a_{ij}(x) + (1 - \chi_k(x))\delta_{ij}, \quad x \in \Omega; \quad i,j = 1,\ldots,n$$

and observe their convergence

$$\|a_{ij}^k - a_{ij}\|_{L^s(\Omega)} \to 0 \quad \text{for} \quad k \to \infty$$

with arbitrary numbers $1 \leq s < +\infty$. The differential operator $\mathcal{L}_k$ possesses the approximate Green's function g^k for $k = 1,2,\ldots$ and Proposition 9.1 implies the following representation:

$$\int_\Omega \left(\sum_{i,j=1}^n a_{ij}^k(x)u_{x_i}(x)g_{x_j}^k(x,y) \right) dx = u(y). \tag{27}$$

Due to Proposition 9.5 the sequence $g_{x_j}^k(x,y)$, $k = 1,2,\ldots$ is bounded in the $L^p(\Omega)$-norm for the given p and all indices $j = 1,\ldots,n$. Then Theorem 8.9

in Section 8 of Chapter 2 allows the transition to a weakly convergent subsequence

$$g_{x_j}^{k'}(x,y) \rightharpoonup D^{e_j}G(x,y) \quad (k' \to \infty) \quad \text{for} \quad j = 1, \ldots, n.$$

Here the limit function $G = G(.,y)$ belongs to the Sobolev space $W^{1,p}(\Omega)$. The sequence $g^k(.,y)$, $k = 1, 2, \ldots$ satisfies the growth conditions from Proposition 9.4 and is therefore uniformly bounded in each compact set $K \subset \overline{\Omega} \setminus \{y\}$. Consequently, we can estimate these functions in the $C^{2+\mu}(K)$-norm by the Schauder estimates derived in Chapter 9. Finally, the limit function G belongs to the class $W_0^{1,p}(\Omega) \cap C^{2+\mu}(\overline{\Omega} \setminus \{y\})$.

We pass to the limit $k' \to \infty$ in (27) and observe the weak convergence of the derivatives and the strong convergence of the coefficients. We use familiar arguments for Hilbert spaces - see the Remark 3.) on weak convergence in Chapter 8, Section 6 - which pertain to these Lebesgue spaces. Then we obtain the representation formula stated in the theorem for test functions.

Finally, we approximate the Sobolev functions $u \in W_0^{1,q}(\Omega)$ with the exponent $q > n$ by test functions and take Theorem 2.2 from Section 2 into account. We then obtain the representation formula even in the Sobolev class.   q.e.d.

Definition 9.7. *We call the function* $G(x,y)$ *from Theorem 9.6 the general-ized Green's function for the operator* $\mathcal{L}$ *on the domain* Ω.

For twice differentiable functions we obtain the following corollary:

Theorem 9.8. *With the generalized Green's function from Theorem 9.6 we have the representation*

$$u(y) = - \int_\Omega \mathcal{L}u(x) \cdot G(x,y)dx, \quad y \in \Omega$$

for all functions $u \in C^2(\overline{\Omega})$ *with* $u = 0$ *on* $\partial\Omega$.

Proof: We integrate the relation

$$\sum_{i,j=1}^n a_{ij}(x)u_{x_i}(x)G_{x_j}(x,y) = \sum_{j=1}^n \frac{\partial}{\partial x_j}\Big(\sum_{i=1}^n a_{ij}(x)u_{x_i}(x)G(x,y)\Big) - G(x,y)\mathcal{L}(u)$$

over the domain Ω_ε via the Gaussian integral theorem. The boundary integrals vanish for $\varepsilon \to 0+$ because of the growth condition, and we evaluate the integral on the left-hand side by Theorem 9.6. This gives us the representation formula stated above.   q.e.d.

Remarks:

1. Originally the Green function for elliptic differential operators in divergence form has been considered by W. Littman, G. Stampacchia, and H.F. Weinberger. Later M. Grüter constructed Green's function for elliptic operators with L^∞-coefficients in the Sobolev space and derived global estimates together with K.-O. Widman. Here we refer the reader to the original paper of M. Grüter and K.-O. Widman quoted at the end of Chapter 8 in Section 9.

2. In his graduate seminar at the University of Göttingen, E. Heinz gave us the present approach to Green's function via the Schauder theory in the winter-semester 1985/86. I am grateful to H.-C. Grunau for an elaborate copy of these beautiful lectures on Green's function and for valuable discussions.

3. One can even derive the familiar growth estimates for the first and second partial derivatives of Green's function near the singularity:

$$|G_{x_i}(x,y)| \le c_3(M,n)|x-y|^{1-n}, \quad i=1,\ldots,n \quad \text{for} \quad x \in \Omega, \quad x \neq y$$

and

$$|G_{x_i x_j}(x,y)| \le c_4(M,n)|x-y|^{-n}, \quad i,j=1,\ldots,n \quad \text{for} \quad x \in \Omega, \quad x \neq y.$$

Here we utilize the weighted Schauder estimates in Theorem 7.9 of Section 7 from Chapter 9 without boundary conditions. We apply them to the generalized Green's function $G(x,y)$ at all midpoints $x \in \Omega \setminus \{y\}$ in the full disc $B(x,R)$ of radius $R = \frac{1}{2}|y-x|$ outside the singularity.

4. We infer the symmetry of the generalized Green's function from the symmetry of the Riemann-Dirichlet bilinear form above. Here we proceed as in Section 1 of Chapter 8 for the ordinary Green's function and utilize the representation formula in Theorem 9.6.

5. When we take only $L^\infty(\Omega)$-coefficients for the differential operator, our approximation method described in the proof of Theorem 9.6 gives us a generalized Green's function even in this situation. Here we approximate the coefficients by $C^{1+\alpha}(\Omega)$-coefficients and control the representation formula in the limit. This generalized Green's function belongs to the Sobolev space in Theorem 9.6, satisfies the given growth condition, and is Hölder continuous outside the singularity - according to the regularity result of de Georgi and Nash from Theorem 5.2 in Section 5. However, differentiability and growth conditions for the derivatives of Green's function cannot be attained in this situation.

10 Spectral Theory of the Laplace-Beltrami Operator

Let $\Omega \subset \mathbb{R}^n$ denote a bounded domain with the $C^{2+\mu}$-boundary $\partial\Omega$ and $0 < \mu < 1$. Here we prescribe the elliptic Riemannian metric

$$ds^2 = \sum_{i,j=1}^{n} g_{ij}(x)dx_i dx_j, \quad x \in \overline{\Omega}$$

of the class $C^{1+\mu}(\overline{\Omega})$ with its Gramian determinant

$$g(x) := \det\big(g_{ij}(x)\big)_{i,j=1,\dots,n}$$

and its inverse matrix

$$\big(g^{ij}(x)\big)_{i,j=1,\dots,n}, \quad x \in \Omega.$$

For the functions $\psi = \psi(x) \in C^{2+\mu}(\overline{\Omega})$ we consider the *Laplace-Beltrami operator*

$$\boldsymbol{\Delta}\psi(x) = \frac{1}{\sqrt{g(x)}} \sum_{i=1}^{n} \frac{\partial}{\partial x_i} \left(\sqrt{g(x)} \sum_{j=1}^{n} g^{ij}(x) \frac{\partial}{\partial x_j} \psi(x) \right) \tag{1}$$

introduced in Section 8 of Chapter 1. Of central interest is the following eigenvalue problem

$$-\boldsymbol{\Delta}\psi(x) = \lambda\psi(x), \quad x \in \Omega \quad \text{and} \quad \psi(x) = 0, \quad x \in \partial\Omega \tag{2}$$

for real eigenvalues $\lambda \in \mathbb{R}$. We recall the *Beltrami operator of first order*

$$\nabla(\phi, \psi) := \sum_{i,j=1}^{n} g^{ij}(x)\phi_{x_i}(x)\psi_{x_j}(x)$$

and obtain the invariant *Riemann-Dirichlet bilinear form*

$$\mathbb{D}(\phi, \psi) := \int_{\Omega} \nabla(\phi, \psi)\sqrt{g(x)}dx = \int_{\Omega} \left\{ \sum_{i,j=1}^{n} \sqrt{g(x)}g^{ij}(x)\phi_{x_i}(x)\psi_{x_j}(x) \right\}dx.$$

Furthermore, we introduce the canonical bilinear form on the Riemannian manifold

$$\mathbb{B}(\phi, \psi) := \int_{\Omega} \{\phi(x)\psi(x)\}\sqrt{g(x)}dx; \quad \phi, \psi \in L^2(\Omega).$$

Now we multiply (2) by an arbitrary test function and we arrive at the *weak eigenvalue equation*

$$\mathbb{D}(\phi, \psi) = \lambda\mathbb{B}(\phi, \psi), \quad \phi \in C_0^{\infty}(\Omega). \tag{3}$$

As described in Section 9, we determine the symmetric Green's function to the elliptic operator

$$\mathcal{L}(\psi) := \sum_{i=1}^{n} \frac{\partial}{\partial x_i} \left(\sum_{j=1}^{n} a_{ij}(x)\psi_{x_j}(x) \right), \quad x \in \Omega \tag{4}$$

with the coefficients

$$a_{ij}(x) := \sqrt{g(x)}g^{ij}(x), \quad x \in \overline{\Omega} \quad \text{for} \quad i,j = 1,\ldots,n.$$

Now we insert $G(.,y)$ into the weak eigenvalue equation and obtain the following identity

$$\lambda \int_\Omega G(x,y)\psi(x)\sqrt{g(x)}dx = \lambda \mathbb{B}(G(.,y),\psi) =$$

$$= \mathbb{D}(G(.,y),\psi) = \int_\Omega \{\sum_{i,j=1}^n a_{ij}(x)G_{x_i}(x,y)\psi_{x_j}(x)\}dx = \psi(y)$$

for all points $y \in \Omega$. Finally, we define the weakly singular integral operator

$$\mathbb{K}\psi(y) := \int_\Omega G(x,y)\psi(x)\sqrt{g(x)}dx, \quad y \in \overline{\Omega}.$$

We have transformed the eigenvalue problem (1) into the equivalent eigenvalue problem

$$\mathbb{K}\psi(y) = \frac{1}{\lambda}\psi(y), \quad y \in \overline{\Omega} \tag{5}$$

for the weakly singular integral operator $\mathbb{K}$.

Now we can proceed as in Chapter 8 in order to study the eigenvalue problem of this integral operator. We only have to integrate with respect to the surface element

$$\sqrt{g(x)}, \quad x \in \Omega$$

over the manifold. This constitutes a positive $C^{1+\mu}(\overline{\Omega})$-function: We obtain Hilbert-Schmidt integral operators with respect to the Riemannian metric ds^2, where their kernels are symmetric and weakly singular. We construct our eigenfunctions and their eigenvalues in the Hilbert space $\mathcal{H}$, endowed with the inner product $\mathbb{B}(.,.)$, via Rellich's spectral theorem. Then we regularize the eigenfunctions with the aid of Schur's theory of iterated kernels. Thus we arrive at the following result, whose complete proof can be taken from the Sections 1, 2, 6, 7, and 9 in Chapter 8.

Theorem 10.1. *To the Laplace-Beltrami operator from above, there exists a complete orthonormal system in $\mathcal{H}$ of eigenfunctions $\psi_k(x) \in C^{2+\mu}(\overline{\Omega})$ satisfying*

$$-\Delta\psi_k(x) = \lambda_k\psi_k(x), \quad x \in \Omega \quad \text{and} \quad \psi_k(x) = 0, \quad x \in \partial\Omega$$

for the eigenvalues λ_k with $k = 1,2,3.\ldots$ such that

$$0 < \lambda_1 \le \lambda_2 \le \lambda_3 \le \ldots \to +\infty$$

holds true.

11 Some Historical Notices to Chapter 10

The concept of weak solutions for partial differential equations was created by D. Hilbert already in 1900. His theory of integral equations from Chapter 8 provides the transition from the classical to the modern approach for partial differential equations.

Before they became widely known under the present name, Sobolev spaces have already been applied by K. Friedrichs and F. Rellich to spectral problems, as described in Chapter 8. Especially, Rellich's selection theorem from 1930 provided the decisive tool treating weak partial differential equations.

About 1957, E. de Georgi and independently J. Nash achieved the breakthrough in the regularity theory from weak to classical solutions. This was substantially simplified by J. Moser in 1960 by his iteration technique, consisting of inverse Hölder inequalities. Finally, W. Littman, G. Stampacchia, and H. Weinberger constructed even the Green's function for weak elliptic differential equations in 1963.

Figure 2.5   PORTRAIT OF K. FRIEDRICHS (1901–1982)
taken from the biography by *C. Reid: Courant, in Göttingen and New York - An Album*; Springer-Verlag, Berlin... (1976).

Chapter 11

Nonlinear Partial Differential Equations

In this chapter we consider geometric partial differential equations, which appear for two-dimensional surfaces in their state of equilibrium. Here we give the differential-geometric foundations in Section 1 and determine in Section 2 the Euler equations of 2-dimensional, parametric functionals. In Section 3 we present the theory of characteristics for quasilinear hyperbolic differential equations, and Section 4 is devoted to the solution of Cauchy's initial value problem with the aid of successive approximation. In Section 5 we treat the Riemannian integration method for linear hyperbolic differential equations. Finally, we prove S. Bernstein's analyticity theorem in Section 6 using ideas of H. Lewy.

1 The Fundamental Forms and Curvatures of a Surface

In the first part of this section, we consider the *differential-geometrically regular surface* on the parameter domain $\Omega \subset \mathbb{R}^2$:

$$\mathbf{x}(u,v) = (x(u,v), y(u,v), z(u,v))^* \; : \; \Omega \to \mathbb{R}^3 \in C^2(\Omega, \mathbb{R}^3),$$

satisfying the condition

$$\mathbf{x}_u(u,v) \wedge \mathbf{x}_v(u,v) \neq \mathbf{0} \qquad \text{for all} \quad (u,v) \in \Omega. \tag{1}$$

Here $\wedge$ denotes the exterior product in $\mathbb{R}^3$. Now the surface $\mathbf{x}$ has the normal

$$\mathbf{N}(u,v) := |\mathbf{x}_u \wedge \mathbf{x}_v(u,v)|^{-1} \mathbf{x}_u \wedge \mathbf{x}_v(u,v) \; : \; \Omega \to S^2 \tag{2}$$

with $S^2 := \{\mathbf{y} \in \mathbb{R}^3 \; : \; |\mathbf{y}| = 1\}$ and the tangential space

$$T_{\mathbf{x}(u,v)} := \left\{ \mathbf{y} \in \mathbb{R}^3 \; : \; \mathbf{y} \cdot \mathbf{N}(u,v) = 0 \right\}. \tag{3}$$

For each point $(u,v) \in \Omega$ we define the linear mapping

F. Sauvigny, *Partial Differential Equations 2*, Universitext,
DOI 10.1007/978-1-4471-2984-4_5, © Springer-Verlag London 2012

$$dx(u,v): \quad \mathbb{R}^2 \to T_{\mathbf{x}(u,v)},$$

$$(du, dv) \mapsto (\mathbf{x}_u, \mathbf{x}_v) \cdot \binom{du}{dv} = \mathbf{x}_u(u,v)\, du + \mathbf{x}_v(u,v)\, dv \tag{4}$$

with its adjoint mapping

$$dx(u,v)^*: \quad (du, dv) \mapsto (du, dv) \cdot \binom{\mathbf{x}_u^*}{\mathbf{x}_v^*} = \mathbf{x}_u(u,v)^*\, du + \mathbf{x}_v(u,v)^*\, dv. \tag{5}$$

We remark that the relation $1 = \mathbf{N}^* \cdot \mathbf{N}$ implies

$$\mathbf{N}^* \cdot \mathbf{N}_u = 0 = \mathbf{N}^* \cdot \mathbf{N}_v \quad \text{in} \quad \Omega. \tag{6}$$

Consequently, we obtain a further linear mapping

$$dN(u,v): \quad \mathbb{R}^2 \to T_{\mathbf{x}(u,v)},$$

$$(du, dv) \mapsto (\mathbf{N}_u, \mathbf{N}_v) \cdot \binom{du}{dv} = \mathbf{N}_u(u,v)\, du + \mathbf{N}_v(u,v)\, dv \tag{7}$$

with the adjoint mapping

$$dN(u,v)^*: \quad (du, dv) \mapsto (du, dv) \cdot \binom{\mathbf{N}_u^*}{\mathbf{N}_v^*} = \mathbf{N}_u(u,v)^*\, du + \mathbf{N}_v(u,v)^*\, dv. \tag{8}$$

We now define three quadratic forms on the space $\mathbb{R}^2$ depending on the point $(u,v) \in \Omega$. *The first fundamental form* is given by

$$\begin{aligned}
I(u,v) &:= dx(u,v)^* \cdot dx(u,v) \\
&= \mathbf{x}_u^* \cdot \mathbf{x}_u(u,v)\, du^2 + 2\mathbf{x}_u^* \cdot \mathbf{x}_v(u,v)\, du\, dv + \mathbf{x}_v^* \cdot \mathbf{x}_v(u,v)\, dv^2 \\
&=: E(u,v)\, du^2 + 2F(u,v)\, du\, dv + G(u,v)\, dv^2,
\end{aligned} \tag{9}$$

and *the second fundamental form* is defined by

$$\begin{aligned}
II(u,v) &:= -dx(u,v)^* \cdot dN(u,v) \\
&= -(\mathbf{x}_u^* \cdot \mathbf{N}_u)\, du^2 - (\mathbf{x}_u^* \cdot \mathbf{N}_v + \mathbf{x}_v^* \cdot \mathbf{N}_u)\, du\, dv - (\mathbf{x}_v^* \cdot \mathbf{N}_v)\, dv^2 \\
&= (\mathbf{N}^* \cdot \mathbf{x}_{uu})\, du^2 + 2(\mathbf{N}^* \cdot \mathbf{x}_{uv})\, du\, dv + (\mathbf{N}^* \cdot \mathbf{x}_{vv})\, dv^2 \\
&=: L(u,v)\, du^2 + 2M(u,v)\, du\, dv + N(u,v)\, dv^2.
\end{aligned} \tag{10}$$

Here we have used that the relation $\mathbf{N}^* \cdot \mathbf{x}_u = 0 = \mathbf{N}^* \cdot \mathbf{x}_v$ implies

$$-\mathbf{N}_u^* \cdot \mathbf{x}_u = \mathbf{N}^* \cdot \mathbf{x}_{uu}, \qquad -\mathbf{N}_u^* \cdot \mathbf{x}_v = \mathbf{N}^* \cdot \mathbf{x}_{uv}, \qquad \text{etc.}$$

Finally, we define *the third fundamental form*

$$\begin{aligned}
III(u,v) &:= dN(u,v)^* \cdot dN(u,v) \\
&= (\mathbf{N}_u^* \cdot \mathbf{N}_u)\, du^2 + 2(\mathbf{N}_u^* \cdot \mathbf{N}_v)\, du\, dv + (\mathbf{N}_v^* \cdot \mathbf{N}_v)\, dv^2 \\
&=: e(u,v)\, du^2 + 2f(u,v)\, du\, dv + g(u,v)\, dv^2.
\end{aligned} \tag{11}$$

The behavior as far as the curvatures of a surface are concerned is determined by the *Weingarten mapping* or alternatively the *shape operator*

$$W(u,v) := -d\mathbf{N}(u,v) \circ (d\mathbf{x}(u,v))^{-1} \; : \; T_{\mathbf{x}(u,v)} \to T_{\mathbf{x}(u,v)}. \qquad (12)$$

The parameters $(u,v) \in \Omega$ being fixed, the mapping $W(u,v)$ attributes the vectors $\mathbf{x}_u \mapsto -\mathbf{N}_u$ and $\mathbf{x}_v \mapsto -\mathbf{N}_v$.

Geometric interpretation:
The tangential vector $\mathbf{y} \in T_{\mathbf{x}(u,v)}$ given, we consider a regular curve

$$\mathbf{x}(t) := \mathbf{x}(u(t), v(t)), -\varepsilon < t < \varepsilon$$

on the surface $\mathbf{x}$ satisfying

$$\mathbf{x}(0) = \mathbf{x}(u(0), v(0)) = \mathbf{x}(u,v) \qquad \text{and} \qquad \mathbf{x}'(0) = \mathbf{y} \in T_{\mathbf{x}(u,v)}.$$

We then observe the curve

$$\mathbf{N}(t) := -\mathbf{N}(u(t), v(t)), -\varepsilon < t < \varepsilon$$

with the tangent vector $\mathbf{N}'(0) \in T_{\mathbf{x}(u,v)}$. The mapping

$$\mathbf{y} = \mathbf{x}'(0) \quad \mapsto \quad \mathbf{N}'(0) =: -\nabla_{\mathbf{y}}\mathbf{N}(u,v) \; : \; T_{\mathbf{x}(u,v)} \to T_{\mathbf{x}(u,v)}$$

is usually denoted as *covariant derivative* of the vector-field $\mathbf{N}$ in direction $\mathbf{y}$. Since this linear mapping coincides with the Weingarten mapping on the basis $\{\mathbf{x}_u, \mathbf{x}_v\}$, the Weingarten mapping is the negative covariant derivative of the normal $\mathbf{N}$ in the direction of the tangential vector $\mathbf{y}$. Consequently, the Weingarten mapping is invariant with respect to positive-oriented parameter transformations.

With respect to the basis $\{\mathbf{x}_u, \mathbf{x}_v\}$ in the tangential space $T_{\mathbf{x}(u,v)}$ the Weingarten map $W(u,v)$ is described by the symmetric matrix

$$\begin{pmatrix} -\mathbf{N}_u \cdot \mathbf{x}_u & -\mathbf{N}_u \cdot \mathbf{x}_v \\ -\mathbf{N}_v \cdot \mathbf{x}_u & -\mathbf{N}_v \cdot \mathbf{x}_v \end{pmatrix}. \qquad (13)$$

Therefore, $W(u,v)$ is a symmetric linear mapping. The latter possesses two real eigenvalues $\kappa_j(u,v)$ belonging to the eigenvectors $\mathbf{e}_j(u,v) \in T_{\mathbf{x}(u,v)}$ with $|\mathbf{e}_j(u,v)| = 1$ for $j = 1, 2$. We obtain the *principal curvatures* with $\kappa_j(u,v)$ attributed to the *principal curvature directions* $\mathbf{e}_j(u,v)$. We summarize

$$W(u,v) \circ \mathbf{e}_j(u,v) = \kappa_j(u,v)\mathbf{e}_j(u,v) \qquad \text{for} \quad j = 1, 2. \qquad (14)$$

Let $\mathbf{y} = \cos\vartheta\, \mathbf{e}_1(u,v) + \sin\vartheta\, \mathbf{e}_2(u,v)$, $0 \leq \vartheta \leq 2\pi$, be an arbitrary tangential vector to the surface $\mathbf{x}(u,v)$. Then we consider the quadratic form

$$Q(\mathbf{y}) := \big(W(u,v) \circ \mathbf{y}\big) \cdot \mathbf{y}$$
$$= \big(W(u,v) \circ (\cos \vartheta \, \mathbf{e}_1 + \sin \vartheta \, \mathbf{e}_2)\big) \cdot (\cos \vartheta \, \mathbf{e}_1 + \sin \vartheta \, \mathbf{e}_2)$$
$$= (\cos \vartheta \, \kappa_1 \mathbf{e}_1 + \sin \vartheta \, \kappa_2 \mathbf{e}_2) \cdot (\cos \vartheta \, \mathbf{e}_1 + \sin \vartheta \, \mathbf{e}_2)$$
$$= \kappa_1(u,v) \cos^2 \vartheta + \kappa_2(u,v) \sin^2 \vartheta.$$

Consequently, we obtain the

Theorem 1.1. (Euler's formula for the normal curvature)
We determine the normal curvature of the surface in the direction $\mathbf{y} = \cos \vartheta \, \mathbf{e}_1(u,v) + \sin \vartheta \, \mathbf{e}_2(u,v)$ *by*

$$Q(\mathbf{y}) = \kappa_1(u,v) \cos^2 \vartheta + \kappa_2(u,v) \sin^2 \vartheta. \qquad (15)$$

In the case $\kappa_1(u,v) \leq \kappa_2(u,v)$, *the normal curvature is minimized in the direction* $\mathbf{e}_1(u,v)$ *and maximized in the direction* $\mathbf{e}_2(u,v)$.

Definition 1.2. *A point* $\mathbf{x}(u,v)$ *of the surface* $\mathbf{x}$ *is called an* umbilical point, *if* $\kappa_1(u,v) = \kappa_2(u,v)$ *is satisfied.*

Definition 1.3. *We define the* Gaussian curvature *of the surface by*

$$K(u,v) := \kappa_1(u,v)\kappa_2(u,v) = \det W(u,v), \qquad (u,v) \in \Omega. \qquad (16)$$

The mean curvature *is given by*

$$H(u,v) := \frac{1}{2}\big(\kappa_1(u,v) + \kappa_2(u,v)\big) = \frac{1}{2} \, \mathrm{tr}\, W(u,v), \qquad (u,v) \in \Omega. \qquad (17)$$

Here det and tr denote the determinant and the trace of a matrix.

With respect to the bases $\{\mathbf{x}_u, \mathbf{x}_v\}$, $\{(1,0),(0,1)\}$, $\{\mathbf{x}_u, \mathbf{x}_v\}$ the Weingarten mapping is described by the matrices

$$\begin{pmatrix} L & M \\ M & N \end{pmatrix} \begin{pmatrix} E & F \\ F & G \end{pmatrix}^{-1} = \frac{1}{EG - F^2} \begin{pmatrix} L & M \\ M & N \end{pmatrix} \begin{pmatrix} G & -F \\ -F & E \end{pmatrix}. \qquad (18)$$

This reveals the following formulas

$$K(u,v) = \frac{LN - M^2}{EG - F^2} \qquad (19)$$

and

$$H(u,v) = \frac{1}{2} \frac{GL - 2FM + EN}{EG - F^2}. \qquad (20)$$

Finally, we show the

Theorem 1.4. *We have the following relation between the three fundamental forms*

$$\begin{pmatrix} e & f \\ f & g \end{pmatrix} - 2H \begin{pmatrix} L & M \\ M & N \end{pmatrix} + K \begin{pmatrix} E & F \\ F & G \end{pmatrix} = \begin{pmatrix} 0 & 0 \\ 0 & 0 \end{pmatrix}. \tag{21}$$

Proof: From the theorem of Hamilton-Cayley we infer that a symmetric matrix represents a zero of its characteristic polynomial. Noting the symmetry of $W(u, v)$ we obtain

$$\begin{aligned} 0 &= W(u,v)^* \circ W(u,v) - 2H(u,v)W(u,v) + K(u,v)\,\mathrm{Id} \\ &= (d\mathbf{N} \circ (d\mathbf{x})^{-1})^* \circ d\mathbf{N} \circ (d\mathbf{x})^{-1} + 2H d\mathbf{N} \circ (d\mathbf{x})^{-1} + K\,\mathrm{Id} \\ &= (d\mathbf{x}^*)^{-1} \circ d\mathbf{N}^* \circ d\mathbf{N} \circ (d\mathbf{x})^{-1} + 2H d\mathbf{N} \circ (d\mathbf{x})^{-1} + K\,\mathrm{Id}. \end{aligned}$$

Applying the operations $d\mathbf{x}^* \circ$ and $\circ d\mathbf{x}$ to this equation, we attain the identity

$$\begin{aligned} 0 &= d\mathbf{N}^* \circ d\mathbf{N} + 2H\,d\mathbf{x}^* \circ d\mathbf{N} + K\,d\mathbf{x}^* \circ d\mathbf{x} \\ &= \mathit{III}(u,v) - 2H\,\mathit{II}(u,v) + K\,\mathit{I}(u,v), \end{aligned}$$

and (21) follows.   q.e.d.

In the second part of this section we investigate *graphs in arbitrary dimensions* $n \geq 2$:

$$\mathbf{z}(x) = \mathbf{z}(x_1, \ldots, x_n) := (x_1, \ldots, x_n, \zeta(x_1, \ldots, x_n)) : \Omega \mapsto \mathbb{R}^{n+1}. \tag{22}$$

Here we defined the height function

$$z = \zeta(x) = \zeta(x_1, \ldots, x_n) \in C^2(\Omega) \tag{23}$$

on the domain $\Omega \subset \mathbb{R}^n$. We determine the tangential vectors

$$\mathbf{z}_{x_i}(x) = (\delta_{i1}, \ldots, \delta_{in}, \zeta_{x_i}(x)), \quad x \in \Omega \quad \text{for} \quad i = 1, \ldots, n. \tag{24}$$

We have the upper unit normal

$$\mathbf{N}(x) := (1 + |\nabla\zeta(x)|^2)^{-\frac{1}{2}} (-\zeta_{x_1}, \ldots, -\zeta_{x_n}, 1), \quad x \in \Omega \tag{25}$$

and the *tangential space*

$$T_{\mathbf{z}(x)} := \{\mathbf{y} \in \mathbb{R}^{n+1} | \mathbf{y} \cdot \mathbf{N}(x) = 0\}. \tag{26}$$

As above we introduce the first fundamental form with the coefficients

$$g_{ij}(x) := \mathbf{z}_{x_i} \cdot \mathbf{z}_{x_j}(x) = \delta_{ij} + \zeta_{x_i}(x)\zeta_{x_j}(x), \quad x \in \Omega \quad \text{for} \quad i, j = 1, \ldots, n. \tag{27}$$

The tangential map

$$d\mathbf{z}(x) : \mathbb{R}^n \mapsto T_{\mathbf{z}(x)} \tag{28}$$

with respect to the bases $\mathbf{e}_i = (\delta_{i1}, \ldots, \delta_{in}) \in \mathbb{R}^n$ and $\mathbf{z}_{x_i}(x) \in T_{\mathbf{z}(x)}$ for $i = 1, \ldots, n$ is given by the matrix

$$\left(g_{ij}(x)\right)_{i,j=1,\ldots,n}.$$

We now consider the family of matrices

$$G^\lambda(x) := \left(\delta_{ij} + \lambda \zeta_{x_i}(x)\zeta_{x_j}(x)\right)_{i,j=1\ldots,n} \tag{29}$$

and observe

$$G^\lambda(x) \circ \nabla\zeta(x) = (1 + \lambda|\nabla\zeta(x)|^2)\nabla\zeta(x) \tag{30}$$

with an arbitrary parameter $\lambda \in \mathbb{R}$. Furthermore, we note that

$$G^\lambda(x) \circ \mathbf{y} = \mathbf{y} \quad \text{for all} \quad \mathbf{y} \in \mathbb{R}^n \quad \text{with} \quad \mathbf{y} \cdot \nabla\zeta(x) = 0. \tag{31}$$

We deduce

$$G^\lambda(x) \circ G^1(x) \circ \nabla\zeta(x) = (1 + \lambda|\nabla\zeta(x)|^2)(1 + |\nabla\zeta(x)|^2)\nabla\zeta(x)$$

and choose λ such that

$$1 = (1 + \lambda|\nabla\zeta(x)|^2)(1 + |\nabla\zeta(x)|^2) \quad \text{or equivalently}$$

$$\frac{1}{1 + |\nabla\zeta(x)|^2} - 1 = \lambda|\nabla\zeta(x)|^2 \quad \text{or equivalently}$$

$$\frac{-1}{1 + |\nabla\zeta(x)|^2} = \lambda \quad \text{holds true.}$$

Introducing the matrix

$$g^{ij}(x) := \delta_{ij} - \frac{\zeta_{x_i}(x)\zeta_{x_j}(x)}{1 + |\nabla\zeta(x)|^2} \quad \text{with} \quad i, j = 1, \ldots, n \tag{32}$$

we obtain

$$g^{ij}(x)g_{jk}(x) = \delta_k^i, \quad x \in \Omega \tag{33}$$

via the Einstein summation convention.
As above we can introduce the second fundamental form with the coefficients

$$h_{ij}(x) := -\mathbf{N}_{x_i} \cdot \mathbf{z}_{x_j}(x) = \mathbf{N} \cdot \mathbf{z}_{x_i x_j}(x) = \left(\mathbf{N}(x) \cdot \mathbf{e}\right)\zeta_{x_i x_j}(x), \quad x \in \Omega \tag{34}$$

for $i, j = 1, \ldots, n$. Here we used the unit vector $\mathbf{e} = (0, \ldots, 0, 1) \in \mathbb{R}^{n+1}$. The linear map

$$- d\mathbf{N}(x) : \mathbb{R}^n \mapsto T_{\mathbf{z}(x)} \tag{35}$$

is represented by the matrix $\left(h_{ij}(x)\right)_{i,j=1,\ldots,n}$ with respect to the canonical bases.

We summarize to the

Theorem 1.5. *The Weingarten mapping for n-dimensional graphs is given by the matrices*

$$W(x) = \big(\mathbf{N}(x) \cdot \mathbf{e}\big)\big(\zeta_{x_i x_j}(x)\big)_{i,j=1,\ldots,n} \circ \Big(\delta_{ij} - \frac{\zeta_{x_i}(x)\zeta_{x_j}(x)}{1 + |\nabla\zeta(x)|^2}\Big)_{i,j=1,\ldots,n}, \quad x \in \Omega \tag{36}$$

with respect to the canonical bases.

Remarks: From this theorem we can deduce curvature equations for graphs in arbitrary dimensions. Especially for the equation

$$\text{trace} \ \ W(x) = 0, \quad x \in \Omega \tag{37}$$

we obtain the quasilinear *n-dimensional minimal surface equation*

$$\sum_{i,j=1}^{n} \Big(\delta_{ij} - \frac{\zeta_{x_i}(x)\zeta_{x_j}(x)}{1 + |\nabla\zeta(x)|^2}\Big)\zeta_{x_i x_j}(x) = 0 \quad \text{in} \ \ \Omega. \tag{38}$$

We now calculate

$$\sqrt{1 + |\nabla\zeta(x)|^2}\,\text{div}\big((1 + |\nabla\zeta(x)|^2)^{-\frac{1}{2}}\nabla\zeta(x)\big) =$$
$$\Delta\zeta(x) - \frac{1}{2(1+|\nabla\zeta(x)|^2)}\big(\nabla\textstyle\sum_{i=1}^{n}\zeta_{x_i}^2\big) \cdot \big(\nabla\zeta(x)\big) =$$
$$\Delta\zeta(x) - \frac{1}{1+|\nabla\zeta(x)|^2}\textstyle\sum_{j=1}^{n}\big(\sum_{i=1}^{n}\zeta_{x_i}\zeta_{x_i x_j}\big)\zeta_{x_j}(x) = \tag{39}$$
$$\Delta\zeta(x) - \frac{1}{1+|\nabla\zeta(x)|^2}\textstyle\sum_{i,j=1}^{n}\zeta_{x_i}\zeta_{x_j}\zeta_{x_i x_j}(x).$$

By the identity (39) we transform (38) into the *minimal surface equation in divergence form*

$$\text{div} \ \Big(\frac{\nabla\zeta(x)}{\sqrt{1 + |\nabla\zeta(x)|^2}}\Big) = 0 \quad \text{in} \ \ \Omega. \tag{40}$$

Geometrically, the arithmetic means of the n principal curvatures vanishes for these graphs at each point.

2 Two-dimensional Parametric Integrals

We consider differential-geometrically regular surfaces on the parameter domain $(u, v) \in \Omega \subset \mathbb{R}^2$, namely

$$\mathbf{x} = \mathbf{x}(u, v) = \big(x_1(u, v), x_2(u, v), x_3(u, v)\big) = \big(x(u, v), y(u, v), z(u, v)\big),$$
$$\mathbf{x} : \Omega \to \mathbb{R}^3 \in C^3(\Omega) \tag{1}$$

satisfying $|\mathbf{x}_u \wedge \mathbf{x}_v(u, v)| > 0$ for all $(u, v) \in \Omega$ and

$$\iint\limits_{\Omega} |\mathbf{x}_u \wedge \mathbf{x}_v(u,v)| \, du \, dv < +\infty.$$

Denoting by $S^2 := \{\mathbf{z} \in \mathbb{R}^3 \ : \ |\mathbf{z}| = 1\}$ the unit sphere in $\mathbb{R}^3$, the normal $\mathbf{X}$ of the surface $\mathbf{x}$ is given as follows:

$$\mathbf{X}(u,v) := |\mathbf{x}_u \wedge \mathbf{x}_v|^{-1} \mathbf{x}_u \wedge \mathbf{x}_v(u,v) : \Omega \to S^2 \in C^2(\Omega). \qquad (2)$$

We consider a density function

$$F = F(\mathbf{x}, \mathbf{p}) = F(x_1, x_2, x_3; p_1, p_2, p_3),$$

$$F : \mathbb{R}^3 \times \mathbb{R}^3 \to \mathbb{R} \in C^2\big(\mathbb{R}^3 \times (\mathbb{R}^3 \setminus \{\mathbf{0}\})\big) \cap C^0(\mathbb{R}^3 \times \mathbb{R}^3),$$

which we assume to be positive-homogeneous of degree 1; that means

$$F(\mathbf{x}, \lambda \mathbf{p}) = \lambda F(\mathbf{x}, \mathbf{p}) \qquad \text{for all} \quad \lambda > 0. \qquad (3)$$

From the relation (3) we obtain the following condition by differentiation with respect to λ at $\lambda = 1$:

$$F_{\mathbf{p}}(\mathbf{x}, \mathbf{p}) \cdot \mathbf{p}^* = F(\mathbf{x}, \mathbf{p}), \quad \mathbf{p} \circ F_{\mathbf{p}\mathbf{p}}(\mathbf{x}, \mathbf{p}) \circ \mathbf{p}^* = 0 \qquad \text{for} \quad \mathbf{p} \neq \mathbf{0}. \qquad (4)$$

Furthermore, $F_{\mathbf{x}}(\mathbf{x}, \lambda \mathbf{p}) = \lambda F_{\mathbf{x}}(\mathbf{x}, \mathbf{p})$ implies

$$F_{\mathbf{x}\mathbf{p}}(\mathbf{x}, \mathbf{p}) \circ \mathbf{p}^* = F_{\mathbf{x}}(\mathbf{x}, \mathbf{p}) \qquad \text{for} \quad \mathbf{p} \neq \mathbf{0}. \qquad (5)$$

Here we have abbreviated $F_{\mathbf{p}} := (F_{p_1}, F_{p_2}, F_{p_3})$, $F_{\mathbf{p}\mathbf{p}} := (F_{p_i p_j})_{i,j=1,2,3}$, etc.

We define the *generalized area integral*

$$A(\mathbf{x}) = \iint\limits_{\Omega} F(\mathbf{x}(u,v), \mathbf{X}(u,v)) |\mathbf{x}_u \wedge \mathbf{x}_v(u,v)| \, du \, dv$$

$$= \iint\limits_{\Omega} F(\mathbf{x}(u,v), \mathbf{x}_u \wedge \mathbf{x}_v(u,v)) \, du \, dv. \qquad (6)$$

Evidently, an arbitrary positive-oriented diffeomorphism

$$f = f(\alpha, \beta) = (u(\alpha, \beta), v(\alpha, \beta)) : \Theta \to \Omega \in C^1(\Theta, \mathbb{R}^2)$$

satisfies the identity

$$A(\mathbf{x}) = A(\mathbf{x} \circ f).$$

Consequently, A represents a parametric functional. We can show that the expression A from (6) gives us the most general two-dimensional parameter-invariant functional in $\mathbb{R}^3$.

Examples:

1. For $F = F(\mathbf{x}, \mathbf{p}) := |\mathbf{p}|$ we obtain the *ordinary area functional*.
2. In the case

$$F = F(\mathbf{x}, \mathbf{p}) = |\mathbf{p}| + \frac{2H}{3}\mathbf{x} \cdot \mathbf{p}, \qquad H \in \mathbb{R},$$

we get the *functional of E. Heinz*

$$A(\mathbf{x}) = \iint_{\Omega} \left\{ |\mathbf{x}_u \wedge \mathbf{x}_v| + \frac{2H}{3}(\mathbf{x}, \mathbf{x}_u, \mathbf{x}_v) \right\} du\, dv, \qquad (7)$$

abbreviating the triple product as follows:

$$(\mathbf{x}, \mathbf{y}, \mathbf{z}) := \mathbf{x} \cdot (\mathbf{y} \wedge \mathbf{z}), \qquad \mathbf{x}, \mathbf{y}, \mathbf{z} \in \mathbb{R}^3.$$

In (7) we have to comprehend H as a Lagrange parameter. Therefore, one minimizes the ordinary area functional with the subsidiary condition of keeping the volume constant:

$$\frac{2H}{3} \iint_{\Omega} (\mathbf{x}, \mathbf{x}_u, \mathbf{x}_v)\, du\, dv = 1.$$

3. When we finally consider

$$F = F(\mathbf{x}, \mathbf{p}) = |\mathbf{p}| + 2\mathbf{Q}(\mathbf{x}) \cdot \mathbf{p},$$

$$\mathbf{Q} : \mathbb{R}^3 \to \mathbb{R}^3 \in C^2(\mathbb{R}^3) \qquad \text{with} \quad \operatorname{div} \mathbf{Q}(\mathbf{x}) = H(\mathbf{x}),$$

we obtain the *functional of S. Hildebrandt*

$$A(\mathbf{x}) = \iint_{\Omega} \left\{ |\mathbf{x}_u \wedge \mathbf{x}_v| + 2\big(\mathbf{Q}(\mathbf{x}), \mathbf{x}_u, \mathbf{x}_v)\big) \right\} du\, dv. \qquad (8)$$

Here one minimizes the ordinary area functional with respect to constant weighted volume as a subsidiary condition:

$$2 \iint_{\Omega} (\mathbf{Q}(\mathbf{x}), \mathbf{x}_u, \mathbf{x}_v)\, du\, dv = 1.$$

We shall now determine the Euler equations of our generalized area integral A: Therefore, we consider the surface varied in the normal direction, when we take an arbitrary test function $\varphi = \varphi(u, v) \in C_0^\infty(\Omega)$, namely

$$\overline{\mathbf{x}}(u, v; t) := \mathbf{x}(u, v) + t\varphi(u, v)\mathbf{X}(u, v) : \Omega \times (-\varepsilon, \varepsilon) \to \mathbb{R}^3. \qquad (9)$$

When we choose the number $\varepsilon > 0$ sufficiently small, these surfaces remain differential-geometrically regular. We calculate

$$\overline{\mathbf{x}}_u = \mathbf{x}_u + t(\varphi\mathbf{X})_u, \qquad \overline{\mathbf{x}}_v = \mathbf{x}_v + t(\varphi\mathbf{X})_v,$$

$$\overline{\mathbf{x}}_u \wedge \overline{\mathbf{x}}_v = \mathbf{x}_u \wedge \mathbf{x}_v + t\{\mathbf{x}_u \wedge (\varphi\mathbf{X})_v + (\varphi\mathbf{X})_u \wedge \mathbf{x}_v\} + t^2(\varphi\mathbf{X})_u \wedge (\varphi\mathbf{X})_v.$$

$$(10)$$

This implies

$$\frac{\partial}{\partial t}A(\overline{\mathbf{x}}) = \frac{\partial}{\partial t}\iint\limits_{\Omega} F(\overline{\mathbf{x}},\overline{\mathbf{x}}_u \wedge \overline{\mathbf{x}}_v)\, du\, dv$$

$$= \iint\limits_{\Omega} \left(F_{\mathbf{x}}(\overline{\mathbf{x}},\overline{\mathbf{x}}_u \wedge \overline{\mathbf{x}}_v)\cdot\mathbf{X}\right)\varphi\, du\, dv$$

$$+ \iint\limits_{\Omega} F_{\mathbf{p}}(\overline{\mathbf{x}},\overline{\mathbf{x}}_u \wedge \overline{\mathbf{x}}_v)\cdot\{\mathbf{x}_u \wedge (\varphi\mathbf{X})_v + (\varphi\mathbf{X})_u \wedge \mathbf{x}_v\}\, du\, dv$$

$$(11)$$

$$+2t\iint\limits_{\Omega} \left(F_{\mathbf{p}}(\overline{\mathbf{x}},\overline{\mathbf{x}}_u \wedge \overline{\mathbf{x}}_v),(\varphi\mathbf{X})_u,(\varphi\mathbf{X})_v\right)\, du\, dv.$$

Then we obtain the Euler equations in the weak form

$$0 = \frac{\partial}{\partial t}A(\overline{\mathbf{x}})\Big|_{t=0}$$

$$= \iint\limits_{\Omega} \{F_{\mathbf{x}}(\mathbf{x},\mathbf{X})\cdot\mathbf{X}\}\,\varphi\,|\mathbf{x}_u \wedge \mathbf{x}_v|\, du\, dv$$

$$+ \iint\limits_{\Omega} \left\{(F_{\mathbf{p}}(\mathbf{x},\mathbf{X}),\mathbf{x}_u,\varphi\mathbf{X})_v + (F_{\mathbf{p}}(\mathbf{x},\mathbf{X}),\varphi\mathbf{X},\mathbf{x}_v)_u\right\}\, du\, dv$$

$$- \iint\limits_{\Omega} \left\{((F_{\mathbf{p}}(\mathbf{x},\mathbf{X}))_v,\mathbf{x}_u,\varphi\mathbf{X}) + ((F_{\mathbf{p}}(\mathbf{x},\mathbf{X}))_u,\varphi\mathbf{X},\mathbf{x}_v)\right\}\, du\, dv$$

$$= \iint\limits_{\Omega} \{\mathbf{X}\circ F_{\mathbf{px}}(\mathbf{x},\mathbf{X})\circ\mathbf{X}^*\}\,\varphi\,|\mathbf{x}_u \wedge \mathbf{x}_v|\, du\, dv$$

$$+ \iint\limits_{\Omega} \left\{(\mathbf{x}_u,F_{\mathbf{px}}(\mathbf{x},\mathbf{X})\circ\mathbf{x}_v,\mathbf{X}) + (F_{\mathbf{px}}(\mathbf{x},\mathbf{X})\circ\mathbf{x}_u,\mathbf{x}_v,\mathbf{X})\right\}\varphi\, du\, dv$$

$$+ \iint\limits_{\Omega} \left\{(\mathbf{x}_u,F_{\mathbf{pp}}(\mathbf{x},\mathbf{X})\circ\mathbf{X}_v,\mathbf{X}) + (F_{\mathbf{pp}}(\mathbf{x},\mathbf{X})\circ\mathbf{X}_u,\mathbf{x}_v,\mathbf{X})\right\}\varphi\, du\, dv.$$

$$(12)$$

We now set

$$2H(\mathbf{x},\mathbf{p}) := \operatorname{div} F_{\mathbf{p}}(\mathbf{x},\mathbf{p}) = \operatorname{tr} F_{\mathbf{px}}(\mathbf{x},\mathbf{p}).$$

($\operatorname{tr} F_{\mathbf{px}}$ denotes the trace of the matrix $F_{\mathbf{px}}$.) Consequently, the following parameter invariant equation holds true:

$$\left(F_{\mathbf{px}}(\mathbf{x}, \mathbf{X}) \circ \mathbf{x}_u, \mathbf{x}_v, \mathbf{X}\right) + \left(\mathbf{x}_u, F_{\mathbf{px}}(\mathbf{x}, \mathbf{X}) \circ \mathbf{x}_v, \mathbf{X}\right) + \left(\mathbf{x}_u, \mathbf{x}_v, F_{\mathbf{px}}(\mathbf{x}, \mathbf{X}) \circ \mathbf{X}\right)$$
$$= 2H(\mathbf{x}, \mathbf{X})(\mathbf{x}_u, \mathbf{x}_v, \mathbf{X}).$$

(13)

Therefore, the weak Euler differential equation (12) appears in the form

$$0 = \iint_{\Omega} \left\{ \left(F_{\mathbf{pp}}(\mathbf{x}, \mathbf{X}) \circ \mathbf{X}_u, \mathbf{x}_v, \mathbf{X}\right) + \left(\mathbf{x}_u, F_{\mathbf{pp}}(\mathbf{x}, \mathbf{X}) \circ \mathbf{X}_v, \mathbf{X}\right) \right.$$
$$\left. +2H(\mathbf{x}, \mathbf{X}) |\mathbf{x}_u \wedge \mathbf{x}_v| \right\} \varphi(u, v)\, du\, dv \qquad \text{for all} \quad \varphi \in C_0^{\infty}(\Omega).$$

(14)

We obtain the Euler equation as follows:

$$0 = \left(F_{\mathbf{pp}}(\mathbf{x}, \mathbf{X}) \circ \mathbf{X}_u, \mathbf{x}_v, \mathbf{X}\right) + \left(\mathbf{x}_u, F_{\mathbf{pp}}(\mathbf{x}, \mathbf{X}) \circ \mathbf{X}_v, \mathbf{X}\right)$$
$$+2H(\mathbf{x}, \mathbf{X}) |\mathbf{x}_u \wedge \mathbf{x}_v| \qquad \text{in} \quad \Omega.$$

(15)

This equation is obviously equivalent to the system

$$\mathbf{0} = \left\{ F_{\mathbf{pp}}(\mathbf{x}, \mathbf{X}) \circ \mathbf{X}_u \right\} \wedge \mathbf{x}_v + \mathbf{x}_u \wedge \left\{ F_{\mathbf{pp}}(\mathbf{x}, \mathbf{X}) \circ \mathbf{X}_v \right\}$$
$$+2H(\mathbf{x}, \mathbf{X}) \mathbf{x}_u \wedge \mathbf{x}_v \qquad \text{in} \quad \Omega.$$

(16)

Following the arguments in Section 3.6 of the book [K] by W. Klingenberg: *Eine Vorlesung über Differentialgeometrie*, we now introduce the lines of principal curvatures as parameters u, v into the surface. We obtain

$$\mathbf{x}_u \cdot \mathbf{x}_v = 0 = \mathbf{X}_u \cdot \mathbf{x}_v = \mathbf{X}_v \cdot \mathbf{x}_u,$$
$$\mathbf{X}_u = -\kappa_1 \mathbf{x}_u, \quad \mathbf{X}_v = -\kappa_2 \mathbf{x}_v \qquad \text{in} \quad \Omega$$

(17)

with the principal curvatures κ_1, κ_2. Furthermore, we define the weight factors

$$\varrho_1(u, v) := |\mathbf{x}_u \wedge \mathbf{x}_v|^{-1} \left(F_{\mathbf{pp}}(\mathbf{x}, \mathbf{X}) \circ \mathbf{x}_u, \mathbf{x}_v, \mathbf{X}\right)$$

(18)

and

$$\varrho_2(u, v) := |\mathbf{x}_u \wedge \mathbf{x}_v|^{-1} \left(\mathbf{x}_u, F_{\mathbf{pp}}(\mathbf{x}, \mathbf{X}) \circ \mathbf{x}_v, \mathbf{X}\right).$$

(19)

Then the relation (15) is transformed into the *quasilinear curvature equation*

$$\varrho_1(u, v)\kappa_1(u, v) + \varrho_2(u, v)\kappa_2(u, v) = 2H(\mathbf{x}(u, v), \mathbf{X}(u, v)) \qquad \text{in} \quad \Omega. \quad (20)$$

The weight factors ϱ_1 and ϱ_2 have the same positive (different) sign if and only if the matrix $F_{\mathbf{pp}}(\mathbf{x}, \mathbf{p})$ is positive-definite (indefinite) on the space orthogonal to $\mathbf{p}$.

Theorem 2.1. *The quasilinear curvature equation (20) represents the Euler equation of the parametric functional (6).*

In the case of Hildebrandt's functional (8) we observe

$$F(\mathbf{x}, \mathbf{p}) = |\mathbf{p}| + 2\mathbf{Q}(\mathbf{x}) \cdot \mathbf{p} = \sqrt{\sum_{k=1}^{3} p_k^2 + 2\sum_{k=1}^{3} q_k(\mathbf{x})p_k}$$

with $\operatorname{div}\mathbf{Q}(\mathbf{x}) = H(\mathbf{x})$. We calculate

$$F_{p_i} = \frac{p_i}{\sqrt{\sum\limits_{k=1}^{3} p_k^2}} + 2q_i(\mathbf{x}), \qquad F_{p_i p_j} = \frac{\delta_{ij}}{\sqrt{\sum\limits_{k=1}^{3} p_k^2}} - \frac{p_i p_j}{\sqrt{\sum\limits_{k=1}^{3} p_k^2}^3}$$

for $i, j = 1, 2, 3$. The weight factors reduce to $\varrho_1(u, v) \equiv 1 \equiv \varrho_2(u, v)$ in Ω and the equation (20) specializes to

$$\frac{1}{2}(\kappa_1(u, v) + \kappa_2(u, v)) = \frac{1}{2}\operatorname{div} F_{\mathbf{p}} = \operatorname{div}\mathbf{Q}(\mathbf{x}) = H(\mathbf{x}) \qquad \text{in} \quad \Omega. \qquad (21)$$

Then the system (16) appears in the form

$$\mathbf{X}_u \wedge \mathbf{x}_v + \mathbf{x}_u \wedge \mathbf{X}_v + 2H(\mathbf{x})\mathbf{x}_u \wedge \mathbf{x}_v = \mathbf{0} \qquad \text{in} \quad \Omega \qquad (22)$$

or equivalently

$$-(\mathbf{X} \wedge \mathbf{x}_v)_u + (\mathbf{X} \wedge \mathbf{x}_u)_v = 2H(\mathbf{x})\mathbf{x}_u \wedge \mathbf{x}_v \qquad \text{in} \quad \Omega. \qquad (23)$$

The equations (23) become transparent if we introduce conformal parameters into the surface as follows:

$$\mathbf{x}_u \cdot \mathbf{x}_v = 0 = |\mathbf{x}_u|^2 - |\mathbf{x}_v|^2 \qquad \text{in} \quad \Omega. \qquad (24)$$

We now observe

$$\mathbf{X} \wedge \mathbf{x}_u = \mathbf{x}_v, \qquad \mathbf{X} \wedge \mathbf{x}_v = -\mathbf{x}_u \qquad \text{in} \quad \Omega. \qquad (25)$$

Inserting (25) into (23), we obtain the *H-surface system*

$$\Delta\mathbf{x}(u, v) = 2H(\mathbf{x})\mathbf{x}_u \wedge \mathbf{x}_v \qquad \text{in} \quad \Omega. \qquad (26)$$

We summarize our considerations to the following

Theorem 2.2. (F. Rellich)
A conformally parametrized surface $\mathbf{x} = \mathbf{x}(u, v) : \Omega \to \mathbb{R}^3$ *with (24) has the prescribed mean curvature* $H = H(\mathbf{x})$ *if and only if* $\mathbf{x}$ *fulfills the H-surface system (26).*

Remark: If the matrix $F_{\mathbf{p}\mathbf{p}}(\mathbf{x}, \mathbf{p})$ is positive-definite on the space orthogonal to $\mathbf{p}$, we can introduce conformal parameters into a weighted first fundamental form. We then obtain the following elliptic system for the mapping

$$\mathbf{y}(u,v) := (\mathbf{x}(u,v), \mathbf{X}(u,v)) : \Omega \to \mathbb{R}^6,$$

namely

$$|\Delta \mathbf{y}(u,v)| \le c|\nabla \mathbf{y}(u,v)|^2 \quad \text{in} \quad \Omega.$$

In this context we refer the reader to

F. Sauvigny: *Curvature estimates for immersions of minimal surface type via uniformization and theorems of Bernstein type.* Manuscripta math. 67 (1990), 69-97.

On the domain $\Omega \subset \mathbb{R}^2$ we now define the surface

$$\mathbf{x}(x,y) := (x, y, \zeta(x,y)), \qquad (x,y) \in \Omega, \tag{27}$$

given as a graph above the x, y-plane. The normal to the surface $\mathbf{x}$ is then represented by

$$\mathbf{X}(x,y) := \frac{1}{\sqrt{1 + |\nabla \zeta(x,y)|^2}}(-\zeta_x, -\zeta_y, 1), \qquad (x,y) \in \Omega, \tag{28}$$

and the surface element by

$$|\mathbf{x}_x \wedge \mathbf{x}_y| = \sqrt{1 + |\nabla \zeta(x,y)|^2} =: \sqrt{\,}. \tag{29}$$

We determine the derivatives

$$\mathbf{x}_x(x,y) = (1, 0, \zeta_x(x,y)), \qquad \mathbf{x}_y(x,y) = (0, 1, \zeta_y(x,y)) \tag{30}$$

and

$$\mathbf{X}_x = \frac{1}{\sqrt{\,}}(-\zeta_{xx}, -\zeta_{xy}, 0) + \lambda_1 \mathbf{X}, \qquad \mathbf{X}_y = \frac{1}{\sqrt{\,}}(-\zeta_{xy}, -\zeta_{yy}, 0) + \lambda_2 \mathbf{X} \tag{31}$$

with certain functions λ_1, λ_2. When we insert the relations (30) and (31) into (15), we get the differential equation

$$0 = \left(F_{\mathbf{pp}}(\mathbf{x}, \mathbf{X}) \circ \begin{pmatrix} -\zeta_{xx} \\ -\zeta_{xy} \\ 0 \end{pmatrix}, \begin{pmatrix} 0 \\ 1 \\ \zeta_y \end{pmatrix}, \begin{pmatrix} -\zeta_x \\ -\zeta_y \\ 1 \end{pmatrix} \right)$$

$$+ \left(\begin{pmatrix} 1 \\ 0 \\ \zeta_x \end{pmatrix}, F_{\mathbf{pp}}(\mathbf{x}, \mathbf{X}) \circ \begin{pmatrix} -\zeta_{xy} \\ -\zeta_{yy} \\ 0 \end{pmatrix}, \begin{pmatrix} -\zeta_x \\ -\zeta_y \\ 1 \end{pmatrix} \right) \tag{32}$$

$$+ 2H(\mathbf{x}, \mathbf{X})\sqrt{1 + |\nabla \zeta(x,y)|^2}^3 \quad \text{in} \quad \Omega.$$

This represents a quasilinear differential equation of the form

$$a(x, y, \zeta(x,y), \nabla \zeta(x,y))\zeta_{xx} + 2b(\ldots)\zeta_{xy} + c(\ldots)\zeta_{yy} + d(\ldots) = 0 \quad \text{in} \quad \Omega. \tag{33}$$

In particular, for Hildebrandt's functional we obtain

$$0 = \begin{vmatrix} -\zeta_{xx} & 0 & -\zeta_x \\ -\zeta_{xy} & 1 & -\zeta_y \\ 0 & \zeta_y & 1 \end{vmatrix} + \begin{vmatrix} 1 & -\zeta_{xy} & -\zeta_x \\ 0 & -\zeta_{yy} & -\zeta_y \\ \zeta_x & 0 & 1 \end{vmatrix} + 2H(\mathbf{x})\sqrt{1 + |\nabla\zeta(x,y)|^2}^3$$

$$= -(1 + \zeta_y^2)\zeta_{xx} + 2\zeta_x\zeta_y\zeta_{xy} - (1 + \zeta_x^2)\zeta_{yy} + 2H(\mathbf{x})\sqrt{1 + |\nabla\zeta(x,y)|^2}^3$$

or equivalently

$$\mathcal{M}\zeta := (1 + \zeta_y^2)\zeta_{xx} - 2\zeta_x\zeta_y\zeta_{xy} + (1 + \zeta_x^2)\zeta_{yy}$$
$$= 2H(\mathbf{x})\sqrt{1 + |\nabla\zeta(x,y)|^2}^3 \quad \text{in} \quad \Omega. \tag{34}$$

Theorem 2.3. (Lagrange, Gauß)
The graph $z = \zeta(x,y)$, $(x,y) \in \Omega$, possesses the prescribed mean curvature $H = H(x,y,z)$ if and only if the function ζ satisfies the nonparametric equation of prescribed mean curvature (34).

Remark: In the case $H \equiv 0$ we obtain the minimal surface equation

$$\mathcal{M}\zeta(x,y) \equiv 0 \quad \text{in} \quad \Omega.$$

Example 2.4. The minimal surface of H. F. Scherk.
With the aid of the ansatz $z = \zeta(x,y) = f(x) + g(y)$ we search for all minimal surfaces of this form satisfying $\zeta(0,0) = 0$, $\nabla\zeta(0,0) = 0$. Inserting into the minimal surface equation we obtain

$$0 = (1 + \zeta_y^2)\zeta_{xx} - 2\zeta_x\zeta_y\zeta_{xy} + (1 + \zeta_x^2)\zeta_{yy}$$
$$= \{1 + (g'(y))^2\}f''(x) + \{1 + (f'(x))^2\}g''(y) \quad \text{in} \quad \Omega.$$

This is equivalent to

$$\frac{f''(x)}{1 + (f'(x))^2} = -\frac{g''(y)}{1 + (g'(y))^2} \quad \text{in} \quad \Omega.$$

Consequently, the condition

$$-\frac{f''(x)}{1 + (f'(x))^2} = a = \frac{g''(y)}{1 + (g'(y))^2}, \qquad a \in \mathbb{R},$$

holds true, and we assume $a > 0$ without loss of generality. We deduce

$$a = -(\arctan f'(x))', \qquad \arctan f'(x) = -ax + b$$

and via $b = 0$ we obtain

$$f'(x) = \tan(-ax), \qquad f(x) = \frac{1}{a}\log\cos(ax).$$

Similarly we comprehend

$$g(y) = -\frac{1}{a}\log\cos(ay)$$

and consequently

$$\zeta(x,y) = f(x) + g(y) = \frac{1}{a}\log\frac{\cos ax}{\cos ay}, \qquad a > 0.$$

This surface is defined on the open square

$$\Omega := \left\{ (x,y) \in \mathbb{R}^2 \ : \ |x| < \frac{\pi}{2a}, \ |y| < \frac{\pi}{2a} \right\}$$

and cannot be extended beyond this domain.

Figure 2.6   GRAPHIC OF SCHERK'S MINIMAL SURFACE

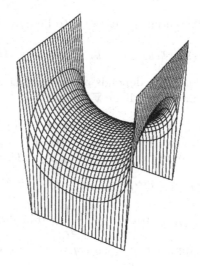

3 Quasilinear Hyperbolic Differential Equations and Systems of Second Order (Characteristic Parameters)

Let the solution $z = \zeta(x,y) : \Omega \to \mathbb{R} \in C^3(\Omega)$ of the quasilinear differential equation

$$\mathcal{L}\zeta(x,y) := a(x,y,\zeta(x,y),\nabla\zeta(x,y))\zeta_{xx}(x,y) + 2b(\ldots)\zeta_{xy} + c(\ldots)\zeta_{yy}$$
$$+d(x,y,\zeta(x,y),\nabla\zeta(x,y)) = 0 \qquad \text{in} \quad \Omega \tag{1}$$

be given on the domain $\Omega \subset \mathbb{R}^2$. Here the coefficients b and c depend on the same quantities as a does. In the sequel, we often use the abbreviations

$$z(x,y) := \zeta(x,y), \quad p(x,y) := \zeta_x(x,y), \quad q(x,y) := \zeta_y(x,y),$$
$$r(x,y) := \zeta_{xx}(x,y), \quad s(x,y) := \zeta_{xy}(x,y), \quad t(x,y) := \zeta_{yy}(x,y) \quad \text{in} \quad \Omega. \tag{2}$$

For a given solution $z = \zeta(x,y)$ of (1) we set

$$a(x,y) := a(x,y,\zeta(x,y),\nabla\zeta(x,y)),$$
$$b(x,y) := b(x,y,\zeta(x,y),\nabla\zeta(x,y)), \tag{3}$$
$$c(x,y) := c(x,y,\zeta(x,y),\nabla\zeta(x,y)) \quad \text{in} \quad \Omega,$$

and obtain the differential equation

$$0 = a(x,y)\zeta_{xx}(x,y) + 2b(x,y)\zeta_{xy}(x,y) + c(x,y)\zeta_{yy}(x,y)$$
$$+ d(x,y,\zeta(x,y),\nabla\zeta(x,y)) \quad \text{in} \quad \Omega. \tag{4}$$

We now assume the differential equation (4) to be hyperbolic, which means

$$a(x,y)c(x,y) - b(x,y)^2 < 0 \quad \text{in} \quad \Omega. \tag{5}$$

We observe that this condition depends on the coefficients $a(x,y,z,p,q), \ldots$ and on the solution ζ and its gradient $\nabla\zeta$ as well.

We now intend to bring the differential equation (1) or equivalently (4) into a form as simple as possible. To this aim we consider the following transformation of variables in the neighborhood $\mathcal{U}(x_0,y_0) \subset \Omega$, namely

$$\xi = \xi(x,y), \; \eta = \eta(x,y) \in C^2(\mathcal{U}(x_0,y_0)),$$

$$\xi_0 = \xi(x_0,y_0), \; \eta_0 = \eta(x_0,y_0), \quad \frac{\partial(\xi,\eta)}{\partial(x,y)} \neq 0 \text{ in } \mathcal{U}(x_0,y_0), \tag{6}$$

with the inverse mapping $x = x(\xi,\eta), \; y = y(\xi,\eta) \in C^2(\mathcal{U}(\xi_0,\eta_0))$.

We calculate

$$z = \zeta(x,y) = z(\xi(x,y),\eta(x,y)), \quad (x,y) \in \mathcal{U}(x_0,y_0),$$
$$\zeta_x = z_\xi\xi_x + z_\eta\eta_x, \quad \zeta_y = z_\xi\xi_y + z_\eta\eta_y,$$
$$\zeta_{xx} = z_{\xi\xi}\xi_x^2 + 2z_{\xi\eta}\xi_x\eta_x + z_{\eta\eta}\eta_x^2 + z_\xi\xi_{xx} + z_\eta\eta_{xx} \tag{7}$$
$$\zeta_{xy} = z_{\xi\xi}\xi_x\xi_y + z_{\xi\eta}(\xi_x\eta_y + \xi_y\eta_x) + z_{\eta\eta}\eta_x\eta_y + z_\xi\xi_{xy} + z_\eta\eta_{xy}$$
$$\zeta_{yy} = z_{\xi\xi}\xi_y^2 + 2z_{\xi\eta}\xi_y\eta_y + z_{\eta\eta}\eta_y^2 + z_\xi\xi_{yy} + z_\eta\eta_{yy}.$$

Therefore, the relation (4) yields the transformed differential equation

$$0 = a(x,y)\zeta_{xx} + 2b(x,y)\zeta_{xy} + c(x,y)\zeta_{yy} + d(x,y,\zeta,\nabla\zeta)$$
$$= A(x,y)z_{\xi\xi} + 2B(x,y)z_{\xi\eta} + C(x,y)z_{\eta\eta} + D(x,y,z,\nabla z) \tag{8}$$

with

$$A(x,y) = a(x,y)\xi_x^2 + 2b(x,y)\xi_x\xi_y + c(x,y)\xi_y^2 =: Q(\xi,\xi),$$
$$B(x,y) = a(x,y)\xi_x\eta_x + b(x,y)(\xi_x\eta_y + \xi_y\eta_x) + c(x,y)\xi_y\eta_y =: Q(\xi,\eta), \tag{9}$$
$$C(x,y) = a(x,y)\eta_x^2 + 2b(x,y)\eta_x\eta_y + c(x,y)\eta_y^2 =: Q(\eta,\eta).$$

The quadratic form

$$Q(\xi,\eta) := (\xi_x,\xi_y) \circ \begin{pmatrix} a(x,y) & b(x,y) \\ b(x,y) & c(x,y) \end{pmatrix} \circ \begin{pmatrix} \eta_x \\ \eta_y \end{pmatrix} \tag{10}$$

is called the *characteristic form of the differential equation (4)*; we finally set $Q(\varphi):=Q(\varphi,\varphi)$. We summarize our relations (9) to the following matrix equation:

$$\begin{pmatrix} A(x,y) & B(x,y) \\ B(x,y) & C(x,y) \end{pmatrix} = \begin{pmatrix} \xi_x & \xi_y \\ \eta_x & \eta_y \end{pmatrix} \circ \begin{pmatrix} a(x,y) & b(x,y) \\ b(x,y) & c(x,y) \end{pmatrix} \circ \begin{pmatrix} \xi_x & \eta_x \\ \xi_y & \eta_y \end{pmatrix}. \tag{11}$$

This implies

$$AC - B^2 = \left(\frac{\partial(\xi,\eta)}{\partial(x,y)}\right)^2 (ac - b^2) < 0, \tag{12}$$

and the transformed equation (8) is hyperbolic as well. Those level curves

$$\Gamma: \quad \varphi(x,y) = \text{const}$$

satisfying

$$Q(\varphi) := Q(\varphi,\varphi) = (a\varphi_x^2 + 2b\varphi_x\varphi_y + c\varphi_y^2)\big|_\Gamma = 0,$$

are the *characteristic curves of the hyperbolic differential equation (4)* (compare Chapter 6, Section 4). Choosing the parameter transformation $\xi = \xi(x,y)$, $\eta = \eta(x,y)$ such that

$$A(x,y) = Q(\xi) = 0, \quad C(x,y) = Q(\eta) = 0 \quad \text{in} \quad \mathcal{U}(x_0,y_0), \tag{13}$$

then the curves $\xi(x,y) = \text{const}$ and $\eta(x,y) = \text{const}$ are the characteristic curves of (4). From the relation (12) we infer the identity

$$|B(x,y)| = \sqrt{b^2 - ac}\left|\frac{\partial(\xi,\eta)}{\partial(x,y)}\right| > 0, \tag{14}$$

and (8) is reduced to the *hyperbolic normal form*

$$z_{\xi\eta}(\xi,\eta) = -\left\{\frac{1}{2B(x,y)}D(x,y,z,p,q)\right\}\bigg|_{\substack{x=x(\xi,\eta) \\ y=y(\xi,\eta)}}. \tag{15}$$

We remind the reader that introducing characteristic parameters ξ, η has already been essential for the treatment of the one-dimensional wave equation $\zeta_{xx} - \zeta_{yy} = 0$ in Chapter 6, Section 5.

We now show the existence of a local parameter transformation (6) with the property (13). The transition to inverse matrices in the relation (11) yields

$$\frac{1}{AC - B^2} \begin{pmatrix} C & -B \\ -B & A \end{pmatrix} = \begin{pmatrix} x_\xi & y_\xi \\ x_\eta & y_\eta \end{pmatrix} \circ \frac{1}{ac - b^2} \begin{pmatrix} c & -b \\ -b & a \end{pmatrix} \circ \begin{pmatrix} x_\xi & x_\eta \\ y_\xi & y_\eta \end{pmatrix}. \quad (16)$$

Taking the equation (12) into account, we deduce

$$(C \, d\xi^2 - 2B \, d\xi \, d\eta + A \, d\eta^2) \left(\frac{\partial(x,y)}{\partial(\xi, \eta)} \right)^2$$

$$= \frac{ac - b^2}{AC - B^2} \, (d\xi, d\eta) \circ \begin{pmatrix} C & -B \\ -B & A \end{pmatrix} \circ \begin{pmatrix} d\xi \\ d\eta \end{pmatrix}$$

$$= (d\xi, d\eta) \circ \begin{pmatrix} x_\xi & y_\xi \\ x_\eta & y_\eta \end{pmatrix} \circ \begin{pmatrix} c & -b \\ -b & a \end{pmatrix} \circ \begin{pmatrix} x_\xi & x_\eta \\ y_\xi & y_\eta \end{pmatrix} \circ \begin{pmatrix} d\xi \\ d\eta \end{pmatrix}$$

$$= (dx, dy) \circ \begin{pmatrix} c & -b \\ -b & a \end{pmatrix} \circ \begin{pmatrix} dx \\ dy \end{pmatrix}$$

$$= c \, dx^2 - 2b \, dx \, dy + a \, dy^2.$$

Therefore, we obtain the transformation formula

$$c(x,y) \, dx^2 - 2b(x,y) \, dx \, dy + a(x,y) \, dy^2$$

$$= \left(\frac{\partial(x,y)}{\partial(\xi, \eta)} \right)^2 \left\{ C(x,y) \, d\xi^2 - 2B(x,y) \, d\xi \, d\eta + A(x,y) \, d\eta^2 \right\}. \quad (17)$$

Since the coefficient matrix is transformed under parameter transformations according to (11), a rotation of the x, y-plane allows us to achieve the condition

$$a(x,y) c(x,y) \neq 0 \quad \text{in} \quad \mathcal{U}(x_0, y_0). \quad (18)$$

We now solve the differential equation

$$0 = a(x,y) \, dy^2 - 2b(x,y) \, dx \, dy + c(x,y) \, dx^2$$

$$= a \left(dy^2 - 2\frac{b}{a} \, dx \, dy + \frac{c}{a} \, dx^2 \right) \quad (19)$$

$$= a(dy - \lambda^+ dx)(dy - \lambda^- dx)$$

with

$$\lambda^\pm := \frac{b \pm \sqrt{b^2 - ac}}{a}. \quad (20)$$

Respecting $\lambda^{\pm} \in C^2(\mathcal{U}(x_0, y_0))$, the solutions of the regular first-order differential equation

$$dy - \lambda^+ dx = 0 \qquad (21)$$

are constructed as level lines $\eta(x, y) = \text{const}$ of a function $\eta \in C^2(\mathcal{U}(x_0, y_0))$. In the same way we find the solutions of

$$dy - \lambda^- dx = 0 \qquad (22)$$

in the form $\xi(x, y) = \text{const}$ for $\xi \in C^2(\mathcal{U}(x_0, y_0))$. On account of $\lambda^+(x_0, y_0) \neq \lambda^-(x_0, y_0)$ the vectors $(1, \lambda^+(x_0, y_0))$ and $(1, \lambda^-(x_0, y_0))$ are linearly independent. The vectors $\nabla\xi(x_0, y_0)$ and $\nabla\eta(x_0, y_0)$, respectively, are orthogonal to them, and we see

$$\frac{\partial(\xi, \eta)}{\partial(x, y)} = \det \begin{pmatrix} \xi_x & \xi_y \\ \eta_x & \eta_y \end{pmatrix} \neq 0 \qquad \text{in} \quad \mathcal{U}(x_0, y_0). \qquad (23)$$

Therefore, the inverse mapping exists as well $x = x(\xi, \eta), y = y(\xi, \eta) \in C^2(\mathcal{U}(\xi_0, \eta_0))$ in a sufficiently small neighborhood $\mathcal{U}(\xi_0, \eta_0)$. Along the ξ-curve $\eta(x, y) = \text{const}$ we have

$$y_\xi - \lambda^+ x_\xi = 0, \qquad (24)$$

and (17) implies $C(x, y) = Q(\eta) = 0$. Along the η-curve $\xi(x, y) = \text{const}$ we have

$$y_\eta - \lambda^- x_\eta = 0, \qquad (25)$$

and (17) yields $A(x, y) = Q(\xi) = 0$. Consequently, we arrive at the following

Theorem 3.1. (Linear hyperbolic differential equations)
For the hyperbolic differential equation with linear principal part (4), (5) given, we have a transformation of variables (6) with

$$Q(\xi) = 0 = Q(\eta) \qquad \text{in} \quad \mathcal{U}(x_0, y_0). \qquad (26)$$

The differential equation appears in the hyperbolic normal form (15) and the parameter transformation $x = x(\xi, \eta)$, $y = y(\xi, \eta)$ satisfies the first-order system (24), (25).

We now consider the case $a = a(x, y, z)$, $b = b(x, y, z)$, $c = c(x, y, z)$ and consequently $\lambda^{\pm} = \lambda^{\pm}(x, y, z)$. The characteristic differential equations (24), (25) now additionally depend on the solution $z = \zeta(x, y)$. Differentiating (24) with respect to η and (25) with respect to ξ, we see

$$y_{\xi\eta} - \lambda^+ x_{\xi\eta} = \lambda_\eta^+ x_\xi = \lambda_x^+ x_\eta x_\xi + \lambda_y^+ y_\eta x_\xi + \lambda_z^+ z_\eta x_\xi \qquad (27)$$

and

$$y_{\xi\eta} - \lambda^- x_{\xi\eta} = \lambda_\xi^- x_\eta = \lambda_x^- x_\xi x_\eta + \lambda_y^- y_\xi x_\eta + \lambda_z^- z_\xi x_\eta, \qquad (28)$$

respectively. The coefficient matrix for this linear system of equations is nonsingular due to $\lambda^+ \neq \lambda^-$, and we can therefore resolve the equations (27), (28) to $x_{\xi\eta}, y_{\xi\eta}$. Then we arrive at the following

Theorem 3.2. *A quasilinear differential equation (1) with the coefficients $a = a(x, y, z)$, $b = b(x, y, z)$, $c = c(x, y, z)$, which is hyperbolic according to (5) with respect to its solution $z = \zeta(x, y)$, appears as the following system in characteristic parameters (24), (25), namely*

$$\mathbf{x}_{\xi\eta}(\xi, \eta) = \mathbf{h}(\xi, \eta, \mathbf{x}(\xi, \eta), \mathbf{x}_\xi(\xi, \eta), \mathbf{x}_\eta(\xi, \eta)) \tag{29}$$

for the vector-valued function $\mathbf{x}(\xi, \eta) := (x(\xi, \eta), y(\xi, \eta), z(\xi, \eta))$.

We now consider the general case

$$a = a(x, y, z, p, q), \qquad b = b(x, y, z, p, q), \qquad c = c(x, y, z, p, q).$$

Noting that $\lambda^\pm = \lambda^\pm(x, y, z, p, q)$ holds true in this situation, the characteristic curves depend on the solution $z = \zeta(x, y)$ and its gradient $\nabla\zeta(x, y)$. The equations (27) and (28) are modified to

$$y_{\xi\eta} - \lambda^+ x_{\xi\eta} = \lambda_x^+ x_\eta x_\xi + \lambda_y^+ y_\eta x_\xi + \lambda_z^+ z_\eta x_\xi + \lambda_p^+ p_\eta x_\xi + \lambda_q^+ q_\eta x_\xi \tag{30}$$

and

$$y_{\xi\eta} - \lambda^- x_{\xi\eta} = \lambda_x^- x_\xi x_\eta + \lambda_y^- y_\xi x_\eta + \lambda_z^- z_\xi x_\eta + \lambda_p^- p_\xi x_\eta + \lambda_q^- q_\xi x_\eta, \tag{31}$$

respectively. In order to obtain a complete system, we derive two additional differential equations of the first order for the functions $p = p(\xi, \eta)$, $q = q(\xi, \eta)$ in characteristic parameters: Let $z = \zeta(x, y)$ be a given solution of (1). The second derivatives $\zeta_{xx}, \zeta_{xy}, \zeta_{yy}$ then satisfy three linear equations

$$\begin{aligned}
a\zeta_{xx} + 2b\zeta_{xy} + c\zeta_{yy} &= -d \\
dx\zeta_{xx} + dy\zeta_{xy} &= dp \\
dx\zeta_{xy} + dy\zeta_{yy} &= dq.
\end{aligned} \tag{32}$$

We refer the reader to the considerations in Chapter 6, Section 4: Posing the Cauchy initial value problem along a characteristic curve $\Gamma \subset \Omega$

$$\begin{aligned}
\mathcal{L}\zeta &= 0 \qquad \text{in} \quad \Omega, \\
\zeta(x, y) &= f(x, y) \qquad \text{on} \quad \Gamma, \\
\frac{\partial\zeta}{\partial\nu}(x, y) &= g(x, y) \qquad \text{on} \quad \Gamma,
\end{aligned} \tag{33}$$

not all the second derivatives $\zeta_{xx}, \zeta_{xy}, \zeta_{yy}$ are determined by the data $\mathcal{L}, f, g$. Since dp and dq are known along a characteristic, the linear system of equations (32) could be resolved to $\zeta_{xx}, \zeta_{xy}, \zeta_{yy}$, if the determinant of the coefficient matrix did not vanish. Therefore, the relation

$$0 = \begin{vmatrix} a & 2b & c \\ dx & dy & 0 \\ 0 & dx & dy \end{vmatrix} = a\,dy^2 - 2b\,dx\,dy + c\,dx^2 \tag{34}$$

is valid along a characteristic, which has already been shown alternatively with the aid of (17). On the other hand, the system of equations (32) possesses the solution $\{\zeta_{xx}, \zeta_{xy}, \zeta_{yy}\}$. Consequently, the relation

$$\text{rank} \begin{pmatrix} a & 2b & c & d \\ dx & dy & 0 & -dp \\ 0 & dx & dy & -dq \end{pmatrix} = 2 \tag{35}$$

holds true along the characteristics. In particular, we obtain

$$0 = \begin{vmatrix} a & c & d \\ dx & 0 & -dp \\ 0 & dy & -dq \end{vmatrix} = a\,dy\,dp + c\,dx\,dq + d\,dx\,dy. \tag{36}$$

Evaluating this equation along the ξ-characteristic, the multiplication by $(a\,dy\,d\xi)^{-1}$ together with relation (21) yields

$$0 = p_\xi + \frac{c\,dx}{a\,dy}q_\xi + \frac{d}{a}x_\xi = p_\xi + \lambda^+\lambda^- \frac{1}{\lambda^+}q_\xi + \frac{d}{a}x_\xi$$

and consequently

$$p_\xi + \lambda^- q_\xi + \frac{d}{a}x_\xi = 0. \tag{37}$$

Along the η-characteristic the relation (36) together with (22) implies the following equation by multiplication with $(a\,dy\,d\eta)^{-1}$, namely

$$0 = p_\eta + \frac{c\,dx}{a\,dy}q_\eta + \frac{d}{a}x_\eta = p_\eta + \lambda^+\lambda^- \frac{1}{\lambda^-}q_\eta + \frac{d}{a}x_\eta$$

and consequently

$$p_\eta + \lambda^+ q_\eta + \frac{d}{a}x_\eta = 0. \tag{38}$$

Finally, the differential equation $dz = p\,dx + q\,dy$ along the ξ-characteristic yields

$$z_\xi - px_\xi - qy_\xi = 0. \tag{39}$$

We now prove the interesting

Theorem 3.3. (Hyperbolic normal form for quasilinear differential equations)

The quasilinear differential equation (1), which is hyperbolic with respect to its solution $z = \zeta(x, y)$ acccording to (5), can be equivalently transformed into the following first-order system by the local parameter transformation (6):

$$y_\xi - \lambda^+ x_\xi = 0, \qquad y_\eta - \lambda^- x_\eta = 0,$$

$$p_\xi + \lambda^- q_\xi + \frac{d}{a}x_\xi = 0, \qquad p_\eta + \lambda^+ q_\eta + \frac{d}{a}x_\eta = 0, \tag{40}$$

$$z_\xi - px_\xi - qy_\xi = 0.$$

For the function $\mathbf{y}(\xi, \eta) := (x(\xi, \eta), y(\xi, \eta), z(\xi, \eta), p(\xi, \eta), q(\xi, \eta))$ *we obtain a hyperbolic system of the second order*

$$\mathbf{y}_{\xi\eta}(\xi, \eta) = \mathbf{h}(\xi, \eta, \mathbf{y}(\xi, \eta), \mathbf{y}_{\xi}(\xi, \eta), \mathbf{y}_{\eta}(\xi, \eta)), \tag{41}$$

where the right-hand side is quadratic in the first derivatives $x_{\xi}, y_{\xi}, \ldots, p_{\eta}, q_{\eta}$.

Proof:

1. Starting from the solution (40) we show the validity of the differential equation (1). The first and second equation from (40) together with the matrix equation

$$\begin{pmatrix} x_{\xi} & x_{\eta} \\ y_{\xi} & y_{\eta} \end{pmatrix} = \begin{pmatrix} \xi_x & \xi_y \\ \eta_x & \eta_y \end{pmatrix}^{-1} = \frac{\partial(x,y)}{\partial(\xi,\eta)} \begin{pmatrix} \eta_y & -\xi_y \\ -\eta_x & \xi_x \end{pmatrix}$$

imply the relations

$$\eta_x + \lambda^+ \eta_y = 0, \quad \xi_x + \lambda^- \xi_y = 0.$$

Therefore, we obtain

$$\begin{aligned}
z_{xx} = p_x &= p_{\xi}\xi_x + p_{\eta}\eta_x \\
&= -\left(\lambda^- q_{\xi} + \frac{d}{a}x_{\xi}\right)\xi_x - \left(\lambda^+ q_{\eta} + \frac{d}{a}x_{\eta}\right)\eta_x \\
&= -(\lambda^+ + \lambda^-)(q_{\xi}\xi_x + q_{\eta}\eta_x) - \lambda^+\lambda^-(q_{\xi}\xi_y + q_{\eta}\eta_y) - \frac{d}{a} \\
&= -\frac{2b}{a}z_{yx} - \frac{c}{a}z_{yy} - \frac{d}{a},
\end{aligned}$$

which reveals that $az_{xx} + 2bz_{xy} + cz_{yy} + d = 0$.

2. Differentiating all the equations of (40) containing only ξ-derivatives with respect to η and vice-versa, we obtain

$$\begin{aligned}
-\lambda^+ x_{\xi\eta} + y_{\xi\eta} &\qquad\qquad\qquad = \ldots \\
-\lambda^- x_{\xi\eta} + y_{\xi\eta} &\qquad\qquad\qquad = \ldots \\
\frac{d}{a}x_{\xi\eta} &\qquad + p_{\xi\eta} + \lambda^- q_{\xi\eta} = \ldots \\
\frac{d}{a}x_{\xi\eta} &\qquad + p_{\xi\eta} + \lambda^+ q_{\xi\eta} = \ldots \\
-px_{\xi\eta} - qy_{\xi\eta} + z_{\xi\eta} &\qquad\qquad\qquad = \ldots
\end{aligned} \tag{42}$$

On the right-hand side only quadratic terms in the first derivatives of x, y, z, p, q appear. We treat (42) as a linear system of equations in the unknowns $x_{\xi\eta}, y_{\xi\eta}, z_{\xi\eta}, p_{\xi\eta}, q_{\xi\eta}$. The coefficient matrix of this system is nonsingular due to

$$\begin{vmatrix} -\lambda^+ & 1 & 0 & 0 & 0 \\ -\lambda^- & 1 & 0 & 0 & 0 \\ \dfrac{d}{a} & 0 & 0 & 1 & \lambda^- \\ \dfrac{d}{a} & 0 & 0 & 1 & \lambda^+ \\ -p & -q & 1 & 0 & 0 \end{vmatrix} = -4\frac{b^2 - ac}{a^2} \neq 0. \tag{43}$$

Therefore, we can resolve the system (42) in the form (41).

q.e.d.

4 Cauchy's Initial Value Problem for Quasilinear Hyperbolic Differential Equations and Systems of Second Order

Theorem 5.1 (d'Alembert) in Chapter 6, Section 5 gives us the solution of Cauchy's initial value problem (briefly CIP) for the one-dimensional wave equation

$$u = u(x,y) \in C^2(\mathbb{R} \times \mathbb{R}, \mathbb{R}),$$

$$\Box u(x,y) := u_{yy}(x,y) - u_{xx}(x,y) = 0 \qquad \text{in } \mathbb{R} \times \mathbb{R}, \tag{1}$$

$$u(x,0) = f(x), \quad \frac{\partial}{\partial y}u(x,0) = g(x) \qquad \text{for all } x \in \mathbb{R},$$

namely

$$u(x,y) = \frac{1}{2}\Big(f(x+y) + f(x-y)\Big) + \frac{1}{2}\int_{x-y}^{x+y} g(s)\,ds, \qquad (x,y) \in \mathbb{R} \times \mathbb{R}. \tag{2}$$

Here we need $f \in C^2(\mathbb{R})$ and $g \in C^1(\mathbb{R})$. Since the problem (1) is uniquely solvable, due to Theorem 4.7 from Section 4 in Chapter 6, we easily deduce the regularity of the solution from d'Alembert's solution formula (2). We obtain the following:

(a) With the assumptions $f \in C^{2+k}(\mathbb{R})$ and $g \in C^{1+k}(\mathbb{R})$ for $k = 0, 1, 2, \ldots,$ we have the regularity $u \in C^{2+k}(\mathbb{R} \times \mathbb{R})$ for the solution.

(b) We now require that the functions f and g can be expanded into convergent power series in a disc of radius $2R \in (0, +\infty)$. With the variable $x = x_1 + ix_2 \in \mathbb{C}$ we have the representations

$$f(x) = \sum_{k=0}^{\infty} a_k x^k, \quad g(x) = \sum_{k=0}^{\infty} b_k x^k \qquad \text{for } x \in \mathbb{C} \text{ with } |x| < 2R. \tag{3}$$

In the dicylinder $Z_R := \{(x, y) \in \mathbb{C}^2 : |x| < R, |y| < R\}$ the function

$$u(x, y) = \frac{1}{2}\Big(f(x+y) + f(x-y)\Big) + \frac{1}{2} \int\limits_{x-y}^{x+y} g(s)\, ds, \qquad (x, y) \in Z_R, \quad (4)$$

then gives us a solution which is holomorphic in Z_R of the following CIP:

$$\frac{\partial^2}{\partial y^2} u(x, y) - \frac{\partial^2}{\partial x^2} u(x, y) = 0 \qquad \text{in} \quad Z_R,$$

$$u(x, 0) = f(x), \quad \frac{\partial}{\partial y} u(x, 0) = g(x) \qquad \text{for all} \quad x \in \mathbb{C} \quad \text{with} \quad |x| < R.$$
$$(5)$$

Here the complex derivatives are denoted by $\frac{\partial}{\partial x}$ and $\frac{\partial}{\partial y}$.

We now perform a rotation about the angle $-\frac{\pi}{4}$ by the mapping

$$\begin{pmatrix} \xi \\ \eta \end{pmatrix} = \begin{pmatrix} \cos(-\frac{\pi}{4}) & -\sin(-\frac{\pi}{4}) \\ \sin(-\frac{\pi}{4}) & \cos(-\frac{\pi}{4}) \end{pmatrix} \circ \begin{pmatrix} x \\ y \end{pmatrix} = \frac{1}{\sqrt{2}} \begin{pmatrix} 1 & 1 \\ -1 & 1 \end{pmatrix} \circ \begin{pmatrix} x \\ y \end{pmatrix}$$

and get the equations

$$\xi = \frac{1}{\sqrt{2}}(x + y), \qquad \eta = \frac{1}{\sqrt{2}}(y - x). \qquad (6)$$

From the wave equation we determine the coefficients of the transformed differential equation with the aid of formula (11) in Section 3 as follows:

$$\begin{pmatrix} A & B \\ B & C \end{pmatrix} = \frac{1}{\sqrt{2}} \begin{pmatrix} 1 & 1 \\ -1 & 1 \end{pmatrix} \circ \begin{pmatrix} -1 & 0 \\ 0 & 1 \end{pmatrix} \circ \frac{1}{\sqrt{2}} \begin{pmatrix} 1 & -1 \\ 1 & 1 \end{pmatrix} = \begin{pmatrix} 0 & 1 \\ 1 & 0 \end{pmatrix}. \qquad (7)$$

By this rotation, the x-axis $y = 0$ - where the Cauchy data are prescribed - is transferred into the secondary diagonal

$$\xi + \eta = 0.$$

The vector $(0, 1)$ is transformed into the unit normal to the secondary diagonal in the direction of the first quadrant, namely $\nu = \frac{1}{\sqrt{2}}(1, 1)$. Therefore, the CIP (1) is transformed into the following CIP:

$u = u(\xi, \eta) \in C^2(\mathbb{R}^2, \mathbb{R})$,

$u_{\xi\eta}(\xi, \eta) = 0 \qquad \text{in} \quad \mathbb{R}^2$,

$u(\xi, -\xi) = f(\sqrt{2}\xi) \quad \text{for} \quad \xi \in \mathbb{R}$,
$$(8)$$

$\frac{\partial}{\partial \nu} u(\xi, -\xi) := \frac{1}{\sqrt{2}}\big(u_\xi(\xi, -\xi) + u_\eta(\xi, -\xi)\big) = g(\sqrt{2}\xi) \qquad \text{for} \quad \xi \in \mathbb{R}.$

The problem (5) is similarly transferred in the case of real-analytic initial values f, g.

We summarize our considerations to the following

Theorem 4.1. *The functions $f = f(\xi) \in C^2(\mathbb{R})$ and $g = g(\xi) \in C^1(\mathbb{R})$ being prescribed, the CIP (8) possesses exactly one solution $u = u(\xi, \eta) \in C^2(\mathbb{R}^2)$. If we assume $f \in C^{2+k}(\mathbb{R})$ and $g \in C^{1+k}(\mathbb{R})$ with an integer $k \in \{0, 1, 2, \ldots\}$, we have $u \in C^{2+k}(\mathbb{R}^2)$. If the functions f and g on $\{\xi \in \mathbb{C} : |\xi| < \sqrt{2}R\}$ can be expanded into convergent power series, then the function $u = u(\xi, \eta)$ is holomorphic in Z_R, the differential equation*

$$\frac{\partial^2}{\partial \xi \, \partial \eta} u(\xi, \eta) = 0 \qquad in \quad Z_R$$

is fulfilled, and the initial conditions in (8) are valid for all $\xi \in \mathbb{C}$ with $|\xi| < R$.

In Section 3 we have transformed a quasilinear hyperbolic differential equation of second order into a hyperbolic system in the normal form. In order to obtain a solution of the CIP for the quasilinear equation, we solve the following CIP:

$$\mathbf{x} = \mathbf{x}(\xi, \eta) \in C^{2+k}(Q_R, \mathbb{R}^n), \qquad Q_R := [-R, R] \times [-R, R], \quad n \in \mathbb{N},$$

$$\mathbf{x}_{\xi\eta}(\xi, \eta) = \mathbf{h}(\xi, \eta, \mathbf{x}(\xi, \eta), \mathbf{x}_\xi(\xi, \eta), \mathbf{x}_\eta(\xi, \eta)) \qquad in \quad Q_R, \tag{9}$$

$$\mathbf{x}(\xi, -\xi) = \mathbf{f}(\xi), \quad \frac{\partial}{\partial \nu}\mathbf{x}(\xi, -\xi) = \mathbf{g}(\xi) \qquad for \quad \xi \in [-R, R]$$

with the Cauchy data

$$\mathbf{f} = \mathbf{f}(\xi) \in C^{2+k}(\mathbb{R}, \mathbb{R}^n) \quad and \quad \mathbf{g} = \mathbf{g}(\xi) \in C^{1+k}(\mathbb{R}, \mathbb{R}^n)$$

and the continuous right-hand side

$$\mathbf{h} = \mathbf{h}(\xi, \eta, \mathbf{x}, \mathbf{p}, \mathbf{q}).$$

Applying Theorem 4.1 to each component function, we find a uniquely determined solution of the CIP

$$\mathbf{y} = \mathbf{y}(\xi, \eta) \in C^{2+k}(\mathbb{R}^2, \mathbb{R}^n),$$

$$\mathbf{y}_{\xi\eta}(\xi, \eta) = 0 \qquad in \quad \mathbb{R}^2, \tag{10}$$

$$\mathbf{y}(\xi, -\xi) = \mathbf{f}(\xi), \quad \frac{\partial}{\partial \nu}\mathbf{y}(\xi, -\xi) = \mathbf{g}(\xi), \qquad \xi \in \mathbb{R}.$$

When we make the transition to

$$\tilde{\mathbf{x}}(\xi, \eta) := \mathbf{x}(\xi, \eta) - \mathbf{y}(\xi, \eta),$$

$$\tilde{\mathbf{h}}(\xi, \eta, \tilde{\mathbf{x}}, \tilde{\mathbf{p}}, \tilde{\mathbf{q}}) := \mathbf{h}(\xi, \eta, \mathbf{y}(\xi, \eta) + \tilde{\mathbf{x}}, \mathbf{y}_\xi(\xi, \eta) + \tilde{\mathbf{p}}, \mathbf{y}_\eta(\xi, \eta) + \tilde{\mathbf{q}}), \tag{11}$$

the problem (9) is equivalently transformed into the CIP

$$\tilde{\mathbf{x}} = \tilde{\mathbf{x}}(\xi, \eta) \in C^{2+k}(Q_R, \mathbb{R}^n),$$

$$\tilde{\mathbf{x}}_{\xi\eta}(\xi, \eta) = \tilde{\mathbf{h}}(\xi, \eta, \tilde{\mathbf{x}}(\xi, \eta), \tilde{\mathbf{x}}_\xi(\xi, \eta), \tilde{\mathbf{x}}_\eta(\xi, \eta)) \qquad \text{in} \quad Q_R, \qquad (12)$$

$$\tilde{\mathbf{x}}(\xi, -\xi) = 0 = \frac{\partial}{\partial \nu}\tilde{\mathbf{x}}(\xi, -\xi) \qquad \text{for} \quad \xi \in [-R, R].$$

In the sequel we suppress $\tilde{}$ in (12) and transform (12) equivalently into an integro-differential-equation: Let the point $(x, y) \in Q_R$ be chosen with $x + y > 0$. We then define the *characteristic triangle to* (x, y) by

$$T(x, y) := \Big\{ (\xi, \eta) \in \mathbb{R}^2 \ : \ -x < -\xi < \eta < y \Big\} \subset Q_R.$$

We confine our considerations to the subset of Q_R above the secondary diagonal. A solution beneath the secondary diagonal is constructed in the same way, defining the characteristic triangle

$$T(x, y) = \Big\{ (\xi, \eta) \in \mathbb{R}^2 \ : \ y < \eta < -\xi < -x \Big\}.$$

We apply the Stokes integral theorem to the Pfaffian form

$$\omega = \mathbf{x}_\eta(\xi, \eta)\, d\eta - \mathbf{x}_\xi(\xi, \eta)\, d\xi, \qquad (\xi, \eta) \in T(x, y).$$

We deduce

$$2\mathbf{x}(x, y) = \int\limits_{\partial T(x,y)} \mathbf{x}_\eta\, d\eta - \mathbf{x}_\xi\, d\xi = \iint\limits_{T(x,y)} d(\mathbf{x}_\eta\, d\eta - \mathbf{x}_\xi\, d\xi)$$

$$= 2 \iint\limits_{T(x,y)} \mathbf{x}_{\xi\eta}(\xi, \eta)\, d\xi\, d\eta$$

$$= 2 \iint\limits_{T(x,y)} \mathbf{h}(\xi, \eta, \mathbf{x}(\xi, \eta), \mathbf{x}_\xi(\xi, \eta), \mathbf{x}_\eta(\xi, \eta))\, d\xi\, d\eta$$

and consequently

$$\mathbf{x}(x, y) = \iint\limits_{T(x,y)} \mathbf{h}(\xi, \eta, \mathbf{x}(\xi, \eta), \mathbf{x}_\xi(\xi, \eta), \mathbf{x}_\eta(\xi, \eta))\, d\xi\, d\eta, \qquad (x, y) \in Q_R. \quad (13)$$

On the contrary, we depart from a given solution of the integro-differential-equation (13) and immediately comprehend $\mathbf{x}(x, -x) = 0$ for all $x \in [-R, R]$. From the representation

$$\mathbf{x}(x, y) = \int\limits_{-y}^{x} \left(\int\limits_{-\xi}^{y} \mathbf{h}(\xi, \eta, \mathbf{x}(\xi, \eta), \mathbf{x}_\xi(\xi, \eta), \mathbf{x}_\eta(\xi, \eta))\, d\eta \right) d\xi$$

we infer the equation

$$\mathbf{x}_x(x,y) = \int\limits_{-x}^{y} \mathbf{h}(x,\eta,\mathbf{x}(x,\eta),\mathbf{x}_\xi(x,\eta),\mathbf{x}_\eta(x,\eta))\, d\eta. \tag{14}$$

This representation implies $\mathbf{x}_x(x,-x) = 0$ for all $x \in [-R,R]$. Furthermore, the representation

$$\mathbf{x}(x,y) = \int\limits_{-x}^{y} \left(\int\limits_{-\eta}^{x} \mathbf{h}(\xi,\eta,\mathbf{x}(\xi,\eta),\mathbf{x}_\xi(\xi,\eta),\mathbf{x}_\eta(\xi,\eta))\, d\xi \right) d\eta$$

yields

$$\mathbf{x}_y(x,y) = \int\limits_{-y}^{x} \mathbf{h}(\xi,y,\mathbf{x}(\xi,y),\mathbf{x}_\xi(\xi,y),\mathbf{x}_\eta(\xi,y))\, d\xi, \tag{15}$$

and consequently $\mathbf{x}_y(x,-x) = 0$ for $x \in [-R,R]$. Finally, we can differentiate (14) with respect to y and (15) with respect to x, and we obtain the fundamental

Theorem 4.2. *The function* $\mathbf{x} = \mathbf{x}(x,y)$ *of the class*

$$C_{xy}(Q_R,\mathbb{R}^n) := \left\{ \mathbf{y} \in C^1(Q_R,\mathbb{R}^n) : \mathbf{y}_{xy} = \mathbf{y}_{yx} \text{ exist continuously in } Q_R \right\}$$

solves the CIP

$$\mathbf{x}_{xy}(x,y) = \mathbf{h}(x,y,\mathbf{x}(x,y),\mathbf{x}_x(x,y),\mathbf{x}_y(x,y)) \quad \text{in} \quad Q_R,$$
$$\mathbf{x}(x,-x) = 0 = \frac{\partial}{\partial \nu}\mathbf{x}(x,-x) \quad \text{for} \quad x \in [-R,R] \tag{16}$$

if and only if $\mathbf{x}$ *satisfies the integro-differential-equation (13).*

For the right-hand side $\mathbf{h} = \mathbf{h}(\xi,\eta,\mathbf{x},\mathbf{p},\mathbf{q})$ we now require the following Lipschitz condition

$$|\mathbf{h}(\xi,\eta,\mathbf{x},\mathbf{p},\mathbf{q}) - \mathbf{h}(\xi,\eta,\tilde{\mathbf{x}},\tilde{\mathbf{p}},\tilde{\mathbf{q}})| \leq L|(\mathbf{x},\mathbf{p},\mathbf{q}) - (\tilde{\mathbf{x}},\tilde{\mathbf{p}},\tilde{\mathbf{q}})|$$
$$\text{for all} \quad (\xi,\eta,\mathbf{x},\mathbf{p},\mathbf{q}),(\xi,\eta,\tilde{\mathbf{x}},\tilde{\mathbf{p}},\tilde{\mathbf{q}}) \in Q_R \times \mathbb{R}^n \times \mathbb{R}^n \times \mathbb{R}^n \tag{17}$$

with the Lipschitz constant $L \in [0,+\infty)$. With this assumption we derive a contraction condition for the *integro-differential-operator*

$$I(\mathbf{x})(x,y) := \iint\limits_{T(x,y)} \mathbf{h}(\xi,\eta,\mathbf{x}(\xi,\eta),\mathbf{x}_\xi(\xi,\eta),\mathbf{x}_\eta(\xi,\eta))\, d\xi\, d\eta, \quad (x,y) \in Q_R.$$

$$\tag{18}$$

For the function $\mathbf{x},\mathbf{y} \in C^1(Q_R,\mathbb{R}^n)$ we set

$$\hat{\mathbf{x}}(x,y) := I(\mathbf{x})(x,y), \quad \hat{\mathbf{y}}(x,y) := I(\mathbf{y})(x,y), \qquad (x,y) \in Q_R.$$

Now we can estimate

$$|\hat{\mathbf{x}}(x,y) - \hat{\mathbf{y}}(x,y)| \le \iint\limits_{T(x,y)} |\mathbf{h}(\xi,\eta,\mathbf{x},\mathbf{x}_\xi,\mathbf{x}_\eta) - \mathbf{h}(\xi,\eta,\mathbf{y},\mathbf{y}_\xi,\mathbf{y}_\eta)| \, d\xi \, d\eta$$

$$\le L \iint\limits_{T(x,y)} \left| (\mathbf{x}(\xi,\eta) - \mathbf{y}(\xi,\eta), \mathbf{x}_\xi - \mathbf{y}_\xi, \mathbf{x}_\eta - \mathbf{y}_\eta) \right| \, d\xi \, d\eta$$

$$\le L \int_0^{x+y} |x+y-\tau| \, \phi(\tau) \, d\tau \tag{19}$$

with

$$\phi(\tau) := \max_{(\xi,\eta)\in T(x,y),\ \xi+\eta=\tau} \left| (\mathbf{x}(\xi,\eta) - \mathbf{y}(\xi,\eta), \mathbf{x}_\xi - \mathbf{y}_\xi, \mathbf{x}_\eta - \mathbf{y}_\eta) \right|. \tag{20}$$

Furthermore, the relation (14) implies the inequality

$$|\hat{\mathbf{x}}_x(x,y) - \hat{\mathbf{y}}_x(x,y)| \le \int_{-x}^{y} |\mathbf{h}(x,\eta,\mathbf{x}(x,\eta),\mathbf{x}_\xi,\mathbf{x}_\eta) - \mathbf{h}(x,\eta,\mathbf{y}(x,\eta),\mathbf{y}_\xi,\mathbf{y}_\eta)| \, d\eta$$

$$\le L \int_{-x}^{y} \left| (\mathbf{x}(x,\eta) - \mathbf{y}(x,\eta), \mathbf{x}_\xi - \mathbf{y}_\xi, \mathbf{x}_\eta - \mathbf{y}_\eta) \right| \, d\eta$$

$$\le L \int_0^{x+y} \phi(\tau) \, d\tau. \tag{21}$$

In the same way we deduce the following estimate via (15):

$$|\hat{\mathbf{x}}_y(x,y) - \hat{\mathbf{y}}_y(x,y)| \le L \int_0^{x+y} \phi(\tau) \, d\tau. \tag{22}$$

We summarize the inequalities (19), (21), and (22) to the following

Theorem 4.3. *For arbitrary functions* $\mathbf{x}, \mathbf{y} \in C^1(Q_R, \mathbb{R}^n)$ *we have the subsequent estimate in* Q_R:

$$\left| \left(\hat{\mathbf{x}}(x,y) - \hat{\mathbf{y}}(x,y), \hat{\mathbf{x}}_x(x,y) - \hat{\mathbf{y}}_x(x,y), \hat{\mathbf{x}}_y(x,y) - \hat{\mathbf{y}}_y(x,y) \right) \right|$$

$$\le L \int_0^{x+y} (2 + |x+y-\tau|)\phi(\tau) \, d\tau.$$

We define the set

$$Q_{R,S} := \left\{ (x, y, \mathbf{x}, \mathbf{p}, \mathbf{q}) \in Q_R \times \mathbb{R}^n \times \mathbb{R}^n \times \mathbb{R}^n \; : \; |\mathbf{x}|, |\mathbf{p}|, |\mathbf{q}| \leq S \right\},$$

and prove the central

Theorem 4.4. *Let the parameter-dependent right-hand side*

$$\mathbf{h} = \mathbf{h}(x, y, \mathbf{x}, \mathbf{p}, \mathbf{q}, \lambda) : Q_{R,S} \times [\lambda_1, \lambda_2] \to \mathbb{R}^n$$

of class $C^1(Q_{R,S} \times [\lambda_1, \lambda_2], \mathbb{R}^n)$ *with* $R > 0$, $S > 0$ *and* $-\infty < \lambda_1 < \lambda_2 < +\infty$ *be given. Then we have a number* $r \in (0, R]$, *such that the following CIP has exactly one solution for all* $\lambda \in [\lambda_1, \lambda_2]$:

$$\mathbf{x} = \mathbf{x}(x, y, \lambda) \in C_{xy}(Q_r, \mathbb{R}^n),$$

$$\mathbf{x}_{xy}(x, y, \lambda) = \mathbf{h}(x, y, \mathbf{x}(x, y, \lambda), \mathbf{x}_x(x, y, \lambda), \mathbf{x}_y(x, y, \lambda), \lambda) \qquad in \quad Q_r, \quad (23)$$

$$\mathbf{x}(x, -x, \lambda) = 0 = \frac{\partial}{\partial \nu} \mathbf{x}(x, -x, \lambda) \qquad for \quad x \in [-r, r].$$

Furthermore, the solution depends differentiably on the parameter as follows:

$$\mathbf{x}(x, y, \lambda) \in C^1(Q_r \times [\lambda_1, \lambda_2], \mathbb{R}^n).$$

Proof:

1. At first, we fix the parameter $\lambda \in [\lambda_1, \lambda_2]$ and construct a solution $\mathbf{x}(x, y, \lambda)$ with the aid of Banach's fixed point theorem. To this aim, we define the Banach space

$$\mathcal{B} := \left\{ \mathbf{y} \in C^1(Q_R, \mathbb{R}^n) \; : \; \mathbf{y}(x, -x) = 0 = \frac{\partial}{\partial \nu} \mathbf{y}(x, -x), \; x \in [-R, R] \right\}$$

endowed with the norm

$$\|\mathbf{y}\| := \sup_{(x,y) \in Q_R} \left| (\mathbf{y}(x, y), \mathbf{y}_\xi(x, y), \mathbf{y}_\eta(x, y)) \right|. \qquad (24)$$

We extend the right-hand side $\mathbf{h} : Q_{R,S} \times [\lambda_1, \lambda_2] \to \mathbb{R}^n$ onto the set

$$Q_R \times \mathbb{R}^n \times \mathbb{R}^n \times \mathbb{R}^n \times [\lambda_1, \lambda_2]$$

in such a way that the Lipschitz condition (17) with a uniform Lipschitz constant $L \geq 0$ is valid for arbitrary $\lambda \in [\lambda_1, \lambda_2]$. Due to Theorem 4.3 we find a sufficiently small $R > 0$ such that the integro-differential-operator $I : \mathcal{B} \to \mathcal{B}$ defined in (18) is contractible. This means, we have a constant $\theta \in [0, 1)$ satisfying

$$\|I(\mathbf{x}) - I(\mathbf{y})\| \leq \theta \|\mathbf{x} - \mathbf{y}\| \qquad for \; all \quad \mathbf{x}, \mathbf{y} \in \mathcal{B}. \qquad (25)$$

Banach's fixed point theorem (see Theorem 1.19 in Chapter 7, Section 1) gives us the existence of a solution $\mathbf{x} = \mathbf{x}(x, y, \lambda) \in \mathcal{B}$ for the integro-differential-equation

$$\mathbf{x}(x, y, \lambda) = \iint\limits_{T(x,y)} \mathbf{h}(\xi, \eta, \mathbf{x}(\xi, \eta, \lambda), \mathbf{x}_\xi(\xi, \eta, \lambda), \mathbf{x}_\eta(\xi, \eta, \lambda), \lambda) \, d\xi \, d\eta \quad (26)$$

for all $\lambda \in [\lambda_1, \lambda_2]$. Parallel to the proof of Theorem 2, we see that $\mathbf{x}(x, y, \lambda) \in C_{xy}(Q_R, \mathbb{R}^n)$ holds true for each $\lambda \in [\lambda_1, \lambda_2]$.

2. We now show that the solution is independent of the extension of the right-hand side $\mathbf{h}$ for sufficiently small $R > 0$: Let $\mathbf{x}$ be a solution of the CIP (23) to the fixed parameter $\lambda \in [\lambda_1, \lambda_2]$, and we set

$$\mathbf{y}(x, y) := I(\mathbf{0})(x, y) = \iint\limits_{T(x,y)} \mathbf{h}(\xi, \eta, 0, 0, 0, \lambda) \, d\xi \, d\eta.$$

We apply Theorem 4.3 to the function

$$\psi(t) := \max_{(x,y) \in Q_R, \ x+y=t} \left| (\mathbf{x}(x, y) - \mathbf{y}(x, y), \mathbf{x}_\xi - \mathbf{y}_\xi, \mathbf{x}_\eta - \mathbf{y}_\eta) \right|$$

and obtain the following estimate

$$\psi(t) \leq A \int\limits_0^t (\psi(\tau) + \|\mathbf{y}\|) \, d\tau \quad (27)$$

with a constant $A > 0$. The comparison lemma (see Proposition 5.1 in Section 5) yields

$$\psi(t) \leq \|\mathbf{y}\| (e^{At} - 1).$$

Choosing $R > 0$ sufficiently small, the inclusion

$$(\mathbf{x}(x, y), \mathbf{x}_\xi(x, y), \mathbf{x}_\eta(x, y)) \in Q_{R,S} \qquad \text{for all} \quad (x, y) \in Q_R$$

is fulfilled.

3. Let two solutions $\mathbf{x} = \mathbf{x}(x, y, \lambda)$ and $\tilde{\mathbf{x}} = \tilde{\mathbf{x}}(x, y, \tilde{\lambda})$ to the parameters λ and $\tilde{\lambda}$ be given. Then we derive an inequality of the form

$$\psi(t) \leq A \int\limits_0^t \left(\psi(\tau) + \varepsilon(\lambda, \tilde{\lambda}) \right) d\tau \quad (28)$$

for the function

$$\psi(t) := \max_{(x,y) \in Q_R, \ x+y=t} \left| (\mathbf{x}(x, y) - \tilde{\mathbf{x}}(x, y), \mathbf{x}_\xi - \tilde{\mathbf{x}}_\xi, \mathbf{x}_\eta - \tilde{\mathbf{x}}_\eta) \right|,$$

as in the proof of Theorem 4.3. Here $\varepsilon(\lambda, \tilde{\lambda}) \to \varepsilon(\lambda, \lambda) = 0$ for $\lambda \to \tilde{\lambda}$ is satisfied. With the comparison lemma from above we infer

$$\psi(t) \le \varepsilon(\lambda, \tilde{\lambda})(e^{At} - 1), \tag{29}$$

which implies the continuous dependence of the solution on the parameter in the C^1-norm. Furthermore, the equation $\varepsilon(\lambda, \lambda) = 0$ gives us the unique solvability of the CIP.

4. In order to show the differentiable dependence on the parameter, we consider the difference quotient as in the theory of ordinary differential equations and observe the limit in the integro-differential-equation.

$$\text{q.e.d.}$$

Remarks:

1. The solution of the CIP (23) is constructed by successive approximation

$$\mathbf{x}^{(0)}(x, y) := 0 \quad \text{in} \quad Q_R,$$

$$\mathbf{x}^{(j+1)}(x, y) := \iint\limits_{T(x,y)} \mathbf{h}\big(\xi, \eta, \mathbf{x}^{(j)}(\xi, \eta), \mathbf{x}_\xi^{(j)}, \mathbf{x}_\eta^{(j)}\big) \, d\xi \, d\eta \quad \text{in} \quad Q_R, \tag{30}$$

for $j = 0, 1, 2, \ldots$

2. Assuming higher regularity of the right-hand side $\mathbf{h}$, we obtain the corresponding higher regularity for the solutions. This statement pertains to the differentiability for the family of solutions with respect to the parameter $\lambda \in [\lambda_1, \lambda_2]$. Again one uses the method of difference quotients indicated in part 4 of the proof above.

Theorem 4.5. Assumptions: *Let the quasilinear differential equation*

$$0 = a(x, y, \zeta(x, y), \zeta_x(x, y), \zeta_y(x, y))\zeta_{xx} + 2b(\ldots)\zeta_{xy} + c(\ldots)\zeta_{yy}$$
$$+ d(x, y, \zeta(x, y), \zeta_x(x, y), \zeta_y(x, y)) = 0 \quad \text{in} \quad \Omega \tag{31}$$

with the coefficients

$$a = a(x, y, z, p, q), \ldots, d = d(x, y, z, p, q) \in C^2(\Omega \times \mathbb{R} \times \mathbb{R} \times \mathbb{R}, \mathbb{R})$$

be given, where $\Omega \subset \mathbb{R}^2$ is an open set. We consider a regular curve

$$\Gamma : x = x(t), \quad y = y(t), \quad t \in [t_0 - T, t_0 + T], \quad \text{in} \quad \Omega$$

with the height function $f = f(t) \in C^3([t_0 - T, t_0 + T], \mathbb{R})$ and the prescribed derivative $g = g(t) \in C^2([t_0 - T, t_0 + T], \mathbb{R})$ in the direction of its normal

$$\nu = \nu(t) := \frac{1}{\sqrt{x'(t)^2 + y'(t)^2}} \big(-y'(t), x'(t) \big).$$

The differential equation (31) is hyperbolic along this stripe, which means

$$a(t)c(t) - b(t)^2 < 0 \qquad \text{for all} \quad t \in [t_0 - T, t_0 + T].$$

Here we have set $a(t) := a(x(t), y(t), f(t), p(t), q(t))$ *etc. with*

$$p(t) := \frac{x'(t)f'(t) - \sqrt{x'(t)^2 + y'(t)^2}\, y'(t)g(t)}{x'(t)^2 + y'(t)^2},$$

$$q(t) := \frac{y'(t)f'(t) + \sqrt{x'(t)^2 + y'(t)^2}\, x'(t)g(t)}{x'(t)^2 + y'(t)^2}.$$

Finally, the curve Γ with respect to this stripe shall represent a noncharacteristic curve for the differential equation (31), which means

$$c(t)x'(t)^2 - 2b(t)x'(t)y'(t) + a(t)y'(t)^2 \neq 0 \qquad \text{for all} \quad t \in [t_0 - T, t_0 + T].$$

Statement: *Then we have a neighborhood $\Theta = \Theta(x^0, y^0)$ of the point $(x^0, y^0) := (x(t_0), y(t_0))$ and a function $\zeta = \zeta(x, y) \in C^2(\Theta)$, which solves the Cauchy initial value problem*

$$a(x, y, \zeta(x, y), \zeta_x, \zeta_y)\zeta_{xx} + 2b(\ldots)\zeta_{xy} + c(\ldots)\zeta_{yy} + d(\ldots) = 0 \qquad in \quad \Theta$$

$$\zeta(x(t), y(t)) = f(t), \quad \frac{\partial}{\partial \nu}\zeta(x(t), y(t)) = g(t) \qquad on \quad \Gamma \cap \Theta.$$

$$(32)$$

Here $\frac{\partial}{\partial \nu}$ denotes the derivative in the direction of the normal ν to the curve Γ. The solution of (32) is uniquely determined.

Remark: We can locally supplement the prescribed noncharacteristic stripe $\{\Gamma, f, g\}$ to a solution of the given differential equation.

Proof: With the aid of Section 3, Theorem 3.3 we introduce characteristic parameters (ξ, η) into the differential equation (31). Differentiating the first-order system once, we obtain a system of the form

$$\mathbf{y}_{\xi\eta}(\xi, \eta) = \mathbf{h}(\xi, \eta, \mathbf{y}(\xi, \eta), \mathbf{y}_\xi(\xi, \eta), \mathbf{y}_\eta(\xi, \eta)). \qquad (33)$$

On account of $a, b, c, d \in C^2$ the function $\mathbf{h}$ belongs to the class C^1 with respect to $\mathbf{y}, \mathbf{y}_\xi, \mathbf{y}_\eta$. From the Cauchy data

$$(x(t), y(t), f(t), g(t)) \qquad \text{with} \quad t_0 - T \leq t \leq t_0 + T$$

we calculate the initial values for $\mathbf{y}$. Since ξ =const and η =const are the characteristic curves, we can transfer the noncharacteristic curve Γ into the secondary diagonal $\xi + \eta = 0$ by the transformation

$$\xi \mapsto \varphi(\xi), \eta \mapsto \psi(\eta).$$

With the aid of Theorem 4.1 we make the transition to homogeneous initial values, and we solve the CIP for the system (33) by Theorem 4.4. Via resubstitution we obtain a solution of the CIP (32) (compare the proof of Section 3, Theorem 3.3). The uniqueness follows from the corresponding statement for the system (33).

q.e.d.

5 Riemann's Integration Method

In this paragraph we shall investigate linear hyperbolic differential equations. Though we established only local solvability in Theorem 4.5 of Section 4, we now shall prove global solvability of the linear Cauchy initial value problem. For the convenience of the reader we supply the preparatory

Proposition 5.1. (Comparison lemma)
The continuous function $f : [\xi - h, \xi + h] \to [0, +\infty)$ satisfies the integral inequality

$$f(x) \le A \int_{\xi}^{x} \left(f(t) + \varepsilon\right) |dt| \qquad \text{for all} \quad x \in [\xi - h, \xi + h]$$

with the constants $A > 0$ and $\varepsilon \ge 0$. Then we have the estimate

$$0 \le f(x) \le \varepsilon \left(e^{A|x-\xi|} - 1\right) = \varepsilon \sum_{k=1}^{\infty} \frac{A^k}{k!} |x - \xi|^k$$

for all $x \in [\xi - h, \xi + h]$.

Proof: We set $M := \max\{f(x) : \xi - h \le x \le \xi + h\}$ and show via complete induction

$$f(x) \le \varepsilon \sum_{k=1}^{n} \frac{A^k}{k!} |x - \xi|^k + M \frac{A^n}{n!} |x - \xi|^n, \qquad x \in [\xi - h, \xi + h].$$

From the integral inequality we deduce

$$f(x) \le MA|x - \xi| + \varepsilon A|x - \xi| \qquad \text{for all} \quad x \in [\xi - h, \xi + h],$$

such that the case $n = 1$ is established. If the estimate above is valid for a number $n \in \mathbb{N}$, we then find

$$f(x) \le \varepsilon A|x - \xi| + A \int_{\xi}^{x} f(t) |dt|$$

$$\le \varepsilon A|x - \xi| + A \int_{\xi}^{x} \left\{ \varepsilon \sum_{k=1}^{n} \frac{A^k}{k!} |x - \xi|^k + M \frac{A^n}{n!} |x - \xi|^n \right\} |dt|$$

$$= \varepsilon A|x - \xi| + \varepsilon \sum_{k=1}^{n} \frac{A^{k+1}}{(k+1)!} |x - \xi|^{k+1} + M \frac{A^{n+1}}{(n+1)!} |x - \xi|^{n+1}$$

$$= \varepsilon \sum_{k=1}^{n+1} \frac{A^k}{k!} |x - \xi|^k + M \frac{A^{n+1}}{(n+1)!} |x - \xi|^{n+1}.$$

We observe

$$\lim_{n \to \infty} \frac{(A|x - \xi|)^{n+1}}{(n+1)!} = 0,$$

and the limit procedure in the estimate above yields

$$f(x) \leq \varepsilon \sum_{k=1}^{\infty} \frac{A^k}{k!} |x - \xi|^k = \varepsilon \left(e^{A|x-\xi|} - 1 \right).$$

q.e.d.

Theorem 5.2. *Let the functions* $f = f(t) \in C_0^2(\mathbb{R})$ *and* $g = g(t) \in C_0^1(\mathbb{R})$ *be given. Furthermore, the coefficient functions* $a = a(x, y)$, $b = b(x, y)$, $c = c(x, y)$, $d = d(x, y)$ *belong to the class* $C_0^1(\mathbb{R}^2)$. *Then the Cauchy initial value problem*

$$u_{xy}(x, y) + a u_x(x, y) + b u_y(x, y) + c u(x, y) = d(x, y) \quad \text{in} \quad \mathbb{R}^2,$$

$$u(x, -x) = f(x), \quad \frac{\partial}{\partial \nu} u(x, -x) = g(x) \quad \text{for} \quad x \in \mathbb{R} \tag{1}$$

possesses exactly one solution. Here the symbol $\frac{\partial}{\partial \nu}$ *again denotes the derivative in the direction of the normal* $\nu = \frac{1}{\sqrt{2}}(1, 1)$.

Proof: We write the differential equation in the form

$$u_{xy} = h(x, y, u, u_x, u_y) := d(x, y) - a(x, y)u_x - b(x, y)u_y - c(x, y)u.$$

Here the right-hand side h globally satisfies a Lipschitz condition as in Section 4, formula (17), with the Lipschitz constant $L \in [0, +\infty)$. We consider a solution $u = u(x, y)$ existing in a neighborhood of the secondary diagonal $x + y = 0$, and we investigate the function

$$\phi(t) := \max_{x+y=t} \left| \left(u(x, y) - v(x, y), u_x(x, y) - v_x(x, y), u_y(x, y) - v_y(x, y) \right) \right|$$

with

$$v(x, y) := I(0)|_{(x,y)}.$$

Here I denotes the integro-differential-operator defined in Section 4, formula (18). From Theorem 4.3 in Section 4 we infer the differential inequality

$$\phi(t) \leq L \int_0^t (2 + T)(\phi(\tau) + K) \, d\tau \quad \text{for all} \quad 0 \leq t \leq T < +\infty$$

with a constant $K > 0$. Proposition 5.1 gives us the estimate

$$\phi(t) \leq K \left(e^{L(2+T)t} - 1 \right) \quad \text{for all} \quad 0 \leq t \leq T < +\infty$$

and consequently

$$\phi(T) \le K\big(e^{L(2+T)T} - 1\big) \qquad \text{for} \quad 0 \le T < +\infty. \tag{2}$$

Therefore, the solution of the Cauchy initial value problem remains bounded in the C^1-norm, and the procedure of successive approximation yields a global solution on $\mathbb{R}^2$.

q.e.d.

We shall now prove an integral representation for the solution of Cauchy's initial value problem (1). This *Riemannian integration method* corresponds to the representation for solutions of Poisson's equation with the aid of Green's function. Together with the linear differential operator

$$\mathcal{L}u(x,y) := u_{xy}(x,y) + a(x,y)u_x(x,y) + b(x,y)u_y(x,y) + c(x,y)u(x,y) \tag{3}$$

we consider the *adjoint differential operator*

$$\mathcal{M}v(x,y) := v_{xy}(x,y) - [a(x,y)v(x,y)]_x - [b(x,y)v(x,y)]_y + c(x,y)v(x,y). \tag{4}$$

The operators $\mathcal{L}$ and $\mathcal{M}$ coincide if and only if $a \equiv 0 \equiv b$ is satisfied.

Proposition 5.3. *We have*

$$v\,\mathcal{L}u - u\,\mathcal{M}v = (-v_y u + auv)_x + (vu_x + bvu)_y. \tag{5}$$

Proof: We calculate

$$
\begin{aligned}
v\,\mathcal{L}u &= vu_{xy} + avu_x + bvu_y + cvu \\
&= (vu_x)_y - v_y u_x + (avu)_x - (av)_x u + (bvu)_y - (bv)_y u + cvu \\
&= (vu_x + bvu)_y + (-v_y u + avu)_x + uv_{xy} - u(av)_x - u(bv)_y + ucv \\
&= (vu_x + bvu)_y + (-v_y u + avu)_x + u\,\mathcal{M}v.
\end{aligned}
$$

q.e.d.

Let the closed, regular arc $\Gamma \subset \mathbb{R}^2$ be given, which represents a noncharacteristic curve for the differential equation (1). This means, the arc Γ never appears parallel to the coordinate axes. Therefore, we can find a continuous, strictly monotonic function $\varphi = \varphi(x) : [x_1, x_2] \to \mathbb{R}$ such that

$$
\begin{aligned}
\Gamma &= \Big\{ (x,y) \in \mathbb{R}^2 : y = \varphi(x),\ x_1 \le x \le x_2 \Big\} \\
&= \Big\{ (x,y) \in \mathbb{R}^2 : x = \varphi^{-1}(y),\ y_2 \le y \le y_1 \Big\}
\end{aligned}
$$

holds true with $y_1 = \varphi(x_1)$ and $y_2 = \varphi(x_2)$. (Without loss of generality we assume $y_2 < y_1$.) Furthermore, $P = (x,y) \notin \Gamma$ represents a fixed point in the square $[x_1, x_2] \times [y_2, y_1]$ above the arc Γ, that means $y > \varphi(x)$. Then we define the *characteristic triangle*

$$T(x,y) := \left\{ (\xi, \eta) \in \mathbb{R}^2 \; : \; \varphi(x) < \varphi(\xi) < \eta < y \right\}.$$

Furthermore, we use the abbreviation

$$\Gamma(x,y) := \Gamma \cap \partial T(x,y).$$

Finally, $\nu = (\nu_1, \nu_2)$ denotes the exterior normal to $T(x,y)$, and we set $A := (\varphi^{-1}(y), y), B := (x, \varphi(x)) \in \Gamma$.

With the aid of the Gaussian integral theorem we integrate (5) over the triangle $T(x,y)$ and obtain

$$\iint\limits_{T(x,y)} (v\,\mathcal{L}u - u\,\mathcal{M}v)\Big|_{(\xi,\eta)} \, d\xi\, d\eta$$

$$= \int\limits_{\partial T(x,y)} \left\{ (-v_y u + auv)\nu_1 + (vu_x + bvu)\nu_2 \right\} d\sigma$$

$$= \int\limits_{\widehat{AB}} \left\{ (-v_y u + auv)\nu_1 + (vu_x + bvu)\nu_2 \right\} d\sigma$$

$$+ \int\limits_{\widehat{BP}} (-v_y + av)u \, d\eta + \int\limits_{\widehat{PA}} (u_x + bu)v \, d\xi$$

$$= \int\limits_{\widehat{AB}} \left\{ (-v_y u + auv)\nu_1 + (vu_x + bvu)\nu_2 \right\} d\sigma$$

$$+ \int\limits_{\widehat{BP}} (-v_y + av)u \, d\eta + \int\limits_{\widehat{PA}} (-v_x + bv)u \, d\xi + \int\limits_{\widehat{PA}} (uv)_x \, d\xi$$

$$= \int\limits_{\widehat{AB}} \left\{ (-v_y u + auv)\nu_1 + (vu_x + bvu)\nu_2 \right\} d\sigma$$

$$+ \int\limits_{\widehat{BP}} (-v_y + av)u \, d\eta + \int\limits_{\widehat{PA}} (-v_x + bv)u \, d\xi$$

$$-u(P)v(P) + u(A)v(A).$$

Here $\widehat{AB} = \Gamma(x,y)$ denotes the positive-oriented arc from A to B on the boundary of $T(x,y)$ between the points A and B.

Definition 5.4. *The function* $v(\xi, \eta) =: R(\xi, \eta; x, y)$ *is called* Riemannian function *if the following conditions are fulfilled:*

1. *The function v satisfies the differential equation $\mathcal{M}v = 0$ in $T(x,y)$.*
2. *We have $v(x,y) = R(x,y;x,y) = 1$.*
3. *Along the arc $\overset{\frown}{BP}$ we have $-v_y + av = 0$, and therefore*

$$v(x,\eta) = \exp\left\{\int_y^\eta a(x,t)\,dt\right\}.$$

4. *Along the arc $\overset{\frown}{PA}$ we have $-v_x + bv = 0$, and therefore*

$$v(\xi,y) = \exp\left\{\int_x^\xi b(t,y)\,dt\right\}.$$

If we can find a Riemannian function, we have the following

Theorem 5.5. (Riemannian integration method)
A solution of the hyperbolic differential equation $\mathcal{L}u(\xi,\eta) = h(\xi,\eta)$ can be represented by the Cauchy data with the aid of the Riemannian function $R(\xi,\eta;x,y)$ as follows: For the point $P = (x,y)$ we have

$$
\begin{aligned}
u(P) = {} & u(A)R(A;P) - \iint\limits_{T(x,y)} R(\xi,\eta;P)h(\xi,\eta)\,d\xi\,d\eta \\
& + \int\limits_{\Gamma(x,y)} \left\{\left(-R_\eta(\xi,\eta;P)u(\xi,\eta) + auR\right)\nu_1 + (Ru_\xi + bRu)\nu_2\right\}d\sigma.
\end{aligned}
\tag{6}
$$

Remark: The problem remains to construct a Riemannian function.

6 Bernstein's Analyticity Theorem

On the unit disc $B := \{(u,v) : u^2 + v^2 < 1\}$ we consider a solution

$$\mathbf{x} = \mathbf{x}(u,v) = \big(x_1(u,v),\ldots,x_n(u,v)\big) : B \to \mathbb{R}^n \in C^3(B,\mathbb{R}^n) \tag{1}$$

of the quasilinear elliptic system

$$\Delta\mathbf{x}(u,v) = \mathbf{F}(u,v,\mathbf{x}(u,v),\mathbf{x}_u(u,v),\mathbf{x}_v(u,v)), \qquad (u,v) \in B. \tag{2}$$

In an open neighborhood $\mathcal{O} \subset \mathbb{R}^{2+3n}$ of the surface

$$\mathcal{F} := \Big\{\big(u,v,\mathbf{x}(u,v),\mathbf{x}_u(u,v),\mathbf{x}_v(u,v)\big) : (u,v) \in B\Big\}$$

the function

$$\mathbf{F} : \mathcal{O} \to \mathbb{R}^n \qquad \text{is assumed to be real-analytic.} \tag{3}$$

At each point $\mathbf{z} \in \mathcal{O}$ we can locally expand the function $\mathbf{F}$ with $2 + 3n$ variables into a power series whose coefficients belong to $\mathbb{R}^n$. This series also converges in the complex variables $u, v, z_1, \ldots, z_n, p_1, \ldots, p_n, q_1, \ldots, q_n \in \mathbb{C}$. This enables us to continue the right-hand side of (2) onto an open set $\mathcal{O}$ in $\mathbb{C}^{2+3n}$ with $\mathcal{F} \subset \mathcal{O}$. Then we obtain

$$\mathbf{F} = \mathbf{F}(u, v, z_1, \ldots, z_n, p_1, \ldots, p_n, q_1, \ldots, q_n) : \mathcal{O} \to \mathbb{C}^n \in C^1(\mathcal{O}, \mathbb{C}^n) \tag{4}$$

without relabeling our function. Then $\mathbf{F}$ satisfies the $2 + 3n$ Cauchy-Riemann equations

$$\mathbf{F}_{\overline{u}} \equiv \mathbf{F}_{\overline{v}} \equiv \mathbf{F}_{\overline{z_1}} \equiv \ldots \equiv \mathbf{F}_{\overline{z_n}} \equiv \mathbf{F}_{\overline{p_1}} \equiv \ldots \equiv \mathbf{F}_{\overline{p_n}} \equiv \mathbf{F}_{\overline{q_1}} \equiv \ldots \equiv \mathbf{F}_{\overline{q_n}} \equiv 0 \tag{5}$$

in $\mathcal{O}$. With the assumptions (3) or equivalently (4)-(5), we shall show that a solution (1) of (2) is real-analytic on the disc B. Then we have an open neighborhood $\mathcal{B} \subset \mathbb{C}^2$ of B such that the following function extended to $\mathcal{B}$

$$\mathbf{x}(u, v) = (x_1(u, v), \ldots, x_n(u, v)) : \mathcal{B} \to \mathbb{C}^n \in C^3(\mathcal{B}, \mathbb{C}^n) \tag{6}$$

satisfies the Cauchy-Riemann equations

$$\frac{\partial}{\partial \overline{u}} x_j(u, v) \equiv 0 \equiv \frac{\partial}{\partial \overline{v}} x_j(u, v), \quad (u, v) \in \mathcal{B}, \qquad \text{for} \quad j = 1, \ldots, n \tag{7}$$

or equivalently

$$\mathbf{x}_{\overline{u}} := (x_{1,\overline{u}}, \ldots, x_{n,\overline{u}}) \equiv 0 \equiv (x_{1,\overline{v}}, \ldots, x_{n,\overline{v}}) =: \mathbf{x}_{\overline{v}} \qquad \text{in} \quad \mathcal{B}. \tag{8}$$

With the aid of ideas from H. Lewy, we shall analytically extend the solution (1) of (2) from B onto $\mathcal{B}$. This is achieved by solving initial value problems for nonlinear hyperbolic differential equations with two variables. Starting from an extension into the variables $(u, v) = (\alpha + i\beta, \gamma + i\delta) \in \mathcal{B}$ for the moment, the system (2) appears in the form

$$\mathbf{x}_{\alpha\alpha}(u, v) + \mathbf{x}_{\gamma\gamma}(u, v) = \mathbf{F}(u, v, \mathbf{x}, \mathbf{x}_\alpha(u, v), \mathbf{x}_\gamma(u, v)) \qquad \text{in} \quad \mathcal{B}. \tag{9}$$

We can write the Cauchy-Riemann equations as follows:

$$\mathbf{x}_\beta(u, v) = i\mathbf{x}_\alpha(u, v) \qquad \text{in} \quad \mathcal{B} \tag{10}$$

and

$$\mathbf{x}_\delta(u, v) = i\mathbf{x}_\gamma(u, v) \qquad \text{in} \quad \mathcal{B}. \tag{11}$$

These imply the Laplace equations

$$\mathbf{x}_{\alpha\alpha}(u, v) + \mathbf{x}_{\beta\beta}(u, v) = 0 \qquad \text{in} \quad \mathcal{B} \tag{12}$$

and

$$\mathbf{x}_{\gamma\gamma}(u,v) + \mathbf{x}_{\delta\delta}(u,v) = 0 \quad \text{in} \quad \mathcal{B}. \tag{13}$$

Inserting (12) and (10) into (9) we obtain

$$-\mathbf{x}_{\beta\beta}(u,v) + \mathbf{x}_{\gamma\gamma}(u,v) = \mathbf{F}(u,v,\mathbf{x},-i\mathbf{x}_\beta,\mathbf{x}_\gamma) \quad \text{in} \quad \mathcal{B}. \tag{14}$$

From (13), (11), and (9) we infer

$$\mathbf{x}_{\alpha\alpha}(u,v) - \mathbf{x}_{\delta\delta}(u,v) = \mathbf{F}(u,v,\mathbf{x},\mathbf{x}_\alpha,-i\mathbf{x}_\delta) \quad \text{in} \quad \mathcal{B}. \tag{15}$$

Now we solve initial value problems for the hyperbolic equations (14) and (15) with initial velocities given by (10) and (11), respectively.

We thus obtain the

Theorem 6.1. (Analyticity theorem of S. Bernstein)
Let the solution $\mathbf{x} = \mathbf{x}(u,v)$ *of the p.d.e. problem (1)-(2) be given with the real-analytic right-hand side (3) or equivalently (4)-(5). Then the function* $\mathbf{x}$ *is real-analytic in* B.

Proof (H. Lewy):

1. Using the notations from above, we start with a solution $\mathbf{x} = \mathbf{x}(\alpha,\gamma)$: $B \to \mathbb{R}^n \in C^3(B,\mathbb{R}^n)$ for the system of differential equations

$$\mathbf{x}_{\alpha\alpha}(\alpha,\gamma) + \mathbf{x}_{\gamma\gamma}(\alpha,\gamma) = \mathbf{F}(\alpha,\gamma,\mathbf{x},\mathbf{x}_\alpha(\alpha,\gamma),\mathbf{x}_\gamma(\alpha,\gamma)) \quad \text{in} \quad B. \tag{16}$$

We consider the Cauchy initial value problem

$$-\mathbf{x}_{\beta\beta}(\alpha,\beta,\gamma) + \mathbf{x}_{\gamma\gamma}(\alpha,\beta,\gamma) = \mathbf{F}(\alpha,\beta,\gamma,\mathbf{x},-i\mathbf{x}_\beta,\mathbf{x}_\gamma) \quad \text{in} \quad \mathcal{B}',$$

$$\mathbf{x}(\alpha,0,\gamma) = \mathbf{x}(\alpha,\gamma) \quad \text{in} \quad B, \tag{17}$$

$$\mathbf{x}_\beta(\alpha,0,\gamma) = i\mathbf{x}_\alpha(\alpha,\gamma) \quad \text{in} \quad B$$

with the parameter α. Here $\mathcal{B}' \subset \mathbb{R}^3$ denotes a suitable open set satisfying $B \subset \mathcal{B}'$. According to Section 4, the problem (17) possesses a locally unique solution $\mathbf{x} = \mathbf{x}(\alpha,\beta,\gamma)$, since the characteristic curves of the differential equation point out of B. We emphasize the differentiable dependence of the solution on the parameter α. We now define $u := \alpha + i\beta$. Taking the Remark 2 following Theorem 4.4 from Section 4 into consideration, we can apply the operator

$$\frac{\partial}{\partial\overline{u}} = \frac{1}{2}\left(\frac{\partial}{\partial\alpha} + i\frac{\partial}{\partial\beta}\right)$$

to the differential equation in (17). For the function

$$\mathbf{y}(\alpha,\beta,\gamma) = \big(y_1(\alpha,\beta,\gamma),\ldots,y_n(\alpha,\beta,\gamma)\big) := \mathbf{x}_{\overline{u}}(\alpha,\beta,\gamma)$$

we obtain the system of differential equations

$$-\mathbf{y}_{\beta\beta}(\alpha,\beta,\gamma) + \mathbf{y}_{\gamma\gamma}(\alpha,\beta,\gamma) = \sum_{j=1}^{n}\left\{\mathbf{F}_{z_j}y_j - i\mathbf{F}_{p_j}y_{j,\beta} + \mathbf{F}_{q_j}y_{j,\gamma}\right\} \quad \text{in } \mathcal{B}'.$$
$$(18)$$

Noting (17) we comprehend

$$\mathbf{y}(\alpha,0,\gamma) = \frac{1}{2}\big(\mathbf{x}_\alpha(\alpha,0,\gamma) + i\mathbf{x}_\beta(\alpha,0,\gamma)\big)$$
$$(19)$$
$$= \frac{1}{2}\big(\mathbf{x}_\alpha(\alpha,\gamma) + ii\mathbf{x}_\alpha(\alpha,\gamma)\big) = 0 \quad \text{in } B.$$

Furthermore, we observe (17) and (16) and calculate

$$\mathbf{y}_\beta(\alpha,0,\gamma) = \frac{1}{2}\big(\mathbf{x}_{\alpha\beta}(\alpha,0,\gamma) + i\mathbf{x}_{\beta\beta}(\alpha,0,\gamma)\big)$$

$$= \frac{1}{2}\big(\mathbf{x}_{\alpha\beta}(\alpha,0,\gamma) + i\mathbf{x}_{\gamma\gamma}(\alpha,0,\gamma) - i\mathbf{F}(\alpha,0,\gamma,\mathbf{x},\mathbf{x}_\alpha,\mathbf{x}_\gamma)\big)$$

$$= \frac{1}{2}\big(\mathbf{x}_{\alpha\beta}(\alpha,0,\gamma) - i\mathbf{x}_{\alpha\alpha}(\alpha,\gamma)\big)$$

$$= \frac{1}{2}\frac{\partial}{\partial\alpha}\big(\mathbf{x}_\beta(\alpha,0,\gamma) - i\mathbf{x}_\alpha(\alpha,\gamma)\big) = 0 \quad \text{in } B,$$

and consequently
$$\mathbf{y}_\beta(\alpha,0,\gamma) = 0 \quad \text{in } B. \tag{20}$$

The homogeneous Cauchy initial value problem (18)-(20) is uniquely solvable by $\mathbf{y}(\alpha,\beta,\gamma) \equiv 0$ in $\mathcal{B}'$, and we see

$$\mathbf{x}_{\overline{u}}(\alpha,\beta,\gamma) \equiv 0 \quad \text{in } \mathcal{B}'. \tag{21}$$

2. We now extend $\mathbf{x}$ from $\mathcal{B}'$ onto $\mathcal{B} \subset \mathbb{C}^2$. In this context we solve the Cauchy initial value problem

$$\mathbf{x}_{\alpha\alpha}(\alpha,\beta,\gamma,\delta) - \mathbf{x}_{\delta\delta}(\alpha,\beta,\gamma,\delta) = \mathbf{F}(\alpha,\beta,\gamma,\delta,\mathbf{x},\mathbf{x}_\alpha,-i\mathbf{x}_\delta) \quad \text{in } \mathcal{B},$$

$$\mathbf{x}(\alpha,\beta,\gamma,0) = \mathbf{x}(\alpha,\beta,\gamma) \quad \text{in } \mathcal{B}',$$

$$\mathbf{x}_\delta(\alpha,\beta,\gamma,0) = i\mathbf{x}_\gamma(\alpha,\beta,\gamma) \quad \text{in } \mathcal{B}'.$$
$$(22)$$

The solution depends differentiably on the parameters β,γ, and higher regularity follows as in Section 4. At first, we consider the function

$$\mathbf{y}(\alpha,\beta,\gamma,\delta) = \big(y_1(\alpha,\beta,\gamma,\delta),\ldots,y_n(\alpha,\beta,\gamma,\delta)\big) := \mathbf{x}_{\overline{u}}(\alpha,\beta,\gamma,\delta).$$

Due to (22), this function satisfies the hyperbolic system

$$\mathbf{y}_{\alpha\alpha}(\alpha,\beta,\gamma,\delta) - \mathbf{y}_{\delta\delta}(\alpha,\beta,\gamma,\delta) = \sum_{j=1}^{n}\left\{\mathbf{F}_{z_j}y_j + \mathbf{F}_{p_j}y_{j,\alpha} - i\mathbf{F}_{q_j}y_{j,\delta}\right\} \quad \text{in } \mathcal{B}.$$
$$(23)$$

Due to (21) we have the initial conditions

$$\mathbf{y}(\alpha,\beta,\gamma,0) = \frac{1}{2}\big(\mathbf{x}_\alpha(\alpha,\beta,\gamma,0) + i\mathbf{x}_\beta(\alpha,\beta,\gamma,0)\big)$$
$$= \mathbf{x}_{\overline{u}}(\alpha,\beta,\gamma) = 0 \quad \text{in} \quad \mathcal{B}' \tag{24}$$

and

$$\mathbf{y}_\delta(\alpha,\beta,\gamma,0) = \frac{1}{2}\big(\mathbf{x}_{\alpha\delta}(\alpha,\beta,\gamma,0) + i\mathbf{x}_{\beta\delta}(\alpha,\beta,\gamma,0)\big)$$
$$= \frac{i}{2}\big(\mathbf{x}_{\alpha\gamma}(\alpha,\beta,\gamma) + i\mathbf{x}_{\beta\gamma}(\alpha,\beta,\gamma)\big) \tag{25}$$
$$= i\frac{\partial}{\partial\gamma}\mathbf{x}_{\overline{u}}(\alpha,\beta,\gamma) = 0 \quad \text{in} \quad \mathcal{B}'.$$

From (23)-(25) we deduce $\mathbf{y}(\alpha,\beta,\gamma,\delta) = 0$ in $\mathcal{B}$ and consequently

$$\mathbf{x}_{\overline{u}}(\alpha,\beta,\gamma,\delta) \equiv 0 \quad \text{in} \quad \mathcal{B}. \tag{26}$$

Finally, we investigate the function

$$\mathbf{z}(\alpha,\beta,\gamma,\delta) = \big(z_1(\alpha,\beta,\gamma,\delta),\ldots,z_n(\alpha,\beta,\gamma,\delta)\big) := \mathbf{x}_{\overline{v}}(\alpha,\beta,\gamma,\delta)$$

and infer the following system of differential equations from (22):

$$\mathbf{z}_{\alpha\alpha}(\alpha,\beta,\gamma,\delta) - \mathbf{z}_{\delta\delta}(\alpha,\beta,\gamma,\delta) = \sum_{j=1}^{n}\big\{\mathbf{F}_{z_j}z_j + \mathbf{F}_{p_j}z_{j,\alpha} - i\mathbf{F}_{q_j}z_{j,\delta}\big\} \quad \text{in} \quad \mathcal{B}. \tag{27}$$

For the function $\mathbf{z}$ we determine the initial conditions

$$\mathbf{z}(\alpha,\beta,\gamma,0) = \frac{1}{2}\big(\mathbf{x}_\gamma(\alpha,\beta,\gamma,0) + i\mathbf{x}_\delta(\alpha,\beta,\gamma,0)\big)$$
$$= \frac{1}{2}\big(\mathbf{x}_\gamma(\alpha,\beta,\gamma) + ii\mathbf{x}_\gamma(\alpha,\beta,\gamma)\big) = 0 \quad \text{in} \quad \mathcal{B}' \tag{28}$$

and

$$\mathbf{z}_\delta(\alpha,\beta,\gamma,0) = \frac{1}{2}\big(\mathbf{x}_{\gamma\delta}(\alpha,\beta,\gamma,0) + i\mathbf{x}_{\delta\delta}(\alpha,\beta,\gamma,0)\big)$$
$$= \frac{1}{2}\big(\mathbf{x}_{\gamma\delta}(\alpha,\beta,\gamma,0) + i\mathbf{x}_{\alpha\alpha}(\alpha,\beta,\gamma,0) - i\mathbf{F}(\alpha,\beta,\gamma,0,\mathbf{x},\mathbf{x}_\alpha,\mathbf{x}_\gamma)\big)$$
$$= \frac{1}{2}\big(\mathbf{x}_{\gamma\delta}(\alpha,\beta,\gamma,0) - i\mathbf{x}_{\gamma\gamma}(\alpha,\beta,\gamma)\big)$$
$$= \frac{\partial}{\partial\gamma}\frac{1}{2}\big(\mathbf{x}_\delta(\alpha,\beta,\gamma,0) - i\mathbf{x}_\gamma(\alpha,\beta,\gamma)\big) = 0 \quad \text{in} \quad \mathcal{B}', \tag{29}$$

using (22) and (16). The equation (16) remains valid in $\mathcal{B}'$ due to (21). From (27)-(29) we infer $\mathbf{z}(\alpha,\beta,\gamma,\delta) \equiv 0$ in $\mathcal{B}$ and finally

$$\mathbf{x}_{\overline{v}}(\alpha, \beta, \gamma, \delta) \equiv 0 \qquad \text{in} \quad \mathcal{B}. \tag{30}$$

Consequently, we have extended the solution $\mathbf{x} = \mathbf{x}(\alpha, \gamma)$ of (16) to a function $\mathbf{x} = \mathbf{x}(\alpha, \beta, \gamma, \delta) : \mathcal{B} \to \mathbb{C}^n$, which is holomorphic in the variables $u = \alpha + i\beta$ and $v = \gamma + i\delta$ due to (26) and (30). Therefore, the function

$$\mathbf{x}(\alpha, \gamma) = \mathbf{x}(\alpha, \beta, \gamma, \delta)|_{\beta = \delta = 0}$$

is real-analytic in α and γ.   q.e.d.

The theorems about holomorphic mappings necessary for the next proof are contained in the beautiful book [GF] of H.Grauert and K. Fritzsche, especially in Chapter 1, Sections 6 and 7.

Theorem 6.2. *On the open set* $\Omega \subset \mathbb{R}^2$ *let us consider a solution* $z = \zeta(x, y) \in C^3(\Omega, \mathbb{R})$ *of the nonparametric H-surface equation*

$$\begin{aligned}
\mathcal{M}\zeta(x, y) &:= (1 + \zeta_y^2)\zeta_{xx}(x, y) - 2\zeta_x\zeta_y\zeta_{xy}(x, y) + (1 + \zeta_x^2)\zeta_{yy}(x, y) \\
&= 2H(x, y, \zeta(x, y))(1 + |\nabla\zeta(x, y)|^2)^{\frac{3}{2}} \qquad \text{in} \quad \Omega.
\end{aligned} \tag{31}$$

The mean curvature $H = H(x, y, z)$ *is assumed to be real-analytic in a three-dimensional open neighborhood of the surface*

$$\mathcal{F} := \Big\{ (x, y, \zeta(x, y)) \; : \; (x, y) \in \Omega \Big\}.$$

Then the solution $z = \zeta(x, y) : \Omega \to \mathbb{R}$ *is real-analytic in* Ω.

Proof: We choose the point $(x^0, y^0) \in \Omega$ arbitrarily and determine $r > 0$ such that the corresponding disc satisfies

$$B_r(x^0, y^0) := \Big\{ (x, y) \in \mathbb{R}^2 \; : \; (x - x^0)^2 + (y - y^0)^2 < r^2 \Big\} \subset\subset \Omega.$$

We consider the C^2-metric

$$ds^2 = (1 + \zeta_x^2)\, dx^2 + 2\zeta_x\zeta_y\, dx\, dy + (1 + \zeta_y^2)\, dy^2 \qquad \text{in} \quad B_r(x^0, y^0)$$

and introduce isothermal parameters via the diffeomorphic mapping

$$f(u, v) = (x(u, v), y(u, v)) : B \to B_r(x^0, y^0) \in C^3(B).$$

The function

$$\mathbf{x}(u, v) := (f(u, v), z(u, v)) \in C^3(B, \mathbb{R}^3) \tag{32}$$

with

$$z(u, v) := \zeta \circ f(u, v), \qquad (u, v) \in B$$

satisfies Rellich's system

$$\Delta\mathbf{x}(u, v) = 2H(\mathbf{x}(u, v))\, \mathbf{x}_u \wedge \mathbf{x}_v(u, v) \qquad \text{in} \quad B. \tag{33}$$

Due to Theorem 6.1, the function $\mathbf{x}$ is real-analytic in B and the mapping $f : B \to B_r(x^0, y^0)$ as well. Since $J_f(u, v) \neq 0$ in B holds true, also the inverse mapping $f^{-1} : B_r(x^0, y^0) \to B$ is real-analytic. Consequently, the function

$$\zeta(x, y) = z \circ f^{-1}(x, y), \qquad (x, y) \in B_r(x^0, y^0), \tag{34}$$

is real-analytic in $B_r(x^0, y^0)$. This holds true in Ω as well, since the point $(x^0, y^0) \in \Omega$ has been chosen arbitrarily.

q.e.d.

Remarks:

a) The introduction of conformal parameters by the uniformization theorem (see Chapter 12, Section 8) can be achieved by various methods: A continuity method is presented in the paper by

F. Sauvigny: *Introduction of isothermal parameters into a Riemannian metric by the continuity method.* Analysis **19** (1999), 235-243.

A variational method is applied by the authors

S. Hildebrandt, H. von der Mosel: *On Lichtenstein's theorem about globally conformal mappings.* Calc. Var. **23** (2005), 415-424.

b) For arbitrary quasilinear, real-analytic, elliptic differential equations in two variables, F. Müller has proved the Bernstein analyticity theorem by the uniformization method used in Theorem 6.2. In this context we refer the reader to the treatises:

F. Müller: *On the continuation of solutions for elliptic equations in two variables.* Ann. Inst. H. Poincaré - Analyse Non Linéaire **19** (2002), 745-776;

F. Müller: *Analyticity of solutions for semilinear elliptic systems of second order.* Calc. Var. and PDE **15** (2002), 257-288.

c) Finally, the reader should consider Hans Lewy's original treatise:

H. Lewy: *Neuer Beweis des analytischen Charakters der Lösungen elliptischer Differentialgleichungen.* Math. Annalen **101** (1929), 609-619.

7 Some Historical Notices to Chapter 11

The Bernstein analyticity theorem represents the first regularity result in the theory of partial differential equations. This question was proposed by D. Hilbert to S. Bernstein, who solved this problem by intricate methods in 1904.

K. Friedrichs and H. Lewy treated the initial value problem for hyperbolic equations in 1927. Their method of successive approximations, which is nowadays established via Banach's fixed point theorem, has been invented already by E. Picard. Two years later in 1929, H. Lewy ingeniously built the bridge from hyperbolic to elliptic equations with his approach to Bernstein's analyticity theorem. We would like to mention that H. Lewy attended lectures of J. Hadamard (1865–1963) during his research visit to Paris in 1930.

In his wonderful book *Partial Differential Equations*, P. Garabedian observed that Lewy's proof is substantially simplified when the principal part of the equation reduces to the Laplacian. With the uniformization theorem for non-analytic Riemannian metrics, we present the decisive tool to investigate arbitrary elliptic equations in the next chapter.

Figure 2.7   PORTRAIT OF H. LEWY (1904–1988)
taken from the biography by *C. Reid: Courant, in Göttingen and New York – An Album*, Springer-Verlag, Berlin... (1976).

Chapter 12

Nonlinear Elliptic Systems

We present a maximum principle of W. Jäger for the H-surface system in Section 1. Then we prove the fundamental gradient estimate of E. Heinz for nonlinear elliptic systems of differential equations in Section 2. Global estimates are established in Section 3. In combination with the Leray-Schauder degree of mapping, we deduce an existence theorem for nonlinear elliptic systems in Section 4. Especially for the system $\Delta \mathbf{x} = 2H\mathbf{x}_u \wedge \mathbf{x}_v$ which was discovered by F. Rellich, this result was proved by E. Heinz already in 1954. In Section 5 we derive an inner distortion estimate for plane nonlinear elliptic systems, which implies a curvature estimate presented in Section 6. In the next Sections 7-8 we introduce conformal parameters into a Riemannian metric and establish a priori estimates up to the boundary in this context. Finally, we explain the uniformization method for quasilinear elliptic differential equations in Section 9.

1 Maximum Principles for the H-surface System

Let $B := \{w = u + iv \in \mathbb{C} : |w| < 1\}$ denote the unit disc. We prescribe the function

$$H = H(w, \mathbf{x}) : \overline{B} \times \mathbb{R}^3 \to \mathbb{R} \in C^0(\overline{B} \times \mathbb{R}^3, \mathbb{R}) \tag{1}$$

with the bounds

$$|H(w, \mathbf{x})| \leq h_0, \quad |H(w, \mathbf{x}) - H(w, \mathbf{y})| \leq h_1 |\mathbf{x} - \mathbf{y}|$$
$$\text{for all} \quad w \in B, \quad \mathbf{x}, \mathbf{y} \in \mathbb{R}^3 \tag{2}$$

and consider *Rellich's H-surface system*

$$\mathbf{x} = \mathbf{x}(u, v) \in C^2(B, \mathbb{R}^3) \cap C^0(\overline{B}, \mathbb{R}^3),$$
$$\Delta \mathbf{x}(u, v) = 2H(w, \mathbf{x}(w)) \, \mathbf{x}_u \wedge \mathbf{x}_v, \qquad w \in B. \tag{3}$$

If a solution of (3) additionally satisfies the relations

F. Sauvigny, *Partial Differential Equations 2*, Universitext,
DOI 10.1007/978-1-4471-2984-4_6, © Springer-Verlag London 2012

$$|\mathbf{x}_u|^2 = |\mathbf{x}_v|^2, \quad \mathbf{x}_u \cdot \mathbf{x}_v = 0 \qquad \text{in} \quad B,$$

which means $\mathbf{x}$ represents a conformally parametrized surface, then $\mathbf{x}$ possesses the prescribed mean curvature $H = H(w, \mathbf{x}(w))$. We now start with two suitable solutions $\mathbf{x}, \mathbf{y}$ of (3), consider the difference function

$$\mathbf{z}(w) := \mathbf{x}(w) - \mathbf{y}(w), \quad w \in \overline{B}$$

and deduce an inequality of the form

$$\sup_{w \in \overline{B}} |\mathbf{z}(w)| \le C(h_0, h_1, \dots) \sup_{w \in \partial B} |\mathbf{z}(w)|. \tag{4}$$

The latter implies unique solvability of the Dirichlet problem for (3) and its stability with respect to perturbations of the boundary values in the C^0-topology.

The special case $H \equiv 0$: Let the two solutions $\mathbf{x}, \mathbf{y}$ of the system (3) with $H \equiv 0$ be given. Then their difference

$$\mathbf{z}(u, v) = \mathbf{x}(u, v) - \mathbf{y}(u, v) \in C^2(B) \cap C^0(\overline{B})$$

is a harmonic function as well. We consider the auxiliary function

$$f(u, v) := |\mathbf{z}(u, v)|^2 = \mathbf{z}(u, v) \cdot \mathbf{z}(u, v) = \mathbf{z}(u, v)^2, \qquad (u, v) \in \overline{B}, \tag{5}$$

and calculate

$$\begin{aligned} \Delta f(u, v) &= \nabla \cdot \nabla f(u, v) = \nabla \cdot \nabla(\mathbf{z} \cdot \mathbf{z}) \\ &= 2\nabla(\mathbf{z} \cdot \nabla \mathbf{z}) = 2\big(|\nabla \mathbf{z}|^2 + \mathbf{z} \cdot \Delta \mathbf{z}\big) \\ &= 2|\nabla \mathbf{z}(u, v)|^2 \ge 0 \qquad \text{in} \quad B. \end{aligned}$$

Here we used the gradient $\quad \nabla = \left(\dfrac{\partial}{\partial u}, \dfrac{\partial}{\partial v}\right) \quad$ and abbreviated the expression $\mathbf{z} \cdot \nabla \mathbf{z} = (\mathbf{z} \cdot \mathbf{z}_u, \mathbf{z} \cdot \mathbf{z}_v) \in \mathbb{R}^2$. The maximum principle for subharmonic functions yields

$$\sup_{w \in \overline{B}} |\mathbf{z}(w)| \le \sup_{w \in \partial B} |\mathbf{z}(w)|. \tag{6}$$

In **the general case** $H \not\equiv 0$, we start with two solutions $\mathbf{x}, \mathbf{y}$ of (3) and consider the difference function $\mathbf{z}(u, v) = \mathbf{x}(u, v) - \mathbf{y}(u, v)$. Then we introduce the *weighted distance function of W. Jäger*

$$F(u, v) := |\mathbf{z}(u, v)|^2 \exp\left\{\frac{1}{2}\Big(\phi(|\mathbf{x}(u, v)|^2) + \phi(|\mathbf{y}(u, v)|^2)\Big)\right\}, \quad w = u + iv \in \overline{B}. \tag{7}$$

Here the symbol

$$\phi = \phi(t) : [0, M^2] \to \mathbb{R} \in C^2([0, M^2]) \tag{8}$$

denotes an auxiliary function, still to be determined, with an appropriate quantity $M > 0$. In the distance function (7), we obviously can consider only *small solutions* satisfying

$$|\mathbf{x}(u, v)| < M, \quad |\mathbf{y}(u, v)| < M \qquad \text{for all} \quad w = u + iv \in \overline{B}. \tag{9}$$

Now we shall choose ϕ in such a way that F fulfills a differential inequality subject to the maximum principle. At first, we note that

$$\nabla e^{\frac{1}{2}(\phi(\mathbf{x}^2)+\phi(\mathbf{y}^2))} = e^{\frac{1}{2}(\phi(\mathbf{x}^2)+\phi(\mathbf{y}^2))}\left[\phi'(\mathbf{x}^2)(\mathbf{x}\cdot\nabla\mathbf{x}) + \phi'(\mathbf{y}^2)(\mathbf{y}\cdot\nabla\mathbf{y})\right] \tag{10}$$

holds true. Then we calculate

$$\nabla F = e^{\frac{1}{2}(\phi(\mathbf{x}^2)+\phi(\mathbf{y}^2))}\left\{\nabla(\mathbf{z}^2) + \mathbf{z}^2\left[\phi'(\mathbf{x}^2)(\mathbf{x}\cdot\nabla\mathbf{x}) + \phi'(\mathbf{y}^2)(\mathbf{y}\cdot\nabla\mathbf{y})\right]\right\}$$

and consequently

$$e^{-\frac{1}{2}(\phi(\mathbf{x}^2)+\phi(\mathbf{y}^2))}\nabla F = \nabla(\mathbf{z}^2) + \mathbf{z}^2\left[\phi'(\mathbf{x}^2)(\mathbf{x}\cdot\nabla\mathbf{x}) + \phi'(\mathbf{y}^2)(\mathbf{y}\cdot\nabla\mathbf{y})\right]. \tag{11}$$

Applying the operator ∇ to this identity, we obtain the subsequent

Proposition 1.1. *The function $F(u, v)$ defined in (7) satisfies the following differential equation in B:*

$$\mathcal{L}F := \left(e^{-\frac{1}{2}(\phi(\mathbf{x}^2)+\phi(\mathbf{y}^2))}F_u\right)_u + \left(e^{-\frac{1}{2}(\phi(\mathbf{x}^2)+\phi(\mathbf{y}^2))}F_v\right)_v$$

$$= \Delta(\mathbf{z}^2) + \frac{1}{2}\mathbf{z}^2\left[\phi'(\mathbf{x}^2)\Delta(\mathbf{x}^2) + \phi'(\mathbf{y}^2)\Delta(\mathbf{y}^2)\right]$$

$$+2\mathbf{z}^2\left[\phi''(\mathbf{x}^2)(\mathbf{x}\cdot\nabla\mathbf{x})^2 + \phi''(\mathbf{y}^2)(\mathbf{y}\cdot\nabla\mathbf{y})^2\right]$$

$$+2(\mathbf{z}\cdot\nabla\mathbf{z})\cdot\left[\phi'(\mathbf{x}^2)(\mathbf{x}\cdot\nabla\mathbf{x}) + \phi'(\mathbf{y}^2)(\mathbf{y}\cdot\nabla\mathbf{y})\right].$$

We intend to choose ϕ in such a way that $\mathcal{L}F \geq 0$ in B holds true. At first, we note that

$$|\Delta\mathbf{x}| \leq 2|H|\,|\mathbf{x}_u \wedge \mathbf{x}_v| \leq h_0|\nabla\mathbf{x}|^2$$

and obtain

$$\Delta(\mathbf{x}^2) = 2(|\nabla\mathbf{x}|^2 + \mathbf{x}\cdot\Delta\mathbf{x}) \geq 2(|\nabla\mathbf{x}|^2 - |\mathbf{x}|\,|\Delta\mathbf{x}|)$$

$$\geq 2|\nabla\mathbf{x}|^2(1 - h_0|\mathbf{x}|),$$

$$\Delta(\mathbf{y}^2) = 2(|\nabla\mathbf{y}|^2 + \mathbf{y}\cdot\Delta\mathbf{y}) \geq 2(|\nabla\mathbf{y}|^2 - |\mathbf{y}|\,|\Delta\mathbf{y}|)$$

$$\geq 2|\nabla\mathbf{y}|^2(1 - h_0|\mathbf{y}|) \qquad \text{in} \quad B.$$

$$\tag{12}$$

Proposition 1.2. *For all $w \in B' := \{\zeta \in B \mid |z(\zeta)| \neq 0\}$ we have*

$$\Delta(\mathbf{z}^2) - 2\left(\frac{\mathbf{z}}{|\mathbf{z}|} \cdot \nabla \mathbf{z}\right)^2 \geq -(h_0^2 + h_1)|\mathbf{z}|^2(|\nabla \mathbf{x}|^2 + |\nabla \mathbf{y}|^2).$$

Proof: We evaluate $\Delta(\mathbf{z}^2) = 2(|\nabla \mathbf{z}|^2 + \mathbf{z} \cdot \Delta \mathbf{z})$ and estimate as follows:

$$|\mathbf{z} \cdot \Delta \mathbf{z}| = |\mathbf{z} \cdot (\Delta \mathbf{x} - \Delta \mathbf{y})| = |\mathbf{z} \cdot \left(2H(w, \mathbf{x})\mathbf{x}_u \wedge \mathbf{x}_v - 2H(w, \mathbf{y})\mathbf{y}_u \wedge \mathbf{y}_v\right)|$$

$$\leq 2|H(w, \mathbf{x})| \, |\mathbf{z} \cdot (\mathbf{x}_u \wedge \mathbf{x}_v - \mathbf{y}_u \wedge \mathbf{y}_v)|$$

$$+ 2|H(w, \mathbf{x}) - H(w, \mathbf{y})| \, |\mathbf{z}| \, |\mathbf{y}_u \wedge \mathbf{y}_v|$$

$$\leq 2h_0|(\mathbf{z}, \mathbf{z}_u, \mathbf{x}_v) + (\mathbf{z}, \mathbf{y}_u, \mathbf{z}_v)| + h_1|\mathbf{z}|^2|\nabla \mathbf{y}|^2$$

$$\leq 2h_0\left(\frac{|\mathbf{z} \wedge \mathbf{z}_u|}{|\mathbf{z}|}|\mathbf{x}_v| \, |\mathbf{z}| + \frac{|\mathbf{z} \wedge \mathbf{z}_v|}{|\mathbf{z}|}|\mathbf{y}_u| \, |\mathbf{z}|\right) + h_1|\mathbf{z}|^2|\nabla \mathbf{y}|^2$$

$$\leq \frac{|\mathbf{z} \wedge \mathbf{z}_u|^2}{|\mathbf{z}|^2} + \frac{|\mathbf{z} \wedge \mathbf{z}_v|^2}{|\mathbf{z}|^2} + h_0^2|\mathbf{z}|^2(|\mathbf{x}_v|^2 + |\mathbf{y}_u|^2) + h_1|\mathbf{z}|^2|\nabla \mathbf{y}|^2$$

$$= |\nabla \mathbf{z}|^2 - \frac{1}{|\mathbf{z}|^2}\{(\mathbf{z} \cdot \mathbf{z}_u)^2 + (\mathbf{z} \cdot \mathbf{z}_v)^2\}$$

$$+ h_0^2|\mathbf{z}|^2(|\mathbf{x}_v|^2 + |\mathbf{y}_u|^2) + h_1|\mathbf{z}|^2|\nabla \mathbf{y}|^2.$$

Interchanging $\mathbf{x}$ and $\mathbf{y}$ we add both inequalities and obtain

$$2|\mathbf{z} \cdot \Delta \mathbf{z}| \leq 2|\nabla \mathbf{z}|^2 - \frac{2}{|\mathbf{z}|^2}(\mathbf{z} \cdot \nabla \mathbf{z})^2 + h_0^2|\mathbf{z}|^2(|\nabla \mathbf{x}|^2 + |\nabla \mathbf{y}|^2)$$

$$+ h_1|\mathbf{z}|^2(|\nabla \mathbf{x}|^2 + |\nabla \mathbf{y}|^2).$$

Finally, we arrive at

$$\Delta(\mathbf{z}^2) \geq 2\frac{1}{|\mathbf{z}|^2}(\mathbf{z} \cdot \nabla \mathbf{z})^2 - (h_0^2 + h_1)|\mathbf{z}|^2(|\nabla \mathbf{x}|^2 + |\nabla \mathbf{y}|^2).$$

q.e.d.

We now combine the Propositions 1.1 and 1.2 with the formula (12) and deduce

$$\mathcal{L}F \geq \Big\{ -(h_0^2 + h_1) + \phi'(\mathbf{x}^2)(1 - h_0|\mathbf{x}|) \Big\} |\mathbf{z}|^2 |\nabla \mathbf{x}|^2$$

$$+ \Big\{ -(h_0^2 + h_1) + \phi'(\mathbf{y}^2)(1 - h_0|\mathbf{y}|) \Big\} |\mathbf{z}|^2 |\nabla \mathbf{y}|^2 + 2 \Big(\frac{\mathbf{z}}{|\mathbf{z}|} \cdot \nabla \mathbf{z} \Big)^2$$

$$+ + 2\sqrt{2} \Big(\frac{\mathbf{z}}{|\mathbf{z}|} \cdot \nabla \mathbf{z} \Big) \cdot \Big\{ \phi'(\mathbf{x}^2)(\mathbf{x} \cdot \nabla \mathbf{x}) + \phi'(\mathbf{y}^2)(\mathbf{y} \cdot \nabla \mathbf{y}) \Big\} \frac{|\mathbf{z}|}{\sqrt{2}}$$

$$+ 2|\mathbf{z}|^2 \Big\{ \phi''(\mathbf{x}^2)(\mathbf{x} \cdot \nabla \mathbf{x})^2 + \phi''(\mathbf{y}^2)(\mathbf{y} \cdot \nabla \mathbf{y})^2 \Big\}$$

$$\geq \psi(|\mathbf{x}|)|\mathbf{z}|^2 |\nabla \mathbf{x}|^2 + \psi(|\mathbf{y}|)|\mathbf{z}|^2 |\nabla \mathbf{y}|^2$$

$$- \frac{1}{2}|\mathbf{z}|^2 |\phi'(\mathbf{x}^2)(\mathbf{x} \cdot \nabla \mathbf{x}) + \phi'(\mathbf{y}^2)(\mathbf{y} \cdot \nabla \mathbf{y})|^2$$

$$+ 2|\mathbf{z}|^2 \Big\{ \phi''(\mathbf{x}^2)(\mathbf{x} \cdot \nabla \mathbf{x})^2 + \phi''(\mathbf{y}^2)(\mathbf{y} \cdot \nabla \mathbf{y})^2 \Big\}.$$

Consequently, we obtain

$$\mathcal{L}F \geq \psi(|\mathbf{x}|)|\mathbf{z}|^2 |\nabla \mathbf{x}|^2 + \psi(|\mathbf{y}|)|\mathbf{z}|^2 |\nabla \mathbf{y}|^2$$

$$+ |\mathbf{z}|^2 \Big\{ (2\phi''(\mathbf{x}^2) - \phi'(\mathbf{x}^2)^2)(\mathbf{x} \cdot \nabla \mathbf{x})^2 + (2\phi''(\mathbf{y}^2) - \phi'(\mathbf{y}^2)^2)(\mathbf{y} \cdot \nabla \mathbf{y})^2 \Big\}$$

$$\tag{13}$$

with the auxiliary function

$$\psi(t) := -(h_0^2 + h_1) + \phi'(t^2)(1 - h_0 t), \qquad t \in [0, M). \tag{14}$$

In formula (13) we additionally required $\phi'(t) \geq 0$ for $t \in [0, M^2)$. Let us determine a function $\phi(t) : [0, M^2) \to \mathbb{R} \in C^2$ such that

$$\phi''(t) \geq \frac{1}{2}\phi'(t)^2 \qquad \text{in} \quad [0, M^2)$$

holds true. This is obviously realized by the function

$$\phi(t) = -2\log(M^2 - t), \qquad t \in [0, M^2),$$

with $\phi'(t) = 2(M^2 - t)^{-1}$, $\phi''(t) = 2(M^2 - t)^{-2} = \frac{1}{2}\phi'(t)^2$ for $t \in [0, M^2)$. Inserting this function ϕ and the corresponding function ψ into (13), we infer that

$$\mathcal{L}F \geq \psi(|\mathbf{x}|)\,|\mathbf{z}|^2\,|\nabla \mathbf{x}|^2 + \psi(|\mathbf{y}|)\,|\mathbf{z}|^2\,|\nabla \mathbf{y}|^2 \qquad \text{in} \quad B,$$

$$\text{if} \qquad |\mathbf{x}| < M, \, |\mathbf{y}| < M \quad \text{in} \quad \overline{B}$$

$$\tag{15}$$

is correct. For all $t \in [0, M)$ we have the estimate

$$\psi(t) = -(h_0^2 + h_1) + 2\frac{1 - h_0 t}{M^2 - t^2}$$

$$= \frac{h_0^2 + h_1}{M^2 - t^2}\left\{-(M^2 - t^2) + 2\frac{1 - h_0 t}{h_0^2 + h_1}\right\}$$

$$= \frac{h_0^2 + h_1}{M^2 - t^2}\left\{t^2 - 2\frac{h_0}{h_0^2 + h_1}t + \left(\frac{h_0}{h_0^2 + h_1}\right)^2\right.$$

$$\left. -\frac{h_0^2}{(h_0^2 + h_1)^2} + \frac{2}{h_0^2 + h_1} - M^2\right\}$$

$$\geq \frac{h_0^2 + h_1}{M^2 - t^2}\left\{\frac{2(h_0^2 + h_1) - h_0^2}{(h_0^2 + h_1)^2} - M^2\right\} = 0,$$

choosing

$$M = \frac{\sqrt{h_0^2 + 2h_1}}{h_0^2 + h_1}. \tag{16}$$

We therefore obtain the following

Theorem 1.3. (Jäger's maximum principle)

The function $H = H(w, \mathbf{x}) \in C^0(\overline{B} \times \mathbb{R}^3)$ is subject to the inequalities (2), and by $\mathbf{x} = \mathbf{x}(u,v)$, $\mathbf{y} = \mathbf{y}(u,v)$ we denote two solutions of the H-surface system (3). We now define

$$F(u,v) := \frac{|\mathbf{x}(u,v) - \mathbf{y}(u,v)|^2}{(M^2 - |\mathbf{x}(u,v)|^2)(M^2 - |\mathbf{y}(u,v)|^2)}, \qquad (u,v) \in \overline{B}. \tag{17}$$

Here we assume $|\mathbf{x}(u,v)| < M$, $|\mathbf{y}(u,v)| < M$ for all $(u,v) \in \overline{B}$ with

$$M = \frac{\sqrt{h_0^2 + 2h_1}}{h_0^2 + h_1}.$$

Statement: *Then the function F satisfies the linear elliptic differential inequality*

$$\mathcal{L}F := \left\{(M^2 - |\mathbf{x}|^2)(M^2 - |\mathbf{y}|^2)F_u\right\}_u + \left\{(M^2 - |\mathbf{x}|^2)(M^2 - |\mathbf{y}|^2)F_v\right\}_v$$

$$\geq 0 \quad in \ \ B.$$

Theorem 1.4. (Geometric maximum principle of E. Heinz)

Let the function $\mathbf{x}(u,v) = (x_1(u,v), \ldots, x_n(u,v)) : \overline{B} \to \mathbb{R}^n \in C^2(B) \cap C^0(\overline{B})$ denote a solution of the differential inequality

$$|\Delta\mathbf{x}(u,v)| \leq a|\nabla\mathbf{x}(u,v)|^2, \qquad (u,v) \in B. \tag{18}$$

The smallness condition

$$|\mathbf{x}(u,v)| \leq M, \qquad (u,v) \in B, \tag{19}$$

may be fulfilled, and we have

$$aM \leq 1 \qquad \text{for the constants} \quad a \in [0,+\infty), \ M \in (0,+\infty). \tag{20}$$

Statement: *Then we infer*

$$\sup_{(u,v)\in B} |\mathbf{x}(u,v)| \leq \sup_{(u,v)\in \partial B} |\mathbf{x}(u,v)|.$$

Proof: The auxiliary function $\quad f(u,v) := |\mathbf{x}(u,v)|^2, \ (u,v) \in \overline{B} \quad$ satisfies the differential inequality

$$\begin{aligned}
\Delta f(u,v) &= 2\Big(|\nabla \mathbf{x}(u,v)|^2 + \mathbf{x}(u,v)\cdot\Delta\mathbf{x}(u,v)\Big) \\
&\geq 2\Big(|\nabla \mathbf{x}(u,v)|^2 - |\mathbf{x}(u,v)|\,|\Delta\mathbf{x}(u,v)|\Big) \\
&\geq 2\Big(|\nabla \mathbf{x}(u,v)|^2 - a|\mathbf{x}(u,v)|\,|\nabla\mathbf{x}(u,v)|^2\Big) \\
&\geq 2|\nabla\mathbf{x}(u,v)|^2(1 - aM) \ \geq \ 0 \qquad \text{in} \quad B.
\end{aligned}$$

The maximum principle for subharmonic functions yields the statement above.

q.e.d.

Remarks:

1. If $|\mathbf{x}(u,v)| \not\equiv M$ on B is valid, we deduce $|\mathbf{x}(u,v)| < M$ for all $(u,v) \in B$.
2. Theorem 1.4 holds true especially for solutions of the H-surface system (3) with $a = h_0$.

Theorem 1.5. (Jäger's estimate)
The function $H = H(w,\mathbf{x})$ satisfies (1) and (2), and we set

$$M := \frac{\sqrt{h_0^2 + 2h_1}}{h_0^2 + h_1}.$$

Furthermore, the symbols $\mathbf{x}$ and $\mathbf{y}$ denote two solutions of the H-surface system (3) such that

$$|\mathbf{x}(u,v)| \leq M, \quad |\mathbf{y}(u,v)| \leq M \qquad \text{for all} \quad (u,v) \in \overline{B}. \tag{21}$$

Additionally, we require $\|\mathbf{x}\|_{C^0(\partial B)} := \sup_{w\in\partial B} |\mathbf{x}(w)| < M$ and $\|\mathbf{y}\|_{C^0(\partial B)} < M$.
Statement: *Then we have the inequality*

$$\frac{|\mathbf{x}(w) - \mathbf{y}(w)|^2}{(M^2 - |\mathbf{x}(w)|^2)(M^2 - |\mathbf{y}(w)|^2)} \leq \frac{\|\mathbf{x} - \mathbf{y}\|_{C^0(\partial B)}^2}{(M^2 - \|\mathbf{x}\|_{C^0(\partial B)}^2)(M^2 - \|\mathbf{y}\|_{C^0(\partial B)}^2)} \tag{22}$$

for all $w \in \overline{B}$.

Proof: We shall apply the geometric maximum principle of E. Heinz to the functions $\mathbf{x}$ and $\mathbf{y}$ with $a = h_0$. In this context we note that $aM \leq 1$ is valid if and only if

$$\frac{h_0^2(h_0^2 + 2h_1)}{(h_0^2 + h_1)^2} \leq 1$$

or equivalently $h_0^4 + 2h_0^2 h_1 \leq h_0^4 + 2h_0^2 h_1 + h_1^2$ is correct - and the latter inequality is evidently fulfilled. Therefore, Theorem 1.4 yields

$$\|\mathbf{x}\|_{C^0(\overline{B})} \leq \|\mathbf{x}\|_{C^0(\partial B)} < M \quad \text{and} \quad \|\mathbf{y}\|_{C^0(\overline{B})} \leq \|\mathbf{y}\|_{C^0(\partial B)} < M.$$

We apply E. Hopf's maximum principle to the auxiliary function $F(u,v)$ from Theorem 1.3 and obtain (22).

q.e.d.

Corollary: In addition to the assumptions of Theorem 1.5 let the inequalities

$$\|\mathbf{x}\|_{C^0(\partial B)} \leq M' < M \quad \text{and} \quad \|\mathbf{y}\|_{C^0(\partial B)} \leq M' < M$$

be satisfied. Then we have a constant $k = k(M, M') > 0$, such that

$$\|\mathbf{x} - \mathbf{y}\|_{C^0(\overline{B})} \leq k(M, M')\|\mathbf{x} - \mathbf{y}\|_{C^0(\partial B)}. \tag{23}$$

Remark: In the original paper of

W. Jäger: *Ein Maximumprinzip für ein System nichtlinearer Differential-gleichungen*; Nachr. Akad. Wiss. Göttingen, II. Math. Phys. Kl. 1976, 157-164.

a maximum principle is derived even for systems of the form

$$\Delta\mathbf{x}(u,v) = \mathbf{F}(u, v, \mathbf{x}(u,v), \nabla\mathbf{x}(u,v)), \qquad (u,v) \in B, \tag{24}$$

under certain structural conditions for the right-hand side.

2 Gradient Estimates for Nonlinear Elliptic Systems

We take a domain $\Omega \subset \mathbb{R}^2$ and consider solutions

$$\mathbf{x} = \mathbf{x}(u,v) = (x_1(u,v), \ldots, x_n(u,v)) \in C^2(\Omega, \mathbb{R}^n) \cap C^0(\overline{\Omega}, \mathbb{R}^n) \tag{1}$$

of the *differential inequality*

$$|\Delta\mathbf{x}(u,v)| \leq a|\nabla\mathbf{x}(u,v)|^2 + b \quad \text{for all} \quad (u,v) \in \Omega \tag{2}$$

with the constants $a, b \in [0, +\infty)$. We require the *smallness condition*

$$|\mathbf{x}(u,v)| \leq M \quad \text{for all} \quad (u,v) \in \overline{\Omega} \tag{3}$$

for the solution of (1), (2) with the constant $M \in (0, +\infty)$.

Remark: The H-surface system, linear systems as well as the Poisson equation are covered by the differential inequality (2).

Now we shall estimate the quantity $|\nabla \mathbf{x}(u, v)|$ from above, in the interior of Ω and on the boundary - with respect to adequate boundary conditions.

Proposition 2.1. *Let* $\mathbf{x} = \mathbf{x}(u, v)$ *denote a solution of (1)-(3). Then the function* $f(u, v) := |\mathbf{x}(u, v)|^2$ *in* Ω *satisfies the differential inequality*

$$\Delta f(u, v) \geq 2(1 - aM)|\nabla \mathbf{x}(u, v)|^2 - 2bM, \qquad (u, v) \in \Omega. \qquad (4)$$

Proof: At first, we have

$$\Delta f(u, v) = 2(|\nabla \mathbf{x}(u, v)|^2 + \mathbf{x} \cdot \Delta \mathbf{x}(u, v)) \quad \text{in} \quad \Omega.$$

Furthermore, the relation (2) yields the inequality

$$|\mathbf{x} \cdot \Delta \mathbf{x}(u, v)| \leq aM|\nabla \mathbf{x}(u, v)|^2 + bM \quad \text{in} \quad \Omega,$$

and we infer (4). q.e.d.

Proposition 2.2. (Inner energy estimate)
Let the condition $aM < 1$ *be satisfied and* $\vartheta \in (0, 1)$ *be chosen. Furthermore, the disc* $B_R(w_0) := \{w \in \mathbb{C} : |w - w_0| < R\}$ *with the center* $w_0 \in \Omega$ *and the radius* $R > 0$ *fulfills the inclusion* $B_R(w_0) \subset \Omega$. *Then all solutions of (1)-(3) satisfy the inequality*

$$\iint\limits_{B_{\vartheta R}(w_0)} |\nabla \mathbf{x}(u, v)|^2 \, du \, dv$$

$$\leq \frac{1}{-\log \vartheta} \left\{ \frac{2\pi M}{1 - aM} \sup_{w \in \partial B_R(w_0)} |\mathbf{x}(w) - \mathbf{x}(w_0)| + \frac{\pi b M R^2}{2(1 - aM)} \right\}. \qquad (5)$$

Proof: Theorem 2.4 from Chapter 5, Section 2 implies that arbitrary functions $\phi \in C^2(\Omega) \cap C^0(\overline{\Omega})$ satisfy the identity

$$\phi(w_0) = \frac{1}{2\pi R} \int\limits_{\partial B_R(w_0)} \phi(w) \, d\sigma(w) - \frac{1}{2\pi} \iint\limits_{B_R(w_0)} \left(\log \frac{R}{|w - w_0|} \right) \Delta\phi(w) \, du \, dv. \qquad (6)$$

Inserting $\phi(w) := |w - w_0|^2 = (u - u_0)^2 + (v - v_0)^2$, $w \in \mathbb{R}^2$ we obtain

$$0 = R^2 - \frac{1}{2\pi} \iint\limits_{B_R(w_0)} \left(\log \frac{R}{|w - w_0|} \right) 4 \, du \, dv$$

and consequently

$$\iint\limits_{B_R(w_0)} \log\frac{R}{|w - w_0|}\,du\,dv = \frac{\pi R^2}{2}. \tag{7}$$

Inserting $\phi = f(u, v) = |\mathbf{x}(u, v)|^2$ into (6), Proposition 2.1 yields

$$\frac{1 - aM}{\pi}\iint\limits_{B_R(w_0)}\Big(\log\frac{R}{|w - w_0|}\Big)|\nabla\mathbf{x}(u, v)|^2\,du\,dv - \frac{bMR^2}{2}$$

$$\leq \frac{1}{2\pi}\iint\limits_{B_R(w_0)}\Big(\log\frac{R}{|w - w_0|}\Big)\Delta f(u, v)\,du\,dv$$

$$= \frac{1}{2\pi R}\int\limits_{\partial B_R(w_0)}\Big(f(w) - f(w_0)\Big)\,d\sigma(w)$$

$$\leq \frac{1}{2\pi R}\int\limits_{\partial B_R(w_0)}|\mathbf{x}(w) - \mathbf{x}(w_0)|\,|\mathbf{x}(w) + \mathbf{x}(w_0)|\,d\sigma(w)$$

$$\leq 2M\sup_{w\in\partial B_R(w_0)}|\mathbf{x}(w) - \mathbf{x}(w_0)|.$$

This implies the estimate

$$\iint\limits_{B_R(w_0)}\Big(\log\frac{R}{|w - w_0|}\Big)|\nabla\mathbf{x}(u, v)|^2\,du\,dv$$

$$\leq \frac{2\pi M}{1 - aM}\sup_{w\in\partial B_R(w_0)}|\mathbf{x}(w) - \mathbf{x}(w_0)| + \frac{\pi bMR^2}{2(1 - aM)}. \tag{8}$$

The inequality now yields (5).   q.e.d.

Proposition 2.3. (Boundary-energy-estimate)
We have the condition $aM < 1$ and choose $\vartheta \in (0, 1)$. The disc $B_R(w_0)$ with the center $w_0 \in \mathbb{R}$ and the radius $R > 0$ satisfies

$$B_R(w_0) \cap \Omega = \Big\{w \in B_R(w_0) : \operatorname{Im} w > 0\Big\} =: H_R(w_0). \tag{9}$$

We set $\partial H_R(w_0) = C_R(w_0) \cup I_R(w_0)$ with

$$C_R(w_0) := \Big\{w \in \partial B_R(w_0) : \operatorname{Im} w \geq 0\Big\},$$

$$I_R(w_0) := [w_0 - R, w_0 + R].$$

For all solutions $\mathbf{x} \in C^1(\overline{\Omega})$ of (1)-(3) satisfying the boundary condition

$$\mathbf{x}(u, 0) = 0 \qquad \text{for all} \quad u \in I_R(w_0)$$

we have the following estimate

$$\iint\limits_{H_{\vartheta R}(w_0)} |\nabla \mathbf{x}(u,v)|^2 \, du \, dv$$

$$\leq \frac{1}{-\log \vartheta} \left\{ \frac{\pi M}{1 - aM} \sup_{w \in C_R(w_0)} |\mathbf{x}(w) - \mathbf{x}(w_0)| + \frac{\pi b M R^2}{4(1 - aM)} \right\}. \tag{10}$$

Proof:

1. With the aid of a reflection we continue $\mathbf{x}$ as follows:

$$\hat{\mathbf{x}}(u,v) := \begin{cases} \mathbf{x}(u,v), & w = u + iv \in \overline{H_R(w_0)} \\ \mathbf{x}(u,-v), & w \in \overline{B_R(w_0)} \setminus \overline{H_R(w_0)} \end{cases}. \tag{11}$$

The function $\hat{\mathbf{x}}(u,v)$ is continuous in $\overline{B_R(w_0)}$ and satisfies the differential inequality (2) in $B_R(w_0) \setminus I_R(w_0)$. However, the function $\mathbf{x}_v(u,v)$ may possess jump discontinuities on the interval $I_R(w_0)$. We consider the function

$$\phi(u,v) := |\hat{\mathbf{x}}(u,v)|^2, \qquad (u,v) \in \overline{B_R(w_0)}, \tag{12}$$

which is subject to the differential inequality (4) in $B_R(w_0) \setminus I_R(w_0)$. Furthermore, we deduce

$$\phi \in C^1(\overline{B_R(w_0)}) \qquad \text{and} \qquad \phi(u,0) = 0 = \phi_v(u,0) \quad \text{in } I_R(w_0). \tag{13}$$

We now show that formula (6) is valid for the function $\phi = |\hat{\mathbf{x}}|^2$ as well. Continuing as in the proof of Proposition 2.2, we then obtain the relation (5) for the reflected function $\hat{\mathbf{x}}(w)$. The estimate (10) finally follows by means of symmetry arguments.

2. Choosing the parameter $0 < \varepsilon < \varepsilon_0$ sufficiently small, we define the sets

$$B_\varepsilon^\pm := \left\{ w \in \mathbb{C} : 0 < \varepsilon < |w - w_0| < R, \ \pm \text{Im} \, w > 0 \right\}$$

and set $r := |w - w_0| \in [0, R]$. At the point w_0 we utilize Green's function

$$\psi(w) = \psi(u,v) = \frac{1}{2\pi} \log \frac{R}{|w - w_0|}$$

$$= \frac{1}{2\pi} (\log R - \log r), \qquad w \in \overline{B_R(w_0)} \setminus \{w_0\}. \tag{14}$$

With the aid of Green's formula we calculate

$$\frac{1}{2\pi} \iint\limits_{B_\varepsilon^\pm} \left(\log \frac{R}{|w - w_0|} \right) \Delta\phi(u,v) \, du \, dv = \iint\limits_{B_\varepsilon^\pm} (\psi \Delta\phi - \phi \Delta\psi) \, du \, dv$$

$$= \int\limits_{\partial B_\varepsilon^\pm} \left(\psi \frac{\partial\phi}{\partial\nu} - \phi \frac{\partial\psi}{\partial\nu} \right) d\sigma \tag{15}$$

for $0 < \varepsilon < \varepsilon_0$. From the boundary conditions (13) and (14) for ϕ and ψ we infer the following identity from (15) by addition

$$\frac{1}{2\pi} \iint\limits_{B_\varepsilon^+ \cup B_\varepsilon^-} \Big(\log \frac{R}{|w - w_0|} \Big) \Delta\phi(u, v) \, du \, dv = \frac{1}{2\pi R} \int\limits_{\partial B_R(w_0)} \phi(w) \, d\sigma(w)$$

$$- \frac{1}{2\pi\varepsilon} \int\limits_{|w - w_0| = \varepsilon} \phi(w) \, d\sigma(w) + \frac{1}{2\pi} \int\limits_{|w - w_0| = \varepsilon} \Big(\log \frac{R}{\varepsilon} \Big) \frac{\partial\phi(w)}{\partial\nu} \, d\sigma(w)$$

with $0 < \varepsilon < \varepsilon_0$. On account of $\phi \in C^1(\overline{B_R(w_0)})$ the transition to the limit $\varepsilon \to 0+$ yields

$$\phi(w_0) = \frac{1}{2\pi R} \int\limits_{\partial B_R(w_0)} \phi(w) \, d\sigma(w)$$

$$- \frac{1}{2\pi} \iint\limits_{B_R(w_0)} \Big(\log \frac{R}{|w - w_0|} \Big) \Delta\phi(u, v) \, du \, dv.$$

Following part 1 of the proof, we now arrive at the statement above.

q.e.d.

With the aid of Proposition 2.2 and 2.3, we now can estimate the oscillation of $\mathbf{x}$ via the Courant-Lebesgue lemma on selected circular lines. Using the Wirtinger operators

$$\frac{\partial}{\partial w} = \frac{1}{2} \Big(\frac{\partial}{\partial u} - i\frac{\partial}{\partial v} \Big) \quad \text{and} \quad \frac{\partial}{\partial \overline{w}} = \frac{1}{2} \Big(\frac{\partial}{\partial u} + i\frac{\partial}{\partial v} \Big)$$

we consider the complex derivative function

$$\mathbf{y}(w) := \mathbf{x}_w(w), \qquad w \in \Omega. \tag{16}$$

The differential inequality (2) can be rewritten into the form

$$4|\mathbf{x}_{w\overline{w}}(w)| \le 4a|\mathbf{x}_w(w)|^2 + b$$

or equivalently

$$|\mathbf{y}_{\overline{w}}(w)| \le a|\mathbf{y}(w)|^2 + \frac{1}{4}b \qquad \text{for all} \quad w \in \Omega. \tag{17}$$

By the oscillation inequalities we now shall estimate the Cauchy integral of the complex derivative function **y** for solutions of (2).

Proposition 2.4. *Given the assumptions of Proposition 2.2, each number $\vartheta \in (0,1)$ admits a number $\lambda = \lambda(\vartheta) \in [\frac{1}{4}, \frac{1}{2}]$, such that the derivative function $\mathbf{y}(w) = \mathbf{x}_w(w)$ for a solution $\mathbf{x}(w)$ of (1)-(3) satisfies the following inequality:*

$$\left| \frac{1}{2\pi i} \int_{\partial B_{\lambda \vartheta R}(w_0)} \frac{\mathbf{y}(w)}{w - w_0} \, dw \right| \leq \frac{8\sqrt{M^2 + \frac{1}{8}bMR^2}}{\sqrt{\log 4}\sqrt{1 - aM}} \frac{1}{\vartheta \sqrt{-\log \vartheta}} \frac{1}{R}$$

$$+ \frac{a}{2}\vartheta R \sup_{w \in B_{\vartheta R}(w_0)} |\mathbf{y}(w)|^2 + \frac{b}{8}\vartheta R. \tag{18}$$

Proof:

1. Proposition 2.2 yields the following estimate for arbitrary $\vartheta \in (0,1)$:

$$\sqrt{\iint_{B_{\vartheta R}(w_0)} |\nabla \mathbf{x}(u,v)|^2 \, du \, dv} \leq \frac{\sqrt{\pi}}{\sqrt{-\log \vartheta}} \frac{2\sqrt{M^2 + \frac{b}{8}MR^2}}{\sqrt{1 - aM}}. \tag{19}$$

On account of the Oscillation lemma of Courant and Lebesgue from Section 5 in Chapter 1, we have a number $\lambda = \lambda(\vartheta) \in [\frac{1}{4}, \frac{1}{2}]$ satisfying

$$\int_{\partial B_{\lambda \vartheta R}(w_0)} |d\mathbf{x}(w)| \leq \frac{4\pi}{\sqrt{\log 4}} \frac{\sqrt{M^2 + \frac{b}{8}MR^2}}{\sqrt{1 - aM}} \frac{1}{\sqrt{-\log \vartheta}}. \tag{20}$$

2. We set $B := B_{\lambda \vartheta R}(w_0)$ and $\varrho := \lambda \vartheta R$. With the aid of the Gaussian integral theorem in the complex form (see Chapter 4, Section 4), we derive the identity

$$\int\limits_{\partial B} \frac{\mathbf{y}(w)}{w - w_0}\, dw - \int\limits_{\partial B} \frac{d\mathbf{x}(w)}{w - w_0}$$

$$= \int\limits_{\partial B} \frac{\mathbf{x}_w(w)}{w - w_0}\, dw - \int\limits_{\partial B} \frac{\mathbf{x}_w(w)\, dw + \mathbf{x}_{\overline{w}}(w)\, d\overline{w}}{w - w_0}$$

$$= -\int\limits_{\partial B} \frac{\mathbf{x}_{\overline{w}}(w)\, d\overline{w}}{w - w_0} = -\frac{1}{\varrho}\left(\overline{\int\limits_{\partial B} \frac{\mathbf{x}_w(w)\, dw}{\left(\frac{w - w_0}{\varrho}\right)}} \right)$$

$$= -\frac{1}{\varrho}\left(\overline{\int\limits_{\partial B} \frac{\mathbf{x}_w(w)\, dw}{\frac{\varrho}{w - w_0}}} \right) = -\frac{1}{\varrho^2}\left(\overline{\int\limits_{\partial B} (w - w_0)\mathbf{x}_w\, dw} \right)$$

$$= -\frac{1}{\varrho^2}\left(\overline{2i \iint\limits_{B} \frac{\partial}{\partial \overline{w}}\left\{ (w - w_0)\mathbf{x}_w(w) \right\} du\, dv} \right)$$

$$= \frac{2i}{\varrho^2} \iint\limits_{B} (\overline{w} - \overline{w_0})\mathbf{x}_{w\overline{w}}(w)\, du\, dv$$

$$= \frac{2i}{\varrho^2} \iint\limits_{B} (\overline{w} - \overline{w_0})\mathbf{y}_{\overline{w}}(w)\, du\, dv.$$

3. We now estimate as follows:

$$\left| \frac{1}{2\pi i}\int\limits_{\partial B} \frac{\mathbf{y}(w)}{w - w_0}\, dw \right| \leq \frac{1}{2\pi}\int\limits_{\partial B} \frac{|d\mathbf{x}(w)|}{|w - w_0|} + \frac{1}{\pi \varrho^2} \iint\limits_{B} |w - w_0||\mathbf{y}_{\overline{w}}(w)|\, du\, dv$$

$$\leq \frac{1}{2\pi\varrho}\int\limits_{\partial B} |d\mathbf{x}(w)| + \varrho \sup_{w \in B} |\mathbf{y}_{\overline{w}}(w)|$$

$$\leq \frac{2}{\pi\vartheta R}\frac{4\pi}{\sqrt{\log 4}}\frac{\sqrt{M^2 + \frac{b}{8}MR^2}}{\sqrt{1 - aM}}\frac{1}{\sqrt{-\log\vartheta}}$$

$$+ \frac{\vartheta R}{2}\left(a \sup_{w \in B} |\mathbf{y}(w)|^2 + \frac{b}{4} \right)$$

$$\leq \frac{8}{\vartheta R}\frac{\sqrt{M^2 + \frac{b}{8}MR^2}}{\sqrt{\log 4}\sqrt{1 - aM}}\frac{1}{\sqrt{-\log\vartheta}}$$

$$+ \frac{a}{2}\vartheta R \sup_{w \in B_{\vartheta R}(w_0)} |\mathbf{y}(w)|^2 + \frac{b}{8}\vartheta R.$$

q.e.d.

Proposition 2.5. *With the assumptions of Proposition 2.3 we consider the reflected gradient function*

$$
\mathbf{z}(w) := \begin{cases} i\mathbf{x}_w(w), & w \in \overline{H_R(w_0)} \\ -i\mathbf{x}_{\overline{w}}(\overline{w}), & w \in \overline{B_R(w_0)} \setminus \overline{H_R(w_0)} \end{cases}. \tag{21}
$$

This function belongs to the class $C^0(\overline{B_R(w_0)}) \cap C^1(B_R(w_0) \setminus I_R(w_0))$ and satisfies the differential inequality

$$
|\mathbf{z}_{\overline{w}}(w)| \le a|\mathbf{z}(w)|^2 + \frac{b}{4} \quad \text{for all} \quad w \in B_R(w_0) \setminus I_R(w_0). \tag{22}
$$

Furthermore, there exists a number $\lambda = \lambda(\vartheta) \in [\frac{1}{4}, \frac{1}{2}]$ for each $\vartheta \in (0,1)$, such that

$$
\left| \frac{1}{2\pi i} \int\limits_{\partial B_{\lambda\vartheta R}(w_0)} \frac{\mathbf{z}(w)}{w - w_0} \, dw \right| \le \frac{8\sqrt{M^2 + \frac{1}{8}bMR^2}}{\sqrt{\log 4}\sqrt{1 - aM}} \frac{1}{\vartheta\sqrt{-\log\vartheta}} \frac{1}{R}
$$
$$
+ \frac{a}{2}\vartheta R \sup_{w \in B_{\vartheta R}(w_0)} |\mathbf{z}(w)|^2 + \frac{b}{8}\vartheta R. \tag{23}
$$

Proof:

1. According to the relation (11) we reflect $\mathbf{x}(u,v)$ and obtain the function $\hat{\mathbf{x}}(u,v)$ satisfying

$$
|\Delta\hat{\mathbf{x}}(u,v)| \le a|\nabla\hat{\mathbf{x}}(u,v)|^2 + b \quad \text{for all} \quad (u,v) \in B_R(w_0) \setminus I_R(w_0). \tag{24}
$$

We observe $\mathbf{x}_u(u,0) = 0 = \operatorname{Im}\mathbf{z}(u,0)$ in $I_R(w_0)$, and the function defined in (21) is continuous in $B_R(w_0)$. Furthermore, we have

$$
\hat{\mathbf{x}}_w(w) = \begin{cases} \mathbf{x}_w(w) = -i\mathbf{z}(w), & w \in H_R(w_0) \\ \mathbf{x}_{\overline{w}}(\overline{w}) = i\mathbf{z}(w), & w \in B_R(w_0) \setminus \overline{H}_R(w_0) \end{cases}, \tag{25}
$$

and (24) yields the inequality (22). In part 1 of the proof for Proposition 2.4 we replace $\mathbf{x}$ by the function $\hat{\mathbf{x}}$, and we deduce

$$
\int\limits_{\partial B_{\lambda\vartheta R}(w_0)} |d\hat{\mathbf{x}}(w)| \le \frac{4\pi}{\sqrt{\log 4}} \frac{\sqrt{M^2 + \frac{b}{8}MR^2}}{\sqrt{1 - aM}} \frac{1}{\sqrt{-\log\vartheta}} \tag{26}
$$

with a suitable $\lambda = \lambda(\vartheta) \in [\frac{1}{4}, \frac{1}{2}]$. The Courant-Lebesgue lemma is namely applicable to the function $\hat{\mathbf{x}}$, whose derivatives might possess jump discontinuities at the interval $I_R(w_0)$.

2. We now follow the arguments in part 2 of the proof for Proposition 2.4, and we additionally set

$$B^\pm := \left\{ w \in B_{\lambda\vartheta R}(w_0) \ : \ \pm \mathrm{Im}\, w > 0 \right\},$$

$$C^\pm := \left\{ w \in \partial B_{\lambda\vartheta R}(w_0) \ : \ \pm \mathrm{Im}\, w \geq 0 \right\}.$$

The calculation there yields

$$\int_{C^\pm} \frac{i\hat{\mathbf{x}}_w(w)}{w - w_0}\, dw - i \int_{C^\pm} \frac{d\hat{\mathbf{x}}(w)}{w - w_0} = \frac{1}{\varrho^2} \left(\overline{\int_{C^\pm} (w - w_0) i\hat{\mathbf{x}}_w(w)\, dw} \right). \qquad (27)$$

We observe that the integrand $(w - w_0) i\hat{\mathbf{x}}_w(w)$, approaching the interval $I_R(w_0)$ from above or from below, is subject to a change of sign, and we infer

$$\int_{C^+} \frac{i\hat{\mathbf{x}}_w(w)}{w - w_0}\, dw + \int_{C^-} \frac{-i\hat{\mathbf{x}}_w(w)}{w - w_0}\, dw - i \int_{C^+} \frac{d\hat{\mathbf{x}}(w)}{w - w_0} + i \int_{C^-} \frac{d\hat{\mathbf{x}}(w)}{w - w_0}$$

$$= \frac{1}{\varrho^2} \left(\overline{\int_{\partial B^+} (w - w_0) i\hat{\mathbf{x}}_w(w)\, dw} \right)$$

$$+ \frac{1}{\varrho^2} \left(\overline{\int_{\partial B^-} (w - w_0)(-i)\hat{\mathbf{x}}_w(w)\, dw} \right)$$

$$= \frac{1}{\varrho^2} \left(\overline{2i \iint_{B^+} (w - w_0) i\hat{\mathbf{x}}_{w\overline{w}}(w)\, du\, dv} \right)$$

$$+ \frac{1}{\varrho^2} \left(\overline{2i \iint_{B^-} (w - w_0)(-i)\hat{\mathbf{x}}_{w\overline{w}}(w)\, du\, dv} \right)$$

$$= -\frac{2}{\varrho^2} \iint_{B^+} (\overline{w - w_0})\hat{\mathbf{x}}_{w\overline{w}}(w)\, du\, dv$$

$$+ \frac{2}{\varrho^2} \iint_{B^-} (\overline{w - w_0})\hat{\mathbf{x}}_{w\overline{w}}(w)\, du\, dv.$$

Taking (25) into account, we obtain the following estimate:

$$\left| \frac{1}{2\pi i} \int_{\partial B_{\lambda\vartheta R}(w_0)} \frac{\mathbf{z}(w)}{w - w_0}\, dw \right| = \frac{1}{2\pi} \left| \int_{C^+} \frac{i\hat{\mathbf{x}}_w(w)}{w - w_0}\, dw + \int_{C^-} \frac{-i\hat{\mathbf{x}}_w(w)}{w - w_0}\, dw \right|$$

$$\leq \frac{1}{2\pi} \int_{\partial B} \frac{|d\hat{\mathbf{x}}(w)|}{|w - w_0|} + \frac{1}{\pi\varrho^2} \iint_B |w - w_0|\, |\mathbf{z}_{\overline{w}}(w)|\, du\, dv.$$

We now deduce the estimate (23) as in part 3 of the proof for Proposition 2.4. Therefore, we replace the function $\mathbf{y}$ by $\mathbf{z}$ and utilize besides (22) the oscillation estimate (26).

q.e.d.

Remark: With adequate assumptions, Proposition 2.5 remains true for the discs with centers $w_0 \in \mathbb{C}$ and $\operatorname{Im} w_0 > 0$ satisfying

$$B_R(w_0) \cap \Omega = \{w \in B_R(w_0) \ : \ \operatorname{Im} w > 0\}.$$

The estimate of Dirichlet's integral in Proposition 2.3 remains correct - in a slightly modified form - also in this situation. We shall use the Propositions 2.3 and 2.5 in the next section, in order to derive a global $C^{1+\alpha}$-estimate.

We are now prepared to prove the profound and central

Theorem 2.6. (Gradient estimate of E. Heinz)
A solution $\mathbf{x} = \mathbf{x}(u,v)$ of the problem (1)-(3) with $aM < 1$ is given on the bounded domain $\Omega \subset \mathbb{R}^2$. We define

$$\delta(w) := dist\{w, \partial\Omega\} = \inf_{\zeta \in \mathbb{C}\setminus\Omega} |\zeta - w|, \quad w \in \Omega \qquad and \qquad d := \sup_{w \in \Omega} \delta(w).$$

Then we have a constant $C = C(a, M, bd^2)$, such that the inequality

$$\delta(w)|\nabla\mathbf{x}(w)| \leq C(a, M, bd^2) \qquad for\ all \ \ w \in \Omega \qquad (28)$$

is satisfied.

Proof:

1. At first, we assume $\mathbf{x} = \mathbf{x}(u,v) \in C^1(\overline{\Omega}, \mathbb{R}^n)$ and consider the continuous function

$$\phi(w) := \delta(w)|\mathbf{y}(w)|, \qquad w \in \Omega \qquad (29)$$

with $\mathbf{y}(w) = \mathbf{x}_w(w)$, $w \in \overline{\Omega}$. Due to the boundary condition $\phi|_{\partial\Omega} = 0$ this function attains its maximum at an interior point $w_0 \in \Omega$. Setting $R := \delta(w_0) > 0$ we obtain $B_R(w_0) \subset \Omega$. We then apply the Proposition 2.4: For an arbitrary number $\vartheta \in (0,1)$ we find a quantity $\lambda = \lambda(\vartheta) \in [\frac{1}{4}, \frac{1}{2}]$, such that

$$R\left|\frac{1}{2\pi i} \int_{\partial B_{\lambda\vartheta R}(w_0)} \frac{\mathbf{y}(w)}{w - w_0}\,dw\right| \leq \frac{c_1(a, M, bd^2)}{\vartheta\sqrt{-\log\vartheta}} + \frac{bd^2}{8}$$

$$+ \frac{a}{2}\vartheta R^2 \sup_{w \in B_{\vartheta R}(w_0)} |\mathbf{y}(w)|^2 \qquad (30)$$

holds true with the constant

$$c_1(a, M, bd^2) := \frac{8\sqrt{M^2 + \frac{1}{8}bMd^2}}{\sqrt{\log 4}\sqrt{1 - aM}}.$$

On the disc $B := B_{\lambda \vartheta R}(w_0)$ of radius $\varrho := \lambda \vartheta R \in (0, R)$ we infer the following integral representation from Theorem 5.5 (Pompeiu, Vekua) in Chapter 4, Section 5:

$$\mathbf{y}(w_0) = \frac{1}{2\pi i} \int\limits_{\partial B} \frac{\mathbf{y}(w)}{w - w_0} \, dw - \frac{1}{\pi} \iint\limits_{B} \frac{\mathbf{y}_{\overline{w}}(w)}{w - w_0} \, du \, dv. \tag{31}$$

The first integral on the right-hand side of (31) has been estimated in (30). We introduce polar coordinates and observe (17); then we obtain the following inequality for the second integral in (31):

$$\frac{R}{\pi} \left| \iint\limits_{B} \frac{\mathbf{y}_{\overline{w}}(w)}{w - w_0} \, du \, dv \right| \leq \frac{R}{\pi} \sup_{w \in B} |\mathbf{y}_{\overline{w}}(w)| \iint\limits_{B} \frac{1}{|w - w_0|} \, du \, dv$$

$$\leq \frac{R}{\pi} \left(a \sup_{w \in B_{\vartheta R}(w_0)} |\mathbf{y}(w)|^2 + \frac{b}{4} \right) 2\pi \frac{1}{2} \vartheta R \tag{32}$$

$$\leq a \vartheta R^2 \sup_{w \in B_{\vartheta R}(w_0)} |\mathbf{y}(w)|^2 + \frac{1}{4} b d^2.$$

2. From (29)-(32) we infer

$$\phi(w_0) = \delta(w_0) |\mathbf{y}(w_0)| = R |\mathbf{y}(w_0)|$$

$$\leq \frac{c_1(a, M, bd^2)}{\vartheta \sqrt{-\log \vartheta}} + \frac{3}{8} b d^2 + \frac{3}{2} a \vartheta R^2 \sup_{w \in B_{\vartheta R}(w_0)} |\mathbf{y}(w)|^2$$

$$\leq \frac{c_1(a, M, bd^2)}{\vartheta \sqrt{-\log \vartheta}} + \frac{3}{8} b d^2 + \frac{3}{2} a \vartheta R^2 \sup_{w \in B_{\vartheta R}(w_0)} \left\{ \frac{\delta(w)}{R - \vartheta R} |\mathbf{y}(w)| \right\}^2$$

$$\leq \frac{c_1(a, M, bd^2)}{\vartheta \sqrt{-\log \vartheta}} + \frac{3}{8} b d^2 + \frac{3a}{2} \frac{\vartheta}{(1 - \vartheta)^2} \sup_{w \in B_{\vartheta R}(w_0)} \phi(w)^2.$$

Consequently, we have the inequality

$$\phi(w_0) \leq \frac{c_1(a, M, bd^2)}{\vartheta \sqrt{-\log \vartheta}} + \frac{3}{8} b d^2 + \frac{3a}{2} \frac{\vartheta}{(1 - \vartheta)^2} \phi(w_0)^2 \tag{33}$$

for all $\vartheta \in (0, 1)$.

3. Taking $\vartheta \in (0, 1)$ we define

$$\alpha(\vartheta) := \frac{3a}{2} \frac{\vartheta}{(1 - \vartheta)^2} > 0 \quad \text{with} \quad \lim_{\vartheta \to 0+} \alpha(\vartheta) = 0$$

and

$$\beta(\vartheta) := \frac{c_1(a, M, bd^2)}{\vartheta \sqrt{-\log \vartheta}} + \frac{3}{8} b d^2 > 0 \quad \text{with} \quad \lim_{\vartheta \to 0+} \beta(\vartheta) = +\infty.$$

Then we deduce

$$\alpha(\vartheta)\beta(\vartheta) = \frac{3ac_1(a, M, bd^2)}{2(1-\vartheta)^2\sqrt{-\log\vartheta}} + \frac{9abd^2\vartheta}{16(1-\vartheta)^2} \to 0, \qquad \vartheta \to 0+.$$

Setting $t := \phi(w_0)$ we obtain the inequality

$$\alpha(\vartheta)t^2 - t + \beta(\vartheta) \geq 0 \qquad \text{for all} \quad \vartheta \in (0, 1). \tag{34}$$

We note the equivalent statement

$$\left(t - \frac{1}{2\alpha(\vartheta)}\right)^2 \geq \frac{1 - 4\alpha(\vartheta)\beta(\vartheta)}{4\alpha(\vartheta)^2} \qquad \text{for all} \quad \vartheta \in (0, 1). \tag{35}$$

There now exists a quantity $\vartheta_0 = \vartheta_0(a, M, bd^2) \in (0, 1)$ satisfying

$$0 < 4\alpha(\vartheta)\beta(\vartheta) \leq \frac{3}{4} \qquad \text{for all} \quad \vartheta \in (0, \vartheta_0], \tag{36}$$

which implies

$$\sqrt{1 - 4\alpha(\vartheta)\beta(\vartheta)} \geq \frac{1}{2} \qquad \text{for all} \quad \vartheta \in (0, \vartheta_0]. \tag{37}$$

When we introduce the functions

$$\chi^{\pm}(\vartheta) := \frac{1 \pm \sqrt{1 - 4\alpha(\vartheta)\beta(\vartheta)}}{2\alpha(\vartheta)}, \qquad \vartheta \in (0, \vartheta_0],$$

the relation (35) yields the subsequent alternative for each number $\vartheta \in (0, \vartheta_0]$, namely

$$t \leq \chi^{-}(\vartheta) \qquad \text{o r} \qquad t \geq \chi^{+}(\vartheta). \tag{38}$$

Since the functions $\chi^{-}(\vartheta) < \chi^{+}(\vartheta)$, $\vartheta \in (0, \vartheta_0]$ depend continuously on the parameter ϑ in the interval $(0, \vartheta_0]$ and the asymptotic behavior $\lim_{\vartheta \to 0+} \chi^{+}(\vartheta) = +\infty$ is correct, we infer

$$t \leq \chi^{-}(\vartheta) \qquad \text{for all} \quad \vartheta \in (0, \vartheta_0].$$

This implies

$$t \leq \chi^{-}(\vartheta_0) = \chi^{-}(\vartheta_0(a, M, bd^2)) =: \frac{1}{2}C(a, M, bd^2) \tag{39}$$

and consequently

$$\sup_{w \in \Omega} \delta(w)|\nabla\mathbf{x}(w)| = 2 \sup_{w \in \Omega} \delta(w)|\mathbf{y}(w)| = 2 \sup_{w \in \Omega} \phi(w)$$

$$= 2\phi(w_0) = 2t \leq C(a, M, bd^2).$$

We thus obtain the desired estimate (28) in the case $\mathbf{x} \in C^1(\overline{\Omega}, \mathbb{R}^n)$.

4. Now we assume only the regularity $\mathbf{x} \in C^2(\Omega) \cap C^0(\overline{\Omega})$. Then we apply the estimate (28) on the set

$$\Omega_\varepsilon := \Big\{ w \in \Omega \, : \, \text{dist}\,\{w, \partial\Omega\} > \varepsilon \Big\} \qquad \text{for} \quad 0 < \varepsilon < \varepsilon_0$$

at first, and we obtain

$$(\delta(w) - \varepsilon)|\nabla\mathbf{x}(w)| \le C(a, M, bd^2) \qquad \text{for all} \quad w \in \Omega_\varepsilon \qquad (40)$$

with an arbitrary parameter $0 < \varepsilon < \varepsilon_0$. The transition to the limit $\varepsilon \to 0+$ yields

$$\delta(w)|\nabla\mathbf{x}(w)| \le C(a, M, bd^2) \qquad \text{for all} \quad w \in \Omega.$$

$$\text{q.e.d.}$$

We consider the compact set $K \subset \mathbb{C}$ and introduce the linear space

$$C^{1+\alpha}(K, \mathbb{R}^n) := \left\{ \mathbf{x} \in C^1(K, \mathbb{R}^n) \, : \, \sup_{\substack{w_1, w_2 \in K \\ w_1 \ne w_2}} \frac{|\nabla\mathbf{x}(w_1) - \nabla\mathbf{x}(w_2)|}{|w_1 - w_2|^\alpha} < +\infty \right\},$$

where $\alpha \in (0, 1)$ has been chosen. When we endow this space with the $C^{1+\alpha}$-Hölder-norm

$$\|\mathbf{x}\|_{C^{1+\alpha}(K)} := \sup_{w \in K} |\mathbf{x}(w)| + \sup_{w \in K} |\nabla\mathbf{x}(w)| + \sup_{\substack{w_1, w_2 \in K \\ w_1 \ne w_2}} \frac{|\nabla\mathbf{x}(w_1) - \nabla\mathbf{x}(w_2)|}{|w_1 - w_2|^\alpha},$$

$$(41)$$

the set $C^{1+\alpha}(K, \mathbb{R}^n)$ becomes a Banach space. We easily infer the following result from Theorem 2.6 above, namely

Theorem 2.7. (Inner $C^{1+\alpha}$-estimate)

Let a solution $\mathbf{x} = \mathbf{x}(u, v)$ *of (1)-(3) with* $aM < 1$ *be given on the bounded domain* $\Omega \subset \mathbb{R}^2$. *We choose an arbitrary number* $\varepsilon > 0$ *and consider the compact set*

$$K_\varepsilon := \Big\{ w \in \Omega \, : \, dist\{w, \partial\Omega\} \ge \varepsilon \Big\},$$

and additionally we fix the exponent $\alpha \in (0, 1)$. *Then we have a constant* $C = C(a, M, b, d, \varepsilon, \alpha) \in (0, +\infty)$, *such that*

$$\|\mathbf{x}\|_{C^{1+\alpha}(K_\varepsilon)} \le C(a, M, b, d, \varepsilon, \alpha). \qquad (42)$$

Proof: At first, we infer the inequality

$$\sup_{w \in K_\varepsilon} |\mathbf{x}(w)| \le M$$

from (3), and Theorem 2.6 yields the gradient estimate

$$|\nabla \mathbf{x}(w)| \leq \frac{2C(a, M, bd^2)}{\varepsilon} \qquad \text{for all} \quad w \in K_{\frac{\varepsilon}{2}}. \tag{43}$$

For all points $w_0 \in K_\varepsilon$ we have the representation

$$\mathbf{x}_w(w_*) = \frac{1}{2\pi i} \int\limits_{\partial B_{\frac{\varepsilon}{2}}(w_0)} \frac{\mathbf{x}_w(w)}{w - w_*} \, dw - \frac{1}{\pi} \iint\limits_{B_{\frac{\varepsilon}{2}}(w_0)} \frac{\mathbf{x}_{w\overline{w}}(w)}{w - w_*} \, du \, dv, \qquad w_* \in B_{\frac{\varepsilon}{2}}(w_0) \tag{44}$$

due to Theorem 5.5 (Pompeiu, Vekua) in Chapter 4, Section 5. On account of (43) the first parameter integral satisfies a Lipschitz condition in the disc $B_{\frac{\varepsilon}{4}}(w_0)$, with a Lipschitz constant depending on a, M, bd^2, ε. Furthermore, the relation (43) implies

$$\sup_{w \in K_{\frac{\varepsilon}{2}}} |\mathbf{x}_{w\overline{w}}(w)| \leq C_1(a, M, b, d, \varepsilon) < +\infty.$$

We apply Theorem 4.12 (Hadamard's estimate) from Section 4 in Chapter 4 to the second parameter integral in $B_{\frac{\varepsilon}{4}}(w_0)$ and get a Hölder condition depending on $a, M, b, d, \varepsilon, \alpha$. Therefore, the relation (44) yields the inequality

$$|\mathbf{x}_w(w_1) - \mathbf{x}_w(w_2)| \leq C_2(a, M, b, d, \varepsilon, \alpha)|w_1 - w_2|^\alpha$$

$$\text{for all} \quad w_1, w_2 \in K_{\frac{3}{4}\varepsilon} \subset \Omega. \tag{45}$$

Finally, we obtain the estimate (42).   q.e.d.

3 Global Estimates for Nonlinear Systems

We continue our considerations from Section 2 and quote these results by the additional symbol *. Let us define the unit disc $E := \{\zeta = \xi + i\eta \; : \; |\zeta| < 1\}$ and consider solutions of the problem

$$\mathbf{x} = \mathbf{x}(\zeta) = (x_1(\xi, \eta), \dots, x_n(\xi, \eta)) \in C^2(E, \mathbb{R}^n) \cap C^1(\overline{E}, \mathbb{R}^n),$$

$$|\Delta\mathbf{x}(\xi, \eta)| \leq a|\nabla\mathbf{x}(\xi, \eta)|^2 + b \qquad \text{for all} \quad (\xi, \eta) \in E,$$

$$|\mathbf{x}(\xi, \eta)| \leq M \qquad \text{for all} \quad (\xi, \eta) \in E, \tag{1}$$

$$\mathbf{x}(\xi, \eta) = 0 \qquad \text{for all} \quad (\xi, \eta) \in \partial E$$

with the constants $a, b \in [0, +\infty)$ and $M \in (0, +\infty)$. We intend to estimate $|\nabla\mathbf{x}|$ in $\overline{E}$ from above and to establish an adequate a priori bound for $\|\mathbf{x}\|_{C^{1+\alpha}(\overline{E})}$. To achieve this aim, we map the unit disc E conformally onto the upper half-plane $\mathbb{C}^+ := \{w = u + iv \; : \; v > 0\}$ with the aid of the following Möbius transformation (compare Example 7.6 in Chapter 4, Section 7):

$$f(\zeta) = \frac{\zeta + i}{i\zeta + 1}, \quad \zeta \in E; \qquad f : \partial E \setminus \{i\} \leftrightarrow \mathbb{R}. \tag{2}$$

We now define the ray

$$S := \left\{ \zeta = -it \,\Big|\, 0 \le t \le 1 \right\} \subset \overline{E}$$

and the interval

$$J := \left\{ w = iv \,\Big|\, 0 \le v \le 1 \right\} \subset \overline{\mathbb{C}^+}.$$

The function

$$f(i\eta) = i\frac{1+\eta}{1-\eta}, \qquad \eta \in [-1,0] \tag{3}$$

then maps the ray S bijectively onto the interval J. The inverse mapping of f is denoted by

$$\zeta = g(w) : \ \mathbb{C}^+ \to E, \ \ \mathbb{R} \to \partial E \setminus \{i\}, \ \ J \to S. \tag{4}$$

With the parameter $\mu \in [0, 2\pi)$ we consider the rotated rays

$$S_\mu := \left\{ \tilde{\zeta} = e^{i\mu}\zeta \,:\, \zeta \in S \right\}$$

and the family of conformal mappings

$$g_\mu(w) := e^{i\mu}g(w), \qquad w \in \mathbb{C}^+. \tag{5}$$

Evidently, we have

$$g_\mu : \mathbb{C}^+ \leftrightarrow E \text{ conformal}, \quad g_\mu : J \leftrightarrow S_\mu \qquad \text{for} \ \ 0 \le \mu < 2\pi. \tag{6}$$

Setting

$$\Omega^+ := \left\{ w \in \mathbb{C}^+ \,:\, \text{dist}\{w, J\} < 1 \right\},$$

we have a constant $\beta \in (0,1)$ such that the distortion estimate

$$\beta \le |g'_\mu(w)| \le \frac{1}{\beta} \qquad \text{for all} \ \ w \in \Omega^+ \ \ \text{and all} \ \ \mu \in [0, 2\pi) \tag{7}$$

is correct. With the aid of arguments from Section 2, we now prove the following

Theorem 3.1. (Global $C^{1+\alpha}$-estimate)

Let $\mathbf{x} = \mathbf{x}(\xi, \eta)$ denote a solution of (1) with $aM < 1$, and let the exponent $\alpha \in (0,1)$ be chosen. Then we have a constant $C = C(a, b, M, \alpha)$ satisfying

$$\|\mathbf{x}\|_{C^{1+\alpha}(\overline{E})} \le C(a, b, M, \alpha). \tag{8}$$

Proof:

1. By the method of Theorem 2.6* we now shall estimate $|\nabla \mathbf{x}|$ in $\overline{E}$ from above. In this context we consider the function

$$\phi(\zeta) := |\mathbf{x}_\zeta(\zeta)| = \frac{1}{2}|\nabla\mathbf{x}(\zeta)|, \qquad \zeta \in \overline{E}, \tag{9}$$

which attains its maximum at a point $\zeta_0 \in \overline{E}$. For this point $\zeta_0 \in \overline{E}$ we find a number $\mu \in [0, 2\pi)$ and a point $w_0 \in J \subset \mathbb{C}^+ \cup \mathbb{R}$ satisfying $g_\mu(w_0) = \zeta_0$. We fix the angle μ and suppress the index. Via the mapping (5) we now introduce new parameters (u, v) into the function $\mathbf{x} = \mathbf{x}(\xi, \eta)$ and reflect $\mathbf{x} \circ g(u, v)$ at the real axis $v = 0$ as follows:

$$\hat{\mathbf{x}}(u, v) := \begin{cases} \mathbf{x} \circ g(u, v), & w = u + iv \in \mathbb{C}^+ \cup \mathbb{R} \\ \mathbf{x} \circ g(u, -v), & w = u + iv \in \mathbb{C}^- := \{\tilde{w} \in \mathbb{C} \,:\, \mathrm{Im}\,\tilde{w} < 0\} \end{cases} . \tag{10}$$

A simple calculation shows

$$\hat{\mathbf{x}}(u, v) \in C^2(\mathbb{C}^+ \cup \mathbb{C}^-) \cap C^1(\mathbb{C}^+ \cup \mathbb{R}) \cap C^0(\mathbb{C}),$$

$$\sup_{u+iv \in \mathbb{C}} |\hat{\mathbf{x}}(u, v)| \leq M, \qquad \hat{\mathbf{x}}(u, 0) = 0 \quad \text{for all } u \in \mathbb{R},$$

$$|\Delta\hat{\mathbf{x}}(u, v)| \leq a|\nabla\hat{\mathbf{x}}(u, v)|^2 + \frac{b}{\beta^2} \qquad \text{for all} \quad w = u + iv \in \Omega^+ \cup \Omega^-, \tag{11}$$

where we abbreviate $\Omega^- := \{w \in \mathbb{C} : \overline{w} \in \Omega^+\}$. We now choose $R = 1$ as fixed and $\vartheta \in (0, 1)$ arbitrarily. Then we estimate the energy as in Proposition 2.3*, namely

$$\iint\limits_{B_\vartheta(w_0)} |\nabla\hat{\mathbf{x}}(u, v)|^2 \, du\, dv.$$

(Please observe the Remark following Proposition 2.5*.)

2. We now make the transition to the reflected complex derivative function

$$\mathbf{z}(w) := \begin{cases} i\hat{\mathbf{x}}_w(w), & w \in \overline{B_1(w_0)} \cap \mathbb{C}^+ \\ -i\overline{\hat{\mathbf{x}}_{\overline{w}}(\overline{w})}, & w \in \overline{B_1(w_0)} \cap \mathbb{C}^- \end{cases} \tag{12}$$

from Proposition 2.5*. This function is continuous in $\overline{B_1(w_0)}$ and satisfies the differential inequality

$$|\mathbf{z}_{\overline{w}}(w)| \leq a|\mathbf{z}(w)|^2 + \frac{b}{4\beta^2} \qquad \text{for all} \quad w \in B_1(w_0) \setminus \mathbb{R}. \tag{13}$$

The integral representation from Theorem 5.5 (Pompeiu, Vekua) in Section 5 of Chapter 4 is valid for $\mathbf{z}$ as well, and we have

$$\mathbf{z}(w_0) = \frac{1}{2\pi i} \int\limits_{\partial B_{\lambda\vartheta}(w_0)} \frac{\mathbf{z}(w)}{w - w_0} \, dw - \frac{1}{\pi} \iint\limits_{B_{\lambda\vartheta}(w_0)} \frac{\mathbf{z}_{\overline{w}}(w)}{w - w_0} \, du\, dv \tag{14}$$

with arbitrary parameters $\vartheta, \lambda \in (0,1)$. In the derivation of this formula we integrate separately on the half-planes $\mathbb{C}^\pm$: Since $\mathbf{z}$ is continuous on the real line $\mathbb{R}$, the curvilinear integrals on the real line annul each other. We use Proposition 2.5* and get a number $\lambda = \lambda(\vartheta) \in [\frac{1}{4}, \frac{1}{2}]$, such that Cauchy's integral of $\mathbf{z}$ can be estimated as follows:

$$\left| \frac{1}{2\pi i} \int\limits_{\partial B_{\lambda\vartheta}(w_0)} \frac{\mathbf{z}(w)}{w - w_0} \, dw \right| \le \frac{c_1(a,b,M)}{\vartheta\sqrt{-\log \vartheta}} + \frac{b}{8\beta^2} + \frac{a}{2}\vartheta \sup_{w \in B_\vartheta(w_0)} |\mathbf{z}(w)|^2 \quad (15)$$

with the constant

$$c_1(a,b,M) := \frac{8\sqrt{M^2 + \frac{b}{8\beta^2} M}}{\sqrt{\log 4}\sqrt{1 - aM}}.$$

Parallel to (32)*, we deduce the following estimate from the differential inequality (13):

$$\left| \frac{1}{\pi} \iint\limits_{B_{\lambda\vartheta}(w_0)} \frac{\mathbf{z}_{\overline{w}}(w)}{w - w_0} \, du\, dv \right| \le a\vartheta \sup_{w \in B_\vartheta(w_0)} |\mathbf{z}(w)|^2 + \frac{b}{4\beta^2}. \quad (16)$$

3. Noting that

$$|\mathbf{z}(w)| = |\hat{\mathbf{x}}_w(w)| = |\mathbf{x}_\zeta \circ g(w)||g'(w)|, \qquad w \in \mathbb{C}^+ \cup \mathbb{R},$$

the relation (7) yields the inequality

$$\beta |\mathbf{z}(w)| \le \phi(g(w)) \le \frac{1}{\beta}|\mathbf{z}(w)| \qquad \text{for all} \quad w \in \overline{B_\vartheta(w_0)} \cap \mathbb{C}^+. \quad (17)$$

We then obtain the following estimate from (14)-(16):

$$\phi(\zeta_0) = \phi(g(w_0)) \le \frac{1}{\beta}|\mathbf{z}(w_0)|$$

$$\le \frac{c_1(a,b,M)}{\beta\vartheta\sqrt{-\log\vartheta}} + \frac{3b}{8\beta^3} + \frac{3a}{2\beta}\vartheta \sup_{w \in B_\vartheta(w_0)} |\mathbf{z}(w)|^2$$

$$\le \frac{c_1(a,b,M)}{\beta\vartheta\sqrt{-\log\vartheta}} + \frac{3b}{8\beta^3} + \frac{3a}{2\beta^3}\vartheta \sup_{w \in \overline{B_\vartheta(w_0)}\cap\mathbb{C}^+} \phi(g(w))^2$$

$$\le \frac{c_1(a,b,M)}{\beta\vartheta\sqrt{-\log\vartheta}} + \frac{3b}{8\beta^3} + \frac{3a}{2\beta^3}\vartheta\phi(\zeta_0)^2.$$

Therefore, the subsequent inequality holds true

$$\phi(\zeta_0) \le \frac{c_1(a,b,M)}{\beta\vartheta\sqrt{-\log\vartheta}} + \frac{3b}{8\beta^3} + \frac{3a}{2\beta^3}\vartheta\phi(\zeta_0)^2 \qquad \text{for all} \quad 0 < \vartheta < 1. \quad (18)$$

4. Parallel to part 3 of the proof for Theorem 2.6*, we obtain the existence of a constant $C_1 = C_1(a, b, M)$ from (18), such that

$$\sup_{\zeta \in E} |\nabla \mathbf{x}(\zeta)| = 2 \sup_{\zeta \in E} \phi(\zeta) \leq C_1(a, b, M) \tag{19}$$

holds true. We apply the representation formula (14) - valid in $\overline{E}$ - to the function $\mathbf{x}_w$. Then we find a constant $C_2 = C_2(a, b, M, \alpha)$ for the number $\alpha \in (0, 1)$ given - as in the proof of Theorem 2.7* - such that

$$|\nabla \mathbf{x}(\zeta_1) - \nabla \mathbf{x}(\zeta_2)| \leq C_2 |\zeta_1 - \zeta_2|^\alpha \qquad \text{for all} \quad \zeta_1, \zeta_2 \in \overline{E} \tag{20}$$

is valid. The statement (8) can now be inferred from the inequalities (19) and (20).

q.e.d.

4 The Dirichlet Problem for Nonlinear Elliptic Systems

We choose the parameters $\alpha \in (0, 1)$ and $M \in (0, +\infty)$ and prescribe periodic boundary values - with the period 2π - on the boundary of the unit disc $B := \{w = u + iv : |w| < 1\}$, namely

$$\begin{aligned} &\mathbf{g} = \mathbf{g}(t) = (g_1(t), \ldots, g_n(t)) : \mathbb{R} \to \mathbb{R}^n \in C_{2\pi}^{2+\alpha}(\mathbb{R}, \mathbb{R}^n), \\ &|\mathbf{g}(t)| \leq M \qquad \text{for all} \quad t \in \mathbb{R}. \end{aligned} \tag{1}$$

Now we concentrate our interest on the Dirichlet problem

$$\begin{aligned} &\mathbf{x} = \mathbf{x}(u, v) = (x_1(u, v), \ldots, x_n(u, v)) \in C^{2+\alpha}(\overline{B}, \mathbb{R}^n), \\ &\Delta \mathbf{x}(u, v) = \mathbf{F}(u, v, \mathbf{x}(u, v), \nabla \mathbf{x}(u, v)) \qquad \text{for all} \quad (u, v) \in B, \\ &|\mathbf{x}(u, v)| \leq M \qquad \text{for all} \quad (u, v) \in B, \\ &\mathbf{x}(\cos t, \sin t) = \mathbf{g}(t) \qquad \text{for all} \quad t \in \mathbb{R}. \end{aligned} \tag{2}$$

As our right-hand side $\mathbf{F}$ we prescribe a homogeneous quadratic polynomial in the first derivatives

$$\nabla \mathbf{x}(u, v) = (x_{1u}(u, v), \ldots, x_{nu}(u, v), x_{1v}(u, v), \ldots, x_{nv}(u, v)).$$

The coefficients are assumed to depend Hölder-continuously on the variables u, v and Lipschitz-continuously on the vector $\mathbf{x}$, and we require that they vanish on the exterior space $|\mathbf{x}| \geq M$. More precisely, we define the function

$$\mathbf{F}(u, v, \mathbf{x}; \mathbf{p}, \mathbf{q}) = (F_1(\ldots), \ldots, F_n(\ldots)) : \overline{B} \times \mathbb{R}^n \times \mathbb{R}^{2n} \times \mathbb{R}^{2n} \to \mathbb{R}^n,$$

$$F_k(u, v, \mathbf{x}; \mathbf{p}, \mathbf{q}) := \sum_{i,j=1}^{2n} f_{ij}^k(u, v, \mathbf{x}) p_i q_j, \qquad k = 1, \ldots, n. \tag{3}$$

Here the coefficients fulfill

$$f_{ij}^k(w, \mathbf{x}) = 0 \qquad \text{for all} \quad w \in \overline{B} \quad \text{and} \quad \mathbf{x} \in \mathbb{R}^n \quad \text{with} \quad |\mathbf{x}| \geq M,$$

$$|f_{ij}^k(w, \mathbf{x})| \leq K \qquad \text{for all} \quad w \in \overline{B} \quad \text{and} \quad \mathbf{x} \in \mathbb{R}^n \quad \text{with} \quad |\mathbf{x}| \leq M,$$

$$|f_{ij}^k(w, \mathbf{x}) - f_{ij}^k(\tilde{w}, \tilde{\mathbf{x}})| \leq L\{|w - \tilde{w}|^\alpha + |\mathbf{x} - \tilde{\mathbf{x}}|\} \tag{4}$$

$$\text{for all} \quad w, \tilde{w} \in \overline{B} \quad \text{and} \quad \mathbf{x}, \tilde{\mathbf{x}} \in \mathbb{R}^n$$

for $i, j = 1, \ldots, 2n$ and $k = 1, \ldots, n$; where the constants $K, L \in [0 + \infty)$ are given. Finally, we use the following function $\mathbf{F}$ as the right-hand side in (2), namely

$$\mathbf{F}(u, v, \mathbf{x}, \mathbf{p}) := \mathbf{F}(u, v, \mathbf{x}; \mathbf{p}, \mathbf{p}), \qquad (u, v) \in \overline{B}, \quad \mathbf{x} \in \mathbb{R}^n, \quad \mathbf{p} \in \mathbb{R}^{2n}.$$

All elliptic systems appearing in differential geometry are of the following form:

$$\Delta \mathbf{x}(u, v) = \mathbf{F}(u, v, \mathbf{x}(u, v); \nabla \mathbf{x}(u, v), \nabla \mathbf{x}(u, v))$$

$$= \mathbf{F}(u, v, \mathbf{x}(u, v), \nabla \mathbf{x}(u, v)), \qquad (u, v) \in B. \tag{5}$$

Fixing $(u, v, \mathbf{x}) \in \overline{B} \times \mathbb{R}^n$, the subsequent mapping

$$(\mathbf{p}, \mathbf{q}) \mapsto \mathbf{F}(w, \mathbf{x}; \mathbf{p}, \mathbf{q}) \tag{6}$$

is bilinear but not necessarily symmetric.

Choosing a parameter $a \in [0, +\infty)$, we now require a *growth condition for the right-hand side* $\mathbf{F}$:

$$|\mathbf{F}(w, \mathbf{x}; \mathbf{p}, \mathbf{p})| \leq a|\mathbf{p}|^2 \text{or equivalently} \sqrt{\sum_{k=1}^n \left(\sum_{i,j=1}^{2n} f_{ij}^k(w, \mathbf{x}) p_i p_j \right)^2} \leq a \sum_{i=1}^{2n} p_i^2$$

$$\text{for all} \quad w \in \overline{B}, \ \mathbf{x} \in \mathbb{R}^n, \ \mathbf{p} = (p_1, \ldots, p_{2n}) \in \mathbb{R}^{2n}. \tag{7}$$

Remarks:

1. On the basis of (3) and (4) we can certainly find a constant a satisfying (7). One should optimize this constant, however. Though the constants K, L from (4) do not enter quantitatively into our later existence result, this is the case for the constant a.

2. If the condition $aM \leq 1$ is fulfilled, then a solution $\mathbf{x} = \mathbf{x}(u, v)$ of (2) is subject to the geometric maximum principle of E. Heinz

$$\sup_{(u,v) \in \overline{B}} |\mathbf{x}(u, v)| \leq \sup_{(u,v) \in \partial B} |\mathbf{x}(u, v)|. \tag{8}$$

In order to solve (2), we make the transition to zero boundary values. In this context we solve the following boundary value problem by potential-theoretic methods (compare Theorem 6.8 in Chapter 9, Section 6):

$$\mathbf{y} = \mathbf{y}(u, v) \in C^{2+\alpha}(\overline{B}, \mathbb{R}^n),$$

$$\Delta \mathbf{y}(u, v) = 0 \qquad \text{for all} \quad (u, v) \in B, \tag{9}$$

$$\mathbf{y}(\cos t, \sin t) = \mathbf{g}(t) \qquad \text{for all} \quad t \in \mathbb{R}.$$

The maximum principle for harmonic functions yields

$$\sup_{(u,v) \in B} |\mathbf{y}(u, v)| \leq M. \tag{10}$$

If $\mathbf{x}$ denotes a solution of (2), we then consider the difference function

$$\mathbf{z}(u, v) := \mathbf{x}(u, v) - \mathbf{y}(u, v), \qquad (u, v) \in \overline{B}, \tag{11}$$

which belongs to the space

$$C_*^{2+\alpha}(\overline{B}) := \left\{ \tilde{\mathbf{z}}(u, v) \in C^{2+\alpha}(\overline{B}, \mathbb{R}^n) \; : \; \tilde{\mathbf{z}}(u, v) = 0 \text{ auf } \partial B \right\}.$$

Now the function $\mathbf{z}$ satisfies the following differential equation:

$$\Delta \mathbf{z}(u, v) = \Delta \mathbf{x}(u, v) = \mathbf{F}(u, v, \mathbf{x}(u, v); \nabla \mathbf{x}(u, v), \nabla \mathbf{x}(u, v))$$

$$= \mathbf{F}(u, v, \mathbf{y}(u, v) + \mathbf{z}(u, v); \nabla \mathbf{y}(u, v) + \nabla \mathbf{z}(u, v), \nabla \mathbf{y}(u, v) + \nabla \mathbf{z}(u, v))$$

$$= \mathbf{F}(u, v, \mathbf{y}(u, v) + \mathbf{z}(u, v); \nabla \mathbf{z}(u, v), \nabla \mathbf{z}(u, v))$$

$$+ \mathbf{F}(u, v, \mathbf{y}(u, v) + \mathbf{z}(u, v); \nabla \mathbf{y}(u, v), \nabla \mathbf{z}(u, v))$$

$$+ \mathbf{F}(u, v, \mathbf{y}(u, v) + \mathbf{z}(u, v); \nabla \mathbf{z}(u, v), \nabla \mathbf{y}(u, v))$$

$$+ \mathbf{F}(u, v, \mathbf{y}(u, v) + \mathbf{z}(u, v); \nabla \mathbf{y}(u, v), \nabla \mathbf{y}(u, v))$$

$$=: \mathbf{G}(u, v, \mathbf{z}(u, v), \nabla \mathbf{z}(u, v)) \qquad \text{for all} \quad (u, v) \in B. \tag{12}$$

Therefore, the function $\mathbf{z}(u, v) \in C_*^{2+\alpha}(\overline{B})$ satisfies an inhomogeneous differential equation with quadratic growth in its gradient. We choose an arbitrary $\varepsilon > 0$, and with the aid of (7) we deduce the following inequality

$$|\Delta \mathbf{z}(u, v)| = |\mathbf{F}(u, v, \mathbf{y}(u, v) + \mathbf{z}(u, v), \nabla \mathbf{y}(u, v) + \nabla \mathbf{z}(u, v))|$$

$$\leq a|\nabla \mathbf{y}(u, v) + \nabla \mathbf{z}(u, v)|^2$$

$$\leq a \left\{ |\nabla \mathbf{y}(u, v)|^2 + 2 \frac{1}{\sqrt{\varepsilon}} |\nabla \mathbf{y}(u, v)| \sqrt{\varepsilon} |\nabla \mathbf{z}(u, v)| + |\nabla \mathbf{z}(u, v)|^2 \right\}$$

$$\leq a(1 + \varepsilon)|\nabla \mathbf{z}(u, v)|^2 + a \left(1 + \frac{1}{\varepsilon} \right) |\nabla \mathbf{y}(u, v)|^2$$

$$\leq a(1 + \varepsilon)|\nabla \mathbf{z}(u, v)|^2 + a \left(1 + \frac{1}{\varepsilon} \right) \sup_{(u,v) \in B} |\nabla \mathbf{y}(u, v)|^2 \tag{13}$$

for all $(u, v) \in B$. Very important is the subsequent

Proposition 4.1. (A priori estimate)

Let the constants $\alpha \in (0,1)$ and $a \in [0, +\infty)$, $M \in (0, +\infty)$ with $2aM < 1$ be chosen. Then we have an a-priori-constant $C_1(a, M, \alpha)$, such that all solutions of the problem

$$\mathbf{z} = \mathbf{z}(u, v) \in C^2(B) \cap C^1(\overline{B}),$$

$$\Delta \mathbf{z}(u, v) = \mathbf{G}(w, \mathbf{z}(w), \nabla \mathbf{z}(w)) \qquad for\ all \quad w \in B, \qquad (14)$$

$$\mathbf{z}(w) = 0 \qquad for\ all \quad w \in \partial B$$

satisfy the following estimate:

$$\|\mathbf{z}\|_{C^{1+\alpha}(\overline{B}, \mathbb{R}^n)} \leq C_1(a, M, \alpha). \qquad (15)$$

Proof:

1. At first, we verify the subsequent statement:

$$\sup_{w \in B} |\mathbf{z}(w)| \leq 2M. \qquad (16)$$

If this were violated, there would exist a point $w_0 \in B$ satisfying

$$2M < |\mathbf{z}(w_0)| \leq |\mathbf{y}(w_0) + \mathbf{z}(w_0)| + |\mathbf{y}(w_0)| \leq |\mathbf{y}(w_0) + \mathbf{z}(w_0)| + M$$

and consequently

$$M < |\mathbf{y}(w_0) + \mathbf{z}(w_0)|.$$

For continuity reasons we can find a disc $B_\varrho(w_0) \subset B$ such that

$$|\mathbf{y}(w) + \mathbf{z}(w)| \geq M \qquad for\ all \quad w \in B_\varrho(w_0). \qquad (17)$$

On account of the assumption (4) for the coefficients f_{ij}^k we infer

$$\Delta \mathbf{z}(w) = \mathbf{F}(w, \mathbf{y}(w) + \mathbf{z}(w); \nabla \mathbf{y}(w) + \nabla \mathbf{z}(w), \nabla \mathbf{y}(w) + \nabla \mathbf{z}(w))$$
$$= 0, \qquad w \in B_\varrho(w_0). \qquad (18)$$

We now consider the function

$$\phi(w) := |\mathbf{z}(w)|^2, \qquad w \in B_\varrho(w_0), \qquad (19)$$

which is subharmonic due to

$$\Delta \phi(w) = 2\big(|\nabla \mathbf{z}(w)|^2 + \mathbf{z}(w) \cdot \Delta \mathbf{z}(w)\big) = 2|\nabla \mathbf{z}(w)|^2 \geq 0 \qquad in \quad B_\varrho(w_0).$$

Choosing $w_0 \in B$ in such a way that

$$|\mathbf{z}(w_0)| = \sup_{w \in B} |\mathbf{z}(w)|$$

holds true, the subharmonic function $\phi(w)$, $w \in B_\varrho(w_0)$ attains its maximum at the interior point w_0. Therefore, we obtain

$$\phi(w) \equiv \phi(w_0) \quad \text{in} \quad \overline{B_\varrho(w_0)}. \tag{20}$$

A continuation argument finally yields

$$\phi(w) \equiv \phi(w_0) \quad \text{in} \quad \overline{B}$$

contradicting the statement $\phi(w) = 0$ on ∂B. Consequently, the relation (16) is satisfied.

2. Formula (13) gives us the differential inequality

$$|\Delta \mathbf{z}(u,v)| \le a(1+\varepsilon)|\nabla \mathbf{z}(u,v)|^2 + b(\varepsilon), \qquad (u,v) \in B, \tag{21}$$

where we have set

$$b(\varepsilon) := a\left(1 + \frac{1}{\varepsilon}\right) \sup_{(u,v)\in B} |\nabla \mathbf{y}(u,v)|^2.$$

We choose $\varepsilon > 0$ so small that $a(1+\varepsilon)\, 2M < 1$ is fulfilled, and Theorem 3.1 from Section 3 yields the following a priori estimate (15) on account of (16) and (21).

q.e.d.

We now transform (14) into an integral equation. We consider the Banach space

$$\mathcal{B} := \left\{ \mathbf{x} \in C^1(\overline{B}, \mathbb{R}^n) \ : \ \mathbf{x}(w) = 0 \text{ on } \partial B \right\}$$

endowed with the norm

$$\|\cdot\| := \|\cdot\|_{C^1(\overline{B},\mathbb{R}^n)},$$

and define the balls

$$\mathcal{B}_N := \left\{ \mathbf{x} \in \mathcal{B} \ : \ \|\mathbf{x}\| < N \right\}$$

with the radii $N > 0$. Taking $0 \le \lambda \le 1$ we investigate the nonlinear integral operators ($\zeta = \xi + i\eta$)

$$\mathbb{V}_\lambda(\mathbf{z})|_w := -\frac{\lambda}{2\pi} \iint_B \log\left|\frac{1 - \overline{w}\zeta}{\zeta - w}\right| \mathbf{G}(\zeta, \mathbf{z}(\zeta), \nabla\mathbf{z}(\zeta))\, d\xi\, d\eta, \qquad w \in B. \tag{22}$$

With the aid of the Leray-Schauder degree of mapping, we shall construct a solution of the nonlinear integral equation $\mathbf{z} = \mathbb{V}_1(\mathbf{z})$. The latter then solves (14), and by the transition (11) we obtain a solution of the problem (2). At first, we need the following

Proposition 4.2. *Green's operator*

$$u(w) \in C^0(\overline{B}) \quad \mapsto \quad \mathbb{L}(u)|_w := -\frac{1}{2\pi} \iint_B \log\left|\frac{1 - \overline{w}\zeta}{\zeta - w}\right| u(\zeta)\, d\xi\, d\eta, \quad w \in B,$$

$$\tag{23}$$

maps the space $C^0(\overline{B})$ *continuously to the space*

$$C_*^{1+\beta}(\overline{B}) := \left\{ v(w) \in C^{1+\beta}(\overline{B}) \;:\; v(w) = 0 \text{ for all } w \in \partial B \right\}$$

for each number $\beta \in (0,1)$. *Therefore, we have a constant* $C_2(\beta)$ *satisfying*

$$\|\mathbb{L}(u)\|_{C^{1+\beta}(\overline{B})} \leq C_2(\beta)\|u\|_{C^0(\overline{B})} \qquad \text{for all } u \in C^0(\overline{B}). \qquad (24)$$

Proof: One should utilize the potential-theoretic estimates from Chapter 9, Section 4 and Theorem 4.12 (Hadamard's estimate) from Section 4 in Chapter 4 for the complex derivative $\frac{\partial}{\partial w} L(u)$.

q.e.d.

Proposition 4.3. *Let the number* $\beta \in (0,1)$ *be chosen arbitrarily. Then the nonlinear integral operator* $\mathbb{V}_\lambda : \mathcal{B} \to C_*^{1+\beta}(\overline{B}, \mathbb{R}^n)$ *is continuous and as the operator* $\mathbb{V}_\lambda : \mathcal{B} \to \mathcal{B}$ *even completely continuous for all* $0 \leq \lambda \leq 1$.

Proof: We observe the following connection for all $0 \leq \lambda \leq 1$, namely

$$\mathbb{V}_\lambda(\mathbf{z}) = \lambda \mathbb{L}(\mathbf{G}(\cdot, \mathbf{z}(\cdot), \nabla \mathbf{z}(\cdot))), \qquad \mathbf{z} \in \mathcal{B}. \qquad (25)$$

On account of (4), the function

$$\mathbf{F}(\cdot, \mathbf{y} + \mathbf{z}, \nabla \mathbf{y} + \nabla \mathbf{z}, \nabla \mathbf{y} + \nabla \mathbf{z}) = \mathbf{G}(\cdot, \mathbf{z}, \nabla \mathbf{z}), \qquad \mathbf{z} \in \mathcal{B}_N \qquad (26)$$

satisfies a Lipschitz condition in the ball $\mathcal{B}_N$ with an arbitrary radius $N > 0$ in all three components, where the Lipschitz constant may depend on N. Therefore, we have a constant $C_3 = C_3(K, L, N)$ satisfying

$$\|\mathbf{G}(\cdot, \mathbf{z}, \nabla \mathbf{z}) - \mathbf{G}(\cdot, \tilde{\mathbf{z}}, \nabla \tilde{\mathbf{z}})\|_{C^0(\overline{B})} \leq C_3(K, L, N)\|\mathbf{z} - \tilde{\mathbf{z}}\|$$
$$\text{for all } \mathbf{z}, \tilde{\mathbf{z}} \in \mathcal{B}_N. \qquad (27)$$

Proposition 4.2 now yields

$$\|\mathbb{V}_\lambda(\mathbf{z}) - \mathbb{V}_\lambda(\tilde{\mathbf{z}})\|_{C^{1+\beta}(\overline{B})} \leq \lambda C_2(\beta) C_3(K, L, N)\|\mathbf{z} - \tilde{\mathbf{z}}\|$$
$$\text{for all } \mathbf{z}, \tilde{\mathbf{z}} \in \mathcal{B}_N. \qquad (28)$$

Consequently, the operator $\mathbb{V}_\lambda : \mathcal{B}_N \to C_*^{1+\beta}(\overline{B})$ is continuous. Furthermore, we infer the following estimate from Proposition 4.2 on account of (26) and (4), namely

$$\|\mathbb{V}_\lambda(\mathbf{z})\|_{C^{1+\beta}(\overline{B})} \leq \lambda C_2(\beta)\|\mathbf{G}(\cdot, \mathbf{z}, \nabla \mathbf{z})\|_{C^0(\overline{B})}$$
$$\leq C_4(K, N, \beta), \qquad \mathbf{z} \in \mathcal{B}_N. \qquad (29)$$

Therefore, the operator $\mathbb{V}_\lambda : \mathcal{B} \to \mathcal{B}$ is completely continuous.   q.e.d.

With the aid of topological methods we now prove the following

Theorem 4.4. *Let the constants $\alpha \in (0,1)$ and $a \in [0,+\infty)$, $M \in (0,+\infty)$ with $aM < \frac{1}{2}$ be chosen. Furthermore, let the boundary values $\mathbf{g}$ from (1) be prescribed, and let the right-hand side $\mathbf{F}$ be defined as in (3) satisfying (4) and the growth condition (7). Then we have a solution $\mathbf{x} = \mathbf{x}(u,v)$ of the Dirichlet problem (2).*

Proof: We choose the quantity $N := C_1(a, M, \alpha) + 1$ with the constant C_1 from Proposition 4.1 as radius for the ball $\mathcal{B}_N$ in the Banach space $\mathcal{B}$. We consider the family of operators

$$\mathrm{Id} - \mathbb{V}_\lambda : \mathcal{B}_N \to \mathcal{B}, \quad \mathbf{z} \mapsto \mathbf{z} - \mathbb{V}_\lambda(\mathbf{z}), \quad 0 \leq \lambda \leq 1. \tag{30}$$

For $\lambda = 0$ the mapping possesses a zero, namely $\mathbf{z} = \mathbf{0} \in \mathcal{B}$. On account of Proposition 4.3 the operator $\mathbb{V}_\lambda : \mathcal{B}_N \to \mathcal{B}$ is completely continuous for each $\lambda \in [0,1]$. Furthermore, the family $\mathbb{V}_\lambda$ depends continuously on the parameter $\lambda \in [0,1]$. We now show that the statement

$$(\mathrm{Id} - \mathbb{V}_\lambda)(\mathbf{z}) \neq \mathbf{0} \quad \text{for all} \quad \mathbf{z} \in \partial\mathcal{B}_N \quad \text{and all} \quad \lambda \in [0,1] \tag{31}$$

is correct. If $\mathbf{z} \in \partial\mathcal{B}_N$ namely would be a zero of $\mathrm{Id} - \mathbb{V}_\lambda$ with a parameter $\lambda \in [0,1]$, we infer

$$\mathbf{z} = \mathbb{V}_\lambda(\mathbf{z}). \tag{32}$$

The solution of this integral equation represents a solution of the Dirichlet problem

$$\begin{aligned}
&\mathbf{z} = \mathbf{z}(u,v) \in C^2(B) \cap C^1(\overline{B}), \\
&\Delta\mathbf{z}(u,v) = \lambda\mathbf{G}(u,v,\mathbf{z}(u,v),\nabla\mathbf{z}(u,v)), \quad (u,v) \in B, \\
&\mathbf{z}(u,v) = 0, \quad (u,v) \in \partial B.
\end{aligned} \tag{33}$$

In (14) we replace $\mathbf{G}(\ldots)$ by $\lambda\mathbf{G}(\ldots)$, and Proposition 4.1 yields an evident contradiction with the inequality

$$\|\mathbf{z}\|_{C^{1+\alpha}(\overline{B},\mathbb{R}^n)} \leq C_1(a,M,\alpha) = N - 1 < N = \|\mathbf{z}\|_{C^1(\overline{B},\mathbb{R}^n)}.$$

Therefore, the relation (31) is fulfilled. Because of the Leray-Schauder theorem (compare Chapter 7, Section 3) the mapping (30) possesses at least one zero $\mathbf{z} = \mathbf{z}(w)$ for each parameter $\lambda \in [0,1]$. Specialized on the parameter $\lambda = 1$, this zero solves the Dirichlet problem (14). Theorem 4.4 from Chapter 9, Section 4 now implies $\mathbf{z} \in C_*^{2+\alpha}(\overline{B},\mathbb{R}^n)$. If $\mathbf{y} = \mathbf{y}(w)$ represents the solution of (9), we obtain a solution of (2) with $\mathbf{x}(u,v) = \mathbf{y}(u,v) + \mathbf{z}(u,v)$, $(u,v) \in \overline{B}$. The property

$$\sup_{(u,v)\in B} |\mathbf{x}(u,v)| \leq M$$

is easily established as in part 1 of the proof for Proposition 4.1.   q.e.d.

We now specialize our result to the H-surface system from Section 1. Taking the boundary values $\mathbf{g}(t)$ from (1) in the case $n = 3$, we consider the Dirichlet problem

$$\mathbf{x} = \mathbf{x}(u,v) = (x_1(u,v), x_2(u,v), x_3(u,v)) \in C^{2+\alpha}(\overline{B}, \mathbb{R}^3),$$

$$\Delta\mathbf{x}(u,v) = 2H(u,v,\mathbf{x}(u,v))\mathbf{x}_u \wedge \mathbf{x}_v(u,v) \quad \text{in} \quad B,$$

$$|\mathbf{x}(u,v)| \le M \quad \text{in} \quad B,$$

$$\mathbf{x}(\cos t, \sin t) = \mathbf{g}(t) \quad \text{for} \quad t \in \mathbb{R}.$$

(34)

Here we prescribe the function $H = H(w, \mathbf{x})$ as follows:

$$H = H(w, \mathbf{x}) : \overline{B} \times \mathbb{R}^3 \to \mathbb{R} \in C^\alpha(\overline{B} \times \mathbb{R}^3) \quad \text{with}$$

$$|H(w, \mathbf{x})| \le h_0, \ |H(w, \mathbf{x}) - H(w, \mathbf{y})| \le h_1|\mathbf{x} - \mathbf{y}|, \quad w \in \overline{B}, \quad \mathbf{x}, \mathbf{y} \in \mathbb{R}^3,$$

$$H(w, \mathbf{x}) = 0, \quad w \in \overline{B}, \quad \mathbf{x} \in \mathbb{R}^3 \quad \text{with} \quad |\mathbf{x}| \ge M.$$

(35)

When we set

$$\mathbf{F}(u, v, \mathbf{x}(u,v), \nabla\mathbf{x}(u,v)) := 2H(u,v,\mathbf{x}(u,v))\mathbf{x}_u \wedge \mathbf{x}_v(u,v),$$

the right-hand side (3) appears in the form

$$\mathbf{F}(u, v, \mathbf{x}; \mathbf{p}, \mathbf{q}) := 2H(w, \mathbf{x})\mathbf{p}' \wedge \mathbf{q}'' \quad \text{with} \quad \mathbf{p}, \mathbf{q} \in \mathbb{R}^6 \quad \text{and}$$

$$\mathbf{p} = (\mathbf{p}', \mathbf{p}'') = (p_1', p_2', p_3', p_1'', p_2'', p_3''), \quad \mathbf{q} = (\mathbf{q}', \mathbf{q}'') = (q_1', q_2', q_3', q_1'', q_2'', q_3'').$$

(36)

We then have the growth condition

$$|\mathbf{F}(w, \mathbf{x}; \mathbf{p}, \mathbf{p})| \le 2|H(w, \mathbf{x})||\mathbf{p}' \wedge \mathbf{p}''| \le h_0(|\mathbf{p}'|^2 + |\mathbf{p}''|^2) = h_0|\mathbf{p}|^2$$

$$\text{for all} \quad w \in \overline{B}, \quad \mathbf{x} \in \mathbb{R}^3, \quad \mathbf{p} \in \mathbb{R}^6.$$

(37)

Theorem 4.4 immediately implies the following

Theorem 4.5. (E. Heinz, H. Werner, S. Hildebrandt)
In the case $h_0 M < \frac{1}{2}$ the Dirichlet problem (34) possesses a solution, with the boundary values (1) and the right-hand side (35).

Remarks:

1. E. Heinz solved the Dirichlet problem (34) for the case $H \equiv \text{const}$ in 1954 by means of the topological method presented here.
2. H. Werner has attained the condition $h_0 M < \frac{1}{2}$.
3. With the aid of variational methods, S. Hildebrandt has solved the Dirichlet problem (34) even in the case $H = H(\mathbf{x})$ and $h_0 M < 1$.
4. Due to Jäger's maximum principle from Section 1, the Dirichlet problem (34) is uniquely solvable in a ball of the radius

$$M := \frac{\sqrt{h_0^2 + 2h_1}}{h_0^2 + h_1}.$$

Therefore, Theorem 4.5 yields an existence result for large h_1 without answering the uniqueness question.

5. According to Section 1, Theorem 1.5 and its corollary, the Dirichlet problem (34) is stable with respect to perturbations of the boundary values in the $C^0(\overline{B}, \mathbb{R}^3)$-norm under the conditions given there. Consequently, we can even solve the Dirichlet problem (34) for continuous boundary values.

We finally note the following

Theorem 4.6. *In the case $H(w, \mathbf{x}) \equiv h_0$ or $H(w, \mathbf{x}) \equiv -h_0$ with $h_0 > 0$ and $h_0 M \leq \frac{1}{2}$, the Dirichlet problem (34) possesses exactly one solution of the regularity class $C^{2+\alpha}(B, \mathbb{R}^3) \cap C^0(\overline{B}, \mathbb{R}^3)$, for the continuous boundary values $\mathbf{g} = \mathbf{g}(t) \in C^0_{2\pi}(\mathbb{R}, \mathbb{R}^n)$ satisfying $|\mathbf{g}(t)| \leq M$, $t \in \mathbb{R}$.*

Proof: We smooth the constant function H at the boundary of the ball $|\mathbf{x}| \leq M$ in such a way that H vanishes for all $|\mathbf{x}| \geq M$. Then we solve (34) for $C^{2+\alpha}$-boundary values and approximate the continuous boundary values $\mathbf{g}$ uniformly with the aid of Theorem 1.5 from Section 1 and Theorem 2.7 from Section 2.

q.e.d.

5 Distortion Estimates for Plane Elliptic Systems

We begin with the important

Theorem 5.1. *Let the radius $R > 0$ be given, and we consider the disc*

$$B_R := \{w = u + iv \in \mathbb{C} : |w| < R\}$$

and the pseudoholomorphic function $f(w) : B_R \to \mathbb{C} \in C^1(B_R, \mathbb{C})$ satisfying

$$|f_{\overline{w}}(w)| \leq M|f(w)|, \qquad w \in B_R, \tag{1}$$

with a constant $M \in [0, +\infty)$. Furthermore, there exists a constant $K \in (0, +\infty)$ such that

$$0 < |f(w)| \leq K, \qquad w \in B_R \tag{2}$$

is fulfilled. Finally, we choose the number $r \in (0, R)$. Then we have the following inequalities for all points $w \in \overline{B}_r$, namely

$$|f(w)| \leq K^{\frac{2r}{R+r}} e^{8MR} |f(0)|^{\frac{R-r}{R+r}} \tag{3}$$

and

$$|f(w)| \geq K^{-\frac{2r}{R-r}} e^{-\frac{8MR(R+r)}{R-r}} |f(0)|^{\frac{R+r}{R-r}}. \tag{4}$$

Proof:

1. The inequality (3) can be transformed into

$$\left|\frac{f(w)}{K}\right| \leq e^{8MR} \left|\frac{f(0)}{K}\right|^{\frac{R-r}{R+r}}, \qquad w \in \overline{B}_r,$$

and (4) is equivalent to

$$\left|\frac{f(w)}{K}\right| \geq e^{-\frac{8MR(R+r)}{R-r}} \left|\frac{f(0)}{K}\right|^{\frac{R+r}{R-r}}, \qquad w \in \overline{B_r}.$$

With $f(w)$ the function $\dfrac{f(w)}{K}$ satisfies the inequality (1) as well. Therefore, is suffices to verify the estimates (3) and (4) only for the case $K = 1$.

2. We define the potential

$$a(w) := \frac{f_{\overline{w}}(w)}{f(w)}, \qquad w \in B_R, \tag{5}$$

and note

$$\|a\|_\infty := \sup_{w \in B_R} |a(w)| \leq M < +\infty.$$

Consequently, the function f satisfies the differential equation

$$\frac{d}{d\overline{w}} f(w) = a(w)f(w), \qquad w \in B_R \tag{6}$$

and is pseudoholomorphic in the sense of Definition 6.1 from Chapter 4, Section 6. From the similarity principle of Bers and Vekua, given there, we infer the representation formula

$$f(w) = e^{\psi(w)} \phi(w), \qquad w \in B_R, \tag{7}$$

with the function ϕ being holomorphic in B_R and the following integral $(\zeta = \xi + i\eta)$

$$\psi(w) := -\frac{1}{\pi} \iint_{B_R} \frac{a(\zeta)}{\zeta - w} \, d\xi \, d\eta, \qquad w \in B_R. \tag{8}$$

We note

$$|\psi(w)| \leq \frac{M}{\pi} \iint_{B_R} \frac{1}{|\zeta - w|} \, d\xi \, d\eta \leq \frac{M}{\pi} 2\pi \, 2R = 4MR, \qquad w \in B_R$$

and obtain

$$e^{-4MR} \leq |e^{\psi(w)}| \leq e^{4MR}, \qquad w \in B_R. \tag{9}$$

Together with (2) and (7), we deduce

$$0 < |\phi(w)| = |e^{-\psi(w)}| |f(w)| \leq e^{4MR}, \qquad w \in B_R. \tag{10}$$

3. We consider the nonnegative harmonic function

$$\chi(w) := 4MR - \log |\phi(w)| \geq 0, \qquad w \in B_R.$$

Theorem 2.5 (Harnack's inequality) from Section 2 in Chapter 5 yields the estimate

$$\frac{R-r}{R+r}\chi(0) \leq \chi(w) \leq \frac{R+r}{R-r}\chi(0), \qquad w \in B_r \tag{11}$$

for all $r \in (0, R)$. We rewrite this inequality into the form

$$\log|\phi(w)| \leq 4MR - \frac{R-r}{R+r}\big(4MR - \log|\phi(0)|\big)$$

$$= \frac{R-r}{R+r}\log|\phi(0)| + \frac{8MRr}{R+r}, \qquad w \in B_r \tag{12}$$

and

$$\log|\phi(w)| \geq 4MR - \frac{R+r}{R-r}\big(4MR - \log|\phi(0)|\big)$$

$$= \frac{R+r}{R-r}\log|\phi(0)| - \frac{8MRr}{R-r}, \qquad w \in B_r, \tag{13}$$

respectively. Via exponentiation we arrive at

$$|\phi(w)| \leq e^{\frac{8MRr}{R+r}}|\phi(0)|^{\frac{R-r}{R+r}}, \qquad w \in B_r \tag{14}$$

and

$$|\phi(w)| \geq e^{-\frac{8MRr}{R-r}}|\phi(0)|^{\frac{R+r}{R-r}}, \qquad w \in B_r. \tag{15}$$

4. From (7) and (9) we infer

$$e^{-4MR}|\phi(w)| \leq |f(w)| \leq e^{4MR}|\phi(w)|, \qquad w \in B_R. \tag{16}$$

Together with (14) we obtain

$$|f(w)| \leq e^{4MR}|\phi(w)| \leq e^{4MR}e^{\frac{8MRr}{R+r}}|\phi(0)|^{\frac{R-r}{R+r}}$$

$$\leq e^{4MR}e^{\frac{8MRr}{R+r}}e^{4MR\frac{R-r}{R+r}}|f(0)|^{\frac{R-r}{R+r}}$$

$$= e^{8MR}|f(0)|^{\frac{R-r}{R+r}}, \qquad w \in B_r$$

and finally the inequality (3) stated. Correspondingly, the relations (16) and (15) imply

$$|f(w)| \geq e^{-4MR}|\phi(w)| \geq e^{-4MR}e^{-\frac{8MRr}{R-r}}|\phi(0)|^{\frac{R+r}{R-r}}$$

$$\geq e^{-4MR}e^{-\frac{8MRr}{R-r}}e^{-4MR\frac{R+r}{R-r}}|f(0)|^{\frac{R+r}{R-r}}$$

$$= e^{-8MR\frac{R+r}{R-r}}|f(0)|^{\frac{R+r}{R-r}}, \qquad w \in B_r,$$

and we get (4).   q.e.d.

Theorem 5.2. (Heinz's inequality)
We take the unit disc $B := \{w = u + iv \in \mathbb{C} : |w| < 1\}$ and consider the plane mapping $\mathbf{z}(u,v) = (x(u,v), y(u,v)) \in C^2(B, \mathbb{R}^2)$. The latter may satisfy the differential inequality

$$|\Delta\mathbf{z}(u,v)| \leq a|\nabla\mathbf{z}(u,v)|^2 + b|\nabla\mathbf{z}(u,v)| \quad in \quad B \tag{17}$$

with the constants $a, b \in [0, +\infty)$, may be subject to the smallness condition

$$|\mathbf{z}(u,v)| \leq m \quad in \quad B \tag{18}$$

with a constant $m \in (0, +\infty)$, and it is positive-oriented as follows:

$$J_{\mathbf{z}}(u,v) := \frac{\partial(x,y)}{\partial(u,v)} > 0 \quad for\ all \quad (u,v) \in B. \tag{19}$$

Finally, we require the condition $am < 1$. Then there exist constants $C^{\pm}(a,b,m,r) > 0$ for each number $r \in (0,1)$, such that

$$C^-(a,b,m,r)|\nabla\mathbf{z}(0)|^{\frac{1+3r}{1-r}} \leq |\nabla\mathbf{z}(w)| \leq C^+(a,b,m,r)|\nabla\mathbf{z}(0)|^{\frac{1-r}{1+3r}}, \ w \in \overline{B_r}. \tag{20}$$

Proof:

1. We take the parameter $\lambda \in (0, +\infty)$ and deduce the following estimate from (17):

$$|\Delta\mathbf{z}(u,v)| \leq a|\nabla\mathbf{z}(u,v)|^2 + 2\lambda|\nabla\mathbf{z}(u,v)|\frac{b}{2\lambda}$$
$$\leq (a + \lambda^2)|\nabla\mathbf{z}(u,v)|^2 + \frac{b^2}{4\lambda^2} \quad in \quad B. \tag{21}$$

Then we choose $\lambda = \lambda(a,m) > 0$ so small that $(a + \lambda^2)m < 1$ holds true. Theorem 2.6 from Section 2 gives us the following estimate in the disc B_R of radius $R := \frac{1+r}{2} \in (r,1)$:

$$|\nabla\mathbf{z}(u,v)| \leq C_1(a,b,m,r), \quad w \in B_R. \tag{22}$$

We obtain the linear differential inequality when we insert into (17), namely

$$|\Delta\mathbf{z}(u,v)| \leq \big(aC_1(a,b,m,r) + b\big)|\nabla\mathbf{z}(u,v)|$$
$$= C_2(a,b,m,r)|\nabla\mathbf{z}(u,v)| \quad in \quad \overline{B_R}. \tag{23}$$

2. We utilize the auxiliary function $f(w) := x_w(w) + iy_w(w) : B \to \mathbb{C}$ and calculate

$$|f(w)|^2 = f(w)\overline{f(w)} = (x_w + iy_w)(x_{\overline{w}} - iy_{\overline{w}})$$
$$= |x_w|^2 + |y_w|^2 - i(x_w y_{\overline{w}} - x_{\overline{w}} y_w)$$
$$= \frac{1}{4}|\nabla\mathbf{z}(w)|^2 - \frac{i}{4}\big\{(x_u - ix_v)(y_u + iy_v) - (x_u + ix_v)(y_u - iy_v)\big\}$$
$$= \frac{1}{4}|\nabla\mathbf{z}(w)|^2 + \frac{1}{2}\frac{\partial(x,y)}{\partial(u,v)} \quad in \quad B.$$

On account of (19), we infer

$$\frac{1}{2}|\nabla \mathbf{z}(w)| < |f(w)| \le \frac{\sqrt{2}}{2}|\nabla \mathbf{z}(w)|, \qquad w \in B. \tag{24}$$

3. The relations (22)-(24) imply the inequalities

$$|f_{\overline{w}}(w)| = \frac{1}{4}|\Delta x(w) + i\Delta y(w)| = \frac{1}{4}|\Delta \mathbf{z}(w)|$$

$$\le \frac{1}{4}C_2(a, b, m, r)|\nabla \mathbf{z}(w)| \tag{25}$$

$$\le \frac{1}{2}C_2(a, b, m, r)|f(w)| \qquad \text{in} \quad B_R$$

and

$$0 < |f(w)| \le \frac{\sqrt{2}}{2}|\nabla \mathbf{z}(w)| \le \frac{\sqrt{2}}{2}C_1(a, b, m, r) \qquad \text{in } B_R. \tag{26}$$

Therefore, the function $f(w)$ is pseudoholomorphic in B_R with the constants

$$M = M(a, b, m, r) := \frac{1}{2}C_2(a, b, m, r),$$

$$K = K(a, b, m, r) := \frac{\sqrt{2}}{2}C_1(a, b, m, r).$$

With the aid of the identities $\frac{R-r}{R+r} = \frac{1-r}{1+3r}$ and $\frac{R+r}{R-r} = \frac{1+3r}{1-r}$, Theorem 5.1 yields the estimate

$$K^{-\frac{2r}{R-r}}e^{-\frac{8MR(R+r)}{R-r}}|f(0)|^{\frac{1+3r}{1-r}} \le |f(w)| \le K^{\frac{2r}{R+r}}e^{8MR}|f(0)|^{\frac{1-r}{1+3r}} \tag{27}$$

in $\overline{B}_r$. Taking (24) into account, we then find the inequality (20) with the a-priori-constants $C^{\pm}(a, b, m, r) > 0$.

$$\text{q.e.d.}$$

The following class of mappings is particularly important for problems in differential geometry:

Definition 5.3. *With the constants $a, b \in [0, +\infty)$ and $N \in (0, +\infty]$ being prescribed, we denote the following class of mappings by the symbol $\Gamma(B, a, b, N)$:*

i) *The function $\mathbf{z}(w) = (x(u, v), y(u, v)) : \overline{B} \to \mathbb{R}^2 \in C^2(B) \cap C^0(\overline{B})$ maps the circumference ∂B topologically and in a positive-oriented way onto the circular line ∂B;*

ii) *The mapping $\mathbf{z}$ is origin-preserving which means $\mathbf{z}(0) = (0, 0)$;*

iii) *We have the condition*

$$J_{\mathbf{z}}(w) = \frac{\partial(x, y)}{\partial(u, v)} > 0 \qquad \text{for all} \quad w = u + iv \in B;$$

iv) The function $\mathbf{z}$ satisfies the differential inequality

$$|\Delta \mathbf{z}(u,v)| \le a|\nabla \mathbf{z}(u,v)|^2 + b|\nabla \mathbf{z}(u,v)| \qquad in \quad B;$$

v) Dirichlet's integral of $\mathbf{z}$ fulfills

$$D(\mathbf{z}) := \iint\limits_B \left(|\mathbf{z}_u(u,v)|^2 + |\mathbf{z}_v(u,v)|^2 \right) du\, dv \le N.$$

Remarks:

1. With the aid of the index-sum formula we easily see that the mapping $\mathbf{z} : \overline{B} \to \overline{B}$ is topological.
2. In the special case $N = +\infty$, we do not require a bound on Dirichlet's integral $D(\mathbf{z})$.
3. E. Heinz has studied this class of mappings and applied it to differential-geometric problems.
4. Linear systems - appearing in the special case $a = 0$ - have already been considered by P. Berg.

We now prove the profound

Theorem 5.4. (Distortion estimate of E. Heinz)

Let the parameters $a, b \in [0, +\infty)$, $N \in (0, +\infty)$, and $r \in (0, 1)$ be chosen. Then we have constants $0 < \Theta(a, b, N, r) \le \Lambda(a, b, N, r) < +\infty$, such that each mapping $\mathbf{z} = \mathbf{z}(w) \in \Gamma(B, a, b, N)$ satisfies the inequality

$$\Theta(a,b,N,r) \le |\nabla \mathbf{z}(w)| \le \Lambda(a,b,N,r) \qquad for\ all \quad w \in \overline{B_r}. \tag{28}$$

Furthermore, the modulus of continuity for the mappings on $\overline{B}$ is estimated according to the formula (29) given below.

Proof:

1. At first, we show the *intermediate statement:* For all functions $\mathbf{z} = \mathbf{z}(w) \in \Gamma(B, a, b, N)$ and all numbers $\delta \in (0, \frac{1}{4})$ we have the estimate

$$|\mathbf{z}(w_1) - \mathbf{z}(w_0)| \le 4\sqrt{\frac{\pi N}{\log \frac{1}{\delta}}} \tag{29}$$

   for all $w_0, w_1 \in \overline{B}$ with $|w_0 - w_1| \le \delta$.

   We assume

$$4\sqrt{\frac{\pi N}{\log \frac{1}{\delta}}} < 2$$

   without loss of generality, since (29) would be trivial otherwise. We choose an arbitrary point $w_0 \in \overline{B}$. Via the Courant-Lebesgue oscillation lemma we find a number $\delta^* \in [\delta, \sqrt{\delta}]$ such that

$$\int_{\substack{w \in B \\ |w - w_0| = \delta^*}} |d\mathbf{z}(w)| \leq 2\sqrt{\frac{\pi N}{\log \frac{1}{\delta}}} \tag{30}$$

holds true. We define the following sets

$$\Omega := \left\{ w \in \overline{B} : |w - w_0| \leq \delta^* \right\}, \qquad \gamma := \left\{ w \in \overline{B} : |w - w_0| = \delta^* \right\}$$

and their topological images $\hat{\Omega} := \mathbf{z}(\Omega)$, $\hat{\gamma} := \mathbf{z}(\gamma)$. Now we distinguish between the following cases:

Case a: $\Omega \subset B$. Then we infer $\partial \hat{\Omega} = \hat{\gamma}$, and the length of $\hat{\gamma}$ satisfies

$$L(\hat{\gamma}) \leq 2\sqrt{\frac{\pi N}{\log \frac{1}{\delta}}}$$

on account of (30). Since the mapping $\mathbf{z}$ is topological, we obtain

$$|\mathbf{z}(w_1) - \mathbf{z}(w_0)| \leq 2\sqrt{\frac{\pi N}{\log \frac{1}{\delta}}} \qquad \text{for all} \quad w_1 \in \Omega. \tag{31}$$

Case b: $\partial \Omega \cap \partial B \neq \emptyset$. Then we have a point $\hat{z} \in \hat{\gamma} \cap \partial B$, and the relation (30) yields

$$\hat{\gamma} \subset K := \left\{ \zeta \in \mathbb{C} : |\zeta - \hat{z}| \leq 2\sqrt{\frac{\pi N}{\log \frac{1}{\delta}}} \right\}.$$

On account of $|\hat{z}| = 1$ and

$$2\sqrt{\frac{\pi N}{\log \frac{1}{\delta}}} < 1$$

the statement $0 \notin K$ holds true; and due to $\delta^* \leq \sqrt{\delta} < \frac{1}{2}$ and $\partial \Omega \cap \partial B \neq \emptyset$ we have $0 \notin \Omega$. Since the mapping $\mathbf{z} : \overline{B} \to \overline{B}$ is topological and origin-preserving, the relation $\hat{\gamma} \subset K$ implies the inclusion $\hat{\Omega} \subset K$. We obtain the following estimate for all points $w_1 \in \Omega$, namely

$$|\mathbf{z}(w_1) - \mathbf{z}(w_0)| \leq |\mathbf{z}(w_1) - \hat{z}| + |\hat{z} - \mathbf{z}(w_0)| \leq 4\sqrt{\frac{\pi N}{\log \frac{1}{\delta}}}. \tag{32}$$

When we additionally note $\delta \leq \delta^*$, the relations (31) and (32) yield the proof of the intermediate statement (29).

2. The function $\mathbf{z} = \mathbf{z}(w) \in \Gamma(B, a, b, N)$ satisfies the differential inequality

$$|\Delta \mathbf{z}(w)| \leq a|\nabla \mathbf{z}(w)|^2 + b|\nabla \mathbf{z}(w)| \leq (a+1)|\nabla \mathbf{z}(w)|^2 + \frac{b^2}{4} \qquad \text{in} \quad B. \tag{33}$$

We now choose the number $r \in (0,1)$ so large that the quantity $\delta := \frac{1-r}{2} > 0$ satisfies both conditions $\delta \in (0, \frac{1}{4})$ and

$$(a+1)\, 4\sqrt{\frac{\pi N}{\log \frac{1}{\delta}}} \leq \frac{1}{2}. \tag{34}$$

We consider an arbitrary point $\tilde{w} \in \overline{B_{1-\delta}} = \overline{B_{\frac{r+1}{2}}}$ and associate the auxiliary function

$$\mathbf{x}(w) := \mathbf{z}(w) - \mathbf{z}(\tilde{w}), \qquad w \in \Omega := \left\{ w \in \overline{B} : |w - \tilde{w}| \leq \delta \right\}. \tag{35}$$

On account of (29) and (33), we then have

$$|\Delta \mathbf{x}(w)| \leq (a+1)|\nabla \mathbf{x}(w)|^2 + \frac{b^2}{4} \qquad \text{in} \quad \overset{\circ}{\Omega},$$

$$\sup_{w \in \Omega} |\mathbf{x}(w)| \leq 4\sqrt{\frac{\pi N}{\log \frac{1}{\delta}}}. \tag{36}$$

When we additionally note (34), the gradient estimate of E. Heinz from Section 2, Theorem 2.6 yields the inequality

$$|\nabla \mathbf{z}(\tilde{w})| = |\nabla \mathbf{x}(\tilde{w})| \leq \tilde{\Lambda}(a, b, N, \delta)$$

$$=: \Lambda(a, b, N, r) \qquad \text{for all} \quad \tilde{w} \in \overline{B_{\frac{r+1}{2}}}. \tag{37}$$

Therefore, we have obtained the estimate (28) from above.

3. We now choose the number $r \in (0,1)$ so large that the quantity $\delta = \frac{1-r}{2}$ satisfies the conditions $\delta \in (0, \frac{1}{8})$ and

$$4\sqrt{\frac{\pi N}{\log \frac{1}{2\delta}}} \leq \frac{1}{2}$$

besides (34). From (29) we deduce

$$|\mathbf{z}(w)| \geq \frac{1}{2} \qquad \text{for all} \quad w \in \mathbb{C} \quad \text{with} \quad r = 1 - 2\delta \leq |w| \leq 1. \tag{38}$$

We now consider the curve $\mathbf{y}(t) := \mathbf{z}(tw_0)$, $0 \leq t \leq 1$ associated with a point $w_0 \in \partial B_r$. Then we calculate

$$\frac{1}{2} \leq |\mathbf{z}(w_0)| = |\mathbf{z}(w_0) - \mathbf{z}(0)| \leq \int_0^1 |\mathbf{y}'(t)|\, dt = |\mathbf{y}'(t_0)| \leq |\nabla \mathbf{z}(t_0 w_0)|$$

with an element $t_0 \in [0,1]$. Therefore, we have a point

$$w_* := t_0 w_0 \in \overline{B_r} \qquad \text{with} \quad |\nabla \mathbf{z}(w_*)| \geq \frac{1}{2}. \tag{39}$$

4. On account of (37), the function $\mathbf{z}$ satisfies the linear differential inequality

$$|\Delta\mathbf{z}(u,v)| \leq (a\Lambda(a,b,N,r) + b)|\nabla\mathbf{z}(u,v)| \quad \text{in} \quad B_{\frac{r+1}{2}}. \qquad (40)$$

We apply Heinz's inequality from Theorem 5.2 (for $a = 0$ and $B \to B_{\frac{r+1}{2}}$, $B_r \to B_r$). Then we obtain the following estimate in $\overline{B_r}$

$$C^-(a,b,N,r)|\nabla\mathbf{z}(0)|^{\varrho-(r)} \leq |\nabla\mathbf{z}(w)| \leq C^+(a,b,N,r)|\nabla\mathbf{z}(0)|^{\varrho+(r)} \qquad (41)$$

with certain exponents $\varrho_\pm(r) > 0$ and constants $C^\pm(a,b,N,r) > 0$. When we additionally take (39) and (41) into account, we find a constant $\Theta(a,b,N,r) > 0$ such that

$$|\nabla\mathbf{z}(w)| \geq \Theta(a,b,N,r) \qquad \text{for all} \quad w \in B_r \qquad (42)$$

holds true for arbitrary mappings $\mathbf{z} \in \Gamma(B,a,b,N)$. Consequently, the estimate from below in (28) has also been proved.

q.e.d.

6 A Curvature Estimate for Minimal Surfaces

We can even prove distortion estimates for the class $\Gamma(B,a,b,+\infty)$ of those mappings without a bound for Dirichlet's integral, if $a \in [0,\frac{1}{2})$ holds true. We confine ourselves to the class $\Gamma(B,0,0,+\infty)$ of one-to-one harmonic mappings on the unit disc B and begin with the

Proposition 6.1. (Continuous boundary behavior)
The harmonic mapping $\mathbf{z} = \mathbf{z}(w)$ *of the class* $\Gamma(B,0,0,+\infty)$ *may satisfy*

$$|\mathbf{z}(e^{i\varphi}) - \mathbf{z}(e^{i\vartheta})| \leq \varepsilon \qquad \text{for all} \quad \varphi \in [\vartheta - \delta, \vartheta + \delta] \qquad (1)$$

with an angle $\vartheta \in [0, 2\pi)$, *a number* $\delta \in (0, \frac{\pi}{2})$, *and a quantity* $\varepsilon > 0$. *Then we have the estimate*

$$|\mathbf{z}(re^{i\vartheta}) - \mathbf{z}(e^{i\vartheta})| \leq \varepsilon + \frac{4}{\sin^2\delta}(1 - r) \qquad \text{for all} \quad r \in (0,1). \qquad (2)$$

Proof: We invoke Poisson's integral formula

$$\mathbf{z}(re^{i\vartheta}) = \frac{1}{2\pi} \int\limits_{-\pi}^{\pi} \frac{1 - r^2}{|e^{i\varphi} - r|^2} \mathbf{z}(e^{i(\vartheta+\varphi)}) \, d\varphi$$

and obtain the following inequality for all $r \in (0,1)$:

$$|\mathbf{z}(re^{i\vartheta}) - \mathbf{z}(e^{i\vartheta})| \le \frac{1}{2\pi} \int_{-\pi}^{\pi} \frac{1-r^2}{|e^{i\varphi}-r|^2} |\mathbf{z}(e^{i(\vartheta+\varphi)}) - \mathbf{z}(e^{i\vartheta})|\, d\varphi$$

$$= \frac{1}{2\pi} \int_{-\pi}^{-\delta} \frac{1-r^2}{|e^{i\varphi}-r|^2} |\mathbf{z}(e^{i(\vartheta+\varphi)}) - \mathbf{z}(e^{i\vartheta})|\, d\varphi$$

$$+ \frac{1}{2\pi} \int_{-\delta}^{\delta} \frac{1-r^2}{|e^{i\varphi}-r|^2} |\mathbf{z}(e^{i(\vartheta+\varphi)}) - \mathbf{z}(e^{i\vartheta})|\, d\varphi$$

$$+ \frac{1}{2\pi} \int_{\delta}^{\pi} \frac{1-r^2}{|e^{i\varphi}-r|^2} |\mathbf{z}(e^{i(\vartheta+\varphi)}) - \mathbf{z}(e^{i\vartheta})|\, d\varphi.$$

Here we have used

$$\frac{1}{2\pi} \int_{-\pi}^{\pi} \frac{1-r^2}{|e^{i\varphi}-r|^2}\, d\varphi = 1 \qquad \text{for all} \quad r \in (0,1).$$

Now we observe $|e^{i\varphi} - r| \ge \sin\delta$ for all $\varphi \in [-\pi, -\delta] \cup [\delta, \pi]$ and all $r \in (0,1)$. We note (1) and infer

$$|\mathbf{z}(re^{i\vartheta}) - \mathbf{z}(e^{i\vartheta})| \le \frac{1}{2\pi} \frac{1-r^2}{\sin^2\delta} 2 \cdot 2\pi + \varepsilon \le 2\frac{(1-r)(1+r)}{\sin^2\delta} + \varepsilon$$

$$\le \frac{4}{\sin^2\delta}(1-r) + \varepsilon \qquad \text{for all} \quad r \in (0,1).$$

q.e.d.

Proposition 6.2. *Let* $\mathbf{z} = \mathbf{z}(w) : \overline{B} \to \overline{B}$ *denote a topological mapping. Then we have an angle* $\vartheta_n \in [0, 2\pi)$ *for each integer* $n \in \mathbb{N}$, *such that*

$$|\mathbf{z}(e^{i\varphi}) - \mathbf{z}(e^{i\vartheta_n})| \le \frac{2\pi}{n} \qquad \text{for all} \quad \varphi \in [\vartheta_n - \frac{\pi}{n}, \vartheta_n + \frac{\pi}{n}]. \tag{3}$$

Proof: We partition the circle ∂B into n arcs $\sigma_1, \ldots, \sigma_n$ of the equal length $\frac{2\pi}{n}$, and denote their images with respect to the topological mapping $\mathbf{z}$ by $\gamma_k := \mathbf{z}(\sigma_k)$ for $k = 1, \ldots, n$. Evidently, their lengths $|\gamma_k|$ fulfill $|\gamma_1| + \ldots + |\gamma_n| = 2\pi$. Therefore, we find an index $m \in \{1, \ldots, n\}$ with the property $|\gamma_m| \le \frac{2\pi}{n}$. If $e^{i\vartheta_n}$ with $\vartheta_n \in [0, 2\pi)$ denotes the center of the arc σ_m, the relation (3) is satisfied.

q.e.d.

In 1952, E. Heinz proved the following remarkable result:

Theorem 6.3. *There exists a universal constant* $\Theta > 0$, *such that each one-to-one harmonic mapping* $\mathbf{z} = \mathbf{z}(w) \in \Gamma(B, 0, 0, +\infty)$ *satisfies the inequality*

$$|\nabla\mathbf{z}(0)| \ge \Theta. \tag{4}$$

Proof: We choose the mapping $\mathbf{z} \in \Gamma(B, 0, 0, +\infty)$ and the integer $n \in \mathbb{N}$. Because of Proposition 6.2, we find an angle $\vartheta_n \in [0, 2\pi)$ satisfying (3). Proposition 6.1 then yields the estimate

$$|\mathbf{z}(re^{i\vartheta_n}) - \mathbf{z}(e^{i\vartheta_n})| \leq \frac{2\pi}{n} + \frac{4}{\sin^2 \frac{\pi}{n}}(1 - r) \qquad \text{for all} \quad r \in (0, 1). \qquad (5)$$

At first, taking the integer $n \in \mathbb{N}$ sufficiently large and afterwards choosing the radius $r \in (0, 1)$ suitably, the right-hand side in (5) becomes less than or equal to $\frac{1}{2}$, and we infer

$$|\mathbf{z}(re^{i\vartheta_n})| \geq |\mathbf{z}(e^{i\vartheta_n})| - |\mathbf{z}(re^{i\vartheta_n}) - \mathbf{z}(e^{i\vartheta_n})| \geq \frac{1}{2}. \qquad (6)$$

As in part 3 of the proof for Theorem 5.4 in Section 5, we then find a point $w_* \in \overline{B_r}$ satisfying

$$|\nabla \mathbf{z}(w_*)| \geq \frac{1}{2}. \qquad (7)$$

We obtain the following estimate via Heinz's inequality from Section 5 Theorem 5.2, namely

$$|\nabla \mathbf{z}(0)| \geq C^+(0, 0, 1, r)^{-\frac{1+3r}{1-r}} |\nabla \mathbf{z}(w_*)|^{\frac{1+3r}{1-r}}$$

$$\geq (2C^+(0, 0, 1, r))^{-\frac{1+3r}{1-r}} =: \Theta, \qquad (8)$$

since the radius $r \in (0, 1)$ has been determined independently of the mapping $\mathbf{z}$ above. $\qquad$ q.e.d.

We now prove the following result with the aid of the uniformization method.

Theorem 6.4. (Curvature estimate of E. Heinz)
Let the radius $R \in (0, +\infty)$ be chosen arbitrarily and the disc

$$B_R := \{z = x + iy \in \mathbb{C} : |z| < R\}$$

be defined. Then we have a universal constant $M \in (0, +\infty)$, such that all solutions of the minimal surface equation

$$z = \zeta(x, y) \in C^{2+\alpha}(\overline{B_R}, \mathbb{R}), \qquad \alpha \in (0, 1),$$

$$\mathcal{M}\zeta(x, y) := (1 + \zeta_y^2)\zeta_{xx} - 2\zeta_x\zeta_y\zeta_{xy} + (1 + \zeta_x^2)\zeta_{yy} = 0 \qquad \text{in} \quad B_R \qquad (9)$$

satisfy the estimate

$$\kappa_1(0, 0)^2 + \kappa_2(0, 0)^2 \leq \frac{1}{R^2} M \qquad (10)$$

for their principal curvatures $\kappa_j(0, 0)$ with $j = 1, 2$ at the point $\mathbf{y}(0, 0)$ of the graph $\mathbf{y}(x, y) := (x, y, \zeta(x, y))$, $(x, y) \in \overline{B_R}$.

Proof:

1. Using the uniformization theorem (compare the subsequent Section 8), we introduce isothermal parameters into the Riemannian metric

$$ds^2 := |\mathbf{y}_x|^2 \, dx^2 + 2(\mathbf{y}_x \cdot \mathbf{y}_y) \, dx \, dy + |\mathbf{y}_y|^2 \, dy^2$$
$$= (1 + \zeta_x^2) \, dx^2 + 2\zeta_x \zeta_y \, dx \, dy + (1 + \zeta_y^2) \, dy^2, \qquad (x, y) \in \overline{B_R}, \tag{11}$$

of the class $C^{1+\alpha}(\overline{B_R})$. We take the uniformizing mapping

$$f(u, v) = x(u, v) + iy(u, v) : \overline{B} \to \overline{B_R} \in C^{2+\alpha}(\overline{B}, \overline{B_R}),$$
$$f(0, 0) = 0, \tag{12}$$

and consider the surface

$$\mathbf{x}(u, v) = \mathbf{y} \circ f(u, v) = (f(u, v), \zeta \circ f(u, v)) = (x(u, v), y(u, v), z(u, v)) \tag{13}$$

of the class $C^{2+\alpha}(\overline{B}, \mathbb{R}^3)$. This surface is subject to the differential equations

$$\Delta \mathbf{x}(u, v) = 0 \quad \text{in} \quad B,$$
$$|\mathbf{x}_u| - |\mathbf{x}_v| = 0 = \mathbf{x}_u \cdot \mathbf{x}_v \quad \text{in} \quad B. \tag{14}$$

In particular, the plane mapping

$$g(u, v) := \frac{1}{R} f(u, v), \qquad (u, v) \in \overline{B}, \tag{15}$$

belongs to the class $\Gamma(B, 0, 0, +\infty)$. Theorem 1 now yields $|\nabla g(0, 0)| \geq \Theta$ and consequently

$$|\nabla f(0, 0)| \geq \Theta R, \tag{16}$$

with the universal constant $\Theta > 0$.

2. The normal to the surface $\mathbf{y}(x, y)$ in direction $\mathbf{e} = (0, 0, 1)$ is denoted by the symbol

$$\mathbf{Y}(x, y) := \frac{1}{\sqrt{1 + |\nabla \zeta(x, y)|^2}} \big(-\zeta_x(x, y), -\zeta_y(x, y), 1 \big), \qquad (x, y) \in \overline{B_R},$$

and we define $\mathbf{X}(u, v) := \mathbf{Y} \circ f(u, v)$, $(u, v) \in \overline{B}$. Due to Theorem 1.4 from Section 1 of Chapter 11, the following mapping

$$\mathbf{X} : B \to S^+ := \Big\{ \mathbf{z} = (z_1, z_2, z_3) \in \mathbb{R}^3 \ : \ |\mathbf{z}| = 1, \ z_3 > 0 \Big\} \tag{17}$$

is antiholomorphic. We now consider the stereographic projection from the south pole $(0, 0, -1)$, more precisely

$$\sigma = \sigma(\mathbf{z}) : S^+ \to B \quad \text{conformal}. \tag{18}$$

The mapping $h(u, v) := \sigma \circ \mathbf{X}(u, v)$, $(u, v) \in \overline{B}$ is antiholomorphic and consequently harmonic. Therefore, we find a constant $\Lambda \in (0, +\infty)$ such that

$$|\nabla \mathbf{X}(0, 0)| \leq \Lambda \tag{19}$$

holds true.

3. We now evaluate via considerations from Chapter 11, Section 1 as follows:

$$\kappa_1(0,0)^2 + \kappa_2(0,0)^2 = -2\kappa_1(0,0)\kappa_2(0,0) = -2K(0,0)$$

$$= 2|K(0,0)| = 2\frac{|\mathbf{X}_u \wedge \mathbf{X}_v(0,0)|}{|\mathbf{x}_u \wedge \mathbf{x}_v(0,0)|}$$

$$= 2\frac{|\nabla \mathbf{X}(0,0)|^2}{|\nabla \mathbf{x}(0,0)|^2} \le 2\frac{|\nabla \mathbf{X}(0,0)|^2}{|\nabla f(0,0)|^2}$$

$$\le 2\frac{\Lambda^2}{\Theta^2 R^2} = \frac{M}{R^2}.$$

Here we have set the quantity $M := 2\frac{\Lambda^2}{\Theta^2}$. q.e.d.

We obtain the following result as a corollary from Theorem 6.4, namely

Theorem 6.5. (S. Bernstein)
Let $z = \zeta(x,y) : \mathbb{R}^2 \to \mathbb{R} \in C^{2+\mu}(\mathbb{R}^2)$ - with $\mu \in (0,1)$ - denote an entire solution of the minimal surface equation $\mathcal{M}\zeta(x,y) = 0$ in $\mathbb{R}^2$. Then we have coefficients $\alpha, \beta, \gamma \in \mathbb{R}$ such that

$$\zeta(x,y) = \alpha x + \beta y + \gamma \qquad in \quad \mathbb{R}^2$$

is satisfied, which means ζ is an affine-linear function.

Proof: We consider the transition to the limit $R \to +\infty$ in the estimate (10) and obtain $\kappa_1(0,0) = 0 = \kappa_2(0,0)$. Since this argument is valid at each point of the minimal graph, we infer

$$\kappa_1(x,y) = 0 = \kappa_2(x,y) \quad in \quad \mathbb{R}^2. \tag{20}$$

Consequently, the surface $\mathbf{y}(x,y) = (x,y,\zeta(x,y))$, $(x,y) \in \mathbb{R}^2$ represents a plane.

q.e.d.

Remarks to Theorem 6.4 and 6.5:

1. We owe the curvature estimate in Theorem 6.4 to:

   E. Heinz: *Über die Lösungen der Minimalflächengleichung.* Nachr. Akad. Wiss. Göttingen, Math.-Phys. Kl. (1952), 51-56.

2. Curvature estimates for surfaces of prescribed mean curvature have been established by:

   F. Sauvigny: *A priori estimates of the principle curvatures for immersions of prescribed mean curvature and theorems of Bernstein-type.* Math. Zeitschrift **205** (1990), 567-582.

3. In his thesis, S. Fröhlich has derived curvature estimates for stable solutions of the Euler equations for parametric elliptic functionals - in particular for relative minima. Here we refer the reader to:

S. Fröhlich: *Curvature estimates for μ-stable G-minimal surfaces and theorems of Bernstein-type.* Analysis **22** (2002), 109-130.

7 Global Estimates for Conformal Mappings with respect to Riemannian Metrics

We define the unit disc $E := \{\mathbf{x} = (x^1, x^2) \in \mathbb{R}^2 : |\mathbf{x}| < 1\}$ in the coordinates (x^1, x^2) and the unit disc $B := \{w = u + iv \in \mathbb{C} : |w| < 1\}$ in the coordinates $u + iv \cong (u, v)$. We prescribe the Riemannian metric

$$
\begin{aligned}
ds^2 &= g_{jk}(x^1, x^2)\, dx^j\, dx^k \\
&= g_{11}(x^1, x^2)\,(dx^1)^2 + 2g_{12}(x^1, x^2)\, dx^1\, dx^2 + g_{22}(x^1, x^2)\,(dx^2)^2
\end{aligned}
\tag{1}
$$

on the disc E. Here we use Einstein's summation convention and require the coefficients to satisfy

$$
\begin{aligned}
g_{jk} &= g_{jk}(x^1, x^2) \in C^{1+\alpha}(E, \mathbb{R}) \qquad \text{for} \quad j, k = 1, 2; \\
g_{12}(x^1, x^2) &= g_{21}(x^1, x^2) \qquad \text{in} \quad E
\end{aligned}
\tag{2}
$$

and

$$
\lambda |\xi|^2 \leq g_{jk}(x^1, x^2)\xi^j \xi^k \leq \frac{1}{\lambda}|\xi|^2
\tag{3}
$$

$$
\text{for all} \quad \xi = (\xi^1, \xi^2) \in \mathbb{R}^2 \quad \text{and} \quad (x^1, x^2) \in E,
$$

with the constants $\alpha, \lambda \in (0, 1)$.

Proposition 7.1. *The C^2-diffeomorphic, positive-orientied mapping*

$$
\mathbf{x} = \mathbf{x}(u, v) = (x^1(u, v), x^2(u, v))^* : \overline{B} \to \overline{E} \in C^2(B, \mathbb{R}^2) \cap C^0(\overline{B}, \overline{E})
$$

may satisfy the weighted conformality relations

$$
x_u^j(u, v) g_{jk}(x^1(u, v), x^2(u, v)) x_v^k(u, v) = 0 \qquad \text{in} \quad B,
\tag{4}
$$

$$
x_u^j(u, v) g_{jk}(x^1, x^2) x_u^k(u, v) = x_v^j(u, v) g_{jk}(x^1, x^2) x_v^k(u, v) \qquad \text{in} \quad B.
\tag{5}
$$

Then the function $\mathbf{x}$ *satisfies the nonlinear elliptic system*

$$
\Delta x^l + \Gamma_{jk}^l(x_u^j x_u^k + x_v^j x_v^k) = 0 \qquad \text{in} \quad B \quad \text{for} \quad l = 1, 2
\tag{6}
$$

where we have used the Christoffel symbols

$$
\Gamma_{jk}^l := \frac{1}{2} g^{li}(g_{ki,x^j} + g_{ij,x^k} - g_{jk,x^i}), \qquad j, k, l = 1, 2
\tag{7}
$$

with the inverse matrix $(g^{jk})_{j,k=1,2} := (g_{jk})^{-1}_{j,k=1,2}$. *Therefore,* $\mathbf{x}$ *represents a harmonic mapping of the disc* $\{B, (\delta_{jk})\}$ *onto the disc* $\{E, (g_{jk})\}$, *with the unit matrix* $(\delta_{jk})_{j,k=1,2}$.

Proof: We derive the equation (4) with respect to the variable v and the equation (5) with respect to the variable u:

$$x^j_{uv} g_{jk} x^k_v + x^j_u g_{jk} x^k_{vv} + x^j_u g_{jk,x^l} x^k_v x^l_v = 0,$$

$$x^j_{uv} g_{jk} x^k_v = x^j_u g_{jk} x^k_{uu} + \frac{1}{2} x^j_u g_{jk,x^l} x^k_u x^l_u - \frac{1}{2} x^l_u g_{jk,x^l} x^j_v x^k_v.$$

When we insert the second equation into the first, we obtain

$$x^j_u g_{jk} \Delta x^k + x^j_u g_{jk,x^l} x^k_v x^l_v + \frac{1}{2} x^j_u g_{jk,x^l} x^k_u x^l_u - \frac{1}{2} x^l_u g_{jk,x^l} x^j_v x^k_v = 0$$

as well as

$$x^j_u g_{jk} \Delta x^k + \frac{1}{2} x^j_u (g_{kj,x^l} + g_{jl,x^k} - g_{lk,x^j})(x^k_u x^l_u + x^k_v x^l_v) = 0.$$

Interchanging the variables u with v in these calculations, we deduce analogously

$$x^j_v g_{jk} \Delta x^k + \frac{1}{2} x^j_v (g_{kj,x^l} + g_{jl,x^k} - g_{lk,x^j})(x^k_u x^l_u + x^k_v x^l_v) = 0.$$

Since the vectors $\mathbf{x}_u$ and $\mathbf{x}_v$ are linearly independent, we obtain

$$g_{jk} \Delta x^k + \frac{1}{2}(g_{kj,x^l} + g_{jl,x^k} - g_{lk,x^j})(x^k_u x^l_u + x^k_v x^l_v) = 0, \qquad j = 1,2.$$

Multiplication by the inverse matrix (g^{ij}) finally yields

$$\delta^i_k \Delta x^k + \frac{1}{2} g^{ij}(g_{kj,x^l} + g_{jl,x^k} - g_{lk,x^j})(x^k_u x^l_u + x^k_v x^l_v) = 0, \qquad i = 1,2$$

and consequently

$$\Delta x^i + \Gamma^i_{lk}(x^k_u x^l_u + x^k_v x^l_v) = 0, \qquad i = 1,2.$$

q.e.d.

We have the following convention for our class of mappings, namely

$$\mathbf{x}(0,0) = (0,0)^*. \tag{8}$$

Furthermore, we define the positive-definite matrix

$$G(x^1, x^2) := (g_{jk}(x^1, x^2))_{j,k=1,2} : E \to \mathbb{R}^{2 \times 2}. \tag{9}$$

Via the principal axes transformation we determine its square root $G^{\frac{1}{2}}(x^1, x^2)$, carrying out this operation for the positive eigenvalues. Then we calculate

$$\left\{|G^{\frac{1}{2}}(\mathbf{x}(u,v))|\,|(\mathbf{x}_u,\mathbf{x}_v)|\right\}^2 = \left|\left(G^{\frac{1}{2}}(\mathbf{x}(u,v))\circ\mathbf{x}_u,\, G^{\frac{1}{2}}(\mathbf{x}(u,v))\circ\mathbf{x}_v\right)\right|^2$$

$$= \left|\left(\begin{array}{c}(G^{\frac{1}{2}}(\mathbf{x})\circ\mathbf{x}_u)^*\\(G^{\frac{1}{2}}(\mathbf{x})\circ\mathbf{x}_v)^*\end{array}\right)\circ\left(G^{\frac{1}{2}}(\mathbf{x})\circ\mathbf{x}_u,\, G^{\frac{1}{2}}(\mathbf{x})\circ\mathbf{x}_v\right)\right|$$

$$= \left|\left(\begin{array}{cc}\mathbf{x}_u^*\circ G(\mathbf{x})\circ\mathbf{x}_u\,,\ \mathbf{x}_u^*\circ G(\mathbf{x})\circ\mathbf{x}_v\\\mathbf{x}_v^*\circ G(\mathbf{x})\circ\mathbf{x}_u\,,\ \mathbf{x}_v^*\circ G(\mathbf{x})\circ\mathbf{x}_v\end{array}\right)\right|$$

$$= \frac{1}{4}\left\{\mathbf{x}_u^*\circ G(\mathbf{x})\circ\mathbf{x}_u + \mathbf{x}_v^*\circ G(\mathbf{x})\circ\mathbf{x}_v\right\}^2 \qquad \text{in} \quad B.$$

This implies

$$|G^{\frac{1}{2}}(\mathbf{x}(u,v))|\,|(\mathbf{x}_u,\mathbf{x}_v)| = \frac{1}{2}\left\{\mathbf{x}_u^*\circ G(\mathbf{x})\circ\mathbf{x}_u + \mathbf{x}_v^*\circ G(\mathbf{x})\circ\mathbf{x}_v\right\} \qquad (10)$$

for all $(u,v)\in B$. With the aid of (3), we obtain

$$\frac{\lambda^2}{2}|\nabla\mathbf{x}(u,v)|^2 \le \frac{\partial(x^1,x^2)}{\partial(u,v)} \le \frac{1}{2\lambda^2}|\nabla\mathbf{x}(u,v)|^2 \qquad \text{for all} \quad (u,v)\in B. \quad (11)$$

We define the discs $E_r := \{\mathbf{x}\in E \,:\, |\mathbf{x}| < r\}$ for the radii $r\in(0,1)$, and similarly $B_r := \{w\in B \,:\, |w| < r\}$. Then we introduce the monotonic function

$$\gamma(r) := \max_{j,k=1,2}\|g_{jk}\|_{C^{1+\alpha}(\overline{E_r})}, \qquad r\in(0,1). \qquad (12)$$

Theorem 7.2. (Inner estimates for conformal mappings)
Associated to the metric (1)-(3), let the mapping

$$\mathbf{x} = \mathbf{x}(u,v) : \overline{B} \to \overline{E} \in C^2(B)\cap C^0(\overline{B})$$

represent a weighted conformal, positive-oriented C^2-diffeomorphism with (4), (5), and (8). For each number $r\in(0,1)$ given, we then have a constant $\Theta = \Theta(r,\lambda,\gamma(\frac{r+1}{2})) > 0$ and a constant $\Lambda = \Lambda(r,\lambda,\alpha,\gamma(\frac{r+1}{2})) < +\infty$, such that the estimates

$$J_{\mathbf{x}}(u,v) = \frac{\partial(x^1,x^2)}{\partial(u,v)} \ge \Theta \qquad \text{for all} \quad (u,v)\in B_r \qquad (13)$$

and

$$\|\mathbf{x}\|_{C^{2+\alpha}(B_r,\mathbb{R}^2)} \le \Lambda \qquad (14)$$

hold true. Furthermore, the class of mappings above is equicontinuous.

Proof: We follow the arguments in the proof of Theorem 5.4 from Section 5. On account of (11), we comprehend

$$D(\mathbf{x}) \leq \frac{2}{\lambda^2} \iint_B \frac{\partial(x^1, x^2)}{\partial(u, v)} \, du \, dv = \frac{2\pi}{\lambda^2}. \tag{15}$$

Therefore, we can estimate the modulus of continuity in $\overline{B}$ parallel to part 1 of the proof quoted above. From (6), (7), (3), and (12) we deduce the subsequent differential inequality for an arbitrary radius $r \in (0,1)$, namely

$$|\Delta\mathbf{x}(u, v)| \leq a|\nabla\mathbf{x}(u, v)|^2 \quad \text{in} \quad B_{\frac{r+1}{2}}, \tag{16}$$

with the constant $a = a(\lambda, \gamma(\frac{r+1}{2})) \in (0, +\infty)$. Then we estimate $|\nabla\mathbf{x}(u, v)|$ in $B_{r+\varepsilon}$ from above for sufficiently small numbers $\varepsilon > 0$, and the transition to a linear differential inequality is possible. On account of (11) we finally obtain the constant Θ from (13), as in part 3 and 4 of the proof quoted above. We further deduce (14) via potential-theoretic estimates.

$$\text{q.e.d.}$$

With the complex derivatives

$$x_w^j = \frac{1}{2}(x_u^j - ix_v^j), \quad x_{\overline{w}}^j = \frac{1}{2}(x_u^j + ix_v^j), \qquad j = 1, 2$$

we rewrite the *weighted conformality relations* into the *complex form*

$$x_w^j(u, v)g_{jk}(x^1(u, v), x^2(u, v))x_w^k(u, v) = 0 \quad \text{in} \quad B. \tag{17}$$

Furthermore, we modify the equation (6) and obtain *harmonic mappings in the complex form*:

$$x_{w\overline{w}}^l + \frac{1}{2}\Gamma_{jk}^l(x_w^j x_{\overline{w}}^k + x_{\overline{w}}^j x_w^k) = 0 \quad \text{in} \quad B; \qquad l = 1, 2. \tag{18}$$

We easily infer the following result from the weighted conformality relation.

Proposition 7.3. (Elimination lemma)
We have the constants $\mu(\lambda) > 1$ and $0 < \mu_1(\lambda) \leq \mu_2(\lambda) < +\infty$, such that all weighted conformal mappings (4) and (5), with respect to the arbitrary Riemannian metric (1)-(3), satisfy the following inequalities:

$$\frac{1}{\mu(\lambda)}|x_w^1(w)| \leq |x_w^2(w)| \leq \mu(\lambda)|x_w^1(w)|, \qquad w \in B, \tag{19}$$

and

$$\mu_1(\lambda)|x_w^1(w)|^2 \leq \frac{\partial(x^1, x^2)}{\partial(u, v)} \leq \frac{1}{2}|\nabla\mathbf{x}(u, v)|^2 \leq \mu_2(\lambda)|x_w^1(w)|^2, \qquad w \in B. \tag{20}$$

Proof:

1. The weighted conformality relation (17) yields

$$g_{11}(x^1, x^2)x_w^1 x_w^1 = -2g_{12}(x^1, x^2)x_w^1 x_w^2 - g_{22}(x^1, x^2)x_w^2 x_w^2 \qquad \text{in} \quad B.$$

With the aid of (3) we deduce

$$\lambda|x_w^1|^2 \le |g_{11}|\,|x_w^1|^2 \le 2|g_{12}|\,|x_w^1|\,|x_w^2| + |g_{22}|\,|x_w^2|^2$$

$$\le 2\left(\sqrt{\frac{\lambda}{2}}|x_w^1|\right)\left(\sqrt{\frac{2}{\lambda}}\frac{|x_w^2|}{\lambda}\right) + \frac{1}{\lambda}|x_w^2|^2$$

$$\le \frac{\lambda}{2}|x_w^1|^2 + \left(\frac{2}{\lambda}\frac{1}{\lambda^2} + \frac{1}{\lambda}\right)|x_w^2|^2$$

$$= \frac{\lambda}{2}|x_w^1|^2 + \frac{2+\lambda^2}{\lambda^3}|x_w^2|^2$$

and consequently

$$|x_w^1|^2 \le \frac{4+2\lambda^2}{\lambda^4}|x_w^2|^2, \qquad w \in B.$$

Similarly, we find

$$|x_w^2|^2 \le \frac{4+2\lambda^2}{\lambda^4}|x_w^1|^2, \qquad w \in B,$$

by resolving the weighted conformality relation (17) with respect to $g_{22}(x^1, x^2)x_w^2 x_w^2$. Setting $\mu(\lambda) := \frac{1}{\lambda^2}\sqrt{4 + 2\lambda^2}$ we obtain (19).

2. We now estimate

$$\frac{1}{2}|\nabla\mathbf{x}(u,v)|^2 = 2\big(|x_w^1(w)|^2 + |x_w^2(w)|^2\big)$$

$$\le 2(1 + \mu(\lambda)^2)|x_w^1(w)|^2$$

$$= \mu_2(\lambda)|x_w^1(w)|^2, \qquad w \in B,$$

with $\mu_2(\lambda) := 2(1 + \mu(\lambda)^2)$. Taking (11) into account, we find

$$\frac{\partial(x^1, x^2)}{\partial(u,v)} \ge \frac{\lambda^2}{2}|\nabla\mathbf{x}(u,v)|^2 = 2\lambda^2\big(|x_w^1(w)|^2 + |x_w^2(w)|^2\big)$$

$$\ge 2\lambda^2\left(1 + \frac{1}{\mu(\lambda)^2}\right)|x_w^1(w)|^2 = \mu_1(\lambda)|x_w^1(w)|^2, \qquad w \in B,$$

with $\mu_1(\lambda) := 2\lambda^2(1 + \frac{1}{\mu(\lambda)^2}) > 0$. Therefore, the relation (20) has been shown.   q.e.d.

We now prove the important

Theorem 7.4. (Global estimates for conformal mappings)
*The metric ds^2 from (1)-(3) with the coefficients $g_{jk}(x^1, x^2) \in C^{1+\alpha}(\overline{E}, \mathbb{R})$
for $j, k = 1, 2$ may be given, and we consider the weighted conformal, positive-
oriented C^2-diffeomorphism*

$$\mathbf{x} = \mathbf{x}(u, v) = (x^1(u, v), x^2(u, v))^* \colon \overline{B} \to \overline{E} \in C^2(B, \mathbb{R}^2) \cap C^1(\overline{B}, \overline{E}) \quad (21)$$

*from (4), (5), and (8). Then we have the regularity property $\mathbf{x} \in C^{2+\alpha}(\overline{B}, \mathbb{R}^2)$
and the following a priori estimates*

$$J_{\mathbf{x}}(u, v) \geq \Theta \qquad \text{for all} \quad (u, v) \in \overline{B} \tag{22}$$

and

$$\|\mathbf{x}\|_{C^{2+\alpha}(\overline{B}, \mathbb{R}^2)} \leq \Lambda, \tag{23}$$

*with the constants $\Theta = \Theta(\lambda, \alpha, \gamma(1)) > 0$ and $\Lambda = \Lambda(\lambda, \alpha, \gamma(1)) < +\infty$; here
the function $\gamma(r)$ is defined in (12).*

Proof:

1. On the circular line ∂E we consider the tangential vector-field

$$\mathbf{t}(x^1, x^2) := (-x^2, x^1)^* \colon \partial E \to \mathbb{R}^2$$

and the constant vector-field $\mathbf{e} = (1, 0)^*$. Furthermore, let

$$\mathbf{a}(x^1, x^2) = (a^1(x^1, x^2), a^2(x^1, x^2))^* \colon \partial E \to \mathbb{R}^2$$

denote a vector-field of length 1 with respect to the metric ds^2, which
means

$$a^j(x^1, x^2) g_{jk}(x^1, x^2) a^k(x^1, x^2) = 1 \qquad \text{on} \quad \partial E. \tag{24}$$

We choose $\mathbf{a}(x^1, x^2)$ such that its oriented angle to the tangential vector
$\mathbf{t}(x^1, x^2)$ in the Riemannian metric coincides with the Euclidean angle
between $\mathbf{e}$ and $\mathbf{t}(x^1, x^2)$. With the symbol

$$\mathbf{b}(x^1, x^2) = (b^1(x^1, x^2), b^2(x^1, x^2))^* \colon \partial E \to \mathbb{R}^2$$

we denote the unit vector-field orthogonal to $\mathbf{a}(x^1, x^2)$ in the Riemannian
metric ds^2, which is oriented as follows

$$\det\left(\mathbf{a}(x^1, x^2), \mathbf{b}(x^1, x^2)\right) = \begin{vmatrix} a^1(x^1, x^2) & b^1(x^1, x^2) \\ a^2(x^1, x^2) & b^2(x^1, x^2) \end{vmatrix} > 0 \qquad \text{on} \quad \partial E. \tag{25}$$

The weighted conformal mapping $\mathbf{x}(u, v)$ then possesses the following *free
boundary condition*

$$\left(\mathbf{x}_u(w), \mathbf{x}_v(w)\right) = \nu(w)\left(\mathbf{a}(\mathbf{x}(w)), \mathbf{b}(\mathbf{x}(w))\right), \qquad w \in \partial B, \tag{26}$$

with the function $\nu(w) \colon \partial B \to (0, +\infty)$. Finally, we find a function

$$\varphi = \varphi(x^1, x^2) : \partial E \to \mathbb{R} \in C^{1+\alpha}(\partial E),$$

such that

$$\begin{pmatrix} a^1(x^1, x^2) \, , \, b^1(x^1, x^2) \\ a^2(x^1, x^2) \, , \, b^2(x^1, x^2) \end{pmatrix} \circ \begin{pmatrix} \cos\varphi(x^1, x^2) \, , \, -\sin\varphi(x^1, x^2) \\ \sin\varphi(x^1, x^2) \, , \, \cos\varphi(x^1, x^2) \end{pmatrix} = \begin{pmatrix} * \ 0 \\ * \ * \end{pmatrix}$$

(27)

holds true on ∂E.

2. We now utilize the Schwarzian integral formula from Theorem 2.2 in Section 2 of Chapter 9, namely

$$F(z) := \frac{1}{2\pi} \int_0^{2\pi} \frac{e^{it} + z}{e^{it} - z} \varphi(e^{it}) \, dt, \qquad z = x^1 + ix^2 \in E, \qquad (28)$$

with the function $\varphi \in C^{1+\alpha}(\partial E)$ defined in part 1. Now the function $F(z)$ is holomorphic in E, and via potential-theoretic methods (compare Theorem 4.6 in Chapter 9, Section 4) we see

$$F(z) \in C^{1+\alpha}(\overline{E}, \mathbb{C}), \qquad \|F\|_{C^{1+\alpha}(\overline{E})} \le C(\alpha) \|\varphi\|_{C^{1+\alpha}(\partial E)}. \qquad (29)$$

Furthermore, F satisfies the boundary condition

$$\operatorname{Re} F(z) = \varphi(z) \qquad \text{for all} \quad z \in \partial E. \qquad (30)$$

The function

$$f(z) := \exp\{iF(z)\}, \qquad z \in \overline{E} \qquad (31)$$

of the class $C^{1+\alpha}(\overline{E}, \mathbb{C} \setminus \{0\})$ is consequently subject to the following boundary condition

$$f(z) = \varrho(z) e^{i\varphi(z)}, \qquad z \in \partial E, \qquad (32)$$

with the positive real function

$$\varrho(z) := e^{-\operatorname{Im} F(z)}, \qquad z \in \partial E. \qquad (33)$$

3. From (32) we deduce the following boundary condition for the function $y(w) := x_w^1(w) f(\mathbf{x}(w)) : \overline{B} \to \mathbb{C}$, namely

$$y(w) = x_w^1(w) f(\mathbf{x}(w)) = \frac{1}{2} \big(x_u^1(w) - i x_v^1(w) \big) \varrho(\mathbf{x}(w)) e^{i\varphi(\mathbf{x}(w))}$$

$$= \frac{\varrho(\mathbf{x}(w))}{2} \big(x_u^1(w) - i x_v^1(w) \big) \big(\cos\varphi(\mathbf{x}(w)) + i \sin\varphi(\mathbf{x}(w)) \big)$$

for all points $w \in \partial B$. From (26) and (27) we infer

$$\operatorname{Im} y(w) = \frac{\varrho(\mathbf{x}(w))}{2}\left(x_u^1(w)\sin\varphi(\mathbf{x}(w)) - x_v^1(w)\cos\varphi(\mathbf{x}(w))\right)$$

$$= \frac{\nu(w)\varrho(\mathbf{x}(w))}{2}\left(a^1(\mathbf{x}(w))\sin\varphi(\mathbf{x}(w)) - b^1(\mathbf{x}(w))\cos\varphi(\mathbf{x}(w))\right)$$

$$= 0 \qquad \text{for all} \quad w \in \partial B.$$

$$(34)$$

Furthermore, we calculate

$$y_{w\overline{w}} = x_{w\overline{w}}^1 f(x^1, x^2) + x_w^1 f_{x^1}(x^1, x^2) x_{\overline{w}}^1 + x_w^1 f_{x^2}(x^1, x^2) x_{\overline{w}}^2 \qquad \text{in} \quad B.$$

Together with the relations (18), (19), and (29), we arrive at the differential inequality

$$|y_{w\overline{w}}(w)| \le a|y(w)|^2, \qquad w \in B, \tag{35}$$

with a constant $a = a(\lambda, \alpha, \gamma(1)) \in (0, +\infty)$.

4. As in Section 3 we transform the unit disc E onto the upper half-plane $\mathbb{C}^+$ via the mapping $g : \mathbb{C}^+ \to E$ and apply the reflection

$$\hat{\mathbf{x}}(w) = (\hat{x}^1(w), \hat{x}^2(w)) := \begin{cases} \mathbf{x} \circ g(w), & \operatorname{Im} w > 0 \\ \mathbf{x} \circ g(\overline{w}), & \operatorname{Im} w < 0 \end{cases}. \tag{36}$$

From (15) we infer a growth condition for Dirichlet's integral of $\hat{\mathbf{x}}(w)$ described in Section 2, Proposition 2.2 and 2.3. Here we utilize the Courant-Lebesgue lemma, estimate the area by the length of the boundary curve via the isoperimetric inequality, and obtain a growth condition for Dirichlet's integral due to (11).

Similarly to Proposition 2.4 and 2.5 in Section 2, we now estimate the oscillation of $\hat{\mathbf{x}}(w)$ on discs in the interior. Then we obtain the following estimates with the notations applied there:

$$2 \int_{\partial B_{\vartheta\lambda(\vartheta)}(w_0)} |\operatorname{Re}(\hat{x}_w^j(w)\,dw)| = \int_{\partial B_{\vartheta\lambda(\vartheta)}(w_0)} |d\hat{x}^j(w)| \le \int_{\partial B_{\vartheta\lambda(\vartheta)}(w_0)} |d\hat{\mathbf{x}}(w)|$$

$$\le \frac{C(\lambda)}{\sqrt{-\log\vartheta}} \qquad \text{for} \quad j = 1, 2. \tag{37}$$

Now we have functions $\varrho^\pm = \varrho^\pm(x^1, x^2) : \mathbb{C}^+ \to \mathbb{C}\setminus\mathbb{R}$ depending on the metric ds^2, such that

$$\hat{x}_w^2(w) = \varrho^\pm(\hat{\mathbf{x}}(w))\hat{x}_w^1(w) \qquad \text{for} \quad w \in \mathbb{C}\setminus\mathbb{R} \quad \text{with} \quad \pm\operatorname{Im} w > 0 \tag{38}$$

holds true (compare the formulas (5) and (6) in Section 9). Consequently, we have

$$2 \int_{\partial B_{\vartheta\lambda(\vartheta)}(w_0)} |\operatorname{Re}(\hat{x}_w^1(w)\,dw)| + 2 \int_{\partial B_{\vartheta\lambda(\vartheta)}(w_0)} |\operatorname{Re}(\varrho^\pm(\hat{\mathbf{x}}(w))\,\hat{x}_w^1(w)\,dw)| \le \frac{2C(\lambda)}{\sqrt{-\log\vartheta}},$$

which implies

$$\int\limits_{\partial B_{\vartheta\lambda(\vartheta)}(w_0)} |\hat{x}_w^1(w)\,dw| \leq \frac{\widetilde{C}(\lambda)}{\sqrt{-\log\vartheta}}. \tag{39}$$

5. Now we consider the reflected derivative function

$$z(w) := \begin{cases} y \circ g(w) = x_w^1(g(w))f(\hat{\mathbf{x}}(w)), & \operatorname{Im} w > 0 \\ \overline{y} \circ g(\overline{w}) = x_{\overline{w}}^1(g(\overline{w}))\overline{f}(\hat{\mathbf{x}}(w)), & \operatorname{Im} w < 0 \end{cases}, \tag{40}$$

and z is continuous according to the boundary condition (34). With the aid of (39) and (29), we then obtain an estimate for the Cauchy integral of $z(w)$ - as described in Proposition 2.4 and 2.5 of Section 2. We apply the method of Theorem 3.1 from Section 3 and find a constant $\widetilde{\Lambda}(\lambda,\alpha,\beta,\gamma(1)) < +\infty$ satisfying

$$\|y\|_{C^{1+\beta}(\overline{B})} \leq \widetilde{\Lambda}(\lambda,\alpha,\beta,\gamma(1)) \qquad \text{for all} \quad \beta \in (0,1) \tag{41}$$

on account of (35). We still observe (19), and the system (6) together with potential-theoretic methods yield the inequality

$$\|\mathbf{x}\|_{C^{2+\alpha}(\overline{B},\mathbb{R}^2)} \leq \Lambda \tag{42}$$

with the a-priori-constant $\Lambda = \Lambda(\lambda,\alpha,\gamma(1))$. Finally, we apply Theorem 5.1 from Section 5 to the nonvanishing function $y(w)$, $w \in \overline{B}$. The methods of proof for Theorem 5.4 from Section 5 provide a constant $\Theta = \Theta(\lambda,\alpha,\gamma(1)) > 0$ satisfying

$$J_{\mathbf{x}}(u,v) \geq \Theta \qquad \text{for all} \quad (u,v) \in \overline{B}, \tag{43}$$

on account of (20). This completes the proof of our theorem.   q.e.d.

Remark: When the condition $g_{jk}(x^1,x^2) = \delta_{jk}$ is valid in the neighborhood of the circular line ∂E for the Riemannian metric, we can reflect the mapping $\mathbf{x}$ at the circumference: Then we do not need the Schwarzian integral formula (28) above.

8 Introduction of Conformal Parameters into a Riemannian Metric

We continue our deliberations from Section 7 and quote those results by adding the symbol *. We shall introduce conformal parameters into the metric ds^2 from (1)*, (2)*, (3)* of the class $C^{1+\alpha}(\overline{E})$. This means to solve the system (4)*, (5)* of the weighted conformality relations and to transfer the metric ds^2 into the isothermal form

$$ds^2 = \sigma(u,v)(du^2 + dv^2) \quad \text{in} \quad \overline{B}, \qquad \sigma(u,v) > 0 \quad \text{in} \quad \overline{B}. \tag{1}$$

At first, we achieve this aim for metrics ds^2 whose coefficients in the $C^{1+\alpha}(\overline{E})$-norm have a sufficiently small deviation from the isothermal metric

$$dr^2 = \varrho(x^1, x^2)\delta_{jk}\, dx^j\, dx^k \quad \text{in} \quad \overline{E},$$
$$\varrho(x^1, x^2) : \overline{E} \to (0, +\infty) \in C^{1+\alpha}(\overline{E}). \tag{2}$$

We define the surface element of ds^2 by

$$g(x^1, x^2) := (\det G(x^1, x^2))^{\frac{1}{2}}$$
$$= \sqrt{g_{11}(x^1, x^2)g_{22}(x^1, x^2) - g_{12}(x^1, x^2)^2} \quad \text{in} \quad \overline{E}. \tag{3}$$

In order to render the subsequent calculations into a more simple form, we set $(x^1, x^2) = (x, y) = z \in \overline{E}$ and

$$G(x^1, x^2) = (g_{jk}(x^1, x^2))_{j,k=1,2} = \begin{pmatrix} a(x,y) & b(x,y) \\ b(x,y) & c(x,y) \end{pmatrix} \quad \text{in} \quad \overline{E}. \tag{4}$$

We shall construct a positive-oriented diffeomorphism

$$w(z) = u(x,y) + iv(x,y) : \overline{E} \to \overline{\Omega} \in C^{2+\alpha}(\overline{E}, \mathbb{C}) \tag{5}$$

onto a bounded, simply connected domain $\Omega \subset \mathbb{C}$ with the inverse mapping

$$z = z(w) = x(u,v) + iy(u,v) : \overline{\Omega} \to \overline{E} \in C^{2+\alpha}(\overline{\Omega}, \mathbb{C}), \tag{6}$$

such that the metric ds^2 is transferred into the isothermal form

$$ds^2 = \sigma(u,v)(du^2 + dv^2) \quad \text{in} \quad \overline{\Omega}. \tag{7}$$

We calculate

$$ds^2 = a\, dx^2 + 2b\, dx\, dy + c\, dy^2$$
$$= \frac{1}{a}\{a^2\, dx^2 + 2ab\, dx\, dy + ac\, dy^2\} \tag{8}$$
$$= \frac{1}{a}\{a\, dx + (b + ig)\, dy\}\{a\, dx + (b - ig)\, dy\}.$$

Now we look for a complex, diffeomorphic primitive function

$$w = w(z) : \overline{E} \to \overline{\Omega} \in C^{2+\alpha}(\overline{E}, \mathbb{C}),$$

such that

$$a\, dx + (b + ig)\, dy = \varrho(z)\, dw \quad \text{in} \quad \overline{E} \tag{9}$$

is correct, with a function $\varrho \in C^{1+\alpha}(\overline{E}, \mathbb{C} \setminus \{0\})$. Then we infer

$$a\,dx + (b - ig)\,dy = \overline{\varrho(z)}\,d\overline{w} \qquad \text{in} \quad \overline{E}, \tag{10}$$

and the relations (8)-(10) provide the desired isothermal form

$$ds^2 = \frac{1}{a}\varrho\,dw\,\overline{\varrho}\,d\overline{w} = \frac{|\varrho(z)|^2}{a(z)}\,dw\,d\overline{w} = \lambda(w)(du^2 + dv^2)$$

$$\text{with} \quad \lambda(w) := \frac{|\varrho(z(w))|^2}{a(z(w))} : \overline{\Omega} \to (0, +\infty) \in C^{1+\alpha}(\overline{\Omega}). \tag{11}$$

The formula (9) is equivalent to the system

$$\varrho(z)\frac{\partial}{\partial x}w(z) = a(z), \quad \varrho(z)\frac{\partial}{\partial y}w(z) = b(z) + ig(z) \qquad \text{in} \quad \overline{E},$$

and consequently to

$$2\varrho\frac{\partial}{\partial z}w = \varrho\frac{\partial}{\partial x}w - i\varrho\frac{\partial}{\partial y}w = a + g - ib,$$

$$2\varrho\frac{\partial}{\partial \overline{z}}w = \varrho\frac{\partial}{\partial x}w + i\varrho\frac{\partial}{\partial y}w = a - g + ib \qquad \text{in} \quad \overline{E},$$

and to the equations

$$\frac{\partial}{\partial \overline{z}}w(z) = \frac{1}{2\varrho(z)}\big(a(z) - g(z) + ib(z)\big),$$

$$\frac{1}{2\varrho(z)} = \frac{1}{a(z) + g(z) - ib(z)}\frac{\partial}{\partial z}w(z) \qquad \text{in} \quad \overline{E}$$

as well. When we insert the second relation into the first, we obtain the following complex equation equivalent to (9), namely

$$\frac{\partial}{\partial \overline{z}}w(z) - \frac{a(z) - g(z) + ib(z)}{a(z) + g(z) - ib(z)}\frac{\partial}{\partial z}w(z) = 0 \qquad \text{in} \quad \overline{E}.$$

We now define

$$q(z) := \frac{a(z) - g(z) + ib(z)}{a(z) + g(z) - ib(z)}, \qquad z \in \overline{E}. \tag{12}$$

We observe

$$q(z) = 0 \quad \Leftrightarrow \quad b(z) = 0, \ a(z) = c(z) \qquad \text{for a point} \quad z \in \overline{E}, \tag{13}$$

and

$$|q(z)| = \sqrt{\frac{(a - g)^2 + b^2}{(a + g)^2 + b^2}} = \sqrt{\frac{(a + g)^2 + b^2 - 4ag}{(a + g)^2 + b^2}}$$

$$= \sqrt{1 - 4\frac{ag}{(a + g)^2 + b^2}} < 1 \qquad \text{for all} \quad z \in \overline{E}. \tag{14}$$

We now have to solve *Beltrami's differential equation in the complex form*

$$\frac{\partial}{\partial \overline{z}} w(z) - q(z) \frac{\partial}{\partial z} w(z) = 0, \qquad z \in \overline{E}. \tag{15}$$

Here we utilize *Hadamard's integral operator* from Definition 4.11 in Section 4 of Chapter 4

$$T_E[f](z) := -\frac{1}{\pi} \iint_E \frac{f(\zeta)}{\zeta - z} \, d\xi \, d\eta, \qquad z \in \overline{E} \tag{16}$$

with $\zeta = \xi + i\eta$. Here the function f lies in the Banach space $\mathcal{B} := C^{1+\alpha}(\overline{E}, \mathbb{C})$ endowed with the norm

$$\|f\| := \sup_{z \in E} \left\{ |f(z)| + |\nabla f(z)| \right\} + \sup_{\substack{z_1, z_2 \in E \\ z_1 \neq z_2}} \frac{|\nabla f(z_1) - \nabla f(z_2)|}{|z_1 - z_2|^\alpha}. \tag{17}$$

In the book of I. N. Vekua [V], namely Theorem 1.33 of Section 8 in Chapter 1, the following inequality is proved by potential-theoretic means:

$$\|T_E[f]\|_{C^{2+\alpha}(\overline{E})} \leq C_1(\alpha) \|f\|, \qquad f \in \mathcal{B}. \tag{18}$$

As in Proposition 5.9 from Chapter 4, Section 5 we define *Vekua's integral operator*

$$\Pi_E[f](z) := \lim_{\varepsilon \to 0+} \left\{ -\frac{1}{\pi} \iint_{\substack{\zeta \in E \\ |\zeta - z| > \varepsilon}} \frac{f(\zeta)}{(\zeta - z)^2} \, d\xi \, d\eta \right\}, \qquad z \in \overline{E}. \tag{19}$$

Due to the Theorem of I. N. Vekua given above, we have the estimate

$$\|\Pi_E[f]\| \leq C_2(\alpha) \|f\|, \qquad f \in \mathcal{B}, \tag{20}$$

with a constant $C_2(\alpha) \in (0, +\infty)$. Proposition 5.11 in Chapter 4, Section 5 provides the identities

$$\frac{\partial}{\partial \overline{z}} \{ T_E[f](z) \} = f(z), \quad \frac{\partial}{\partial z} \{ T_E[f](z) \} = \Pi_E[f](z), \qquad z \in \overline{E}. \tag{21}$$

In order to prove (20), we apply Theorem 4.7 from Chapter 9, Section 4 to the function $\frac{\partial}{\partial z} f$. We recall the identity

$$T_E\left[\frac{\partial}{\partial \zeta} f \right](z) = \Pi_E[f](z) - \frac{1}{2\pi i} \int_{\partial E} \frac{f(\zeta)}{\zeta^2(\zeta - z)} \, d\zeta, \qquad z \in E$$

from Proposition 5.12 in Chapter 4, Section 5; and we still have to estimate the curvilinear integral in the $C^{1+\alpha}(\overline{E})$-norm. The latter represents a holomorphic function in E, attaining certain Cauchy principal values over ∂E as boundary values, according to Theorem 2.1 from Chapter 9, Section 2 of Plemelj. We control them with the aid of Proposition 4.3 from Chapter 9, Section 4 and note Theorem 4.6 there. Then we can estimate the curvilinear integral in the $C^{1+\alpha}(\overline{E})$-norm and obtain (20). We use (21) and have shown (18) as well.

We now propose the *ansatz of L. Ahlfors and I. N. Vekua* for the solution of Beltrami's differential equation (15), namely

$$W(z) = z + T_E[f](z), \qquad z \in \overline{E}, \qquad \text{for} \quad f \in \mathcal{B}. \tag{22}$$

When we insert (22) into (15), we arrive at *Tricomi's integral equation* for $f \in \mathcal{B}$

$$f(z) - q(z)\Pi_E[f](z) = q(z), \qquad z \in \overline{E}, \tag{23}$$

with the aid of (21). We now consider the operator

$$\mathbb{L}f := q(z) + q(z)\Pi_E[f](z), \qquad z \in \overline{E}, \qquad \text{for} \quad f \in \mathcal{B}. \tag{24}$$

If the condition

$$\|q\|C_2(\alpha) < 1 \tag{25}$$

is fulfilled, the operator $\mathbb{L}$ on $\mathcal{B}$ becomes contracting. On account of (20), we have the following inequality for two elements $f_1, f_2 \in \mathcal{B}$, namely

$$
\begin{aligned}
\|\mathbb{L}f_1 - \mathbb{L}f_2\| &= \|q\,\Pi_E[f_1 - f_2]\| \\
&\leq \|q\|\,\|\Pi_E[f_1 - f_2]\| \\
&\leq \|q\|C_2(\alpha)\|f_1 - f_2\|.
\end{aligned} \tag{26}
$$

Given the assumption (25), the operator $\mathbb{L} : \mathcal{B} \to \mathcal{B}$ possesses exactly one fixed point $f \in \mathcal{B}$ with $\mathbb{L}f = f$, because of Banach's fixed point theorem. Now the function $f \in \mathcal{B}$ satisfies Tricomi's integral equation (23). We then obtain a solution of the differential equation (15) with the function $W(z)$ from (22), and the relation (18) implies $W \in C^{2+\alpha}(\overline{E})$. Furthermore, we infer the estimate

$$\|f\| \leq \frac{\|q\|}{1 - \|q\|C_2(\alpha)} \tag{27}$$

for the fixed point $f = \mathbb{L}f$ from (20) and (25). Due to (18), we can estimate the $C^{2+\alpha}(\overline{E})$-norm for the perturbation of the identity caused by $T_E[f](z)$ in (22). When we assume $\|q\|$ to be sufficiently small, the mapping

$$W(z) : \overline{E} \to \overline{\Omega} \in C^{2+\alpha}(\overline{E}, \mathbb{C}) \tag{28}$$

represents a positive-oriented diffeomorphism onto the Jordan domain $\Omega \subset \mathbb{C}$, with the $C^{2+\alpha}$-boundary $\partial\Omega$ constituting a Jordan curve. With the aid of results in Chapter 4, Sections 7-8, we now transform the set $\overline{\Omega}$ conformally onto the unit disc $\overline{B}$ via the mapping $X(w) : \overline{\Omega} \to \overline{B}$, such that $X \circ W(0) = 0$

is fulfilled. The mapping $X^{-1} : \overline{B} \to \overline{\Omega}$ belongs to the class $C^{1,1}(\overline{B})$ due to Theorem 8.7 in Chapter 4, Section 8. Then Theorem 7.4 from Section 7 implies

$$(X \circ W)^{-1} = W^{-1} \circ X^{-1} : \overline{B} \to \overline{E} \in C^{2+\alpha}(\overline{B}, \mathbb{C}),$$
$$(X \circ W)^{-1}(0) = 0.$$

We summarize our considerations to the following

Theorem 8.1. (Stability for conformal mappings)
The metric ds^2 from (1)*, (2)*, (3)* satisfies the following inequality with respect to the metric (2), namely

$$\|g_{jk} - \varrho\delta_{jk}\|_{C^{1+\alpha}(\overline{E})} < \delta \quad for \quad j, k = 1, 2 \tag{29}$$

with a sufficiently small number $\delta = \delta(\alpha, \varrho) > 0$. Then we have a weighted conformal diffeomorphism $\mathbf{x}(u, v) = (x^1(u, v), x^2(u, v)) \in C^{2+\alpha}(\overline{B}, \overline{E})$ which satisfies (4)*, (5)*, (8)*. Therefore, the metric ds^2 appears in the isothermal form (1).

By a nonlinear continuity method we now prove the uniformization theorem, which is of central significance for differential geometry, complex analysis, and the theory of partial differential equations. Already C. F. Gauß could conformally map analytic surface patches in the small, and L. Lichtenstein locally mapped differentiable surface patches conformally. Conformal mappings in the large have been constructed by P. Koebe in the analytic situation, and in the nonanalytic case C. B. Morrey, E. Heinz, L. Ahlfors, and I. N. Vekua attained similar results by different methods.

Theorem 8.2. (Uniformization theorem)
Let the Riemannian metric ds^2 from (1)*, (2)*, (3)* with the coefficients $g_{jk} \in C^{1+\alpha}(\overline{E})$ for $j, k = 1, 2$ be prescribed. Then we have a diffeomorphism $\mathbf{x} = \mathbf{x}(u, v) \in C^{2+\alpha}(\overline{B}, \overline{E})$ satisfying (4)*, (5)*, (8)*, which transfers the metric ds^2 into the isothermal form

$$ds^2 = \sigma(u, v)(du^2 + dv^2) \quad in \quad \overline{B}, \tag{30}$$

with the surface element $\sigma = \sigma(u, v) \in C^{1+\alpha}(\overline{B}, (0, +\infty))$.

Proof: We deform the metric ds^2 into the Euclidean metric via

$$ds^2(\tau) := g_{jk}^{(\tau)}(x^1, x^2) \, dx^j \, dx^k \quad in \quad \overline{E}, \qquad 0 \le \tau \le 1, \qquad with$$

$$g_{jk}^{(\tau)}(x^1, x^2) := (1 - \tau)\delta_{jk} + \tau g_{jk}(x^1, x^2), \qquad (x^1, x^2) \in \overline{E}, \qquad j, k = 1, 2. \tag{31}$$

For the parameter $\tau = 0$, the metric $ds^2(0) = \delta_{jk} \, dx^j \, dx^k$ is already isothermal. With the aid of Theorem 8.1, we then find a maximal number $\tau^* \in (0, 1]$, such that all metrics $ds^2(\tau)$ with $0 \le \tau < \tau^*$ can be transferred into the isothermal form. Theorem 7.4* now implies that also the metric $ds^2(\tau^*)$ can be transferred into the isothermal form, with the aid of the diffeomorphism $\mathbf{x} \in C^{2+\alpha}(\overline{B}, \overline{E})$ satisfying (4)*, (5)*, (8)*. If the inequality $\tau^* < 1$ were true, we could – due to Theorem 8.1 – transfer the metrics $ds^2(\tau)$ for all parameters $\tau^* \leq \tau < \tau^* + \varepsilon$ – with a sufficiently small number $\varepsilon > 0$ – into the isothermal form. Since the number $\tau^* \in (0,1]$ has been chosen maximal, the identity $\tau^* = 1$ holds true. Consequently, the metric

$$ds^2 = ds^2(1) = g_{jk}(x^1, x^2)\, dx^j\, dx^k$$

can be transferred into the isothermal form as described above.   q.e.d.

We finally note the following

Theorem 8.3. *For each Riemannian metric ds^2 from (1)*, (2)*, (3)* we have a $C^{2+\alpha}(B)$-diffeomorphism $\mathbf{x} = \mathbf{x}(u,v)$ satisfying (4)*,(5)*, (8)* which transfers ds^2 into the isothermal form*

$$ds^2 = \sigma(u,v)(du^2 + dv^2)   in   B \tag{32}$$

with the surface element $\sigma = \sigma(u,v) \in C^{1+\alpha}(B, (0, +\infty))$.

Proof: For all radii $r \in (0,1)$, we introduce isothermal parameters into the metric ds^2 on E_r – according to Theorem 8.2. With the aid of Theorem 7.2*, we then find a solution of (32) by approximation.

        q.e.d.

Remark to Theorem 8.3: We can derive this theorem alternatively by approximation with metrics being Euclidean at the boundary. In this context we refer the reader to the Remark following Theorem 7.4*.

9 The Uniformization Method for Quasilinear Elliptic Differential Equations

We consider the Jordan domain $\Omega \subset \mathbb{R}^2$ with the $C^{2+\alpha}$-boundary-curve $\partial\Omega$, and we investigate the quasilinear elliptic differential equation

$$a(x,y,z,p,q)r + 2b(x,y,z,p,q)s + c(x,y,z,p,q)t + d(x,y,z,p,q) = 0   in   \Omega$$
$$with   ac - b^2 > 0.$$
$$\tag{1}$$

Here we use the familiar symbols of G. Monge

$$p = z_x(x,y),   q = z_y(x,y),   r = z_{xx}(x,y),   s = z_{xy}(x,y),   t = z_{yy}(x,y)$$
$$\tag{2}$$

for the derivatives of a function $z = z(x,y) : \overline{\Omega} \to \mathbb{R} \in C^{2+\alpha}(\overline{\Omega})$. With adequate assumptions, we introduce isothermal parameters into the metric

$$ds^2 = c\,dx^2 - 2b\,dx\,dy + a\,dy^2 \tag{3}$$

via the uniformizing mapping

$$x + iy = f(w) = f(u,v) : \overline{B} \to \overline{\Omega}. \tag{4}$$

Here we apply the uniformization theorem from Section 8. The deliberations from Chapter 11, Section 3 can formally be repeated by substitution of the characteristic parameters ξ, η with the complex parameters $w, \overline{w}$. We now define the functions

$$\lambda^{\pm}(u,v) := \left.\frac{b \pm i\sqrt{ac - b^2}}{a}\right|_{x+iy=f(u,v)}. \tag{5}$$

Then we obtain the following system of first order, associated with the differential equation (1), in the same way as in the theory of characteristics quoted above:

$$y_w - \lambda^+ x_w = 0, \qquad y_{\overline{w}} - \lambda^- x_{\overline{w}} = 0,$$

$$p_w + \lambda^- q_w + \frac{d}{a} x_w = 0, \qquad p_{\overline{w}} + \lambda^+ q_{\overline{w}} + \frac{d}{a} x_{\overline{w}} = 0, \tag{6}$$

$$z_w - p x_w - q y_w = 0$$

(with $z = z \circ f(w)$ etc.). In these equations the derivatives with respect to w and $\overline{w}$, respectively, only appear separately. Therefore, we differentiate the equations containing $\frac{\partial}{\partial w}$ with respect to $\overline{w}$, and the equations containing $\frac{\partial}{\partial \overline{w}}$ are derived with respect to w. We obtain a linear system of equations for the functions $x_{w\overline{w}}, y_{w\overline{w}}, z_{w\overline{w}}, p_{w\overline{w}}, q_{w\overline{w}}$, which we can resolve to these quantities as in Chapter 11, Section 3. We introduce the function

$$\mathbf{x}(w) = \mathbf{x}(u,v) = (x(u,v), y(u,v), z(u,v), p(u,v), q(u,v)) \qquad \text{in } \overline{B}, \tag{7}$$

and obtain a system

$$\Delta \mathbf{x}(w) = \mathbf{\Phi}(\mathbf{x}(u,v), \mathbf{x}_u(u,v), \mathbf{x}_v(u,v)), \qquad w = u + iv \in B, \tag{8}$$

with quadratic growth in the gradient. One can deduce results for the differential equation (1) via the system (8) combined with the equations of first order (6). Estimates for the uniformizing mapping f then guarantee the independence of the parametrization.

We remark that the system (8) is deduced by real differentiation from the differential equation (1) in the original papers of F. Müller, which have been quoted in Chapter 11, Section 6.

10 Some Historical Notices to Chapter 12

We owe to C.F. Gauß, already in 1827, the introduction of isothermal parameters in the small for real-analytic surfaces. L. Lichtenstein developed his ideas for conformal mappings between nonanalytic surfaces from 1911 to 1916. T. Carleman made about 1930 the profound observation that the class of pseudoholomorphic functions share the property of isolated zeroes with the much smaller class of holomorphic functions. This fact was utilized by P. Hartman and A. Wintner in 1953 for the investigation of singularities on nonanalytic surfaces.

The significance of isothermal parameters for surfaces of prescribed mean curvature was revealed by F. Rellich: In 1938 he established his H-surface-system. Then E. Heinz solved the Dirichlet problem for this system by topological methods in 1954. Moreover, he developed the profound theory of nonlinear elliptic systems, in 1956/57, presented here. With his pioneering paper from 1952, E. Heinz also initiated the still flourishing study of curvature estimates.

In 1976, W. Jäger presented a uniqueness result for the Dirichlet problem of nonlinear elliptic systems with his well-known maximum principle. We should note that existence and regularity questions are quite well understood today. However, the study of the entire set of solutions and their classification remains a great challenge for the theory of nonlinear partial differential equations.

Figure 2.8   PORTRAIT OF F. RELLICH (1906–1955)
taken from the biography by *C. Reid: Courant, in Göttingen and New York – An Album*, Springer-Verlag, Berlin... (1976).

Chapter 13

Boundary Value Problems from Differential Geometry

On the basis of the theory for nonlinear elliptic systems from Chapter 12, we shall now present applications to differential geometric boundary value problems. At first, we solve the Dirichlet problem for the nonparametric equation of prescribed mean curvature by a continuity method in Section 1. Then we consider in Section 2 and Section 3 minimal graphs on Riemannian surfaces, solving a semi-free boundary value problem.

Later in Section 4, we shall provide a solution to Plateau's problem for surfaces of constant mean curvature by direct variational methods. In Section 5 we discuss the famous question, which class of closed surfaces with constant mean curvature contains only spheres? Finally, we give an outlook to Monge-Ampère equations in Section 7 and study the nonvanishing of the Jacobian for one-to-one solutions of the associate Lewy-Heinz system in Section 6.

1 The Dirichlet Problem for Graphs of Prescribed Mean Curvature

With the aid of the uniformization method, we now shall solve Dirichlet's problem for the nonparametric equation of prescribed mean curvature. We choose the bound $0 < h_0 < +\infty$, define the disc

$$\Omega_0 := \left\{ (x, y) \in \mathbb{R}^2 \; : \; 4h_0^2(x^2 + y^2) \leq 1 \right\},$$

and fix a Hölder exponent $\alpha \in (0, 1)$.

Assumption D_1: The bounded domain $\Omega \subset \Omega_0$ may have a regular $C^{2+\alpha}$-Jordan-curve $\partial \Omega$ as its boundary, whose curvature satisfies the inequality $\kappa(x, y) \geq 2h_0$ for all points $(x, y) \in \partial \Omega$. Furthermore, let the condition $(0, 0) \in \Omega$ be fulfilled.

F. Sauvigny, *Partial Differential Equations 2*, Universitext,
DOI 10.1007/978-1-4471-2984-4_7, © Springer-Verlag London 2012

Figure 2.9   GRAPHIC OF A $(2h_0)$-CONVEX DOMAIN

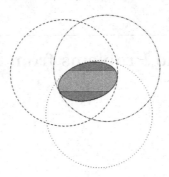

Assumption D_2: On the circular cylinder

$$\mathcal{Z} := \Big\{ (x, y, z) \in \mathbb{R}^3 \; : \; (x, y) \in \Omega_0 \Big\}$$

we prescribe the mean curvature

$$H = H(x, y, z) : \mathcal{Z} \to \mathbb{R} \in C^{1+\alpha}(\mathcal{Z})$$

with the following properties:

– We have a height $z_0 \in \mathbb{R}$ and a number $H_0 \in [-h_0, +h_0]$, such that the relation

$$H(x, y, z) = H_0 \qquad \text{for all} \quad (x, y, z) \in \mathcal{Z} \quad \text{with} \quad z \le z_0 \qquad (1)$$

holds true.
– We require the monotonicity

$$\frac{\partial}{\partial z} H(x, y, z) \ge 0 \qquad \text{for all} \quad (x, y, z) \in \mathcal{Z}. \qquad (2)$$

– Finally, we assume the bound

$$|H(x, y, z)| \le h_0 \qquad \text{for all} \quad (x, y, z) \in \mathcal{Z}. \qquad (3)$$

According to Section 2 in Chapter 6, the following problem possesses at most one solution.

Definition 1.1. *The height representation $g : \partial\Omega \to \mathbb{R} \in C^0(\partial\Omega, \mathbb{R})$ being given as continuous function, we consider a solution $z = \zeta(x, y) \in C^2(\Omega) \cap C^0(\overline{\Omega})$ of Dirichlet's problem $\mathcal{P}(g)$ for the nonparametric equation of prescribed mean curvature*

$$\mathcal{M}\zeta(x, y) := (1 + \zeta_y^2)\zeta_{xx} - 2\zeta_x\zeta_y\zeta_{xy} + (1 + \zeta_x^2)\zeta_{yy}$$

$$= 2H(x, y, \zeta(x, y))\big(1 + |\nabla\zeta(x, y)|^2\big)^{\frac{3}{2}} \qquad in \quad \Omega \qquad (4)$$

and

$$\zeta(x,y) = g(x,y) \qquad \text{for all} \quad (x,y) \in \partial\Omega. \tag{5}$$

We additionally set

$$\|g\|_{C^0(\partial\Omega)} := \sup_{(x,y)\in\partial\Omega} |g(x,y)|.$$

Proposition 1.2. (R. Finn)
A solution $\zeta \in \mathcal{P}(g)$ of our problem with the boundary distribution $g \in C^0(\partial\Omega, \mathbb{R})$ satisfies the following estimates:

(a) $\quad |\zeta(x,y)| \le \|g\|_{C^0(\partial\Omega)} + \dfrac{1}{h_0} \qquad \text{for all} \quad (x,y) \in \Omega,$

(b) $\quad \displaystyle\iint_\Omega \sqrt{1 + |\nabla\zeta(x,y)|^2}\, dx\, dy \le 3|\Omega| + \big(2h_0|\Omega| + |\partial\Omega|\big)\|g\|_{C^0(\partial\Omega)}.$

Here the symbols $|\Omega|$ and $|\partial\Omega|$ denote the area of the domain Ω and the length of the curve $\partial\Omega$, respectively.

Proof:

(a) We consider the spherical graphs

$$\eta^\pm(x,y) := \pm\|g\|_{C^0(\partial\Omega)} \pm \sqrt{\frac{1}{h_0^2} - (x^2 + y^2)}, \qquad (x,y) \in \Omega_0. \tag{6}$$

These fulfill the differential inequalities

$$\mathcal{M}\eta^\pm(x,y) = \pm 2h_0\big(1 + |\nabla\eta^\pm(x,y)|^2\big)^{\frac{3}{2}}$$
$$\begin{array}{c}\ge\\[-4pt]\le\end{array} 2H(x,y,\eta^\pm(x,y))\big(1 + |\nabla\eta^\pm(x,y)|^2\big)^{\frac{3}{2}} \qquad \text{in} \quad \Omega. \tag{7}$$

We now deduce a differential inequality for the function

$$\phi(x,y) := \zeta(x,y) - \eta^\pm(x,y) \quad \text{in} \quad \Omega$$

as in Section 2 from Chapter 6, taking (2) into account. Then the maximum principle yields

$$\eta^-(x,y) \le \zeta(x,y) \le \eta^+(x,y) \qquad \text{in} \quad \Omega. \tag{8}$$

This implies the estimate (a).

(b) We rewrite (4) into the divergence form, abbreviate $\sqrt{} := \sqrt{1 + |\nabla\zeta|^2}$, and obtain

$$\zeta\frac{\partial}{\partial x}\left(\frac{\zeta_x}{\sqrt{}}\right) + \zeta\frac{\partial}{\partial y}\left(\frac{\zeta_y}{\sqrt{}}\right) = 2H(x, y, \zeta)\zeta.$$

Then we integrate over the domain Ω as follows:

$$2\iint_\Omega \zeta(x, y)H(x, y, \zeta(x, y))\,dx\,dy$$

$$= \iint_\Omega \left\{\frac{\partial}{\partial x}\left(\zeta\frac{\zeta_x}{\sqrt{}}\right) + \frac{\partial}{\partial y}\left(\zeta\frac{\zeta_y}{\sqrt{}}\right)\right\}\,dx\,dy - \iint_\Omega \frac{|\nabla\zeta|^2}{\sqrt{}}\,dx\,dy$$

$$= \int_{\partial\Omega} \zeta\left(\frac{\zeta_x}{\sqrt{}}\,dy - \frac{\zeta_y}{\sqrt{}}\,dx\right) - \iint_\Omega \sqrt{}\,dx\,dy + \iint_\Omega \frac{1}{\sqrt{}}\,dx\,dy.$$

We note (a) and estimate

$$\iint_\Omega \sqrt{1 + |\nabla\zeta|^2}\,dx\,dy$$

$$= \int_{\partial\Omega} \zeta\left(\frac{\zeta_x}{\sqrt{}}\,dy - \frac{\zeta_y}{\sqrt{}}\,dx\right) + \iint_\Omega \frac{1}{\sqrt{}}\,dx\,dy - 2\iint_\Omega \zeta H(x, y, \zeta)\,dx\,dy$$

$$\leq \|g\|_{C^0(\partial\Omega)}|\partial\Omega| + |\Omega| + 2\left(\|g\|_{C^0(\partial\Omega)} + \frac{1}{h_0}\right)h_0|\Omega|,$$

and (b) is shown as well.   q.e.d.

Via Theorem 8.3 from Section 8 in Chapter 12, we now introduce conformal parameters into the graph $\zeta \in \mathcal{P}(g)$ by the uniformizing mapping

$$f = f(u, v) : \overline{B} \to \overline{\Omega} \in C^2(B) \cap C^0(\overline{B}) \quad \text{diffeomorphic,}$$
$$f(0, 0) = (0, 0). \tag{9}$$

Then the function

$$\mathbf{x}(u, v) := \big(f(u, v), \zeta(f(u, v))\big), \qquad (u, v) \in \overline{B} \tag{10}$$

is an H-surface in the following sense:

Definition 1.3. *A nonconstant solution* $\mathbf{x} \in C^2(B, \mathbb{R}^3)$ *of the system*

$$\Delta\mathbf{x}(u, v) = 2H(\mathbf{x}(u, v))\mathbf{x}_u \wedge \mathbf{x}_v(u, v) \qquad \text{in} \quad B,$$
$$|\mathbf{x}_u(u, v)|^2 - |\mathbf{x}_v(u, v)|^2 = 0 = \mathbf{x}_u \cdot \mathbf{x}_v(u, v) \qquad \text{in} \quad B \tag{11}$$

is called an H-surface. This surface is called immersed *or* free of branch points *if the condition*

$$E(u,v) := |\mathbf{x}_u \wedge \mathbf{x}_v(u,v)| > 0 \qquad \text{for all} \quad (u,v) \in B$$

is valid.

Proposition 1.4. *The normal* $\mathbf{X}(u,v) \in C^{2+\alpha}(B)$ *to the immersed H-surface* $\mathbf{x}$ *satisfies the following differential equation*

$$\Delta\mathbf{X}(u,v) + 2\Big(2E(H(\mathbf{x}))^2 - EK - E(\nabla H(\mathbf{x}) \cdot \mathbf{X})\Big)\mathbf{X} = -2E\nabla H(\mathbf{x}) \quad \text{in} \quad B,$$
$$(12)$$

with the notations from Chapter 11, Section 1.

Proof: From the Weingarten fundamental equations (compare [BL]) in conformal parameters

$$\mathbf{X}_u = -\frac{L}{E}\mathbf{x}_u - \frac{M}{E}\mathbf{x}_v, \qquad \mathbf{X}_v = -\frac{M}{E}\mathbf{x}_u - \frac{N}{E}\mathbf{x}_v$$

we infer the identities

$$(\mathbf{X} \wedge \mathbf{X}_v)_u - (\mathbf{X} \wedge \mathbf{X}_u)_v = 2\mathbf{X}_u \wedge \mathbf{X}_v = 2\frac{LN - M^2}{E^2}\mathbf{x}_u \wedge \mathbf{x}_v = 2EK\mathbf{X}$$

and

$$\mathbf{X} \wedge \mathbf{X}_u = -\mathbf{X}_v - 2H(\mathbf{x})\mathbf{x}_v, \qquad \mathbf{X} \wedge \mathbf{X}_v = \mathbf{X}_u + 2H(\mathbf{x})\mathbf{x}_u.$$

On account of

$$\big\{H(\mathbf{x}(u,v))\big\}_u = \nabla H(\mathbf{x}) \cdot \mathbf{x}_u, \qquad \big\{H(\mathbf{x}(u,v))\big\}_v = \nabla H(\mathbf{x}) \cdot \mathbf{x}_v$$

we obtain

$$\begin{aligned}
2EK\mathbf{X} &= (\mathbf{X} \wedge \mathbf{X}_v)_u - (\mathbf{X} \wedge \mathbf{X}_u)_v \\
&= \mathbf{X}_{uu} + 2(\nabla H \cdot \mathbf{x}_u)\mathbf{x}_u + 2H\mathbf{x}_{uu} + \mathbf{X}_{vv} + 2(\nabla H \cdot \mathbf{x}_v)\mathbf{x}_v + 2H\mathbf{x}_{vv} \\
&= \Delta\mathbf{X} + 4EH^2\mathbf{X} + 2\big((\nabla H \cdot \mathbf{x}_u)\mathbf{x}_u + (\nabla H \cdot \mathbf{x}_v)\mathbf{x}_v\big).
\end{aligned}$$
$$(13)$$

Now we expand

$$\nabla H = \left(\nabla H \cdot \frac{\mathbf{x}_u}{|\mathbf{x}_u|}\right)\frac{\mathbf{x}_u}{|\mathbf{x}_u|} + \left(\nabla H \cdot \frac{\mathbf{x}_v}{|\mathbf{x}_v|}\right)\frac{\mathbf{x}_v}{|\mathbf{x}_v|} + (\nabla H \cdot \mathbf{X})\mathbf{X}$$

and deduce

$$(\nabla H \cdot \mathbf{x}_u)\mathbf{x}_u + (\nabla H \cdot \mathbf{x}_v)\mathbf{x}_v = E\nabla H - E(\nabla H \cdot \mathbf{X})\mathbf{X}. \qquad (14)$$

The formulas (13) and (14) imply the differential equation (12).   q.e.d.

Theorem 1.5. (Compactness of graphs)
With the assumptions (D_1) and (D_2), let the boundary distributions

$$g_k \in C^0(\partial\Omega, \mathbb{R}) \quad for \quad k = 1, 2, \ldots$$

be given, and our problem may possess a solution $\zeta_k \in \mathcal{P}(g_k)$ for each function g_k. Furthermore, let the sequence $\{g_k\}_{k=1,2,\ldots}$ converge uniformly on $\partial\Omega$ towards the limit function

$$g(x) := \lim_{k\to\infty} g_k(x) \ \in \ C^0(\partial\Omega, \mathbb{R}).$$

Then also the limit problem $\mathcal{P}(g)$ possesses a solution ζ.

Proof:

1. As described in (9)-(10), we introduce conformal parameters into the graphs ζ_k by the uniformizing mappings $f_k = f_k(u,v) : \overline{B} \to \overline{\Omega}$. Then we obtain the immersed H-surfaces

$$\mathbf{x}_k(u,v) := \big(f_k(u,v), \zeta_k(f_k(u,v))\big) =: \big(f_k(u,v), z_k(u,v)\big), \ (u,v) \in \overline{B}.$$
(15)

   Because of Proposition 1.2 of R. Finn, this sequence has a uniformly bounded Dirichlet's integral. Via the Courant-Lebesgue lemma combined with the geometric maximum principle of E. Heinz, we prove that the sequence of functions $\{\mathbf{x}_k\}_{k=1,2,\ldots}$ is equicontinuous on the domain $\overline{B}$. The Arzelà-Ascoli theorem allows the transition to a uniformly convergent subsequence on $\overline{B}$, converging towards an H-surface

$$\mathbf{x}(u,v) = \big(f(u,v), z(u,v)\big) : \overline{B} \to \mathbb{R}^3 \in C^2(B) \cap C^0(\overline{B}), \qquad (16)$$

   on account of Theorem 2.7 in Section 2 of Chapter 12.

2. Since the surface $\mathbf{x}_k$ is conformally parametrized, we can eliminate the third component according to Proposition 7.3 from Section 7 in Chapter 12; more precisely

$$|\nabla z_k(u,v)|^2 \leq |\nabla f_k(u,v)|^2 \quad in \quad B \quad for \quad k = 1, 2, \ldots \qquad (17)$$

   We then obtain the sequence of plane mappings

$$
\begin{aligned}
&f_k(u,v) : \overline{B} \to \overline{\Omega} \in C^2(B) \cap C^0(\overline{B}) \qquad \text{diffeomorphic,} \\
&|\Delta f_k(u,v)| \leq c_1|\nabla f_k(u,v)|^2 \quad in \quad B, \\
&f_k(0,0) = (0,0), \\
&D(f_k) \leq c_2 \qquad for \quad k = 1, 2, \ldots
\end{aligned}
$$
(18)

   with the constants c_1, c_2 and Dirichlet's integral $D(f_k)$ of f_k. The distortion estimate of E. Heinz from Section 5 of Chapter 12 provides a constant $\Theta(c_1, c_2, r) > 0$ for each radius $r \in (0,1)$, such that the inequality

$$|\nabla f_k(u, v)| \geq \Theta(c_1, c_2, r) \qquad \text{for all points} \quad (u, v) \in B_r \qquad (19)$$

is satisfied on the disc $B_r := \{(u, v) \in B : u^2 + v^2 < r^2\}$. Here we replace the image domain B with the domain Ω in the proof of Theorem 5.4 from Section 5 of Chapter 12. On account of (19), we find the estimate

$$|\nabla f(u, v)| > 0 \qquad \text{in} \quad B \qquad (20)$$

for the limit mapping, and the H-surface $\mathbf{x}$ from (16) is immersed.

3. With the normal $\mathbf{X}(u, v)$ for the surface $\mathbf{x}(u, v)$ we associate the auxiliary function

$$\phi(u, v) := \mathbf{X}(u, v) \cdot \mathbf{e} \geq 0, \qquad (u, v) \in B, \qquad (21)$$

where the vector $\mathbf{e} := (0, 0, 1)$ appears. Introducing the potential-function

$$q(u, v) := 2\Big(2E(H(\mathbf{x}))^2 - EK - E(\nabla H(\mathbf{x}) \cdot \mathbf{X})\Big),$$

Proposition 1.4, together with (2), yields the differential inequality

$$\Delta\phi(u, v) + q(u, v)\phi(u, v) \leq 0 \qquad \text{in} \quad B. \qquad (22)$$

Via multiplication by a nonnegative test function and integration, we arrive at the *weak differential inequality*

$$\iint\limits_{B} \Big\{ \nabla\phi(u, v) \cdot \nabla\psi(u, v) - q(u, v)\phi(u, v)\psi(u, v) \Big\} \, du \, dv \geq 0 \qquad (23)$$

for all $\psi \in C_0^\infty(B)$ with $\psi \geq 0$ in B.

Now Moser's inequality (compare Theorem 5.1 from Chapter 10, Section 5) pertains to solutions of these differential inequalities, and the function ϕ is subject to the principle of unique continuation. Consequently, the function ϕ on B must vanish if at least one zero appears in B. Since the case $\phi \equiv 0$ in B is evidently excluded, we infer

$$\phi(u, v) > 0 \qquad \text{for all} \quad (u, v) \in B$$

and finally

$$J_f(u, v) := \frac{\partial(x, y)}{\partial(u, v)} > 0 \qquad \text{in} \quad B. \qquad (24)$$

Consequently, the function $f : \overline{B} \to \overline{\Omega}$ represents a diffeomorphism of the class $C^2(B) \cap C^0(\overline{B})$, when we additionally observe the following: The boundary mapping $f|_{\partial B}$ is weakly monotonic, at first. However, this function cannot develop intervals where it remains constant. Otherwise, we could easily derive the statement $\mathbf{x}_w(w) \equiv 0$ in B, with the aid of the conformality relations and the similarity principle: This is impossible, of course! With the function

$$\zeta(x, y) := z\big(f^{-1}(x, y)\big), \qquad (x, y) \in \overline{\Omega}$$

we finally obtain a solution of the problem $\mathcal{P}(g)$.   q.e.d.

Proposition 1.6. (Geometric maximum principle of S. Hildebrandt)
The auxiliary function $\phi(u,v) := x(u,v)^2 + y(u,v)^2$, $(u,v) \in \overline{B}$, associated with the H-surface $\mathbf{x}(u,v) = (x(u,v), y(u,v), z(u,v)) : \overline{B} \to \mathcal{Z}$, satisfies the differential inequality

$$\Delta\phi(u,v) \geq 0 \quad in \quad B.$$

Proof: We calculate

$$\Delta\phi(u,v) = 2\big(|\nabla x|^2 + |\nabla y|^2 + x\,\Delta x + y\,\Delta y\big)$$
$$= 2\big(|\nabla x|^2 + |\nabla y|^2 + 2H(\mathbf{x})(x,y,0) \cdot \mathbf{x}_u \wedge \mathbf{x}_v\big).$$

Since the surface $\mathbf{x}$ is represented in conformal parameters, we infer

$$|\nabla z|^2 \leq |\nabla x|^2 + |\nabla y|^2 \quad in \quad B$$

and consequently

$$|2H(\mathbf{x})(x,y,0) \cdot \mathbf{x}_u \wedge \mathbf{x}_v| \leq 2h_0 \frac{1}{2h_0} \frac{1}{2}\big(|\nabla x|^2 + |\nabla y|^2 + |\nabla z|^2\big)$$

$$\leq |\nabla x|^2 + |\nabla y|^2 \quad in \quad B.$$

We summarize our considerations to the inequality $\Delta\phi(u,v) \geq 0$ in B. q.e.d.

With a fundamental boundary regularity result of S. Hildebrandt, J. C. C. Nitsche, F. Tomi, and E. Heinz we prove the following

Theorem 1.7. (Regularity of graphs)
With the assumptions (D_1) and (D_2) being given, let $\zeta \in \mathcal{P}(g)$ denote a solution of our problem to the boundary distribution $g \in C^{2+\alpha}(\partial\Omega, \mathbb{R})$. Then we have the regularity statement $\zeta = \zeta(x,y) \in C^{2+\alpha}(\overline{\Omega})$.

Proof: We investigate the H-surface

$$\mathbf{x}(u,v) = (f(u,v), \zeta(f(u,v)), \quad (u,v) \in \overline{B}$$

again, which belongs to the regularity class $C^2(B) \cap C^0(\overline{B})$. From [DHKW] 7.3, Theorem 2 we infer the regularity statement $\mathbf{x} \in C^{2+\alpha}(\overline{B})$; compare also the Chapter 2 in [DHT1]. We already know the condition

$$J_f(u,v) > 0 \quad in \quad B$$

for the Jacobian of the uniformizing mapping, and we intend to establish this estimate on the closed disc ∂B as well. Let the point $w_0 \in \partial B$ be chosen arbitrarily, and let $x_0 := x(w_0)$, $y_0 := y(w_0)$ be defined. Via a translation of the domain $\Omega \subset \Omega_0$, we can achieve the condition $(x_0, y_0) \in \partial\Omega_0$. According to the boundary point lemma of E. Hopf, now Proposition 1.2 implies the following inequality for the auxiliary function ϕ in polar coordinates $w = re^{i\vartheta}$, namely

$$0 < \frac{1}{2}\frac{\partial \phi}{\partial r}\bigg|_{w_0} = (xx_r + yy_r)\bigg|_{w_0}. \tag{25}$$

Since the function $\phi(\vartheta) := \phi(\cos\vartheta, \sin\vartheta)$ attains its maximum at the point ϑ_0, we infer the identity

$$0 = \frac{1}{2}\frac{\partial \phi}{\partial \vartheta}\bigg|_{\vartheta_0} = (xx_\vartheta + yy_\vartheta)\bigg|_{w_0}. \tag{26}$$

The relation (25) implies that w_0 does not represent a branch point, more precisely

$$|\mathbf{x}_\vartheta(w_0)|^2 = |\mathbf{x}_r(w_0)|^2 > 0. \tag{27}$$

Furthermore, we have a number $K > 0$ satisfying

$$z_\vartheta^2 \leq K(x_\vartheta^2 + y_\vartheta^2). \tag{28}$$

From (27) and (28) we infer

$$0 < (x_\vartheta^2 + y_\vartheta^2 + z_\vartheta^2)\bigg|_{w_0} \leq (1+K)(x_\vartheta^2 + y_\vartheta^2)\bigg|_{w_0}$$

and consequently

$$(x_\vartheta^2 + y_\vartheta^2)\bigg|_{w_0} > 0. \tag{29}$$

Since the mapping f is positive-oriented, we find a parameter $\lambda > 0$ satisfying

$$x_\vartheta(w_0) = -\lambda y(w_0), \qquad y_\vartheta(w_0) = \lambda x(w_0),$$

on account of (26). This implies

$$(x_r y_\vartheta - x_\vartheta y_r)\bigg|_{w_0} = \lambda(xx_r + yy_r)\bigg|_{w_0} > 0$$

and equivalently

$$J_f(w_0) > 0.$$

Therefore, the function $f : \overline{B} \to \overline{\Omega}$ represents a $C^{2+\alpha}(\overline{B})$-diffeomorphism, and the height function

$$\zeta(x,y) := z(f^{-1}(x,y)), \quad (x,y) \in \overline{\Omega}$$

belongs to the regularity class $C^{2+\alpha}(\overline{\Omega})$.   q.e.d.

The following result was initiated by considerations of J.C.C. Nitsche for minimal surfaces.

Proposition 1.8. (Stability of graphs) *For the boundary distribution $g \in C^{2+\alpha}(\partial\Omega)$ let $\zeta = \zeta(x,y) \in \mathcal{P}(g)$ denote a solution of the class $C^{2+\alpha}(\overline{\Omega})$. Then we have a quantity $\varepsilon = \varepsilon(\zeta) > 0$, such that all boundary distributions*

$$\tilde{g} \in C^{2+\alpha}(\partial\Omega) \quad \text{satisfying} \quad \|\tilde{g} - g\|_{C^{2+\alpha}(\partial\Omega)} \leq \varepsilon$$

possess a solution of the problem $\mathcal{P}(\tilde{g})$.

Proof: We solve the problem $\mathcal{P}(\tilde{g})$ via perturbation with a function $\eta(x,y) \in C^{2+\alpha}(\overline{\Omega})$. In this context we have to ascertain that besides the function ζ the perturbed function $\zeta + \eta$ satisfies the differential equation (4) as well: From the identity

$$0 = \left(1 + (\zeta_y + \eta_y)^2\right)(\zeta_{xx} + \eta_{xx}) - 2(\zeta_x + \eta_x)(\zeta_y + \eta_y)(\zeta_{xy} + \eta_{xy})$$
$$+\left(1 + (\zeta_x + \eta_x)^2\right)(\zeta_{yy} + \eta_{yy}) - 2H(x,y,\zeta+\eta)\left(1 + |\nabla(\zeta+\eta)|^2\right)^{\frac{3}{2}} \tag{30}$$

we deduce the following differential equation - ordered with respect to the degree of homogeneity in $\eta, \eta_x, \ldots, \eta_{yy}$, namely

$$\mathcal{L}\eta(x,y) = \phi(\eta) \qquad \text{in} \quad \Omega. \tag{31}$$

Here the symbol

$$\mathcal{L}\eta := (1 + \zeta_y^2)\eta_{xx} - 2\zeta_x\zeta_y\eta_{xy} + (1 + \zeta_x^2)\eta_{yy}$$
$$+a(x,y)\eta_x + b(x,y)\eta_y + c(x,y)\eta$$

denotes a linear elliptic differential operator, with coefficients depending on the quantities $\zeta, \zeta_x, \ldots, \zeta_{yy}$. We observe the condition $c(x,y) \leq 0$ in Ω, according to (2). The right-hand side is quadratic and of a higher order in $\eta, \eta_x, \ldots, \eta_{yy}$ and consequently satisfies the *contraction condition*

$$\|\phi(\eta_1) - \phi(\eta_2)\|_{C^\alpha(\overline{\Omega})} \leq C(\varrho)\|\eta_1 - \eta_2\|_{C^{2+\alpha}(\overline{B})}$$
$$\text{for all} \quad \eta_j \in C^{2+\alpha}(\overline{\Omega}) \quad \text{with} \quad \|\eta_j\|_{C^{2+\alpha}(\overline{\Omega})} \leq \varrho \quad \text{and} \quad j = 1,2. \tag{32}$$

Here the property $C(\varrho) \to 0$ for $\varrho \to 0+$ is correct, and we note $\phi(0) = 0$. With the aid of the Schauder theory from Section 6 in Chapter 9, we now solve the linear problem

$$\mathcal{L}\eta = \omega \qquad \text{in} \quad \Omega,$$
$$\eta = \psi \qquad \text{on} \quad \partial\Omega \tag{33}$$

uniquely by a function $\eta \in C^{2+\alpha}(\overline{\Omega})$ – for each right-hand side $\omega \in C^\alpha(\overline{\Omega})$ and all boundary values $\psi \in C^{2+\alpha}(\partial\Omega)$. For vanishing boundary values $\psi \equiv 0$ on $\partial\Omega$, we set

$$C_*^{2+\alpha}(\overline{\Omega}) := \left\{\eta \in C^{2+\alpha}(\overline{\Omega}) \,:\, \eta = 0 \text{ on } \partial\Omega\right\}.$$

We use the symbol $\mathcal{L}_0 := \mathcal{L}|_{C_*^{2+\alpha}(\overline{\Omega})}$ for the restriction of the operator $\mathcal{L}$ to the subspace $C_*^{2+\alpha}(\overline{\Omega})$. Then the operator

$$\omega = \mathcal{L}_0(\eta) : C_*^{2+\alpha}(\overline{\Omega}) \to C^\alpha(\overline{\Omega}) \tag{34}$$

is invertible, and we have Schauder's estimate

$$\|\mathcal{L}_0^{-1}(\omega)\|_{C^{2+\alpha}(\overline{\Omega})} \leq C\|\omega\|_{C^\alpha(\overline{\Omega})} \qquad \text{for all} \quad \omega \in C^\alpha(\overline{\Omega}), \tag{35}$$

according to Theorem 5.2 in Section 5 from Chapter 9. Given sufficiently small boundary values satisfying $\|\psi\|_{C^{2+\alpha}(\partial\Omega)} \leq \varepsilon$, we solve

$$\mathcal{L}\eta_0 = 0 \quad \text{in} \quad \Omega,$$
$$\eta_0 = \psi \quad \text{on} \quad \partial\Omega. \tag{36}$$

Here we have to estimate the solution by its boundary values with respect to the $C^{2+\alpha}$-norm: At first, we estimate the solution $\eta_0(x,y)$ in the C^0-norm by its boundary values, according to Theorem 1.3 (Uniqueness and stability) from Chapter 6, Section 1. With the aid of the Schauder theory from Section 7 of Chapter 9, we secondly estimate the solution in the $C^{2+\alpha}$-norm by its boundary values. Here we locally straighten the boundary of the domain, and then we can extend the boundary values into the ambient space – without augmenting their $C^{2+\alpha}$-norm. Subtracting the extended boundary values, we get an inhomogeneous differential equation with zero boundary values, from which we gain our Schauder estimates. Now we iterate

$$\mathcal{L}\eta_{k+1} = \phi(\eta_k) \quad \text{in} \quad \Omega,$$
$$\eta_{k+1} = \psi \quad \text{on} \quad \partial\Omega \tag{37}$$

for $k = 0, 1, 2, \ldots$. With the aid of (32) and (35), we see that the sequence $\{\eta_k\}_{k=1,2,\ldots}$ converges towards a solution $\eta \in C^{2+\alpha}(\overline{\Omega})$ of (31) in the Banach space $C^{2+\alpha}(\overline{\Omega})$; here we choose the number $\varepsilon > 0$ sufficiently small.   q.e.d.

Theorem 1.9. (Quasilinear Dirichlet problem)
With the assumptions (D_1) and (D_2), we take an arbitrary boundary function $g \in C^0(\partial\Omega, \mathbb{R})$. Then the Dirichlet problem $\mathcal{P}(g)$ for the nonparametric equation of prescribed mean curvature possesses exactly one solution.

Proof: We note the condition (1) and find a spherical graph $\eta(x,y) : \overline{\Omega} \to \mathbb{R} \in C^{2+\alpha}(\overline{\Omega})$ of constant mean curvature H_0, such that the differential equation (4) is fulfilled with the boundary values

$$f(x,y) := \eta(x,y), \qquad (x,y) \in \partial\Omega.$$

We solve the problem $\mathcal{P}(g_\lambda)$ for the family of boundary values

$$g_\lambda(x,y) := f(x,y) + \lambda\big(g(x,y) - f(x,y)\big), \qquad (x,y) \in \partial\Omega, \tag{38}$$

with $0 \leq \lambda \leq 1$ and $g \in C^{2+\alpha}(\partial\Omega, \mathbb{R})$. This has already been done for the start parameter $\lambda = 0$, and the solvability is – due to Proposition 1.8 – an open, and – due to Theorem 1.5 – also a closed property. Consequently, the Dirichlet problem $\mathcal{P}(g_\lambda)$ is solvable for all $0 \leq \lambda \leq 1$. Considering the terminal parameter, the problem $\mathcal{P}(g)$ possesses a solution $\zeta \in C^{2+\alpha}(\overline{\Omega})$. With the aid of Theorem 1.5, we then comprehend the solvability of Dirichlet's problem for continuous boundary values. The uniqueness has already been shown in Section 2 of Chapter 6.   q.e.d.

Remarks:

1.) This approach to the Dirichlet problem is contained in the 2001 paper [S5] by Friedrich Sauvigny. The differential equation for the normal of an H-surface from Proposition 1.4 had already been derived in Section 1 of the treatise [S3] from 1992 – even for branched surfaces with prescribed mean curvature. The latter paper [S3] was based on the dissertation of the author, presented in 1981 to the Fachbereich Mathematik der Georg-August-Universität Göttingen.

2.) This method was utilized by Matthias Bergner to obtain results even for nonconvex and multiply connected domains as well as for mean curvature functions which depend additionally on the normal. Please see
BERGNER, MATTHIAS: *Das Dirichlet-Problem für Graphen von vorgeschriebener mittlerer Krümmung*. Shaker-Verlag Aachen; Dissertation 2006 an der Technischen Universität Darmstadt.

3.) Claudia Szerement has generalized these methods to equations of minimal surface type, satisfying a weighted homogeneous relation between their principal curvatures. Here we refer the reader to the dissertation by
CLAUDIA SZEREMENT: *Lösung des Dirichletproblems für G-minimale Graphen mit der Kontinuitäts- und Approximationsmethode*. Dissertation 2011 an der BTU Cottbus.

2 Winding Staircases and Stable Minimal Graphs on a Riemann Surface

Let us regard the following beautiful winding staircase:

Figure 2.10   THE WINDING STAIRCASE OF THE CENTER FOR INFORMATION, COMMUNICATION, AND MEDIA WITHIN THE BRANDENBURGIAN TECHNICAL UNIVERSITY AT COTTBUS, constructed by the bureau of architects Herzog & de Meuron from Basel. The use of this photograph is authorized by the *Informations-, Kommunikations- und Medien- Zentrum der Brandenburgischen Technischen Universität Cottbus.*

The central form of this staircase, if we ignore the single steps, represents a minimal surface with a multiple projection onto the plane. It appears as graph over a circular ring: The height yields a *linear function*, depending on the circumference of the exterior circle for this ring. We interpret the spiral, which bounds the staircase from the exterior, as a fixed curve Γ: Its terminal points are situated on the circular cylinder $\mathcal{Z}$ which touches the winding staircase from the interior. We realize experimentally Γ by a thin metal wire and $\mathcal{Z}$ by a synthetic glass material such that Γ meets $\mathcal{Z}$ vertically at two points from outside. Then we immerse this configuration $< \Gamma, \mathcal{Z} >$ into a soap liquid, and a soap pellicle is installed which minimizes the area and is bounded by the thin wire Γ. Moreover, this minimal surface possesses a trace on the cylinder $\mathcal{Z}$ and meets vertically this support surface $\mathcal{Z}$. Now we are going to perturb the fixed boundary Γ, and we examine the stability of the minimal surface within this configuration $< \Gamma, \mathcal{Z} >$. The particular inspiration of this staircase consists of visualizing *a minimal graph with a multiple projection onto a plane* in its entire extension: We have a model of a Riemann surface before our eyes!

In the geometric situation of our winding staircase, a minimal graph over a Riemannian surface $\mathcal{R}$ appears. This surface covers the Gaussian plane $\mathbb{C}' := \mathbb{C} \setminus \{0\}$ infinitely often and has the origin as branch point; we shall construct $\mathcal{R}$ in Section 3 by explicit formulas and speak of *the etale plane*. As usual in topology, we use here the French word **étale** in the English form **etale**, in order to indicate that the surface is given *in its complete extension*. We begin with a pull-back of the problem, from $\mathcal{R}$ to the Gaussian plane $\mathbb{C}$, under the exponential function lifted to the etale plane. With the aid of so-called *strong solutions* we solve the stability problem of our minimal graph under a small perturbation for the fixed boundary values later in Theorem 2.5. We shall introduce isothermal parameters into our minimal graph in Section 3 and establish the compactness of the mixed boundary value problem. Via a nonlinear continuity method we solve the mixed boundary value problem for all continuous boundary distributions. Finally, we present a criterion whether our surface is embedded into the Euclidean space.

We begin our considerations with the complex exponential function

$$x+iy = g(s+it) = g(s,t) := \exp(s+it) = e^s \left(\cos t + i\sin t\right), \ s \in \mathbb{R}, \ t \in \mathbb{R}. \quad (1)$$

This mapping transforms topologically the Gaussian plane of the (s,t)-points onto the Riemannian surface, covering universally the plane outside the singular point 0. We address this Riemannian surface as *the etale plane*, with the origin as its branch point, furnishing a conformal diffeomorphism in the local coordinates $x + iy$. With the positive, real quantities s_0, t_0 we introduce the open rectangle

$$R(s_0,t_0) := \{s + it \in \mathbb{C} : 0 < s < s_0, \ 0 < t < t_0\}, \quad (2)$$

which is transformed under the exponential function onto a multiply covered circular ring. We obtain the surface

$$A(s_0, t_0) := g\Big(R(s_0, t_0)\Big) = \{e^s \left(\cos t + i \sin t\right) \in \mathbb{C}' : 0 < s < s_0, 0 < t < t_0\} \tag{3}$$

in the etale plane, with the interior radius 1 and exterior radius $r_0 := \exp(s_0) > 1$, where the angle covers the interval $0 < t < t_0$. Now we consider the *graphs over the etale plane*

$$\mathbf{Z}(s, t) := \Big(g(s, t), \zeta(s, t)\Big) = g(s, t) + \zeta(s, t)\,\mathbf{e}, \ (s, t) \in R(s_0, t_0) \tag{4}$$

with the unit vector $\mathbf{e} = (0, 0, 1)$ in positive z-direction and a *height function*

$$\zeta = \zeta(s, t) \in C^2(R(s_0, t_0)). \tag{5}$$

By elementary differential geometric calculations, we derive the following

Proposition 2.1. (Etale minimal surface equation)
Our surface $\mathbf{Z}(s, t)$ in (4) represents a minimal surface if and only if the following quasilinear, elliptic partial differential equation

$$\begin{aligned}
&(e^{2s} + \zeta_t^2(s, t))\zeta_{ss}(s, t) - 2\zeta_s(s, t)\zeta_t(s, t)\zeta_{st}(s, t) + (e^{2s} + \zeta_s^2(s, t))\zeta_{tt}(s, t) \\
&+|\nabla\zeta(s, t)|^2\zeta_s(s, t) = 0, \quad (s, t) \in R(s_0, t_0)
\end{aligned} \tag{6}$$

is satisfied; here $\nabla := (\dfrac{\partial}{\partial s}, \dfrac{\partial}{\partial t})$ denotes the gradient.

Proof: Within the following calculations, we identify the Gaussian complex plane with the real vector space $\mathbb{R}^2 \times \{0\}$, and partial differentiation yields

$$\mathbf{Z}_s(s, t) = g(s, t) + \zeta_s(s, t)\,\mathbf{e}, \quad \mathbf{Z}_t(s, t) = ig(s, t) + \zeta_t(s, t)\,\mathbf{e}.$$

We designate the exterior and interior products in the Euclidean space $\mathbb{R}^3$ by the symbols $\wedge$ or $\cdot$ respectively, and we calculate

$$\begin{aligned}
\mathbf{Z}_s(s, t) \wedge \mathbf{Z}_t(s, t) &= \Big(g(s, t) + \zeta_s(s, t)\,\mathbf{e}\Big) \wedge \Big(ig(s, t) + \zeta_t(s, t)\,\mathbf{e}\Big) \\
&= e^{2s}\,\mathbf{e} - \zeta_s(s, t)g(s, t) - \zeta_t(s, t)ig(s, t).
\end{aligned} \tag{7}$$

We obtain *the surface element*

$$\begin{aligned}
W(s, t) := |\mathbf{Z}_s(s, t) \wedge \mathbf{Z}_t(s, t)| &= |e^{2s}\,\mathbf{e} - \zeta_s(s, t)g(s, t) - \zeta_t(s, t)ig(s, t)| \\
&= e^s\sqrt{e^{2s} + {\zeta_s}^2(s, t) + {\zeta_t}^2(s, t)} > 0, \quad (s, t) \in R(s_0, t_0).
\end{aligned}$$

Let us determine the coefficients of *the first fundamental form*

$$d\sigma^2 = |\mathbf{Z}_s|^2 ds^2 + 2(\mathbf{Z}_s \cdot \mathbf{Z}_t)dsdt + |\mathbf{Z}_t|^2 dt^2 \quad \text{with}$$

$$|\mathbf{Z}_s(s,t)|^2 = |g(s,t) + \zeta_s(s,t)\,\mathbf{e}|^2 = |g(s,t)|^2 + \zeta_s^2(s,t) = e^{2s} + \zeta_s^2(s,t),$$

$$\mathbf{Z}_s \cdot \mathbf{Z}_t \Big|_{(s,t)} = \Big(g(s,t) + \zeta_s(s,t)\,\mathbf{e}\Big) \cdot \Big(ig(s,t) + \zeta_t(s,t)\,\mathbf{e}\Big) = \zeta_s\zeta_t \Big|_{(s,t)}, \tag{8}$$

$$|\mathbf{Z}_t(s,t)|^2 = |ig(s,t) + \zeta_t(s,t)\,\mathbf{e}|^2 = |ig(s,t)|^2 + \zeta_t^2(s,t) = e^{2s} + \zeta_t^2(s,t).$$

In general, the Riemannian metric $d\sigma^2$ does not appear in the isothermal form. Let us now determine the second partial derivatives

$$\mathbf{Z}_{ss}(s,t) = g(s,t) + \zeta_{ss}(s,t)\,\mathbf{e},$$

$$\mathbf{Z}_{st}(s,t) = ig(s,t) + \zeta_{st}(s,t)\,\mathbf{e}, \tag{9}$$

$$\mathbf{Z}_{tt}(s,t) = -g(s,t) + \zeta_{tt}(s,t)\,\mathbf{e}.$$

Our surface $Z(s,t)$ in (4) represents a minimal surface if and only if its mean curvature $H(s,t)$ vanishes. Consequently, the denominator in the classical formula for H becomes zero (see Section 1 of Chapter 11)

$$|\mathbf{Z}_t|^2 \Big(\mathbf{Z}_s, \mathbf{Z}_t, \mathbf{Z}_{ss}\Big)(s,t) - 2(\mathbf{Z}_s \cdot \mathbf{Z}_t)\Big(\mathbf{Z}_s, \mathbf{Z}_t, \mathbf{Z}_{st}\Big)(s,t)$$

$$+|\mathbf{Z}_s|^2 \Big(\mathbf{Z}_s, \mathbf{Z}_t, \mathbf{Z}_{tt}\Big)(s,t) = 0, \quad (s,t) \in R(s_0, t_0). \tag{10}$$

Here the symbol $(.,.,.)$ denotes the determinant or the triple product in the Euclidean space. We evaluate these triple products by the formulas (7) and (9) in the equation (10), and then we substitute the coefficients (8) of the first fundamental form into this relation. Thus we arrive at the p.d.e.

$$(e^{2s} + \zeta_t^2)\Big(e^{2s}\zeta_{ss} - \zeta_s\,e^{2s}\Big)\Big|_{(s,t)} - 2\zeta_s\zeta_t\Big(e^{2s}\zeta_{st} - \zeta_t\,e^{2s}\Big)\Big|_{(s,t)}$$

$$+(e^{2s} + \zeta_s^2(s,t))\Big(e^{2s}\zeta_{tt}(s,t) + \zeta_s(s,t)e^{2s}\Big) = 0, \quad (s,t) \in R(s_0, t_0).$$

We obtain the equivalent *etale minimal surface equation*

$$0 = (e^{2s} + \zeta_t^2(s,t))\Big(\zeta_{ss}(s,t) - \zeta_s(s,t)\Big)$$

$$-2\zeta_s(s,t)\zeta_t(s,t)\Big(\zeta_{st}(s,t) - \zeta_t(s,t)\Big) + (e^{2s} + \zeta_s^2(s,t))\Big(\zeta_{tt}(s,t) + \zeta_s(s,t)\Big)$$

$$= (e^{2s} + \zeta_t^2(s,t))\zeta_{ss}(s,t) - 2\zeta_s(s,t)\zeta_t(s,t)\zeta_{st}(s,t) + (e^{2s} + \zeta_s^2(s,t))\zeta_{tt}(s,t)$$

$$+|\nabla\zeta(s,t)|^2\zeta_s(s,t), \quad (s,t) \in R(s_0, t_0).$$

q.e.d.

When we specialize our p.d.e. (6) to solutions $\zeta = \zeta(t)$, $0 < t < t_0$ independent of the parameter s, we obtain $0 = e^{2s}\zeta_{tt}(t)$, $0 < t < t_0$. This differential equation possesses the solutions

$$\zeta(t) := \alpha + \beta t, \ 0 < s < s_0, \ 0 < t < t_0 \quad \text{with the constants} \quad \alpha, \beta \in \mathbb{R}. \quad (11)$$

In the form

$$Z(s,t) := \Big(e^s \cos t, \ e^s \sin t, \ \alpha + \beta t\Big), \quad 0 < s < s_0, \ 0 < t < t_0 \qquad (12)$$

there appears a *helicoid over the circular ring* $A(s_0, t_0)$, which we identify as our staircase above. We immediately see that the function (11) solves the central boundary value problem for linear boundary values:

Definition 2.2. *On the closed interval*

$$I = I(s_0, t_0) := \{(s_0, t) \in \mathbb{R} \times \mathbb{R} : 0 \leq t \leq t_0\}$$

at the right boundary of the rectangle $R(s_0, t_0)$, *we prescribe the continuous Dirichlet boundary values* $\Phi = \Phi(t) : I \to \mathbb{R} \in C^0(I, \mathbb{R})$. *Then we look for a classical solution*

$$\zeta = \zeta(s,t) \in C^2(R(s_0,t_0)) \cap C^1(\overline{R(s_0,t_0)} \setminus I(s_0,t_0)) \cap C^0(\overline{R(s_0,t_0)}) \quad (13)$$

of the p.d.e.

$$(e^{2s} + \zeta_t^2(s,t))\zeta_{ss}(s,t) - 2\zeta_s(s,t)\zeta_t(s,t)\zeta_{st}(s,t) + (e^{2s} + \zeta_s^2(s,t))\zeta_{tt}(s,t)$$

$$= -|\nabla\zeta(s,t)|^2 \zeta_s(s,t), \quad (s,t) \in R(s_0,t_0), \tag{14}$$

with the Dirichlet conditions

$$\zeta(s_0,t) = \Phi(t), \ 0 \leq t \leq t_0; \quad \zeta(s,0) = \Phi(0) \ and \ \zeta(s,t_0) = \Phi(t_0), \ 0 \leq s \leq s_0 \tag{15}$$

as well as the Neumann condition

$$\frac{\partial}{\partial s}\zeta(s,t)\Big|_{s=0} = 0, \quad 0 \leq t \leq t_0 \tag{16}$$

on the boundary of the rectangle $R(s_0, t_0)$. *We name (13) – (16) the* mixed boundary value problem $\mathcal{P}(\Phi)$ *or briefly* mixed b.v.p..

Remark 1: For our region $R(s_0, t_0)$ we define *the domain reflected w.r.t. the s-axis* as

$$R(s_0, -t_0) := \{(s,t) \in \mathbb{R} \times \mathbb{R} : 0 < s < s_0, \ -t_0 < t < 0\}.$$

Via a translation we generate homogeneous boundary values on the s-axis, and we consider the *uneven reflection of the graph w.r.t. the s-axis* by

$$\zeta^{**}(s,t) := \zeta(s,t), (s,t) \in R(s_0,t_0) \, ; \ \zeta^{**}(s,t) := -\zeta(s,-t), (s,t) \in R(s_0,-t_0). \tag{17}$$

From the p.d.e. (14) we infer that our prescription (17) extends the solution to the entire rectangle

$$R(s_0, |t_0|) := \{(s,t) \in \mathbb{R} \times \mathbb{R} : 0 < s < s_0, \, |t| < t_0\}$$

solving this p.d.e there. An iteration of the reflections w.r.t. the linear boundary of the graph yields a solution of our p.d.e. (14) on *the entire stripe*

$$R(s_0, |\infty|) := \{(s,t) \in \mathbb{R} \times \mathbb{R} : 0 < s < s_0\}.$$

The reflection concerning the variable s is more difficult, since the p.d.e. (14) does not remain invariant under the transformation $s \to (-s)$. We have to symmetrize our differential operator w.r.t. this substitution and introduce *the symmetric rectangle*

$$\Omega = R(|s_0|, t_0) := \{(s,t) \in \mathbb{R} \times \mathbb{R} : |s| < s_0, \, 0 < t < t_0\}. \tag{18}$$

We utilize the Sobolev space $W^{2,p}(\Omega)$ for the exponents $1 < p < \infty$, whose second weak derivatives lie in the Lebesgue space $L^p(\Omega)$. Our subsequent considerations are based on the theory of *strong solutions* presented in Chapter 9 of the book [GT] by D. Gilbarg and N. Trudinger, where the equations only hold *almost everywhere*.

Definition 2.3. *For the functions*

$$\zeta = \zeta(s,t) \in W^{2,p}(\Omega) \cap C^1(\Omega) \tag{19}$$

we introduce the symmetrized operator

$$\mathcal{Q}\zeta(s,t) := \left(e^{2|s|} + \zeta_t^2(s,t)\right)\zeta_{ss}(s,t) - 2\left(\zeta_s(s,t)\zeta_t(s,t)\right)\zeta_{st}(s,t)$$

$$+ \left(e^{2|s|} + \zeta_s^2(s,t)\right)\zeta_{tt}(s,t) + \chi(s)\zeta_s(s,t)\left(\nabla\zeta(s,t) \cdot \nabla\zeta(s,t)\right), \quad (s,t) \in \Omega \text{ a.e.}, \tag{20}$$

where we use the jump function

$$\chi(s) := \begin{cases} +1, & s \in (0, +\infty) \\ 0, & s = 0 \\ -1, & s \in (-\infty, 0). \end{cases} \tag{21}$$

Remark 2: This quasilinear operator is elliptic; it has continuous coefficients for the second order terms but only bounded coefficients for the first order terms in the class $L^\infty(\Omega)$. Therefore, the theory of strong solutions is necessary on the symmetric rectangle Ω. In the *left half-plane* $\mathbb{C}_- := \{s + it \in \mathbb{C} : s < 0\}$ the p.d.e.

$$\mathcal{Q}\zeta = 0 \quad \text{a.e. in} \quad \Omega \tag{22}$$

does not conserve the character of a minimal graph. However, a *virtual graph* $\zeta : R(-s_0, t_0) \to \mathbb{R}$ appears over the *domain reflected w.r.t. the t-axis*

$$R(-s_0, t_0) := \{(s, t) \in \mathbb{R} \times \mathbb{R} : -s_0 < s < 0, 0 < t < t_0\}. \qquad (23)$$

One can extend each solution ζ of the mixed b.v.p. $\mathcal{P}(\Phi)$ by an *even reflection w.r.t. the (t, z)-plane*

$$\zeta^*(s, t) := \zeta(s, t), \ (s, t) \in R(s_0, t_0) \text{ and } \zeta^*(s, t) := \zeta(-s, t), \ (s, t) \in R(-s_0, t_0) \qquad (24)$$

on the domain $R(-s_0, t_0)$ reflected w.r.t. the t-axis. One easily verifies that our function satisfies *the symmetrized p.d.e.*

$$\mathcal{Q}\zeta^*(s, t) = 0, \quad (s, t) \in \Omega \quad \text{a.e..} \qquad (25)$$

We observe that the function ζ^* of (24) solves the following boundary value problem.

Definition 2.4. *On the closed interval $I = I(s_0, t_0)$ at the right boundary of our rectangle $\Omega = R(|s_0|, t_0)$, we prescribe the continuous Dirichlet boundary values $\Phi = \Phi(t) : I \to \mathbb{R} \in C^0(I, \mathbb{R})$. Then we consider a solution*

$$\zeta^* = \zeta^*(s, t) \in W^{2,p}(\Omega) \cap C^1(\Omega) \cap C^0(\overline{\Omega}) \qquad (26)$$

of the symmetrized p.d.e.

$$\mathcal{Q}\zeta^*(s, t) = 0, \quad (s, t) \in \Omega \quad a.e. \qquad (27)$$

with the symmetric Dirichlet conditions

$$\zeta^*(\pm s_0, t) = \Phi(t), 0 \leq t \leq t_0; \ \zeta^*(s, 0) = \Phi(0), \zeta^*(s, t_0) = \Phi(t_0), -s_0 \leq s \leq s_0 \qquad (28)$$

on the boundary of the rectangle Ω. The combination (26), (27), (28) represents the symmetrized boundary value problem $\mathcal{P}_*(\Phi)$.

Theorem 2.5. (Stability of the mixed boundary value problem)
The following assertions are valid:

1. *For each function $\Phi \in C^0(I, \mathbb{R})$ prescribed on the boundary, the mixed boundary value problem $\mathcal{P}(\Phi)$ of Definition 2.2 possesses at most one solution ζ. Moreover, a solution of $\mathcal{P}(\Phi)$ depends continuously on the boundary values Φ in the C^0-topology, according to the estimate (32) below.*
2. *If a solution $\zeta \in \mathcal{P}(\Phi)$ already exists for the differentiable function*

$$\Phi : I \to \mathbb{R} \in C^2(I, \mathbb{R}) \qquad (29)$$

prescribed on the boundary, we have exactly one solution $\xi = \zeta + \eta$ of the problem $\mathcal{P}(\Psi)$ for all the perturbed boundary values

$$\Psi = \Psi(t) : I \to \mathbb{R} \in C^2(I, \mathbb{R}) \quad with \quad ||\Psi - \Phi||_{C^2(I,\mathbb{R})} \leq \epsilon. \tag{30}$$

The existence is established by an iteration, where the number $\epsilon = \epsilon(\Phi) > 0$ has to be chosen sufficiently small. If $2 < p < \infty$ denotes an arbitrary Sobolev exponent, the $W^{2,p}(\Omega)$-norm of the symmetrized perturbation function $\eta^* \in W^{2,p}(\Omega)$ is majorized by $||\Psi - \Phi||_{C^2(I,\mathbb{R})}$.

Proof:

1.) Let us consider the solutions $\zeta_j \in \mathcal{P}(\Phi_j)$ for their boundary values $\Phi_j \in C^0(I, \mathbb{R})$ with the respective index $j = 1, 2$. Since these functions solve the same quasilinear p.d.e., their difference $\zeta = \zeta_1 - \zeta_2$ satisfies a linear elliptic p.d.e., accessible to the maximum principle (compare Chapter 6, Section 2). More precisely, we obtain the equations

$$\zeta(s,t) := \zeta_1(s,t) - \zeta_2(s,t) \in$$

$$C^2(R(s_0, t_0)) \cap C^1(\overline{R(s_0, t_0)} \setminus I(s_0, t_0)) \cap C^0(\overline{R(s_0, t_0)}),$$

$$\mathcal{L}(\zeta) = 0 \quad in \quad R(s_0, t_0) \quad with \ a \ linear \ elliptic \ operator \ \mathcal{L},$$

such that the Dirichlet conditions $\quad \zeta(s_0, t) = \Phi_1(t) - \Phi_2(t), \ 0 \leq t \leq t_0;$

$$\zeta(s, 0) = \Phi_1(0) - \Phi_2(0), \ \zeta(s, t_0) = \Phi_1(t_0) - \Phi_2(t_0), \ 0 \leq s \leq s_0;$$

and the Neumann condition $\dfrac{\partial}{\partial s} \zeta(s,t)\Big|_{s=0} = 0, \ 0 \leq t \leq t_0$ are satisfied.

$$\tag{31}$$

The maximum principle of E. Hopf and, in particular, his boundary point lemma for the Neumann condition imply that the function ζ cannot attain its maximum – neither in the domain $R(s_0, t_0)$ nor on the free boundary $\{(0, s) : 0 < s < s_0\}$ (see Chapter 6, Section 1). Then we derive the estimate

$$\sup\{|\zeta_1(s,t) - \zeta_2(s,t)| : (s,t) \in R(s_0, t_0)\} \leq \sup\{|\Phi_1(t) - \Phi_2(t)| : t \in I(s_0, t_0)\},$$
$$\tag{32}$$

which shows the uniqueness of $\mathcal{P}(\Phi)$ as well as the continuous dependence on the boundary values Φ.

2.) We reflect the solution ζ of the mixed b.v.p. $\mathcal{P}(\Phi)$ with its boundary values (29) according to the Remark 2, and we obtain a solution $\zeta^* \in \mathcal{P}_*(\Phi)$ of the symmetrized problem from Definition 2.4. We utilize the *ansatz of perturbation* $\zeta^* + \eta$ with the function

$$\eta \in W^{2,p}(\Omega) \cap C^1(\Omega) \tag{33}$$

and solve the symmetrized problem $\mathcal{P}_*(\Psi)$ for the perturbed boundary values (30). In this context we determine the *perturbation equation* (34) as follows: We arrange the terms in this equation according to the powers of derivatives $\eta_s, \eta_t, \eta_{ss}, \eta_{st}, \eta_{tt}$, which appear. The sum of the terms of order 0 vanishes, because ζ^* is a solution of our p.d.e.; the terms of first order represent *the* *linear operator* $\mathcal{L}$; the terms of order 2 and 3 are combined to *the nonlinear operator* $\mathcal{K}$. Then we calculate

$$0 = \mathcal{Q}\left[\zeta^*(s,t) + \eta(s,t)\right]$$

$$= \left(e^{2|s|} + (\zeta_t^*(s,t) + \eta_t(s,t))^2\right)(\zeta_{ss}^*(s,t) + \eta_{ss}(s,t))$$

$$-2[\zeta_s^*(s,t) + \eta_s(s,t)][\zeta_t^*(s,t) + \eta_t(s,t)](\zeta_{st}^*(s,t) + \eta_{st}(s,t))$$

$$+\left(e^{2|s|} + (\zeta_s^*(s,t) + \eta_s(s,t))^2\right)(\zeta_{tt}^*(s,t) + \eta_{tt}(s,t))$$

$$+\chi(s)[\zeta_s^*(s,t)) + \eta_s(s,t)]\Big([\nabla\zeta^*(s,t) + \nabla\eta(s,t)] \cdot [\nabla\zeta^*(s,t) + \nabla\eta(s,t)]\Big)$$

$$= \left(e^{2|s|} + (\zeta_t^*(s,t))^2\right)\eta_{ss}(s,t) - 2\zeta_s^*\zeta_t^*\eta_{st}(s,t) + \left(e^{2|s|} + (\zeta_s^*(s,t))^2\right)\eta_{tt}(s,t)$$

$$+2\zeta_t^*(s,t)\zeta_{ss}^*(s,t)\eta_t(s,t) - 2\zeta_{st}^*\Big(\zeta_s^*\eta_t(s,t) + \zeta_t^*\eta_s(s,t)\Big)$$

$$+2\zeta_s^*(s,t)\zeta_{tt}^*(s,t)\eta_s(s,t) + 2\chi(s)\zeta_s^*(s,t)\Big(\nabla\zeta^*(s,t) \cdot \nabla\eta(s,t)\Big)$$

$$+\chi(s)\Big(\nabla\zeta^*(s,t) \cdot \nabla\zeta(s,t)\Big)\eta_s(s,t) - \mathcal{K}(\eta)$$

$$=: \mathcal{L}(\eta) - \mathcal{K}(\eta) \quad \text{for almost all} \quad (s,t) \in \Omega.$$
$$(34)$$

3.) The linear operator $\mathcal{L}$ possesses only $\chi \in L^\infty(\Omega)$ as a discontinuous coefficient, but all the others are continuous. With the aid of [GT] Theorem 9.15, one can solve the Dirichlet problem

$$\mathcal{L}(\eta) = \omega \quad \text{a.e. in} \quad \Omega \quad \text{with} \quad \eta \in W^{2,p}(\Omega) \quad \text{and} \quad (\eta - \varphi) \in W_0^{1,2}(\Omega)$$

for all right-hand sides $\omega \in L^p(\Omega)$ and for all boundary values

$$\varphi = \varphi(t) \in C^2(I, \mathbb{R}) \quad \text{constantly extended in the variable } s \text{ on} \quad \Omega.$$
$$(35)$$

The result [GT] Lemma 9.17 shows that the inverse operator $\mathcal{L}^{-1}$ exists on this space with homogeneous boundary values, and this operator is bounded w.r.t. the given norms:

A solution of $\mathcal{L}(\eta) = \omega$ a.e. in Ω with $\eta \in W^{2,p}(\Omega) \cap W_0^{1,2}(\Omega)$

satisfies the estimate $\|\eta\|_{W^{2,p}(\Omega)} \le \|\omega\|_{L^p(\Omega)}$ for all $\omega \in L^p(\Omega)$.
$$(36)$$

The Sobolev space $W^{2,p}(\Omega)$ is continuously embedded in the Hölder space $C^{1,\mu}(\Omega)$ with the Hölder exponent $\mu = 1 - \frac{2}{p} \in (0,1)$. This embedding theorem of Morrey from Theorem 6.10 in Section 6 of Chapter 10 is also contained in [GT] Theorem 7.26. Finally, [GT] Corollary 9.18 induces that the boundary values (35) are continuously attained.

4.) The nonlinear operator $\mathcal{K}$ represents a sum with homogeneous terms of second or third degree in $\eta_s, \eta_t, \eta_{ss}, \eta_{st}, \eta_{tt}$, where the coefficients in $L^\infty(\Omega)$ depend on the initial solution ζ and its continuation ζ^* by reflection. We observe that, at most, one of the second derivatives $\eta_{ss}, \eta_{st}, \eta_{tt}$ appears as factor in each term, and that the $W^{2,p}(\Omega)$-norm dominates the $C^1(\Omega)$-norm. Then we arrive at *the contraction inequality*

$$||\mathcal{K}(\eta_1) - \mathcal{K}(\eta_2)||_{L^p(\Omega)} \leq C(\rho)||\eta_1 - \eta_2||_{W^{2,p}(\Omega)}$$

$$(37)$$

for all $\eta_j \in W^{2,p}(\Omega)$ satisfying $||\eta_j||_{W^{2,p}(\Omega)} \leq \rho$ and $j = 1, 2$

with $C(\rho) \to 0+$ for $\rho \to 0+$ and $\mathcal{K}(0) = 0$.

5.) Now we follow the proof of Proposition 1.8 from Section 1, where we replace the Schauder estimates by the above results for strong solutions. For each Ψ from (30), we solve the problem

$$\mathcal{L}(\eta_0) = 0 \text{ a.e. in } \Omega \text{ with } \eta_0 \in W^{2,p}(\Omega) \cap C^1(\overline{\Omega}) \text{ and } \eta_0 = \Psi - \Phi \text{ on } \partial\Omega. \quad (38)$$

We observe that our solution depends continuously on the boundary values, w.r.t. the given norms. Then we solve *the iterative equations*

$$\mathcal{L}(\eta_{k+1}) = \mathcal{K}(\eta_k) \text{ a.e. in } \Omega \text{ with } \eta_{k+1} \in W^{2,p}(\Omega) \cap C^1(\overline{\Omega})$$

$$(39)$$

$$\text{and} \quad \eta_{k+1} = \Psi - \Phi \quad \text{on} \quad \partial\Omega \quad \text{for} \quad k = 1, 2, \dots .$$

With the properties (35), (36), (37) one immediately comprehends that the sequence $\{\eta_k\}_{k=1,2,\dots}$ converges within the Banach space $W^{2,p}(\Omega) \cap C^1(\overline{\Omega})$ to the unique solution

$$\mathcal{L}(\eta) = \mathcal{K}(\eta) \text{ a.e. in } \Omega \text{ with } \eta \in W^{2,p}(\Omega) \cap C^1(\overline{\Omega}) \text{ and } \eta = \Psi - \Phi \text{ on } \partial\Omega. \quad (40)$$

One obtains a solution of the symmetrized problem

$$\xi = \xi(s,t) := \zeta^*(s,t) + \eta(s,t) \in \mathcal{P}_*(\Psi) \quad (41)$$

for the perturbed boundary values (30). Alternatively, the Banach fixed point theorem yields the existence of such a solution.

6.) Let us consider *the reflected function*

$$\xi^- = \xi^-(s,t) := \xi(-s,t), \quad (s,t) \in \Omega, \quad (42)$$

which satisfies the same condition at the boundary as ξ.

For all $s > 0$ we calculate:

$$\mathcal{Q}\xi^-(s,t) = \left(e^{2s} + \xi_t^2(-s,t)\right)\xi_{ss}(-s,t) - 2\left(\xi_s(-s,t)\xi_t(-s,t)\right)\xi_{st}(-s,t)$$

$$+\left(e^{2s} + \xi_s^2(-s,t)\right)\xi_{tt}(-s,t) + \chi(-s)\xi_s(-s,t)\left(\nabla\xi(-s,t) \cdot \nabla\xi(-s,t)\right)$$

$$= \mathcal{Q}\xi\Big|_{(-s,t)} = 0 \quad \text{for almost all} \quad (s,t) \in R(s_0,t_0).$$

$$(43)$$

Then we evaluate for all $s < 0$ as follows:

$$\mathcal{Q}\xi^-(s,t) = \left(e^{-2s} + \xi_t^2(-s,t)\right)\xi_{ss}(-s,t) - 2\left(\xi_s(-s,t)\xi_t(-s,t)\right)\xi_{st}(-s,t)$$

$$+\left(e^{-2s} + \xi_s^2(-s,t)\right)\xi_{tt}(-s,t) + \chi(-s)\xi_s(-s,t)\left(\nabla\xi(-s,t)\cdot\nabla\xi(-s,t)\right)$$

$$= \mathcal{Q}\xi\Big|_{(-s,t)} = 0 \quad \text{for almost all} \quad (s,t) \in R(-s_0,t_0).$$

$$(44)$$

Consequently, the function ξ^- yields a solution of the symmetrized problem $\mathcal{P}_*(\Psi)$ as well as the function ξ. We know from [GT] Theorem 9.5 that the Dirichlet problem possesses at most one strong solution for linear operators, with continuous or bounded coefficients of first or second order. Taking the considerations of part 1.) into account, this statement remains valid for our quasilinear problem $\mathcal{P}_*(\Psi)$. Thus we obtain the identity

$$\xi^-(s,t) \equiv \xi(s,t), \quad (s,t) \in \Omega,$$

$$(45)$$

and the Neumann condition

$$\frac{\partial}{\partial s}\xi(s,0)\Big|_{s=0} = 0, \quad 0 < s < s_0$$

$$(46)$$

follows. By a local reconstruction, one can establish the classical regularity:

$$\xi \in C^2(R(s_0,t_0)).$$

$$(47)$$

Finally, the restriction

$$\xi\Big|_{\overline{R(s_0,t_0)}}$$

$$(48)$$

solves the perturbed problem $\mathcal{P}(\Psi)$ with mixed boundary values.   q.e.d.

3 Mixed Boundary Value Problem for Minimal Graphs over the Etale Plane

At first, we shall explicitly construct the etale plane. To better imagine this, we prescribe the *level of the leaves* for our etale plane

$$-\infty \le \theta_{-\infty} < \ldots < \theta_{-2} < \theta_{-1} < \theta_0 < \theta_1 < \theta_2 < \ldots < \theta_\infty \le \infty.$$

For $n = 0, \pm 1, \pm 2, \pm 3, \ldots$ we define **the n-th leaf**

$$\mathcal{R}_n := \left\{\mathbf{x} = (x,y,\theta_n) \in \mathbb{R}^3 : x = r\cos t, y = r\sin t; 0 < r < \infty, 0 \le t < 2\pi\right\}.$$

Thus we obtain **the etale plane**

$$\mathcal{R} := \bigcup_{n\in\mathbb{Z}} \mathcal{R}_n$$

over the plane $\{(x, y, z) \in \mathbb{R}^3 : z = \theta_\infty\}$ with the surjective **etale projection**

$$\Theta : \mathcal{R} \to \mathbb{C}' := \mathbb{C} \backslash \{0\} \quad \text{defined by} \quad \mathcal{R} \ni (x, y, \theta_n) = \mathbf{x} \to \Theta(\mathbf{x}) = (x, y) \in \mathbb{C}'.$$

Each point $\mathbf{x} \in \mathcal{R}$ is situated on exactly one leaf and allows the representation

$$\mathbf{x} = (r \cos t, y = r \sin t, \theta_n) \quad \text{with the uniquely determined numbers:}$$

$$0 < r < \infty, \quad 0 \le t < 2\pi, \quad n \in \mathbb{Z}.$$

Consequently, we can define **the etale angle**

$$\tau(\mathbf{x}) := 2\pi n + t \in \mathbb{R}, \quad \mathbf{x} \in \mathcal{R}$$

as well as **the etale radius**

$$\rho(\mathbf{x}) := r \in (0, +\infty), \quad \mathbf{x} \in \mathcal{R},$$

which together characterize the points in $\mathcal{R}$ uniquely. The etale projection Θ is not injective; however, it furnishes an infinite covering of the pointed plane $\mathbb{C}'$. More precisely, to each point $(x, y) \in \mathbb{C}'$ we associate **the fiber**

$$\Theta^{-1}\big((x, y)\big) = \{(x, y, \theta_n) : n \in \mathbb{Z}\}.$$

For each point $\mathbf{x}_0 \in \mathcal{R}$ one introduces **the etale neighborhood**

$$\mathcal{V}(\mathbf{x}_0) := \left\{ \mathbf{x} \in \mathcal{R} : |\tau(\mathbf{x}) - \tau(\mathbf{x}_0)| < \frac{\pi}{2} \right\}.$$

The restriction of the etale projection to this neighborhood

$$\Theta_{\mathbf{x}_0} : \mathcal{V}(\mathbf{x}_0) \to V(x_0, y_0) := \left\{ (x, y) \in \mathbb{C}' \,\middle|\, (x, y) \cdot (x_0, y_0) > 0 \right\} \qquad (1)$$

constitutes a bijection onto the half-plane $V(x_0, y_0)$. Therefore, its inverse transformation

$$\Theta_{\mathbf{x}_0}^{-1} : V(x_0, y_0) \to \mathcal{V}(\mathbf{x}_0) \qquad (2)$$

represents a **lifting of $V(x_0, y_0)$ onto $\mathcal{R}$ around the point $\mathbf{x}_0 \in \mathcal{R}$**. For our Riemannian surface $\mathcal{R}$ we obtain a **conformal atlas** with **the family of charts**

$$\left\{ \Theta_{\mathbf{x}_0}^{-1} : \quad \mathbf{x}_0 \in \mathcal{R} \right\}. \qquad (3)$$

The change of charts is conformal, and the leaves $\mathcal{R}_n$ are continuously glued together near **the slits**

$$\mathcal{F}_n := \left\{ \mathbf{x} = (r, 0, \theta_n) \in \mathbb{R}^3 : 0 < r < \infty \right\} \quad \text{for} \quad n = 0, \pm 1, \pm 2, \pm 3, \dots.$$

The continuous family of **the etale circles with radius** r

$$C_r := \Big\{ \mathbf{x} \in \mathcal{R} : \rho(\mathbf{x}) = r \Big\}, \quad 0 < r < \infty$$

sweeps out the etale plane as well as the **etale half-lines to the parameter** t

$$\mathcal{D}_t := \Big\{ \mathbf{x} \in \mathcal{R} : \tau(\mathbf{x}) = t \Big\}, \quad -\infty < t < +\infty;$$

their combination represents an *orthogonal net on our surface* $\mathcal{R}$. We define **the lifted circular ring**

$$\mathcal{A}(r,t) := \Big\{ \mathbf{x} \in \mathcal{R} : 1 < \rho(\mathbf{x}) < r, \, 0 < \tau(\mathbf{x}) < t \Big\} \quad \text{with} \quad r > 1 \quad \text{and} \quad t > 0,$$

bounded by certain segments on the straight lines or circular contours C_1, C_r, $\mathcal{D}_0$, $\mathcal{D}_t$. The **etale unit circle** C_1 is contained in *the unit cylinder*

$$\mathcal{Z} := \{ (x, y, z) \in \mathbb{R}^3 : x^2 + y^2 = 1 \},$$

where we introduce *the normal vector field*

$$\mathcal{N} : \mathcal{Z} \to \mathbb{R}^2 \times \{0\} \quad \text{with} \quad \mathcal{N}(\mathbf{x}) := (x, y, 0) \quad \text{for all} \quad \mathbf{x} = (x, y, z) \in \mathcal{Z}.$$

Now we lift the exponential function onto this Riemannian surface and obtain globally its inverse.

Definition 3.1. *Let us consider the discrete function*

$$\theta : \mathbb{R} \to \{ \theta_n : n \in \mathbb{Z} \} \quad \text{with} \quad \theta(t) := \theta_n \quad \text{if} \quad t \in \Big[2\pi n, 2\pi (n+1) \Big), n \in \mathbb{Z}.$$

We introduce **the lifted exponential function** $Exp_\mathcal{R} : \mathbb{C} \to \mathcal{R}$ *satisfying*

$$Exp_\mathcal{R}(s + it) := \Big(exp(s + it), \theta(t) \Big) = \Big(e^s \cos t, e^s \sin t, \theta(t) \Big), \, s + it \in \mathbb{C},$$

which yields a conformal mapping from the Gaussian plane $\mathbb{C}$ *onto the etale plane* $\mathcal{R}$. *In particular, this mapping transforms conformally the plane rectangle* $R(s_0, t_0)$ *of the lengths* $s_0 > 0$ *and* $t_0 > 0$ *onto the lifted circular ring* $\mathcal{A}(r_0, t_0)$ *with the radius* $r_0 := \exp(s_0) > 1$.

The inverse function of $Exp_\mathcal{R}$ *defines* **the etale logarithm** $Log_\mathcal{R} : \mathcal{R} \to \mathbb{C}$ *satisfying*

$$Log_\mathcal{R}(\mathbf{x}) := s + it, \, \mathbf{x} \in \mathcal{R} \quad \text{if} \quad Exp_\mathcal{R}(s + it) = \mathbf{x}, \, s + it \in \mathbb{C},$$

which represents a conformal mapping of $\mathcal{R}$ *onto* $\mathbb{C}$.

With these functions we lift the mixed b.v.p. of Definition 2.2 in Section 2 on the Riemannian surface. At first, we lift the boundary values

$$\Phi = \Phi(t) : I = I(s_0, t_0) \to \mathbb{R} \in C^0(I, \mathbb{R})$$

on **the fixed boundary** by

$$\widetilde{\Phi}(\mathbf{x}) := \Phi\Big(\tau(\mathbf{x})\Big) - \theta\Big(\tau(\mathbf{x})\Big) \quad \text{if} \quad \mathbf{x} \in \partial\mathcal{A}(r_0, t_0) \setminus \mathcal{C}_1. \tag{4}$$

Secondly, we lift the solution

$$\zeta \in C^2(R(s_0, t_0)) \cap C^1(\overline{R(s_0, t_0)} \setminus I(s_0, t_0)) \cap C^0(\overline{R(s_0, t_0)})$$

of the mixed b.v.p. $\mathcal{P}(\Phi)$ from Definition 2.2 in Section 2 to **the nonpara-metric solution** $\ \widetilde{\zeta}: \overline{\mathcal{A}(r_0, t_0)} \to \mathbb{R}\ $ satisfying

$$\widetilde{\zeta} \in C^2(\mathcal{A}(r_0, t_0)) \cap C^1\Big(\overline{\mathcal{A}(r_0, t_0)} \setminus \mathcal{C}_{r_0}\Big) \cap C^0(\overline{\mathcal{A}(r_0, t_0)} \setminus \{\mathcal{F}_1 \cup \mathcal{F}_2 \cup \ldots\}),$$

$$\widetilde{\zeta}(\mathbf{x}) := (\zeta - \theta) \circ \Big(\mathrm{Log}_\mathcal{R}(\mathbf{x})\Big) = \zeta\Big(\mathrm{Log}_\mathcal{R}(\mathbf{x})\Big) - \theta\Big(\mathrm{Log}_\mathcal{R}(\mathbf{x})\Big), \ \mathbf{x} \in \overline{\mathcal{A}(r_0, t_0)}. \tag{5}$$

When we lift the graphs $Z = Z(s, t): \overline{\mathcal{A}(s_0, t_0)} \to \mathbb{R}$ onto the etale plane, using the prescription (5) on $\mathcal{R}$, we calculate

$$\widetilde{Z}(\mathbf{x}) := Z \circ \mathrm{Log}_\mathcal{R}(\mathbf{x}) = \Big[(g(s,t), \theta(t)) + \Big(\zeta(s,t) - \theta(t)\Big)\mathbf{e}\Big] \circ \mathrm{Log}_\mathcal{R}(\mathbf{x})$$

$$= \Big[\mathrm{Exp}_\mathcal{R}(s + it) + \Big(\zeta(s,t) - \theta(t)\Big)\mathbf{e}\Big] \circ \mathrm{Log}_\mathcal{R}(\mathbf{x}) = \mathbf{x} + \widetilde{\zeta}(\mathbf{x})\,\mathbf{e}, \ \mathbf{x} \in \overline{\mathcal{A}(r_0, t_0)}. \tag{6}$$

One obtains a graph over the leaves in our etale plane, and the representation (6) justifies the classification *nonparametric* for the solution $\widetilde{\zeta}$ in (5).

We use the sign $[t] \in \mathbb{N}_0 := \{0, 1, 2, \ldots\}$ for all real numbers $t > 0$, indicating the *maximal number* $n \in \mathbb{Z}$ which satisfies $n \le t$. Then we infer from (5) the following equations

$$\widetilde{\zeta}(\mathbf{x}) := \zeta\Big(\mathrm{Log}_\mathcal{R}(x, y, \theta_n)\Big) - \theta_n \quad \text{if} \quad \mathbf{x} = (x, y, \theta_n) \in \overline{\mathcal{A}(r_0, t_0)} \cap \mathcal{R}_n$$

$$\text{on the leaves of the order} \quad n = 0, \ldots, n_0 := \left[\frac{t_0}{2\pi}\right] \in \mathbb{N}_0. \tag{7}$$

In the *first case* where $t_0 \ne 2\pi k$ is valid for all $k \in \mathbb{N}$, the intersection $\mathcal{A}(r_0, t_0) \cap \{\mathcal{R}_{n_0} \setminus \mathcal{F}_{n_0}\}$ yields a nonvoid open set. In the *second case* where $t_0 \in 2\pi\mathbb{N}$ is an entire multiple of 2π, the above intersection $\overline{\mathcal{A}(r_0, t_0)} \cap \mathcal{R}_{n_0}$ reduces to a straight closed segment on the slit $\mathcal{F}_{n_0}$.

For $n = 0, 1, \ldots, n_0 - 1$, **the open sheet** $\mathcal{G}_n := \mathcal{A}(r_0, t_0) \cap \{\mathcal{R}_n \setminus \mathcal{F}_n\}$ is conformally transformed under the etale projection Θ onto the plane domain

$$\Omega_n := \{(x, y) \in \mathbb{C} \setminus [0, +\infty) : 1 < x^2 + y^2 < r_0^2\}, \tag{8}$$

with a slit on the positive x-axis.

In the first case, **the sheet** $\mathcal{G}_{n_0} := \mathcal{A}(r_0, t_0) \cap \{\mathcal{R}_{n_0} \setminus \mathcal{F}_{n_0}\}$ is transformed under the projection Θ onto the following *sector of the circular ring*

$$\Omega_{n_0} := \Big\{ (x, y) \in \mathbb{C} : x = r \cos t, \, y = r \sin t, \, 1 < r < r_0, \, 0 < t < (t_0 - 2\pi n_0) \Big\}. \tag{9}$$

In the second case, the set $\mathcal{G}_{n_0}$ is void and we set $\Omega_{n_0} = \emptyset$.
These liftings on $\mathcal{R}$ are achieved by the mappings

$$\overset{n}{\Theta}(x, y) := \Theta \Big|_{\mathcal{G}_n}^{-1}(x, y) = (x, y, \theta_n) : \Omega_n \to \mathcal{G}_n, \quad n = 0, \ldots, n_0, \tag{10}$$

where we simply omit the mapping $\overset{n_0}{\Theta}$ in the second case.

Due to the identities (7), the function $\widetilde{\zeta} : \overline{\mathcal{A}(r_0, t_0)} \to \mathbb{R}$ from (5) possesses a *jump of the height* $(\theta_n - \theta_{n-1})$ at the transition over the slit $\mathcal{F}_n$ for $n = 1, \ldots, n_0$. The differentiation of $\widetilde{\zeta}$ is possible by a pull-back with the aid of the charts (1) and (2) in the complex atlas (3); however, this is impossible on the slits! This differentiation only concerns the variables x and y, but does not affect the last discrete variable θ_n. We employ **the gradient**

$$\operatorname{grad} \widetilde{\zeta}(\mathbf{x}) := \Big(\widetilde{\zeta}_x, \widetilde{\zeta}_y \Big) \Big|_{\mathbf{x}}, \, \mathbf{x} = (x, y, \theta_n) \in \mathcal{G}_n, \quad n = 0, 1, \ldots, n_0,$$

where the sheet $\mathcal{G}_{n_0}$ does not appear in the second case. Moreover, we define **the divergence**

$$\operatorname{div} \widetilde{h}(\mathbf{x}) := \Big(\widetilde{f}_x + \widetilde{g}_y \Big) \Big|_{\mathbf{x}}, \, \mathbf{x} = (x, y, \theta_n) \in \mathcal{A}(r_0, t_0)$$

of the vector fields

$$\widetilde{h} = \widetilde{h}(\mathbf{x}) := \Big(\widetilde{f}(\mathbf{x}), \widetilde{g}(\mathbf{x}) \Big) \in C^1(\mathcal{A}(r_0, t_0), \mathbb{R}^2).$$

Above every sheet $\mathcal{G}_n$, $n = 0, \ldots, n_0$, the function $\widetilde{\zeta}$ represents a nonparametric minimal graph, and we obtain the p.d.e. in divergence form

$$\operatorname{div} \left(\frac{\operatorname{grad} \widetilde{\zeta}(x, y, \theta_n)}{\sqrt{1 + |\operatorname{grad} \widetilde{\zeta}(x, y, \theta_n)|^2}} \right) = 0, \, \mathbf{x} = (x, y, \theta_n) \in \mathcal{G}_n, \, n = 0, \ldots, n_0. \tag{11}$$

The *upward normal* of $\widetilde{\zeta}$, defined by

$$\widetilde{N}(\mathbf{x}) := \{ 1 + |\operatorname{grad} \widetilde{\zeta}(\mathbf{x})|^2 \}^{-\frac{1}{2}} \Big(\operatorname{grad} \widetilde{\zeta}(\mathbf{x}), 1 \Big) \Big|_{\mathbf{x}}, \, \mathbf{x} = (x, y, \theta_n) \in \mathcal{G}_n$$

for $n = 0 \ldots, n_0$, can be continuously extended over the slits onto the topological space $\overline{\mathcal{A}(r_0, t_0)} \setminus \mathcal{C}_{r_0}$. Consequently, the equations (11) remain valid over the slits, and we speak of **the equation for minimal graphs in divergence form on** $\mathcal{A}(r_0.t_0)$.

The Neumann condition (16) from Section 2 for the solution ζ is transformed into the following **free boundary condition** for the solution $\widetilde{\zeta}$

$$\mathcal{N}(\mathbf{x}) \cdot \widetilde{N}(\mathbf{x}) = 0 \quad \text{for all} \quad \mathbf{x} = (x, y, \theta_n) \in \partial\mathcal{G}_n \cap \mathcal{C}_1, \, n = 0, \ldots, n_0, \quad (12)$$

and remains continuous over the slits. The equation (12) shows that the minimal graph meets our unit cylinder $\mathcal{Z}$ orthogonally at the **lifted free boundary** $\partial\mathcal{A}(r_0, t_0) \cap \mathcal{C}_1$.

We remind the reader that the *fixed boundary condition*, equivalent to the equations (15) from Section 2, appears in the form

$$\widetilde{\zeta}(\mathbf{x}) = \widetilde{\Phi}(\mathbf{x}), \quad \mathbf{x} \in \partial\mathcal{A}(r_0, t_0) \setminus \mathcal{C}_1, \tag{13}$$

on the **lifted fixed boundary** $\partial\mathcal{A}(r_0, t_0) \setminus \mathcal{C}_1$.

Definition 3.2. *For the boundary values $\widetilde{\Phi}$ in (4) being given, a solution (5) or (7) of the nonparametric equation (11) for our minimal graphs, with the free (12) and fixed (13) boundary conditions respectively, belongs to the class $\widetilde{\mathcal{P}}(\widetilde{\Phi})$. We address $\widetilde{\mathcal{P}}(\widetilde{\Phi})$ as our* lifted mixed boundary value problem, *or briefly* lifted mixed b.v.p.

Remark: The preceding considerations show the equivalence of the problems $\mathcal{P}(\Phi)$ and $\widetilde{\mathcal{P}}(\widetilde{\Phi})$.

Definition 3.3. *With the plane domains Ω_n for $n = 0, \ldots, n_0$ in (8) and (9), the* nonparametric area formula

$$\Sigma\left(\widetilde{\zeta}, \mathcal{A}(r_0, t_0)\right) := \sum_{n=0}^{n_0} \int_{\Omega_n} \sqrt{1 + |grad\,\widetilde{\zeta}(x, y, \theta_n)|^2} \quad dx\,dy$$

determines the area of the graph $\widetilde{\zeta} : \overline{\mathcal{A}(r_0, t_0)} \to \mathbb{R} \in C^1(\mathcal{A}(r_0, t_0))$.

The ingenious method by R. Finn from Proposition 1.2 in Section 1 allows an estimate of the area for our solutions of the lifted mixed b.v.p.

Proposition 3.4. (Etale area estimate)
One can bound the area of a solution $\widetilde{\zeta}$ for the lifted mixed b.v.p. $\widetilde{\mathcal{P}}(\widetilde{\Phi})$ by their values $\widetilde{\Phi}$ in (4) on the fixed boundary as follows:

$$\Sigma\left(\widetilde{\zeta}, \mathcal{A}(r_0, t_0)\right) \leq \sum_{n=0}^{n_0} |\Omega_n| + (r_0 - 1)|\Phi(0) - \theta_0|$$

$$+(r_0 - 1)\sum_{n=0}^{n_0-1} (\theta_{n+1} - \theta_n) + (r_0 - 1)|\Phi(t_0) - \theta_{n_0}| + \sum_{n=0}^{n_0} \alpha_n. \tag{14}$$

Within this estimate $|\Omega_n|$ *denotes the* plane area *of the domain* Ω_n, *and the* cylindrical areas *are given by*

$$\alpha_n := \int_{\mathcal{H}_n} |\widetilde{\Phi}(\mathbf{x})| \, d\sigma(\mathbf{x}) = \int_{\mathcal{H}_n} |\Phi(\tau(\mathbf{x})) - \theta_n| \, d\sigma(\mathbf{x})$$

for $n = 0, \ldots, n_0$. *Here our* n-*th integral extends over the circular arc* $\mathcal{H}_n :=$ $\partial \mathcal{G}_n \cap \mathcal{C}_{r_0}$ *in positive orientation, with* $d\sigma(\mathbf{x})$ *as its element of arc.*

Proof:

1.) We are going to multiply the p.d.e. (11) with the solution and integrate this product separately on each sheet $\mathcal{G}_n$. For $n = 0, \ldots, n_0$ we pull back the integration from the sheet $\mathcal{G}_n$ into the plane domain Ω_n of (8) and (9) under the projection (10). The boundary of the sheet, being traversed in a positive orientation, yields the following decomposition:

$$\partial \mathcal{G}_n = \partial^+ \mathcal{G}_n \cup \mathcal{H}_n \cup \partial^- \mathcal{G}_n \cup (\partial \mathcal{G}_n \cap \mathcal{C}_1) \quad \text{for} \quad n = 0, \ldots, n_0. \tag{15}$$

The right-hand segments $\partial^+ \mathcal{G}_n$ and $\partial^- \mathcal{G}_n$ on the corresponding half-lines are traversed in a positive or a negative direction, respectively. Moreover, the circular segments $\mathcal{H}_n$ and $\partial \mathcal{G}_n \cap \mathcal{C}_1$ are traversed on the circle $\mathcal{C}_{r_0}$ in a positive direction and on the unit circle $\mathcal{C}_1$ in a negative direction, respectively. On the sheet $\mathcal{G}_n$ with the oriented boundary $\partial \mathcal{G}_n$, we can apply the Stokes integral theorem: The latter can be reduced to the Gaussian integral theorem by the retraction (10) and a subsequent exhaustion of the domains Ω_n, $n = 0, \ldots, n_0$, which possess slits and singular boundaries.

2.) Let us perform a multiplication of the p.d.e. (11) by the solution $\widetilde{\zeta}$ and a partial differentiation.

$$0 = \widetilde{\zeta}(x, y, \theta_n) \operatorname{div} \left(\frac{\operatorname{grad} \widetilde{\zeta}(x, y, \theta_n)}{\sqrt{1 + |\operatorname{grad} \widetilde{\zeta}(x, y, \theta_n)|^2}} \right)$$

$$= \operatorname{div} \left(\frac{\widetilde{\zeta}(x, y, \theta_n) \operatorname{grad} \widetilde{\zeta}(x, y, \theta_n)}{\sqrt{1 + |\operatorname{grad} \widetilde{\zeta}(x, y, \theta_n)|^2}} \right) - \frac{|\operatorname{grad} \widetilde{\zeta}(x, y, \theta_n)|^2}{\sqrt{1 + |\operatorname{grad} \widetilde{\zeta}(x, y, \theta_n)|^2}}$$

$$= \operatorname{div} \left(\frac{\widetilde{\zeta}(x, y, \theta_n) \operatorname{grad} \widetilde{\zeta}(x, y, \theta_n)}{\sqrt{1 + |\operatorname{grad} \widetilde{\zeta}(x, y, \theta_n)|^2}} \right) - \sqrt{1 + |\operatorname{grad} \widetilde{\zeta}(x, y, \theta_n)|^2}$$

$$+ \frac{1}{\sqrt{1 + |\operatorname{grad} \widetilde{\zeta}(\mathbf{x})|^2}} \quad \text{for all} \quad \mathbf{x} = (x, y, \theta_n) \in \mathcal{G}_n, \quad n = 0, \ldots, n_0.$$

Now we arrive at the central estimate:

$$\sqrt{1 + |\operatorname{grad} \widetilde{\zeta}(x, y, \theta_n)|^2} \leq 1 + \operatorname{div} \left(\frac{\widetilde{\zeta}(x, y, \theta_n) \operatorname{grad} \widetilde{\zeta}(x, y, \theta_n)}{\sqrt{1 + |\operatorname{grad} \widetilde{\zeta}(x, y, \theta_n)|^2}} \right) \tag{16}$$

$$\text{for all} \quad \mathbf{x} = (x, y, \theta_n) \in \mathcal{G}_n, \quad n = 0, \ldots, n_0.$$

3.) For $n = 0, \ldots, n_0$ we introduce Pfaffian forms

$$\omega_n(x, y) := \frac{\widetilde{\zeta}(x, y, \theta_n)\widetilde{\zeta}_x(x, y, \theta_n)}{\sqrt{1 + |\operatorname{grad}\widetilde{\zeta}(x, y, \theta_n)|^2}}\, dy - \frac{\widetilde{\zeta}(x, y, \theta_n)\widetilde{\zeta}_y(x, y, \theta_n)}{\sqrt{1 + |\operatorname{grad}\widetilde{\zeta}(x, y, \theta_n)|^2}}\, dx$$

$$\text{for all} \quad \mathbf{x} = (x, y, \theta_n) \in \mathcal{G}_n.$$

$$(17)$$

Integration of our inequality (16), with the aid of the Stokes integral theorem, yields the following result

$$\int_{\Omega_n} \sqrt{1 + |\operatorname{grad}\widetilde{\zeta}(x, y, \theta_n)|^2}\, dx dy \leq |\Omega_n| + \int_{\mathcal{G}_n} d\,\omega_n$$

$$(18)$$

$$= |\Omega_n| + \int_{\partial\mathcal{G}_n} \omega_n(x, y, \theta_n) \quad \text{for} \quad n = 0, \ldots, n_0.$$

Now we evaluate the curvilinear integrals at the right-hand side in (18). We introduce *the vector field*

$$\widetilde{M}(\mathbf{x}) = (\widetilde{l}(\mathbf{x}), \widetilde{m}(\mathbf{x})) := \{1 + |\operatorname{grad}\widetilde{\zeta}(\mathbf{x})|^2\}^{-\frac{1}{2}} \left(\widetilde{\zeta}_x(\mathbf{x}), \widetilde{\zeta}_y(\mathbf{x})\right)$$

$$\text{for all} \quad \mathbf{x} = (x, y, \theta_n) \in \mathcal{G}_n, n = 0 \ldots, n_0,$$

$$(19)$$

which represents the plane projection of the unit normal and satisfies

$$|\widetilde{M}(\mathbf{x})| < 1 \quad \text{for all} \quad \mathbf{x} = (x, y, \theta_n) \in \mathcal{G}_n, n = 0 \ldots, n_0. \qquad (20)$$

4.) We parametrize the circular segment $\partial\mathcal{G}_n \cap \mathcal{C}_1$ of the free boundary by

$$\mathbf{x}(t) = (x(t), y(t), \theta_n) := (\cos t, \sin t, \theta_n), \quad 0 \leq t < \beta_n,$$

where $\beta_n := 2\pi$ for $n = 0, \ldots, n_0 - 1$ and $\beta_{n_0} := (t_0 - 2\pi n_0)$. Utilizing the negative orientation as well as the free boundary condition (12), we obtain

$$\int_{\partial\mathcal{G}_n \cap \mathcal{C}_1} \omega_n = -\int_0^{\beta_n} \widetilde{\zeta}(\mathbf{x}(t))\left\{\widetilde{l}(\mathbf{x}(t))y'(t) - \widetilde{m}(\mathbf{x}(t))x'(t)\right\} dt$$

$$(21)$$

$$= -\int_0^{\beta_n} \widetilde{\zeta}(\mathbf{x}(t))\left\{\widetilde{N}(\mathbf{x}(t)) \cdot \mathcal{N}(\mathbf{x}(t))\right\} dt = 0 \quad \text{for} \quad n = 0, \ldots, n_0.$$

Via the parametrization

$$\mathbf{x}(t) = (x(t), y(t), \theta_n) := (r_0 \cos t, r_0 \sin t, \theta_n), \quad 0 \leq t < \beta_n,$$

we estimate the integrals along the circle $\mathcal{C}_{r_0}$. With the aid of (20) and (13), we comprehend

$$\left| \int_{\mathcal{H}_n} \omega_n \right| = \left| \int_0^{\beta_n} \widetilde{\zeta}\left(\mathbf{x}(t)\right)\left\{\widetilde{l}\left(\mathbf{x}(t)\right)y'(t) - \widetilde{m}\left(\mathbf{x}(t)\right)x'(t)\right\} dt \right|$$

$$\le \int_0^{\beta_n} \left|\widetilde{\zeta}\left(\mathbf{x}(t)\right)\right|\left|\widetilde{l}\left(\mathbf{x}(t)\right)y'(t) - \widetilde{m}\left(\mathbf{x}(t)\right)x'(t)\right| dt$$

$$\le \int_0^{\beta_n} \left|\widetilde{\zeta}\left(\mathbf{x}(t)\right)\right|\left|\widetilde{M}\left(\mathbf{x}(t)\right)\right|\sqrt{x'(t)^2 + y'(t)^2}\, dt \tag{22}$$

$$\le \int_0^{\beta_n} \left|\widetilde{\Phi}\left(\mathbf{x}(t)\right)\right|\sqrt{x'(t)^2 + y'(t)^2}\, dt$$

$$= \int_{\mathcal{H}_n} \left|\widetilde{\Phi}\left(\mathbf{x}\right)\right| d\sigma(\mathbf{x}) \quad \text{for} \quad n = 0, \ldots, n_0.$$

5.) By the method above, we arrive at the corresponding inequalities

$$\left| \int_{\partial^+ \mathcal{G}_0} \omega_0 \right| \le (r_0 - 1)|\widetilde{\Phi}(0)| \quad \text{and} \quad \left| \int_{\partial^- \mathcal{G}_{n_0}} \omega_{n_0} \right| \le (r_0 - 1)|\widetilde{\Phi}(t_0)|. \tag{23}$$

Finally, we control the integrals in transition over the slits $\mathcal{F}_{n+1}$ for $n = 0, \ldots, n_0 - 1$. The vector field $\widetilde{M}$ behaves continuously in transition over these straight segments; however, the function $\widetilde{\zeta}$ possesses a jump of the height $(\theta_{n+1} - \theta_n)$ near the slits. The symbols $\widetilde{\zeta}^-$ and $\widetilde{\zeta}^+$, respectively, designate the continuous extensions of the function $\widetilde{\zeta}$ on the sheet $\mathcal{G}_n$ or $\mathcal{G}_{n+1}$ up to the slit $\mathcal{F}_{n+1}$. We utilize the parametrization

$$\mathbf{x}(r) = (x(r), y(r), \theta_{n+1}) := (r, 0, \theta_{n+1}), \quad 1 \le r < r_0$$

on the slit $\mathcal{F}_{n+1}$, and we traverse one segment $\partial^- \mathcal{G}_n$ in a negative direction and the other $\partial^+ \mathcal{G}_{n+1}$ in a positive direction. Thus we see

$$\int_{\partial^- \mathcal{G}_n} \omega_n + \int_{\partial^+ \mathcal{G}_{n+1}} \omega_{n+1}$$

$$= \int_1^{r_0} \widetilde{\zeta}^-\left(\mathbf{x}(r)\right)\widetilde{m}(\mathbf{x}(r))dr - \int_1^{r_0} \widetilde{\zeta}^+\left(\mathbf{x}(r)\right)\widetilde{m}(\mathbf{x}(r))dr$$

$$= \int_1^{r_0} \left(\widetilde{\zeta}^-\left(\mathbf{x}(r)\right) - \widetilde{\zeta}^+\left(\mathbf{x}(r)\right)\right)\widetilde{m}(\mathbf{x}(r))dr$$

with the aid of 17) and (19). Taking (20) into account, we arrive at the estimate

$$\left| \int_{\partial^- \mathcal{G}_n} \omega_n + \int_{\partial^+ \mathcal{G}_{n+1}} \omega_{n+1} \right| \le \int_1^{r_0} \left|\widetilde{\zeta}^+\left(\mathbf{x}(r)\right) - \widetilde{\zeta}^-\left(\mathbf{x}(r)\right)\right||\widetilde{m}(\mathbf{x}(r))|dr \tag{24}$$

$$\le (r_0 - 1)(\theta_{n+1} - \theta_n) \quad \text{for} \quad n = 0, \ldots, n_0 - 1.$$

6.) By means of Definition 3.3 we add (18) from 0 to n_0 and observe (15) as well as (21). Thus we obtain

$$\Sigma\left(\widetilde{\zeta}, \mathcal{A}(r_0, t_0)\right) \leq \sum_{n=0}^{n_0} |\Omega_n| + \sum_{n=0}^{n_0} \int_{\partial \mathcal{G}_n} \omega_n \leq \sum_{n=0}^{n_0} |\Omega_n| +$$

$$+ \int_{\partial^+ \mathcal{G}_0} \omega_0 + \sum_{n=0}^{n_0-1} \left(\int_{\partial^- \mathcal{G}_n} \omega_n + \int_{\partial^+ \mathcal{G}_{n+1}} \omega_{n+1} \right) + \int_{\partial^- \mathcal{G}_{n_0}} \omega_{n_0} + \sum_{n=0}^{n_0} \int_{\mathcal{H}_n} \omega_n.$$

With the aid of (23), (24), and (22) we achieve our assertion (14), recalling (4) at the fixed boundary.

7.) We remark that the continuity of our vector field $\widetilde{M}$, up to the boundary $\mathcal{C}_{r_0}$, was necessary for the estimate (22). Under our general conditions, we estimate

$$\Sigma\left(\widetilde{\zeta}, \mathcal{A}(r, t_0)\right) \quad \text{for sufficiently large} \quad 1 < r < r_0,$$

and then control the passage $r \to r_0-$ to the limit.   q.e.d.

Finally, we prove the central

Theorem 3.5. (Solution of the mixed boundary value problem)
For each function $\Phi \in C^0(I, \mathbb{R})$ prescribed at the boundary, the mixed b.v.p. $\mathcal{P}(\Phi)$ of the Definition 2.2 in Section 2 possesses exactly one solution ζ. Lifting Φ to the function $\widetilde{\Phi} : \partial \mathcal{A}(r_0, t_0) \setminus \mathcal{C}_1 \to \mathbb{R}$ according to formula (4), we can lift the solution $\zeta = \zeta(s, t) \in \mathcal{P}(\Phi)$ onto the lifted circular ring to a solution $\widetilde{\zeta}(x, y, \theta_n) \in \widetilde{\mathcal{P}}(\widetilde{\Phi})$ of the equivalent mixed b.v.p. from Definition 3.2.

Proof:

1.) At first, we solve the mixed b.v.p. $\mathcal{P}(\Phi)$ for all the functions

$$\Phi \in C^{2+\mu}(I, \mathbb{R}) \quad \text{with a Hölder exponent} \quad \mu \in (0, 1) \tag{25}$$

by *the continuity method*: Let us consider *the homotopy* of the fixed boundary values

$$\Phi[., \lambda] = \Phi[t, \lambda] := \lambda \Phi(t) : I \times [0, 1] \to \mathbb{R} \in C^{2+\mu}(I \times [0, 1], \mathbb{R}), \tag{26}$$

with the deformation parameter $\lambda \in [0, 1]$. Evidently, the problem

$$\mathcal{P}\left(\Phi[., \lambda]\right), \quad 0 \leq \lambda \leq 1 \tag{27}$$

is solved by the function $\zeta(s, t) \equiv 0$, $(s, t) \in \overline{R(s_0, t_0)}$ for the parameter $\lambda = 0$. If we have already found a solution for the parameter $\lambda_* \in [0, 1)$, our Theorem 2.5 from Section 2 demonstrates the existence of a solution (27) for all $\lambda \in (0, 1)$ satisfying $|\lambda - \lambda_*| \leq \epsilon$; with a sufficiently small number $\epsilon = \epsilon(\lambda_*) > 0$. Now we consider the maximal $\lambda_* \in (0, 1]$, such that the problem (27) is solvable for all $0 \leq \lambda < \lambda_*$. Due to the *Compactness statement* below, this problem possesses a solution for the parameter λ_* as well. Thus the case $\lambda_* = 1$ enters necessarily, and the problem $\mathcal{P}(\Phi)$ is solved for all boundary values (25).

Then we solve the problem $\mathcal{P}(\Phi)$ for all the functions $\Phi \in C^0(I, \mathbb{R})$ which are only continuous on the boundary, invoking the *compactness statement* once more.

2.) **Compactness statement:** *If the problem $\mathcal{P}(\Psi_k)$ possesses a solution ζ_k for each $k = 1, 2, \ldots$ and their boundary values $\Psi_k \in C^0(I, \mathbb{R})$ converge uniformly to the function $\Psi \in C^0(I, \mathbb{R})$, the problem $\mathcal{P}(\Psi)$ also possesses a solution ζ. When we assume the boundary regularity $\Psi \in C^{2+\mu}(I, \mathbb{R})$, the solution $\zeta \in \mathcal{P}(\Psi)$ belongs to the class $C^{2+\mu}(\overline{R(s_0, t_0)})$.*

In order to establish our statement, we consider *the pulled-back surfaces*

$$\mathbf{Z}_k(s, t) := \Big(\exp(s + it), \zeta_k(s, t) \Big) = \exp(s + it) + \zeta_k(s, t)\, \mathbf{e}\,, (s, t) \in \overline{R(s_0, t_0)}. \tag{28}$$

With the aid of the uniformization theorems from Section 7 and Section 8 of Chapter 12, we introduce isothermal parameters into the first fundamental forms of (28). In this context, we use the *semi-disc*

$$B := \{w = u + iv = (u, v) \in \mathbb{C} : |w| < 1,\ v > 0\}$$

as our parameter domain, which is bounded by the *semi-circle*

$$C := \{w = u + iv = (u, v) \in \mathbb{C} : |w| = 1,\ v > 0\}$$

and the *unit interval*

$$J := \{w = u + iv = (u, v) \in \mathbb{C} : -1 < u < +1,\ v = 0\}.$$

The *uniformizing mappings* $s_k + it_k = s_k(u, v) + it_k(u, v) : \overline{B} \to \overline{R(s_0, t_0)}$, which satisfy

$$\frac{\partial(s_k, t_k)}{\partial(u, v)} > 0 \quad \text{for all} \quad (u, v) \in B \cup J \tag{29}$$

for their Jacobians, constitute $C^2(B \cup J)$-diffeomorphisms and furnish homeomorphisms up to the closure of the domains. In the sequel, we shall tacitly understand that the identities involving the index k are valid for $k = 1, 2, \ldots$.

3.) Now we introduce the functions

$$f_k(u, v) = x_k(u, v) + iy_k(u, v) := \exp\Big(s_k(u, v) + it_k(u, v) \Big),\, (u, v) \in \overline{B},$$

$$\text{with} \quad \tau_k(u, v) := \arg\Big(x_k(u, v) + iy_k(u, v) \Big),\, (u, v) \in \overline{B},$$

$$\text{and} \quad z_k(u, v) := \zeta_k(s_k(u, v), t_k(u, v)),\, (u, v) \in \overline{B}; \tag{30}$$

here the argument-function arg measures the angle of the vector $x + iy$ to the positive x-axis w.r.t. the continuous extension along arbitrary paths in B. Thus we obtain the regular, parametric minimal surfaces

$$\mathbf{X}_k = \mathbf{X}_k(u,v) := (x_k(u,v), y_k(u,v), z_k(u,v)) = (f_k(u,v), z_k(u,v)), \quad (31)$$

satisfying the following conditions (32) – (36):

$$\mathbf{X}_k = \mathbf{X}_k(u,v) : \overline{B} \to \mathbb{R}^3 \in C^2(B) \cap C^1(B \cup J) \cap C^0(\overline{B}); \quad (32)$$

$$\Delta\mathbf{X}_k = 0, \quad \left|\frac{\partial}{\partial u}\mathbf{X}_k\right| = \left|\frac{\partial}{\partial v}\mathbf{X}_k\right|, \quad \frac{\partial}{\partial u}\mathbf{X}_k \cdot \frac{\partial}{\partial v}\mathbf{X}_k = 0 \quad \text{in} \quad B; \quad (33)$$

With a function $\lambda_k : J \to \mathbb{R} \in C^0(J)$ we have:

$$\frac{\partial}{\partial v}\mathbf{X}_k(u,0) = \lambda_k(u)\mathcal{N}(\mathbf{X}_k(u,0)), \; u \in J; \quad (34)$$

There is a weakly increasing, continuous function $T_k : [0,\pi] \to [0,\pi]$

$$\text{with} \quad T_k(\frac{\pi}{2}l) = \frac{\pi}{2}l \quad \text{for} \quad l = 0,1,2 \quad (35)$$

$$\text{such that} \quad \mathbf{X}_k(e^{it}) = \Gamma_k\left(e^{iT_k(t)}\right), \quad 0 \le t \le \pi.$$

Here we have used the *representation for the fixed boundary curves*

$\Gamma_k = \Gamma_k(e^{it}) : \overline{C} \to \mathbb{R}^3_* := \{(x,y,z) \in \mathbb{R}^3 : x^2 + y^2 > 0\}$ defined by $\Gamma_k(e^{it})$:

$$= \left(\frac{3}{\pi}(r_0 - 1)t + 1, 0, \Psi_k(0)\right) \quad \text{for all} \quad 0 \le t \le \frac{\pi}{3};$$

$$= \left(r_0\cos(\frac{3t_0}{\pi}t - t_0), r_0\sin(\frac{3t_0}{\pi}t - t_0), \Psi_k(\frac{3t_0}{\pi}t - t_0)\right) \quad \text{for all} \quad \frac{\pi}{3} \le t \le \frac{2\pi}{3};$$

$$= \left(\left[\frac{3}{\pi}(1 - r_0)t + 3r_0 - 2\right]\cos t_0, \left[\frac{3}{\pi}(1 - r_0)t + 3r_0 - 2\right]\sin t_0, \Psi_k(t_0)\right)$$

$$\text{for all} \quad \frac{2\pi}{3} \le t \le \pi.$$

$$(36)$$

We observe that the trace $\|\Gamma_k\| := \left\{\Gamma_k(e^{it}) \in \mathbb{R}^3 : 0 \le t \le \pi\right\}$ of this mapping coincides with the graph of the boundary values, lifted onto the etale plane,

$$\left\{\mathbf{y} \in \mathbb{R}^3 : \mathbf{y} = \mathbf{x} + \widetilde{\Psi}_k(\mathbf{x})\,\mathbf{e}, \; \mathbf{x} \in \overline{\partial\mathcal{A}(r_0,t_0) \setminus \mathcal{C}_1}\right\} \quad (37)$$

and set $\Gamma_k(+1) = (1, 0, \Psi_k(0)) =: P_k$, $\Gamma_k(-1) = (\cos t_0, \sin t_0, \Psi_k(t_0)) =: Q_k$.

4.) Let us lift our solutions ζ_k onto the circular ring $\overline{\mathcal{A}(s_0,t_0)}$, and we obtain the nonparametric solutions $\widetilde{\zeta}_k \in \mathcal{P}(\widetilde{\Psi}_k)$. With the aid of Proposition 3.4, we estimate their area $\Sigma\left(\widetilde{\zeta}_k, \mathcal{A}(r_0,t_0)\right)$ by their boundary values $\widetilde{\Psi}_k$. On account of the conformal parametrization, Dirichlet's integral coincides with the double area of the surface $\mathbf{X}_k$. Thus we obtain a uniform bound for the *Dirichlet integrals* of the mappings $\mathbf{X}_k$ and their plane projections f_k in

$$D(f_k) \leq D(\mathbf{X}_k) := \int_B |\nabla \mathbf{X}_k(u,v)|^2 \, du dv = 2\Sigma\left(\widetilde{\zeta}_k \,, \mathcal{A}(r_0,t_0)\right) \leq c \tag{38}$$

$$\text{for} \quad k = 1, 2, \ldots \quad \text{with a constant} \quad c = c(\Psi) \in (0,\infty).$$

Then we lift the plane mapping f_k onto the etale plane by the prescription

$$F_k : \overline{B} \to \overline{\mathcal{A}(s_0,t_0)} \text{ with } F_k(u,v) := \left(x_k(u,v), y_k(u,v), \theta(\tau_k(u,v)) \right), (u,v) \in \overline{B}. \tag{39}$$

This function F_k represents a $C^2(B \cup J)$-diffeomorphism and a homeomorphism up to the closure of the domains. With the aid of the Oscillation lemma by R. Courant and H. Lebesgue (see Section 5 in Chapter 1), one easily derives a modulus of continuity for the class of mappings

$$\left\{ F_k \in C^0\left(\overline{B} \,, \overline{\mathcal{A}(s_0,t_0)} \right) : k = 1, 2, \ldots \right\}. \tag{40}$$

This estimate is achieved via the bound (38) for the functions f_k and the homeomorphic character of our mappings F_k on the lifted circular ring. The derivation of this modulus poses an interesting exercise for our readers.

5.) The gradient estimate of E. Heinz (see Section 2 and Section 3 of Chapter 12) allows the selection of a subsequence $1 \leq j(1) < j(2) < j(3) < \ldots$ such that the functions $\{\mathbf{X}_{j(k)}\}_{k=1,2,\ldots}$ converge in the space $C^{2+\mu}(B \cup J) \cap C^0(\overline{B})$ to *a solution* $\mathbf{X} = \mathbf{X}(u,v)$ *of the asymptotic parametric problem* (31) – (36) for the limit boundary distribution Ψ, by simply omitting the index k in these identities. For the necessary regularity questions at the fixed and free boundaries, we refer our readers to the book [DHT1] *Regularity Theory of Minimal Surfaces* by U. Dierkes, S. Hildebrandt, and A. Tromba.

6.) Finally, we have to guarantee the existence of the inverse transformation for our mapping

$$F(u,v) := \lim_{k \to \infty} F_{j(k)}(u,v), \quad (u,v) \in \overline{B}.$$

As in the proof of Theorem 1.5 from Section 1, we observe the last component of the normal $\mathbf{N}_{j(k)}$ for the minimal surface $\mathbf{X}_{j(k)}$, which remains positive during the passage to the limit within the domain B. The same phenomenon is valid on the free boundary J, where a Neumann condition has been derived for the last component of the normal. Please see Proposition 1 in the following paper by S. HILDEBRANDT AND F. SAUVIGNY: *Uniqueness of stable minimal surfaces with partially free boundaries*; Journal Math.Soc.Japan **47** (1995) 423-440 or consult Chapter 2 in the book [DHT2]. Via the boundary point lemma by E. Hopf, we arrive at the statement

$$\mathbf{N}(u,v) \cdot \mathbf{e} > 0 \quad \text{for all} \quad (u,v) \in B \cup J \tag{41}$$

for the unit normal $\mathbf{N}(u,v)$ of our asymptotic parametric solution $\mathbf{X}(u,v)$. Using the index-sum formula (see Section 1 in Chapter 3), we deduce that our *asymptotic lifted plane mapping* $F : \overline{B} \to \overline{\mathcal{A}(s_0, t_0)}$, which is associated with $\mathbf{X}(u, v)$, represents a homeomorphism with the inverse $F^{-1} : \overline{\mathcal{A}(s_0, t_0)} \to \overline{B}$ and moreover a $C^2(B \cup J)$-diffeomorphism. With the function

$$\widetilde{\zeta}(x, y, \theta_n) := z\Big(F^{-1}(x, y, \theta_n)\Big), \quad (x, y, \theta_n) \in \mathcal{G}_n \quad \text{for} \quad n = 0, \ldots, n_0 \quad (42)$$

we obtain a solution of the problem $\mathcal{P}(\widetilde{\Psi})$ equivalent to the problem $\mathcal{P}(\Psi)$. Thus our compactness statement above is completely proved.

7.) By the methods from the Proof of Theorem 1.7 in Section 1, we deduce *the transversality to the boundary* for our asymptotic parametric solution

$$\mathbf{X}(u, v) = (x(u, v), y(u, v), z(u, v)) \quad \text{with} \quad \frac{\partial(x, y)}{\partial(u, v)} > 0 \quad \text{for all} \quad (u, v) \in C,$$
$$(43)$$

in case of differentiable boundary data $\Psi \in C^{2+\mu}(I, \mathbb{R})$. Here we mention that a reflection over the straight fixed boundary transforms the regularity question into an interior problem, which is solvable as above. Therefore, the nonparametric solution is differentiable up to the fixed boundary, according to $\zeta \in C^{2+\mu}(\overline{R(s_0, t_0)})$.   q.e.d.

Remarks: The investigations of Section 2 and Section 3 are based on the paper [S6] by Friedrich Sauvigny, which we recommend for further study. Here we would also like to refer our readers to the subsequent article:

F. Sauvigny: *Mixed boundary value problems for minimal graphs on a Riemann surface and winding staircases.* Milan Journal of Mathematics **79** (2011), 311-325.

4 Plateau's Problem with Constant Mean Curvature

Given the radius $M > 0$, we define the ball

$$K := \Big\{(x, y, z) \in \mathbb{R}^3 \ : \ x^2 + y^2 + z^2 \leq M^2\Big\}.$$

Within K we take a rectifiable Jordan curve $\Gamma \subset K$, where we fix three points $\mathbf{p}_j \in \Gamma$ for $j = 1, 2, 3$. We define the nonvoid *class of admissible functions*

$$\mathcal{Z}(\Gamma) := \left\{ \mathbf{x} = \mathbf{x}(u, v) : \overline{B} \to K \ : \ \begin{array}{l} \mathbf{x} \in C^2(B) \cap C^0(\overline{B}) \cap W^{1,2}(B), \\ \mathbf{x} : \partial B \to \Gamma \text{ weakly monotonic,} \\ \mathbf{x}(e^{\frac{2\pi i}{3}j}) = \mathbf{p}_j, \ j = 1, 2, 3 \end{array} \right\}.$$

Besides the *generalized area*

$$A(\mathbf{x}) := \iint\limits_B \Big\{|\mathbf{x}_u \wedge \mathbf{x}_v| + \frac{2H}{3}(\mathbf{x}, \mathbf{x}_u, \mathbf{x}_v)\Big\} \, du \, dv \quad (1)$$

from Chapter 11, Section 2, we consider the *energy functional of E. Heinz*

$$E(\mathbf{x}) := \iint\limits_B \left\{ \left(|\mathbf{x}_u|^2 + |\mathbf{x}_v|^2 \right) + \frac{4H}{3}(\mathbf{x}, \mathbf{x}_u, \mathbf{x}_v) \right\} du\, dv \tag{2}$$

for $\mathbf{x} \in \mathcal{Z}(\Gamma)$, assuming $H \in [-\frac{1}{2M}, +\frac{1}{2M}]$. Dirichlet's integral

$$D(\mathbf{x}) := \iint\limits_B \left(|\mathbf{x}_u|^2 + |\mathbf{x}_v|^2 \right) du\, dv$$

has the following relationship to this energy functional:

$$E(\mathbf{x}) \geq \frac{2}{3} D(\mathbf{x}) \qquad \text{for all} \quad \mathbf{x} \in \mathcal{Z}(\Gamma). \tag{3}$$

Furthermore, we observe

$$2A(\mathbf{x}) \leq E(\mathbf{x}) \qquad \text{for all} \quad \mathbf{x} \in \mathcal{Z}(\Gamma), \tag{4}$$

where equality is exactly attained in the case of conformal parametrization

$$|\mathbf{x}_u| = |\mathbf{x}_v|, \quad \mathbf{x}_u \cdot \mathbf{x}_v = 0 \qquad \text{in} \quad B. \tag{5}$$

This fact is based on the inequality

$$\sqrt{EG - F^2} \leq \sqrt{EG} \leq \frac{1}{2}(E + G)$$

for the coefficients of the first fundamental form

$$d\mathbf{x}^2 = E\, du^2 + 2F\, du\, dv + G\, dv^2$$

associate to the surface.

We owe the following result to T. Radó and C. B. Morrey:

Proposition 4.1. (Almost conformal parameters)
Let the function $\mathbf{x} = (x(u,v), y(u,v), z(u,v)) \in \mathcal{Z}(\Gamma)$ *and the quantity* $\varepsilon > 0$ *be given. Then we have a parameter transformation* $f(\alpha, \beta) : \overline{B} \to \overline{B}$ *which is topological, such that the surface* $\mathbf{y}(\alpha, \beta) := \mathbf{x} \circ f(\alpha, \beta) \in \mathcal{Z}(\Gamma)$ *is admissible and the estimate*

$$\frac{1}{2} E(\mathbf{y}) \leq A(\mathbf{y}) + \varepsilon \tag{6}$$

is fulfilled.

Proof: Since the second summand in $2A$ and E are parameter-invariant (with respect to orientation-preserving reparametrizations), we have only to investigate the case $H = 0$. Given the number $\delta > 0$ we define the extended mapping

$$\tilde{\mathbf{x}}(u, v) = \big(x(u, v), y(u, v), z(u, v); \delta u, \delta v\big) : \overline{B} \to \mathbb{R}^5 \qquad (7)$$

with the first fundamental form

$$\tilde{E} = \tilde{\mathbf{x}}_u \cdot \tilde{\mathbf{x}}_u = E + \delta^2, \quad \tilde{F} = \tilde{\mathbf{x}}_u \cdot \tilde{\mathbf{x}}_v = F, \quad \tilde{G} = \tilde{\mathbf{x}}_v \cdot \tilde{\mathbf{x}}_v = G + \delta^2$$

and the surface element

$$\tilde{E}\tilde{G} - \tilde{F}^2 = EG - F^2 + \delta^2(E + G) + \delta^4 > 0.$$

According to Section 8 of Chapter 12, we introduce isothermal parameters into the regular surface $\tilde{\mathbf{x}}(u, v)$ with the aid of the positive-oriented mapping

$$f(\alpha, \beta) = \big(u(\alpha, \beta), v(\alpha, \beta)\big) : \overline{B} \to \overline{B}.$$

The surface

$$\tilde{\mathbf{y}}(\alpha, \beta) := \tilde{\mathbf{x}} \circ f(\alpha, \beta) = \big(\mathbf{x} \circ f(\alpha, \beta), \delta f(\alpha, \beta)\big) = \big(\mathbf{y}(\alpha, \beta), \delta f(\alpha, \beta)\big) : \overline{B} \to \mathbb{R}^5$$

satisfies

$$\tilde{\mathbf{y}}_\alpha \cdot \tilde{\mathbf{y}}_\beta = 0 = |\tilde{\mathbf{y}}_\alpha|^2 - |\tilde{\mathbf{y}}_\beta|^2 \quad \text{in} \quad B,$$

and the transformation formula for multiple integrals yields

$$D(\mathbf{y}) + \delta^2 D(f) = D(\tilde{\mathbf{y}}) = 2 \iint\limits_{B} \sqrt{\tilde{E}\tilde{G} - \tilde{F}^2} \, d\alpha \, d\beta$$

$$= 2 \iint\limits_{B} \sqrt{(EG - F^2) + \delta^2(E + G) + \delta^4} \, du \, dv$$

$$\leq 2 \iint\limits_{B} \sqrt{EG - F^2} \, du \, dv + 2\delta \iint\limits_{B} \sqrt{E + G} \, du \, dv + 2\pi\delta^2.$$

$$(8)$$

The quantity $\varepsilon > 0$ given, we can find a number $\delta > 0$ and an associate parameter transformation f, such that the inequality (6) holds true for the function $\mathbf{y} = \mathbf{x} \circ f$.

                                                                                        q.e.d.

Proposition 4.2. (Minimal property)
Let the function $\mathbf{x}(u, v) \in \mathcal{Z}(\Gamma)$ denote a solution of the H-surface system

$$\Delta\mathbf{x}(u, v) = 2H \, \mathbf{x}_u \wedge \mathbf{x}_v(u, v) \quad \text{in} \quad B$$

with $H \in [-\frac{1}{2M}, +\frac{1}{2M}]$. Then all admissible functions $\mathbf{y}(u, v) \in \mathcal{Z}(\Gamma)$ satisfying $\mathbf{y}(u, v) = \mathbf{x}(u, v)$ on ∂B realize the inequality

$$E(\mathbf{y}) \geq E(\mathbf{x}). \qquad (9)$$

Proof: With the aid of the Gaussian integral theorem we easily verify the following identity:

$$E(\mathbf{x} + \mathbf{z}) = E(\mathbf{x}) + \iint_B \left\{ |\nabla \mathbf{z}|^2 + \frac{4H}{3}(3\mathbf{x} + \mathbf{z}, \mathbf{z}_u, \mathbf{z}_v) \right\} du\, dv \tag{10}$$

for all $\mathbf{z} \in C_0^\infty(B, \mathbb{R}^3)$.

Therefore, we develop

$$E(\mathbf{x} + \mathbf{z}) = \iint_B \left\{ |\nabla \mathbf{x}|^2 + 2\nabla(\mathbf{z} \cdot \nabla \mathbf{x}) + |\nabla \mathbf{z}|^2 - 2\mathbf{z} \cdot \Delta \mathbf{x} \right\} du\, dv$$

$$+ \frac{4H}{3} \iint_B (\mathbf{x} + \mathbf{z}, \mathbf{x}_u + \mathbf{z}_u, \mathbf{x}_v + \mathbf{z}_v)\, du\, dv$$

$$= E(\mathbf{x}) + \iint_B \left\{ |\nabla \mathbf{z}|^2 + \frac{4H}{3}(\mathbf{x} + \mathbf{z}) \cdot \mathbf{z}_u \wedge \mathbf{z}_v - 4H(\mathbf{x}_u, \mathbf{x}_v, \mathbf{z}) \right\} du\, dv$$

$$+ \frac{4H}{3} \iint_B \left\{ (\mathbf{z}, \mathbf{x}_u, \mathbf{x}_v) + (\mathbf{x} + \mathbf{z}, \mathbf{x}_u, \mathbf{z}_v) + (\mathbf{x} + \mathbf{z}, \mathbf{z}_u, \mathbf{x}_v) \right\} du\, dv$$

$$= E(\mathbf{x}) + \iint_B \left\{ |\nabla \mathbf{z}|^2 + \frac{4H}{3}(\mathbf{x} + \mathbf{z}) \cdot \mathbf{z}_u \wedge \mathbf{z}_v \right\} du\, dv$$

$$+ \frac{4H}{3} \iint_B \left\{ (\mathbf{x} + \mathbf{z}, \mathbf{x}_u, \mathbf{z})_v + (\mathbf{x} + \mathbf{z}, \mathbf{z}, \mathbf{x}_v)_u \right\} du\, dv$$

$$- \frac{4H}{3} \iint_B \left\{ (\mathbf{z}_v, \mathbf{x}_u, \mathbf{z}) + (\mathbf{z}_u, \mathbf{z}, \mathbf{x}_v) \right\} du\, dv$$

$$= E(\mathbf{x}) + \iint_B \left\{ |\nabla \mathbf{z}|^2 + \frac{4H}{3}(3\mathbf{x} + \mathbf{z}) \cdot \mathbf{z}_u \wedge \mathbf{z}_v \right\} du\, dv$$

$$- \frac{4H}{3} \iint_B \left\{ (\mathbf{z}_v, \mathbf{x}, \mathbf{z})_u + (\mathbf{z}_u, \mathbf{z}, \mathbf{x})_v \right\} du\, dv$$

$$= E(\mathbf{x}) + \iint_B \left\{ |\nabla \mathbf{z}|^2 + \frac{4H}{3}(3\mathbf{x} + \mathbf{z}) \cdot \mathbf{z}_u \wedge \mathbf{z}_v \right\} du\, dv.$$

Via a well-known approximation procedure, we can insert the function

$$\mathbf{z} = \mathbf{y} - \mathbf{x} \quad \text{with} \quad |\mathbf{x} + \mathbf{z}| \leq M \text{ on } B$$

into (10). From the condition $|H|M \leq \frac{1}{2}$ we then infer the inequality (9).

q.e.d.

We owe the following result for surfaces of constant mean curvature to E. Heinz:

Theorem 4.3. (Plateau's problem)
Let the parameter $H \in [-\frac{1}{2M}, +\frac{1}{2M}]$ be given. Then the variational problem

$$A(\mathbf{x}) \to minimum, \qquad \mathbf{x} \in \mathcal{Z}(\Gamma) \tag{11}$$

possesses a solution $\mathbf{x} \in \mathcal{Z}(\Gamma)$, representing an H-surface with the curve Γ as its boundary.

Proof: We define the number

$$a := \inf_{\mathbf{x} \in \mathcal{Z}(\Gamma)} A(\mathbf{x}) \in (0, +\infty)$$

and choose a minimal sequence $\{\mathbf{x}_n\}_{n=1,2,\ldots} \subset \mathcal{Z}(\Gamma)$ with

$$\lim_{n \to \infty} A(\mathbf{x}_n) = a. \tag{12}$$

Via Proposition 4.1 we make the transition to a sequence $\{\mathbf{y}_n\}_{n=1,2,\ldots} \subset \mathcal{Z}(\Gamma)$ satisfying

$$\frac{1}{2} E(\mathbf{y}_n) \le A(\mathbf{x}_n) + \frac{1}{n}, \qquad n = 1, 2, \ldots. \tag{13}$$

Using Theorem 4.6 from Section 4 of Chapter 12, we can uniquely extend the continuous boundary values of $\mathbf{y}_n$ to a solution of Rellich's system, namely

$$\Delta \mathbf{z}_n(u, v) = 2H\, (\mathbf{z}_n)_u \wedge (\mathbf{z}_n)_v(u, v) \qquad in \quad B,$$
$$\mathbf{z}_n = \mathbf{y}_n \qquad on \quad \partial B. \tag{14}$$

Proposition 4.2 together with (13) yield the inequality

$$\frac{1}{2} E(\mathbf{z}_n) \le A(\mathbf{x}_n) + \frac{1}{n}, \qquad n = 1, 2, \ldots \tag{15}$$

On account of (3) the sequence $\{\mathbf{z}_n\}_n$ possesses a uniformly bounded Dirichlet's integral. With the aid of the Courant-Lebesgue lemma we see that the boundary values $\mathbf{z}_n|_{\partial B}$, $n = 1, 2, \ldots$ are equicontinuous. W. Jäger's maximum principle from Section 1 of Chapter 12 allows the transition to a uniformly convergent subsequence on the closed disc $\overline{B}$. Due to Theorem 2.7 in Section 2 of Chapter 12, we find a function $\mathbf{z}(u, v) \in \mathcal{Z}(\Gamma)$ in the limit satisfying

$$\Delta \mathbf{z}(u, v) = 2H\, \mathbf{z}_u \wedge \mathbf{z}_v \qquad in \quad B. \tag{16}$$

On account of the convergence in $C^1(B)$ we infer the following inequality from (15), namely

$$a \le \frac{1}{2} E(\mathbf{z}) \le a \le A(\mathbf{z}), \tag{17}$$

which implies $A(\mathbf{z}) = \frac{1}{2}E(\mathbf{z})$. Consequently, the surface $\mathbf{z}$ is conformally parametrized and represents a H-surface.

q.e.d.

On each disc $B_r(w_0) \subset\subset B$ with the center $w_0 \in B$, our H-surface satisfies the differential inequality

$$|\mathbf{x}_{w\overline{w}}(w)| \leq c|\mathbf{x}_w(w)| \qquad \text{in} \quad B_r(w_0) \tag{18}$$

with a number $c = c(w_0, r) > 0$. By a well-known result [HW] of P. Hartman and A. Wintner (compare Section 3.1 in [DHT1]), we establish **the asymptotic expansion**

$$\mathbf{x}_w(w) = \mathbf{a}(w - w_0)^n + o(|w - w_0|^n), \qquad w \to w_0 \tag{19}$$

with the nonvanishing complex vector $\mathbf{a} = \mathbf{a}(w_0) \in \mathbb{C}^3 \setminus \{\mathbf{0}\}$ and the integer $n = n(w_0) \in \mathbb{N} \cup \{0\}$.

In the recent lecture notes [Mu] by Frank Müller, it has been shown that this asymptotic expansion can be achieved by an adequate vector-valued version of the similarity principle from Bers and Vekua (compare Theorem 6.3 from Section 6 in Chapter 4).

Only in the special case $n = n(w_0) = 0$ of (19) the surface is differential-geometrically regular at the point w_0.

Definition 4.4. *A point $w_0 \in B$, where the asymptotic expansion (19) with the positive $n(w_0) \in \mathbb{N}$ is valid, is called* branch point $\mathbf{x}_0 = \mathbf{x}(w_0)$ *of the H-surface. We call $n(w_0) \in \mathbb{N}$ the order of the branch point w_0.*

The branch points are obviously isolated, and the surface is not regular in the differential-geometric sense there. Let us represent the nonvanishing vector $\mathbf{a}$ from the asymptotic expansion (19) at a branch point w_0 in the form $\mathbf{a} = \mathbf{b} - i\mathbf{c}$ with the real vectors $\mathbf{b} \in \mathbb{R}^3$ and $\mathbf{c} \in \mathbb{R}^3$. Then we infer

$$|\mathbf{b}| = |\mathbf{c}| > 0 \,, \mathbf{b} \cdot \mathbf{c} = 0$$

from the conformality relations for our H-surface. From (19) we deduce the following limit relation

$$\mathbf{X}(u, v) := |\mathbf{x}_u \wedge \mathbf{x}_v(w)|^{-1} \mathbf{x}_u \wedge \mathbf{x}_v(w)$$

$$= |\mathbf{x}_w \wedge \mathbf{x}_{\overline{w}}(w)|^{-1} \mathbf{x}_w \wedge \mathbf{x}_{\overline{w}}(w) \tag{20}$$

$$\to |\mathbf{b} \wedge \mathbf{c}|^{-1} \mathbf{b} \wedge \mathbf{c} \quad \text{for } w = u + iv \to w_0, \, w \neq w_0$$

for *the unit normal* $\mathbf{X}(u, v) : B \to S^2$ *of our branched H-surface*. In particular, the normal can be uniquely and continuously extended into the branch points! We leave the easy proofs for these statements to our readers as an exercise. An integration of the asymptotic expansion (19) is performed within the formulas (23) and (24) in the following Section 5.

Remarks:

1.) The regularity of H-surfaces, especially at the boundary, is intensively studied in the beautiful *Grundlehren*-books [DHKW] of U. Dierkes, S. Hildebrandt, A. Küster, and O. Wohlrab on *Minimal Surfaces I and II.*

2.) If the boundary curve Γ is real-analytic, the solution can be analytically continued beyond the boundary as an H-surface, due to the result of F. MÜLLER: *Analyticity of solutions for semilinear elliptic systems of second order.* Calc. Var. and PDE **15** (2002), 257-288.

3.) According to a theorem of Alt-Gulliver-Osserman one can exclude the branch points for the solutions of the above variational problem a posteriori. However, the desire remains to solve the variational problem (11) directly in the class

$$\mathcal{Z}^*(\Gamma) := \Big\{ \mathbf{x} \in \mathcal{Z}(\Gamma) \; : \; |\mathbf{x}_u \wedge \mathbf{x}_v(u,v)| > 0 \text{ for all } (u,v) \in B \Big\}. \quad (21)$$

4.) Only under very restrictive conditions on the boundary contour Γ, the Plateau problem is uniquely solvable. To classify the solution set for Plateau's problem with arbitrary contours more precisely, this question has been and still is a great challenge for differential geometers!

5.) For a profound study of the very interesting theory of minimal surfaces, we would like to recommend our new book *Minimal Surfaces* [DHS] by U. Dierkes, S. Hildebrandt, and F. Sauvigny. This is succeeded by the new treatises [DHT1] and [DHT2] by U. Dierkes, S. Hildebrandt, and A. Tromba on *Regularity of Minimal Surfaces* and *Global Analysis of Minimal Surfaces.*

6.) Furthermore, we recommend studying the pioneering monograph *Vorlesungen über Minimalflächen* by J.C.C. Nitsche [N], published already in 1975.

5 Closed Surfaces of Constant Mean Curvature

When we look for a solution of **the isoperimetric problem within the Euclidean space**, finding a closed surface under *all closed surfaces* with given area to bound the maximal volume, the Calculus of Variations tells us that this surface has necessarily constant mean curvature. This raises the following question: In which class is the sphere the only closed surface of constant mean curvature?

Definition 5.1. *Under* **a closed embedded surface** $\mathcal{M}$ *we comprehend a 2-dimensional, compact and connected, orientable C^3-manifold in $\mathbb{R}^3$ without boundary, which does not possess self-intersections. More precisely,* $\mathcal{M}$

represents a compact and connected set of $\mathbb{R}^3$, *which is locally a zero-set
for a nondegenerate* C^3*-function and possesses a continuous unit-normal-field*
$\mathcal{N} : \mathcal{M} \to S^2$.

With the aid of the *reflection principle by A. D. Alexandrov*, we shall give a
first answer to the question above. In this context, we need a geometric version
of the maximum principle.

Proposition 5.2. (Touching principle)
On the C^1*-domain* $\Omega \subset \mathbb{R}^2$, *we have solutions* $\zeta_j = \zeta_j(x,y) \in C^2(\Omega) \cap C^1(\overline{\Omega})$
of the H*-graph-equation*

$$(1 + \zeta_y^2)\zeta_{xx} - 2\zeta_x\zeta_y\zeta_{xy} + (1 + \zeta_x^2)\zeta_{yy} = 2H\left(1 + |\nabla\zeta(x,y)|^2\right)^{\frac{3}{2}}, \ (x,y) \in \Omega \quad (1)$$

for $j = 1,2$ *and constant* $H > 0$ *satisfying*

$$\zeta_1(x,y) \leq \zeta_2(x,y), \quad (x,y) \in \Omega. \tag{2}$$

Furthermore, there exists an interior point

$$(x_0, y_0) \in \Omega \quad with \quad \zeta_1(x_0, y_0) = \zeta_2(x_0, y_0) \tag{3}$$

or alternatively *a boundary point*

$$(x_0, y_0) \in \partial\Omega \ with \zeta_1(x_0, y_0) = \zeta_2(x_0, y_0) \ and \frac{\partial}{\partial\nu}\zeta_1(x_0, y_0) = \frac{\partial}{\partial\nu}\zeta_2(x_0, y_0)$$
$$\tag{4}$$

for the derivative w.r.t. the exterior normal $\nu : \partial\Omega \to S^1$.
Then the identity $\zeta_1(x,y) \equiv \zeta_2(x,y)$, $(x,y) \in \Omega$ *holds true.*

Proof: The difference of the solutions $\zeta_1 - \zeta_2 \leq 0$ satisfies a linear elliptic
differential equation; and the strict maximum principle or the boundary point
lemma by E. Hopf yield the statement above (see Section 1 and Section 2 in
Chapter 6). q.e.d.

Theorem 5.3. (A. D. Alexandrov)
The only closed embedded surfaces $\mathcal{M}$ *of constant mean curvature* $H > 0$ *are
the spheres of radius* H^{-1}.

Proof: Choose an arbitrary directional vector $\mathbf{e} \in \mathbb{R}^3$ with $|\mathbf{e}| = 1$. Orthogonal
to $\mathbf{e}$, we consider a support-plane $\mathcal{E}$ for the surface $\mathcal{M}$, which we move into the
interior of the closed embedded surface. Thus *a cap* $\widehat{\mathcal{M}}$ is cut out by $\mathcal{E}$ on $\mathcal{M}$,
which we reflect at this plane $\mathcal{E}$ to *the reflected cap* $\widehat{\mathcal{M}}^*$. By *the residual cap*
$\widecheck{\mathcal{M}} := \mathcal{M} \setminus \widehat{\mathcal{M}}$ we mean the complement of our cap in $\mathcal{M}$. At first, the small
reflected cap $\widehat{\mathcal{M}}^*$ remains contained in the interior of $\mathcal{M}$. However, moving
the plane continuously into the interior of $\mathcal{M}$, we will arrive at the point where

the reflected cap wants to protrude from the interior of $\mathcal{M}$. We distinguish between the following two situations possible:

Case 1: The reflected cap $\widehat{\mathcal{M}}^*$ touches the residual cap $\widetilde{\mathcal{M}}$ in an interior point $\mathbf{x}_0$. Then we can represent the surfaces nonparametrically as described in the first situation (2) and (3) of Proposition 5.2. In a small neighborhood of $\mathbf{x}_0$, the two surfaces must coincide, and a continuation argument yields $\widehat{\mathcal{M}}^* = \widetilde{\mathcal{M}}$. Consequently, $\mathcal{E}$ is a symmetry plane for our surface $\mathcal{M}$.

Case 2: The reflected cap $\widehat{\mathcal{M}}^*$ touches the residual cap $\widetilde{\mathcal{M}}$ at a boundary point $\mathbf{x}_0$ *of $\widehat{\mathcal{M}}^*$:* Then we can represent the surfaces nonparametrically as described in the second situation (2) and (4) of Proposition 5.2. In a small neighborhood of $\mathbf{x}_0$, the two surfaces must coincide, and a continuation argument yields $\widehat{\mathcal{M}}^* = \widetilde{\mathcal{M}}$. Consequently, $\mathcal{E}$ is a symmetry plane for our surface $\mathcal{M}$.

In both cases, the plane $\mathcal{E}$ constitutes a symmetry plane for our closed surface $\mathcal{M}$, where the directional vector $\mathbf{e}$ has been chosen arbitrarily. Therefore, the closed surface $\mathcal{M}$ must be a sphere.   q.e.d.

Remark: Since a *soap bubble* $\mathcal{M}$ does not have self-intersections in physical experiments, this surface $\mathcal{M}$ represents necessarily a sphere.

For a more thorough study of our central question we need the following observation, already known to O. Bonnet: *If we vary a surface $\mathbf{x}$ of constant mean curvature $H > 0$ in the normal direction $\mathbf{X}$ with the distance H^{-1}, a dual surface $\mathbf{z}$ of the same mean curvature H appears!* In Lemma 1 of Section 2.) from the paper [S4] by Friedrich Sauvigny, it has been observed that even the conformal parametrization is transferred to the parallel surface, rendering the analytical treatment within this paragraph more transparent. Furthermore, we would also like to recommend the instructive Section 5.2 in our joint book on minimal surfaces [DHS] by U. Dierkes, S. Hildebrandt, and F. Sauvigny. However, we offer a derivation of this result here for the convenience of the reader.

Proposition 5.4. (The dual H-surface)

For the constant $H > 0$ let the immersed H-surface $\mathbf{x}(u, v)$, $(u, v) \in B$ on the open unit disc B with the associate unit normal $\mathbf{X}(u, v)$, $(u, v) \in B$ be given. By the parallel surface in the distance H^{-1}, namely

$$\mathbf{z}(u, v) := \mathbf{x}(u, v) + \frac{1}{H} \mathbf{X}(u, v), (u, v) \in B, \tag{5}$$

we obtain a dual H-surface $\mathbf{z}$ with the unit normal

$$\mathbf{Z}(u, v) = -\mathbf{X}(u, v), u, v) \in B, \tag{6}$$

which could degenerate to a single point. If $\mathbf{z}$ is nonconstant, the branch points of this surface are exactly the umbilical points for the original surface $\mathbf{x}$, due to the identity (8) below.

Proof: In order to simplify our calculations, we consider *the rescaled surface*

$$\mathbf{y}(u,v) := H\mathbf{z}(u,v) = H\mathbf{x}(u,v) + \mathbf{X}(u,v), \ (u,v) \in B. \tag{7}$$

We recall the identity (see Theorem 1.4 in Section 1 of Chapter 9) between the first, second, and third fundamental forms of the H-surface $\mathbf{x}$ and use those notations, e.g. K for the Gaussian curvature of $\mathbf{x}$. Thus we deduce

$$\begin{pmatrix} \mathbf{y}_u^2 & \mathbf{y}_u \cdot \mathbf{y}_v \\ \mathbf{y}_v \cdot \mathbf{y}_u & \mathbf{y}_v^2 \end{pmatrix}$$

$$= \begin{pmatrix} \mathbf{X}_u^2 & \mathbf{X}_u \cdot \mathbf{X}_v \\ \mathbf{X}_v \cdot \mathbf{X}_u & \mathbf{X}_v^2 \end{pmatrix} + 2H \begin{pmatrix} \mathbf{x}_u \cdot \mathbf{X}_u & \mathbf{x}_u \cdot \mathbf{X}_v \\ \mathbf{x}_v \cdot \mathbf{X}_u & \mathbf{x}_v \cdot \mathbf{X}_v \end{pmatrix} + H^2 \begin{pmatrix} \mathbf{x}_u^2 & \mathbf{x}_u \cdot \mathbf{x}_v \\ \mathbf{x}_v \cdot \mathbf{x}_u & \mathbf{x}_v^2 \end{pmatrix}$$

$$= \begin{pmatrix} e & f \\ f & g \end{pmatrix} - 2H \begin{pmatrix} L & M \\ M & N \end{pmatrix} + K \begin{pmatrix} E & F \\ F & G \end{pmatrix} + (H^2 - K) \begin{pmatrix} \mathbf{x}_u^2 & \mathbf{x}_u \cdot \mathbf{x}_v \\ \mathbf{x}_v \cdot \mathbf{x}_u & \mathbf{x}_v^2 \end{pmatrix}$$

$$= E(H^2 - K) \begin{pmatrix} 1 & 0 \\ 0 & 1 \end{pmatrix}.$$

This matrix-identity yields the conformality relations for our surface $\mathbf{y}$ with

$$\mathbf{y}_u \cdot \mathbf{y}_v = 0, \quad |\mathbf{y}_u|^2 = |\mathbf{y}_v|^2 = E(H^2 - K) \quad \text{in} \quad B$$

and consequently for our surface $\mathbf{z}$ with

$$\mathbf{z}_u \cdot \mathbf{z}_v = 0, \quad |\mathbf{z}_u|^2 = |\mathbf{z}_v|^2 = E\left(1 - \frac{K}{H^2}\right) \quad \text{in} \quad B. \tag{8}$$

Furthermore, we calculate with the aid of the Weingarten fundamental equations (compare the Proof of Proposition 1.4 in Section 1)

$$\mathbf{y}_u \wedge \mathbf{y}_v = \mathbf{X}_u \wedge \mathbf{X}_v + H(\mathbf{X}_u \wedge \mathbf{x}_v + \mathbf{x}_u \wedge \mathbf{X}_v) + H^2 \mathbf{x}_u \wedge \mathbf{x}_v$$

$$= EK\mathbf{X} - 2EH^2\mathbf{X} + EH^2\mathbf{X} = -E(H^2 - K)\mathbf{X} = -|\mathbf{y}_u \wedge \mathbf{y}_v|\mathbf{X}.$$

This implies

$$\mathbf{z}_u \wedge \mathbf{z}_v = -|\mathbf{z}_u \wedge \mathbf{z}_v|\mathbf{X} \quad \text{in} \quad B \tag{9}$$

and the identity (6), in particular. Finally, we establish

$$\Delta \mathbf{y} = \Delta \mathbf{X} + H\Delta \mathbf{x} = (-4EH^2 + 2EK)\mathbf{X} + 2EH^2\mathbf{X}$$

$$= 2E(K - H^2)\mathbf{X} = 2\mathbf{y}_u \wedge \mathbf{y}_v$$

and therefore the system

$$\Delta \mathbf{z} = 2H\mathbf{z}_u \wedge \mathbf{z}_v \quad \text{in} \quad B. \tag{10}$$

The equations (8) and (10) show that $\mathbf{z}$ represents an H-surface. The asymptotic expansion (19) from Section 4 yields the further properties stated above. q.e.d.

Definition 5.5. *Under* **a closed convex surface** *or equivalently* **an ovaloid** $\mathcal{M}$ *we comprehend a closed embedded surface, whose Gaussian curvature satisfies* $K(\mathbf{x}) > 0$ *for all points* $\mathbf{x} \in \mathcal{M}$.

With the aid of our dual H-surface, we prove the following classical result by the maximum method.

Theorem 5.6. (H. Liebmann)
The only closed convex surfaces $\mathcal{M}$ of constant mean curvature $H > 0$ are the spheres of radius H^{-1}.

Proof:

1.) Due to classical results of J. Hadamard, the surface $\mathcal{M}$ bounds a convex set and is homeomorphic to the sphere S^2 and consequently of genus 0. Theorem 6.5 (Jordan, Brouwer) from Section 6 in Chapter 3 tells us that $\mathcal{M}$ separates the space $\mathbb{R}^3$ into *a convex interior and an exterior domain*. The unit normal to the surface $\mathcal{N} : \mathcal{M} \to S^2$ furnishes a C^2-diffeomorphism by Hadamard's observation; an eventual substitution of $\mathcal{N}$ with $-\mathcal{N}$ guarantees that the normal is directed into the convex interior of $\mathcal{M}$ globally. For these properties of the ovaloids, we refer the reader to Section 6.2 in the book [K] by W. Klingenberg and to the *Grundlehren* [BL], § 102.

2.) Now we consider **Minkowski's support-function to the surface** $\mathcal{M}$, namely

$$\mu : \mathcal{M} \to \mathbb{R} \in C^2(\mathcal{M}) \quad \text{defined by} \quad \mu(\mathbf{x}) := \mathcal{N}(\mathbf{x}) \cdot \mathbf{x}, \quad \mathbf{x} \in \mathcal{M}. \qquad (11)$$

Via a translation of the coordinate system, we can ascertain that we have the equation

$$\mathbf{x}_0 + \frac{1}{H} \mathcal{N}(\mathbf{x}_0) = 0, \qquad (12)$$

at least for one point $\mathbf{x}_0 \in \mathcal{M}$. The Minkowski support-function satisfies the inequality

$$\mu(\mathbf{x}) \leq 0, \quad \mathbf{x} \in \mathcal{M} \qquad (13)$$

for convexity reasons. Let us consider the nonnegative function

$$\phi = \phi(\mathbf{x}) : \mathcal{M} \to \mathbb{R} \in C^2(\mathcal{M}) \text{ defined by } \phi(\mathbf{x}) := \left(\mathbf{x} + \frac{1}{H} \mathcal{N}(\mathbf{x}) \right)^2, \mathbf{x} \in \mathcal{M}. \qquad (14)$$

From (12) we infer the condition $\phi(\mathbf{x}_0) = 0$, and the continuous function ϕ attains its maximum on the compact surface $\mathcal{M}$ at one point $\mathbf{x}_1 \in \mathcal{M}$, at least.

3.) In a neighborhood of this point $\mathbf{x}_1$, we introduce isothermal parameters (see Section 7 and Section 8 of Chapter 12) by the following mapping:

$$B \ni (u, v) \to \mathbf{x}(u, v) \in \mathcal{M} \,; (0,0) \to \mathbf{x}_1 \,; \mathbf{X}(u, v) = \mathcal{N}(\mathbf{x}(u, v)), (u, v) \in B. \tag{15}$$

Here the unit normal $\mathbf{X}(u, v)$ of the H-surface $\mathbf{x}(u, v)$ points to the interior direction $\mathcal{N}(\mathbf{x}(u, v))$ of our closed convex surface $\mathcal{M}$. We now utilize the dual H-surface

$$\mathbf{z}(u, v) := \mathbf{x}(u, v) + \frac{1}{H} \mathbf{X}(u, v) \,, (u, v) \in B$$

from Proposition 5.4 above and consider the function

$$\Phi(u, v) := \Big(\mathbf{z}(u, v) \Big)^2 = \phi(\mathbf{x}(u, v)), (u, v) \in B. \tag{16}$$

Parallel to the geometric maximum principle of E. Heinz (see Theorem 1.4 in Section 1 of Chapter 12), we calculate with the aid of (8) – (10) as follows:

$$\Delta \Phi(u, v) = 2 \Big(|\nabla \mathbf{z}(u, v)|^2 + \mathbf{z}(u, v) \cdot \Delta \mathbf{z}(u, v) \Big)$$

$$= 4 \Big(|\mathbf{z}_u \wedge \mathbf{z}_v(u, v)| + H \mathbf{z}(u, v) \cdot \mathbf{z}_u \wedge \mathbf{z}_v(u, v) \Big)$$

$$= 4 \Big(|\mathbf{z}_u \wedge \mathbf{z}_v(u, v)| + H \Big(\mathbf{x}(u, v) + \frac{1}{H} \mathbf{X}(u, v) \Big) \cdot \Big(-|\mathbf{z}_u \wedge \mathbf{z}_v| \mathbf{X}(u, v) \Big) \Big)$$

$$= 4 \Big(|\mathbf{z}_u \wedge \mathbf{z}_v(u, v)| - H |\mathbf{z}_u \wedge \mathbf{z}_v(u, v)| \Big(\mathbf{x}(u, v) \cdot \mathbf{X}(u, v) \Big) - |\mathbf{z}_u \wedge \mathbf{z}_v(u, v)| \Big)$$

$$= -4H |\mathbf{z}_u \wedge \mathbf{z}_v(u, v)| \Big(\mathbf{x}(u, v) \cdot \mathbf{X}(u, v) \Big)$$

$$= -4H |\mathbf{z}_u \wedge \mathbf{z}_v(u, v)| \Big(\mathbf{x}(u, v) \cdot \mathcal{N}(\mathbf{x}(u, v)) \Big)$$

$$= -4H |\mathbf{z}_u \wedge \mathbf{z}_v(u, v)| \, \mu(\mathbf{x}(u, v)) \geq 0 \,, \quad (u, v) \in B.$$

Here we have taken the inequality (13) into account.

4.) Therefore, the function Φ is subharmonic in B and attains its maximum at the interior point $(0,0) \in B$. Thus Φ is constant on B by the maximum principle. A continuation argument on the connected surface $\mathcal{M}$ shows that $\phi : \mathcal{M} \to \mathbb{R}$ is constant and even vanishes, due to the condition $\phi(\mathbf{x}_0) = 0$ from above. Thus we arrive at the identity

$$0 = \phi(\mathbf{x}) = \Big(\mathbf{x} + \frac{1}{H} \mathcal{N}(\mathbf{x}) \Big)^2 , \mathbf{x} \in \mathcal{M} \tag{17}$$

or equivalently

$$\mathbf{x} = -\frac{1}{H} \mathcal{N}(\mathbf{x}), \mathbf{x} \in \mathcal{M}. \tag{18}$$

This shows that our ovaloid is necessarily a sphere of radius H^{-1}.   q.e.d.

Definition 5.7. *Under* **a closed immersed surface** $\mathcal{M}$ *we comprehend a 2-dimensional, compact and connected, orientable C^3-manifold in $\mathbb{R}^3$ without boundary. More precisely, $\mathcal{M}$ is a compact and connected set of $\mathbb{R}^3$, which is locally representable by charts of an orientable C^3-atlas, and consequently possesses a continuous unit-normal-field $\mathcal{N} : \mathcal{M} \to S^2$.*

From § 97 and § 98 in [BL] we recall the following notion: *The genus $g \in \mathbb{N}_0$ of a closed immersed surface* indicates that $\mathcal{M}$ can be globally parametrized over a sphere with g handles being attached. With the aid of our dual H-surface, we prove the subsequent well-known result of Heinz Hopf by the index method (compare § 100 in [BL]).

Theorem 5.8. (H. Hopf)
The only closed immersed surfaces $\mathcal{M}$ of genus 0 with constant mean curvature $H > 0$ are the spheres of radius H^{-1}.

Proof:

1.) For a tangential vector-field $\mathcal{T}$ along the surface $\mathcal{M}$, we look at its *isolated singularities* $\mathbf{x}_l \in \mathcal{M}$. Then we attribute an index $ind(\mathcal{T}, \mathbf{x}_l) \in \mathbb{Z}$ to this singular point. Invoking the profound *index-sum formula of Poincaré*, the sum of the indices for all these singularities on the surface is independent of the vector-field and equals $2 - 2g$. Thus we have the identity

$$\sum_{l=1}^{L} ind(\mathcal{T}, \mathbf{x}_l) = 2 - 2g, \tag{19}$$

where the situations $L = 0, 1, 2, 3, \ldots$ might occur – and $L = 0$ means the absence of singularities. Here we refer the reader to §99 in the *Grundlehren* [BL] by W. Blaschke and K. Leichtweiss. This profound result is based on the *Gauss-Bonnet theorem* – the central result in differential geometry!

2.) Let us now regard the mapping

$$\mathcal{F} : \mathcal{M} \to \mathcal{M}^* \quad \text{defined by} \quad \mathcal{F}(\mathbf{x}) := \mathbf{x} + \frac{1}{H} \mathcal{N}(\mathbf{x}), \quad \mathbf{x} \in \mathcal{M} \tag{20}$$

onto the set

$$\mathcal{M}^* := \left\{ \mathbf{z} = \mathbf{x} + \frac{1}{H} \mathcal{N}(\mathbf{x}) \,\Big|\, \mathbf{x} \in \mathcal{M} \right\}.$$

About an arbitrary point $\mathbf{x}_1 \in \mathcal{M}$ we can locally introduce conformal parameters as described in (15) above. To the immersed H-surface $\mathbf{x}(u, v)$, we consider its dual surface $\mathbf{z}(u, v)$ from Proposition 5.4 and observe

$$\mathbf{z}(u, v) = \mathcal{F}(\mathbf{x}(u, v)), \quad (u, v) \in B. \tag{21}$$

The H-surface $\mathbf{z}(u, v)$ is not necessarily immersed and might even degenerate to a single point. If the last case does not enter, the asymptotic expansion from (19) in Section 4 yields

$$\mathbf{z}_w(w) = \mathbf{a}\,w^n + o\big(|w|^n\big),\ w = u + iv \to 0 \text{ with } \mathbf{a} = \mathbf{b} - i\mathbf{c} \in \mathbb{C}^3\backslash\{\mathbf{0}\},\ n \in \mathbb{N}\cup\{0\}. \tag{22}$$

The conformality relations (8) imply $|\mathbf{b}| = |\mathbf{c}| > 0$, $\mathbf{b}\cdot\mathbf{c} = 0$, and an integration of (22) yields

$$
\begin{aligned}
\mathbf{z}(w) &= \mathbf{x}_1 + \frac{1}{n+1}\Big(\mathbf{a}\,w^{n+1} + \overline{\mathbf{a}}\,\overline{w}^{n+1}\Big) + o\big(|w|^{n+1}\big)\\
&= \mathbf{x}_1 + \frac{2}{n+1}\mathrm{Re}\Big(\mathbf{a}\,w^{n+1}\Big) + o\big(|w|^{n+1}\big),\quad w = u + iv \to 0.
\end{aligned}
\tag{23}
$$

Using polar coordinates $w = \rho(\cos\varphi + i\sin\varphi)$, we deduce the asymptotic expansion

$$
\begin{aligned}
\mathbf{z}(w) &= \mathbf{x}_1 + \frac{2}{n+1}\mathrm{Re}\Big((\mathbf{b} - i\mathbf{c})\,\rho^{n+1}[\cos(n+1)\varphi + i\sin(n+1)\varphi]\Big) + o(\rho^{n+1})\\
&= \mathbf{x}_1 + \frac{2}{n+1}\Big(\mathbf{b}\cos(n+1)\varphi + \mathbf{c}\sin(n+1)\varphi\Big)\rho^{n+1} + o(\rho^{n+1}),\ \rho \to 0 + .
\end{aligned}
\tag{24}
$$

From the property (6) we infer the following representation

$$\frac{\mathbf{b}\wedge\mathbf{c}}{|\mathbf{b}\wedge\mathbf{c}|} = -\mathcal{N}(\mathbf{x}_1). \tag{25}$$

3.) Let us now introduce *the tangential space to $\mathcal{M}$ at the point* $\mathbf{x}_1$ with

$$T_{\mathcal{M}}(\mathbf{x}_1) := \{\mathbf{y} \in \mathbb{R}^3 : \mathbf{y}\cdot\mathcal{N}(\mathbf{x}_1) = 0\}. \tag{26}$$

From the identities (24) and (25) we infer that the *linear differentials* of the mappings $\mathbf{x} : B \to \mathcal{M}$ and $\mathbf{z} : B \to \mathcal{M}^*$ at the origin $(0,0)$ both map into the same tangent plane $T_{\mathcal{M}}(\mathbf{x}_1)$ according to the prescriptions

$$\partial\mathbf{x}|_{(0,0)} : \mathbb{R}^2 \to T_{\mathcal{M}}(\mathbf{x}_1)\text{ with }\partial\mathbf{x}(s,t) := \mathbf{x}_u(0,0)s + \mathbf{x}_v(0,0)t,\ (s,t) \in \mathbb{R}^2\,;$$

$$\partial\mathbf{z}|_{(0,0)} : \mathbb{R}^2 \to T_{\mathcal{M}}(\mathbf{x}_1)\text{ with }\partial\mathbf{z}(s,t) := \mathbf{z}_u(0,0)s + \mathbf{z}_v(0,0)t,\ (s,t) \in \mathbb{R}^2. \tag{27}$$

Moreover, the mapping $\mathbf{z}(u,v)$ reverses the orientation of $\mathbf{x}(u,v)$ on $T_{\mathcal{M}}(\mathbf{x}_1)$ in linear approximation, according to (6). Finally, the dual H-surface $\mathbf{z}(u,v)$ might possess a branch point of the order $n \in \mathbb{N}$ at the origin $(0,0)$, according to the formula (22) above. Therefore, the set $\mathcal{M}^*$ constitutes *a branched immersion of constant mean curvature H in $\mathbb{R}^3$.*

4.) With the aid of the representation (21), we consider *the linear differential of the mapping* (20) in the following notion:

$$
\begin{aligned}
\partial\mathcal{F}\Big|_{\mathbf{x}_1} &: T_{\mathcal{M}}(\mathbf{x}_1) \to T_{\mathcal{M}}(\mathbf{x}_1)\quad\text{defined by}\\
\partial\mathcal{F}\Big|_{\mathbf{x}_1}(\mathbf{y}) &:= \partial\mathbf{z}|_{(0,0)} \circ (\partial\mathbf{x}|_{(0,0)})^{-1}(\mathbf{y}),\ \mathbf{y} \in T_{\mathcal{M}}(\mathbf{x}_1).
\end{aligned}
\tag{28}
$$

Here the inverse linear transformation $(\partial\mathbf{x}|_{(0,0)})^{-1} = (\partial\mathbf{x}^{-1})|_{\mathbf{x}_1}$ exists, because the H-surface $\mathbf{x}$ is immersed.

The differential $\partial\mathcal{F}\big|_{\mathbf{x}_1}$ vanishes if and only if the dual H-surface $\mathbf{z}(u,v)$ has a branch point at the origin; this is exactly the case, when our original H-surface $\mathbf{x}(u,v)$ has the umbilical point $\mathbf{x}_1 = \mathbf{x}(0,0)$ (compare the identity (8) above). Consequently, we call $\mathbf{x}_1$ a **singular point of the mapping** $\mathcal{F}$ if its linear differential from (28) degenerates to the null-transformation.

If $\mathbf{x}_1$ is such a singular point, we consider the mapping $\mathbf{z} \circ \mathbf{x}^{-1}$ in a small neighborhood of $\mathbf{x}_1$ on $\mathcal{M}$. Then we determine the winding number of this mapping projected onto the tangential plane $T_\mathcal{M}(\mathbf{x}_1)$, which we orient by the normal $\mathcal{N}(\mathbf{x}_1)$. Taking the conditions (24), (25) into account as well as the orientation (15) for our mappings $\mathbf{z}$ and $\mathbf{x}$, respectively, we can determine the winding number which we diminish by 1. Thus we obtain **the index** $ind(\mathbf{x}_1, \mathcal{F})$ **of the map** (20) **at the singular point** $\mathbf{x}_1$ by

$$ind(\mathbf{x}_1, \mathcal{F}) := -n \in -\mathbb{N} \quad \text{with the integer } n \in \mathbb{N} \text{ from (22) or (24).} \quad (29)$$

5.) Let us now determine the index at the singular point $\mathbf{x}_1 = \mathbf{x}(0,0)$ alternatively: For a sufficiently small $\varepsilon > 0$, we introduce *the directional field*

$$\mathbf{t}_\vartheta := \partial\mathcal{F}\big|_{\mathbf{x}(w_\vartheta)} \Big(\cos\vartheta\, \mathbf{x}_u(w_\vartheta) + \sin\vartheta\, \mathbf{x}_v(w_\vartheta) \Big) \in T_\mathcal{M}(\mathbf{x}(w_\vartheta)),\ 0 \leq \vartheta \leq 2\pi,$$

with the circuit $\quad w_\vartheta := \varepsilon(\cos\vartheta + i\sin\vartheta) \in \mathbb{C} \setminus \{0\} \quad \text{for} \quad 0 \leq \vartheta \leq 2\pi.$
$$(30)$$

We consider the projection $\widehat{\mathbf{t}_\vartheta},\ 0 \leq \vartheta \leq 2\pi$ of this directional field onto the tangential plane $T_\mathcal{M}(\mathbf{x}_1)$ oriented by the normal $\mathcal{N}(\mathbf{x}_1)$. This nonvanishing directional field $\widehat{\mathbf{t}_\vartheta}$ covers the plane in *a Riemannian sector of oriented total angle* $2\pi\mu$ for $0 \leq \vartheta \leq 2\pi$, where the multiple $\mu \in \mathbb{Z}$ appears as a *signed covering number*.

From (28) combined with (22) and (25) we deduce that this index equals $-n$, where $n \in \mathbb{N}$ appears in the asymptotic expansion of the branch point for $\mathbf{z}(u,v)$ at the origin. Now we introduce *the tangent bundle*

$$T_\mathcal{M} := \Big\{ (\mathbf{x}_0; \mathbf{y}) \Big| \mathbf{x}_0 \in \mathcal{M},\ \mathbf{y} \in T_\mathcal{M}(\mathbf{x}_0) \Big\},$$

and via (28) we define locally *the tangential vector-field* as follows:

$$\mathcal{T} : T_\mathcal{M} \to T_\mathcal{M} \quad \text{satisfying}$$

$$\mathcal{T}(\mathbf{x}_0, \mathbf{y}) := \partial\mathcal{F}\big|_{\mathbf{x}_0} (\mathbf{y}) \quad \text{for} \quad \mathbf{x}_0 \in \mathcal{M},\ \mathbf{y} \in T_\mathcal{M}(\mathbf{x}_0). \quad (31)$$

The investigations above show that our index (29) at a singular point $\mathbf{x}_1$ of the mapping (20) is identical to the index $ind(\mathcal{T}, \mathbf{x}_1)$ of the tangential mapping $\mathcal{T}$ at the singularity $\mathbf{x}_1$.

6.) If the assertion of the Theorem were false, the mapping (20) could not be constant. We look at the finitely many isolated singularities $\{x_1, \ldots, x_L\}$, with $L \in \mathbb{N}$, of the mapping $\mathcal{F} : \mathcal{M} \to \mathcal{M}^*$ on this compact closed surface $\mathcal{M}$. Due to the observation in part 5.), the index at such a singularity is always negative and equals the $ind(\mathcal{T}, x_k)$ of the tangential vector-field $\mathcal{T}$ for $k = 1, \ldots, L$. Since our surface $\mathcal{M}$ has genus 0, Poincaré's index-sum formula (19) reveals the evident contradiction

$$0 > \sum_{l=1}^{L} ind(\mathcal{T}, x_l) = 2 - 2g = 2, \tag{32}$$

Therefore, only the sphere of radius $\dfrac{1}{H}$ remains as a closed, immersed surface of genus 0 with constant mean curvature H.   q.e.d.

Remarks:

1. For about 36 years the intriguing question remained open, as to whether closed immersed surfaces of higher genus with constant mean curvature exist or not? To the great surprise of the interested community of mathematicians, H. C. Wente answered this question affirmatively by constructing an immersed closed surface with constant mean curvature of the topological type of a torus, the so-called *Wente torus*. This is achieved in the treatise H. C. WENTE: *Counterexample to a question of H. Hopf.* Pacific Journal of Mathematics **121** (1986), 193-243.
2. Under an adequate stability condition, J. L. Barbosa and M. do Carmo showed that those *stable* closed immersed surfaces of constant mean curvature are necessarily spheres. These closed immersed surfaces comprise the relative and absolute minima of the variational problem associated with the isoperimetric problem. Here we refer our readers to J. L. Barbosa, M. do Carmo: *Stability of hypersurfaces with constant mean curvature.* Math. Zeitschrift **185** (1984), 339-353.

6 One-to-one Solutions of Lewy-Heinz Systems

The uniformizing mappings for the elliptic Monge-Ampère equation in the subsequent Section 7 motivate to study the following class of mappings, refining the results from Section 5 in Chapter 12. Although this *Lewy-Heinz class of one-to-one mappings* deserves an independent interest, the reader should consider this paragraph together with the next as an entity.

Definition 6.1. *With the constants* $a_0, a_1 \in [0, +\infty)$ *and* $N \in (0, +\infty]$ *being prescribed, we denote the following* **Lewy-Heinz class of mappings** *by the symbol* $\Gamma_*(B, a_0, a_1, N)$:

i) The function $\mathbf{z}(w) = (x(u,v), y(u,v)) : \overline{B} \to \mathbb{R}^2 \in C^2(B) \cap C^0(\overline{B})$ maps the closed disc $\overline{B}$ homeomorphically and positive-oriented onto $\overline{B}$;

ii) The mapping $\mathbf{z}$ is origin-preserving, which means $\mathbf{z}(0) = (0,0)$;

iii) The function $\mathbf{z}$ satisfies the system of differential equations

$$\Delta x(u,v) = h_1(x(u,v), y(u,v))|\nabla x(u,v)|^2 + h_2(x,y)(\nabla x \cdot \nabla y)(u,v)$$

$$+h_3(x(u,v), y(u,v))|\nabla y(u,v)|^2 + h_4(x(u,v), y(u,v))\frac{\partial(x,y)}{\partial(u,v)},$$

$$\Delta y(u,v) = \tilde{h}_1(x(u,v), y(u,v))|\nabla x(u,v)|^2 + \tilde{h}_2(x,y)(\nabla x \cdot \nabla y)(u,v) \quad (1)$$

$$+\tilde{h}_3(x(u,v), y(u,v))|\nabla y(u,v)|^2 + \tilde{h}_4(x(u,v), y(u,v))\frac{\partial(x,y)}{\partial(u,v)}$$

$$\text{for all} \quad w = u + iv \in B,$$

which we call the **Lewy-Heinz system**.

iv) The coefficient functions $h_1 = h_1(x,y), \ldots, h_4 = h_4(x,y)$ and $\tilde{h}_1 = \tilde{h}_1(x,y), \ldots, \tilde{h}_4 = \tilde{h}_4(x,y)$ belong to the class $C^0(\overline{B})$ and satisfy the following estimates

$$|h_1(z)|, \ldots, |h_4(z)|, |\tilde{h}_1(z)|, \ldots, |\tilde{h}_4(z)| \leq a_0 \quad \text{for all} \quad z = x + iy \in \overline{B}. \quad (2)$$

For the combinations $\omega_1 := \tilde{h}_1$, $\omega_2 := h_1 - \tilde{h}_2$, $\omega_3 := h_2 - \tilde{h}_3$, $\omega_4 := h_3$ we have the following **structure conditions**

$$|\omega_l(z_1) - \omega_l(z_2)| \leq a_1 |z_1 - z_2| \quad \text{for all} \quad z_1, z_2 \in \overline{B} \text{ and } \quad l = 1, \ldots, 4. \quad (3)$$

v) Dirichlet's integral of $\mathbf{z}$ fulfills

$$D(\mathbf{z}) := \iint\limits_{B} \left(|\mathbf{z}_u(u,v)|^2 + |\mathbf{z}_v(u,v)|^2 \right) du\, dv \leq N.$$

Within the comprising class $\Gamma_*(B, a_0, a_1) := \Gamma_*(B, a_0, a_1, +\infty)$ we require no bound on Dirichlet's integral. An arbitrary element $\mathbf{z} \in \Gamma_*(B, a_0, a_1)$ is addressed as **one-to-one solution of the Lewy-Heinz system**.

The Lewy-Heinz system (1), together with the structure conditions (2) and (3), is obviously invariant under translations. By an explicit calculation (compare [Sc], Lemma 8.1.1), we derive the invariance of the Lewy-Heinz system under rotations, and the structure conditions above appear! As usual, the symbol $*$ denotes the transposition of matrices. For the Jacobian in the open unit disc B, we equivalently use the symbols

$$J_{\mathbf{z}}(u,v) = \frac{\partial(x,y)}{\partial(u,v)}\bigg|_{(u,v)} = [\nabla x(u,v), \nabla y(u,v)], \quad (u,v) \in B,$$

where the bracket $[.,.]$ denotes the familiar alternating bilinear form of the determinant.

Proposition 6.2. (Rotational invariance)

For $a_0, a_1 \in (0, +\infty)$ there are positive constants $\hat{a}_0 = \hat{a}_0(a_0)$ and $\hat{a}_1 = \hat{a}_1(a_1)$ such that the following holds: With each solution

$$\mathbf{z} = \mathbf{z}(u,v) = \begin{pmatrix} x(u,v) \\ y(u,v) \end{pmatrix} \in \Gamma_*(B, a_0, a_1)$$

the **rotated solution**

$$\mathbf{z}' = \mathbf{z}'(u,v) = \begin{pmatrix} \xi(u,v) \\ \eta(u,v) \end{pmatrix} := T^* \circ \mathbf{z}(u,v), \ (u,v) \in \overline{B}$$

belongs to the class $\Gamma_*(B, \hat{a}_0, \hat{a}_1)$ for all rotational matrices

$$T := \begin{pmatrix} \alpha & -\beta \\ \beta & \alpha \end{pmatrix} \quad \text{with} \quad \alpha, \beta \in \mathbb{R} \quad \text{and} \quad \alpha^2 + \beta^2 = 1. \tag{4}$$

The coefficients $h'_1, \ldots, h'_4, \tilde{h}'_1, \ldots, \tilde{h}'_4$ of the Lewy-Heinz system for the rotated solution are given in (6) and (7) below.

Proof: 1.) We take the transformation identities

$$\xi = \alpha x + \beta y, \ \eta = -\beta x + \alpha y \quad \text{and} \quad x = \alpha\xi - \beta\eta, \ y = \beta\xi + \alpha\eta \tag{5}$$

from the orthogonal transformation

$$\begin{pmatrix} \xi \\ \eta \end{pmatrix} = T^* \circ \begin{pmatrix} x \\ y \end{pmatrix} = \begin{pmatrix} \alpha & \beta \\ -\beta & \alpha \end{pmatrix} \circ \begin{pmatrix} x \\ y \end{pmatrix} = \begin{pmatrix} \alpha x + \beta y \\ -\beta x + \alpha y \end{pmatrix},$$

and their inverse transformation

$$\begin{pmatrix} x \\ y \end{pmatrix} = T \circ \begin{pmatrix} \xi \\ \eta \end{pmatrix} = \begin{pmatrix} \alpha & -\beta \\ \beta & \alpha \end{pmatrix} \circ \begin{pmatrix} \xi \\ \eta \end{pmatrix} = \begin{pmatrix} \alpha\xi - \beta\eta \\ \beta\xi + \alpha\eta \end{pmatrix}.$$

2.) With the identities (5) we calculate as follows:

$$\Delta\xi(u,v) = \alpha\Delta x(u,v) + \beta\Delta y(u,v)$$

$$= (\alpha h_1 + \beta\tilde{h}_1)|\nabla x|^2 + (\alpha h_2 + \beta\tilde{h}_2)(\nabla x \cdot \nabla y) + (\alpha h_3 + \beta\tilde{h}_3)|\nabla y|^2$$

$$+ (\alpha h_4 + \beta\tilde{h}_4)[\nabla x, \nabla y]$$

$$= (\alpha h_1 + \beta \tilde{h}_1)\left(\alpha^2|\nabla\xi|^2 - 2\alpha\beta(\nabla\xi \cdot \nabla\eta) + \beta^2|\nabla\eta|^2\right)$$

$$+(\alpha h_2 + \beta \tilde{h}_2)\left(\alpha\beta|\nabla\xi|^2 + (\alpha^2 - \beta^2)(\nabla\xi \cdot \nabla\eta) - \alpha\beta|\nabla\eta|^2\right)$$

$$+(\alpha h_3 + \beta \tilde{h}_3)\left(\beta^2|\nabla\xi|^2 + 2\alpha\beta(\nabla\xi \cdot \nabla\eta) + \alpha^2|\nabla\eta|^2\right)$$

$$+(\alpha h_4 + \beta \tilde{h}_4)[\nabla\xi, \nabla\eta]$$

$$= \left(\alpha^3 h_1 + \alpha^2\beta(\tilde{h}_1 + h_2) + \alpha\beta^2(\tilde{h}_2 + h_3) + \beta^3\tilde{h}_3\right)|\nabla\xi|^2$$

$$+\left(\alpha^3 h_2 + \alpha^2\beta(-2h_1 + \tilde{h}_2 + 2h_3) + \alpha\beta^2(-2\tilde{h}_1 - h_2 + 2\tilde{h}_3) - \beta^3\tilde{h}_2\right)$$
$$\cdot(\nabla\xi \cdot \nabla\eta)$$

$$+\left(\alpha^3 h_3 + \alpha^2\beta(\tilde{h}_3 - h_2) + \alpha\beta^2(-\tilde{h}_2 + h_1) + \beta^3\tilde{h}_1\right)|\nabla\eta|^2$$

$$+(\alpha h_4 + \beta\tilde{h}_4)[\nabla\xi, \nabla\eta]$$

$$= h_1'|\nabla\xi|^2 + h_2'(\nabla\xi \cdot \nabla\eta) + h_3'|\nabla\eta|^2 + h_4'[\nabla\xi, \nabla\eta].$$

Here *the transformed coefficient functions* are defined by

$$h_1' := \alpha^3 h_1 + \alpha^2\beta(\tilde{h}_1 + h_2) + \alpha\beta^2(\tilde{h}_2 + h_3) + \beta^3\tilde{h}_3$$

$$h_2' := \alpha^3 h_2 + \alpha^2\beta(-2h_1 + \tilde{h}_2 + 2h_3) + \alpha\beta^2(-2\tilde{h}_1 - h_2 + 2\tilde{h}_3) - \beta^3\tilde{h}_2$$

$$h_3' := \alpha^3 h_3 + \alpha^2\beta(\tilde{h}_3 - h_2) + \alpha\beta^2(-\tilde{h}_2 + h_1) + \beta^3\tilde{h}_1 \tag{6}$$

$$h_4' := \alpha h_4 + \beta\tilde{h}_4.$$

3.) Correspondingly, we calculate

$$\Delta\eta(u, v) = -\beta\Delta x(u, v) + \alpha\Delta y(u, v)$$

$$= (-\beta h_1 + \alpha\tilde{h}_1)|\nabla x|^2 + (-\beta h_2 + \alpha\tilde{h}_2)(\nabla x \cdot \nabla y) + (-\beta h_3 + \alpha\tilde{h}_3)|\nabla y|^2$$

$$+(-\beta h_4 + \alpha\tilde{h}_4)[\nabla x, \nabla y]$$

$$= (-\beta h_1 + \alpha\tilde{h}_1)\left(\alpha^2|\nabla\xi|^2 - 2\alpha\beta(\nabla\xi \cdot \nabla\eta) + \beta^2|\nabla\eta|^2\right)$$

$$+(-\beta h_2 + \alpha\tilde{h}_2)\left(\alpha\beta|\nabla\xi|^2 + (\alpha^2 - \beta^2)(\nabla\xi \cdot \nabla\eta) - \alpha\beta|\nabla\eta|^2\right)$$

$$+(-\beta h_3 + \alpha\tilde{h}_3)\left(\beta^2|\nabla\xi|^2 + 2\alpha\beta(\nabla\xi \cdot \nabla\eta) + \alpha^2|\nabla\eta|^2\right)$$

$$+(-\beta h_4 + \alpha\tilde{h}_4)[\nabla\xi, \nabla\eta]$$

$$= \left(\alpha^3 \tilde{h}_1 + \alpha^2\beta(-h_1 + \tilde{h}_2) + \alpha\beta^2(-h_2 + \tilde{h}_3) - \beta^3 h_3\right)|\nabla\xi|^2$$

$$+\left(\alpha^3 \tilde{h}_2 + \alpha^2\beta(-h_2 - 2\tilde{h}_1 + 2\tilde{h}_3) + \alpha\beta^2(-\tilde{h}_2 + 2h_1 - 2h_3) - \beta^3 h_2\right)$$

$$\cdot(\nabla\xi \cdot \nabla\eta)$$

$$+\left(\alpha^3 \tilde{h}_3 + \alpha^2\beta(-h_3 - \tilde{h}_2) + \alpha\beta^2(h_2 + \tilde{h}_1) - \beta^3 h_1\right)|\nabla\eta|^2$$

$$+(-\beta h_4 + \alpha\tilde{h}_4)[\nabla\xi, \nabla\eta]$$

$$= \tilde{h}'_1|\nabla\xi|^2 + \tilde{h}'_2(\nabla\xi \cdot \nabla\eta) + \tilde{h}'_3|\nabla\eta|^2 + \tilde{h}'_4[\nabla\xi, \nabla\eta].$$

Here *the transformed coefficient functions* are defined by

$$\tilde{h}'_1 := \alpha^3 \tilde{h}_1 + \alpha^2\beta(-h_1 + \tilde{h}_2) + \alpha\beta^2(-h_2 + \tilde{h}_3) - \beta^3 h_3$$

$$\tilde{h}'_2 := \alpha^3 \tilde{h}_2 + \alpha^2\beta(-h_2 - 2\tilde{h}_1 + 2\tilde{h}_3) + \alpha\beta^2(-\tilde{h}_2 + 2h_1 - 2h_3) - \beta^3 h_2$$

$$\tilde{h}'_3 := \alpha^3 \tilde{h}_3 + \alpha^2\beta(-h_3 - \tilde{h}_2) + \alpha\beta^2(h_2 + \tilde{h}_1) - \beta^3 h_1$$ (7)

$$\tilde{h}'_4 := -\beta h_4 + \alpha\tilde{h}_4.$$

4.) From the transformation laws (6) and (7) for the coefficients, we deduce that they remain bounded by a positive constant $\hat{a}_0 = \hat{a}_0(a_0)$, according to (2). Moreover, we can directly control that the structure conditions (3) remain valid for the transformed system with a positive constant $\hat{a}_1 = \hat{a}_1(a_1)$. This completes the proof of our lemma.   q.e.d.

Proposition 6.3. (Nonlinear combination)
For a sufficiently small $\delta = \delta(a_0, a_1) > 0$, there exists a $C^2((-\delta, +\delta), \mathbb{R})$-function $y = g(x)$, $x \in (-\delta, +\delta)$ satisfying $g(0) = 0 = g'(0)$, such that the function

$$f(x, y) := y - g(x), \ (x, y) \in (-\delta, +\delta) \times \mathbb{R} \ with \ f(0, 0) = 0 \ and \ \nabla f(0, 0) = (0, 1)$$

has the following property: When we insert an arbitrary solution

$$\mathbf{z}(u, v) = (x(u, v), y(u, v)) \in \Gamma_*(B, a_0, a_1)$$

of the Lewy-Heinz system into this function f, **the nonlinear combination**

$$\Phi(u, v) := f(x(u, v), y(u, v)) = y(u, v) - g(x(u, v)), \ (u, v) \in K_\epsilon$$

satisfies the differential inequality

$$|\Delta\Phi(u, v)| \le c\Big(|\nabla\Phi(u, v)| + |\Phi(u, v)|\Big), \ (u, v) \in K_\epsilon$$ (8)

with a constant $c = c(a_0, a_1) > 0$, in a sufficiently small disc

$$K_\epsilon := \{w = u + iv \in \mathbb{C} : |w| < \epsilon\}$$

of radius $\epsilon > 0$ about the origin. Here the function g depends on the coefficient functions $h_1, \ldots, \tilde{h}_4$ according to (11) below, but the length of its existence interval δ as well as the constant c in the differential inequality above depend only on the quantities a_0 and a_1 prescribed.

Proof: Let us calculate with the aid of the Lewy-Heinz system (1) as follows:

$$\Delta\Phi(u,v) = \Delta y(u,v) - \Delta\{g(x(u,v))\} = \Delta y(u,v) - \nabla\left(g'(x)x_u, g'(x)x_v\right)$$

$$= \Delta y(u,v) - g'(x)\Delta x(u,v) - g''(x)|\nabla x(u,v)|^2$$

$$= (\tilde{h}_1 - g'h_1 - g'')|\nabla x|^2 + (\tilde{h}_2 - g'h_2)(\nabla x \cdot \nabla y) + (\tilde{h}_3 - g'h_3)|\nabla y|^2$$

$$+ (\tilde{h}_4 - g'h_4)[\nabla x, \nabla y].$$

We observe $\nabla\Phi(u,v) = \nabla y(u,v) - g'(x(u,v))\nabla x(u,v)$ and insert the identity

$$\nabla y(u,v) = \nabla\Phi(u,v) + g'(x(u,v))\nabla x(u,v)$$

into the p.d.e. above. Thus we obtain the equation (10) below, by collecting the terms with $\nabla\Phi$ to a function

$$\Psi(u,v) : K_\epsilon \to \mathbb{R} \quad \text{satisfying} \quad |\Psi(u,v)| \leq \tilde{c}|\nabla\Phi(u,v)|, \ (u,v) \in K_\epsilon \qquad (9)$$

with a constant $\tilde{c} = \tilde{c}(a_0, a_1) > 0$ as follows:

$$\Delta\Phi(u,v) = \Psi(u,v)$$

$$+ \left\{\tilde{h}_1 - g'h_1 - g'' + g'\tilde{h}_2 - (g')^2h_2 + (g')^2\tilde{h}_3 - (g')^3h_3\right\}|\nabla x|^2$$

$$= \Psi(u,v) + \left\{\omega_1(x(u,v), y(u,v)) - g'(x(u,v))\omega_2(x(u,v), y(u,v)) \right. \qquad (10)$$

$$- \left(g'(x(u,v))\right)^2 \omega_3(x(u,v), y(u,v)) - \left(g'(x(u,v))\right)^3 \omega_4(x(u,v), y(u,v))$$

$$\left. - g''(x(u,v))\right\}|\nabla x(u,v)|^2 \quad \text{for all} \quad (u,v) \in K_\epsilon.$$

For a sufficiently small $\delta > 0$, we shall now solve *the initial value problem* for the following ordinary differential equation, briefly o.d.e., namely

$$y = g(x) \in C^2((-\delta, +\delta), \mathbb{R}) \quad \text{satisfying} \quad g(0) = 0 = g'(0) \quad \text{and}$$

$$g''(x) = \omega_1(x, g(x)) - g'(x)\omega_2(x, g(x)) - \left(g'(x)\right)^2 \omega_3(x, g(x)) \qquad (11)$$

$$- \left(g'(x)\right)^3 \omega_4(x, g(x)), \quad x \in (-\delta, +\delta).$$

The right-hand side of the o.d.e. is bounded and Lipschitz-continuous in the variables $y = g(x)$ and $p = g'(x)$, due to the structure conditions (2) and (3), where the Lipschitz constant depends on a_0 and a_1. Therefore, the initial value problem (11) can be uniquely solved, on the given existence interval of controlled half-length $\delta = \delta(a_0, a_1) > 0$, by succesive approximation.

Now we insert the function $x = x(u, v)$ into (11), and together with (10) we obtain the following differential equation involving the function Ψ from above:

$$\Delta\Phi(u, v) = \Psi(u, v) + \Big\{\omega_1(x(u, v), y(u, v)) - \omega_1(x(u, v), g(x(u, v)))\Big\}$$

$$-g'(x(u, v))\Big\{\omega_2(x(u, v), y(u, v)) - \omega_2(x(u, v), g(x(u, v)))\Big\}$$

$$-\big(g'(x(u, v))\big)^2\Big\{\omega_3(x(u, v), y(u, v)) - \omega_3(x(u, v), g(x(u, v)))\Big\} \tag{12}$$

$$-\big(g'(x(u, v))\big)^3\Big\{\omega_4(x(u, v), y(u, v)) - \omega_4(x(u, v), g(x(u, v)))\Big\}$$

$$=: \Psi(u, v) + \chi(u, v) \quad \text{for all} \quad (u, v) \in K_\epsilon.$$

Since the first derivative of g can be estimated by the data a_0 and a_1, the structure conditions yield

$$|\chi(u, v)| \le \hat{c}|y(u, v) - g(x(u, v))| = \hat{c}|\Phi(u, v)| \quad \text{in} \quad K_\epsilon \tag{13}$$

with a constant $\hat{c} = \hat{c}(a_o, a_1) > 0$. Recalling (9) together with (13), the differential equation (12) yields the stated differential inequality (8) from above, with a constant $c = c(a_0, a_1) > 0$.   q.e.d.

Remarks:

a) This central differential inequality (8), in order to exclude the vanishing of the Jacobian for certain one-to-one mappings in Theorem 6.4 below, has been invented by E. Heinz. We have followed the presentation in [Sc], Proposition 8.1.2 here.

b) The projection of Rellich's H-surface system to the plane is *not* accessible to this method, since the third component is involved on the right-hand side of this system! In the treatise [S3] of my dissertation, I have proposed a completely different method to prove the one-to-one property of the plane mapping; here we also refer to the investigations in Section 1.

c) The fundamental idea, to combine the vanishing of the Jacobian in plane harmonic mappings with the critical points of first order for an adequate auxiliary function, originates from the ingenious paper [Kn] by Hellmuth Kneser.

d) At this point, I gratefully acknowledge the good fortune in having learned from his son, Professor Martin Kneser the differential geometry as well as higher algebra. Wonderful lectures by M. Kneser on these topics and a motivating seminar on differential geometry in the years 1972 – 1975 at Göttingen gave me a decisively scientific orientation!

e) H. Kneser's paper [Kn] is the starting point as well for the dissertation by J. JOST: *Univalency of harmonic mappings between surfaces*. Journal reine angew. Math. **342** (1981), 141-153.
Here the diffemorphic character of harmonic mappings between 2-dimensional Riemannian manifolds is studied.

f) H. Kneser's ideas from the paper [Kn] are also present in the joint investigations by S. Hildebrandt and F. Sauvigny, which are described in Chapter 2 of the new book [DHT2] on *Global Analysis of Minimal Surfaces*.

For holomorphic functions, we have the well-known fact that injective mappings necessarily have a nonvanishing Jacobian. Now we can show that one-to-one solutions of the Lewy-Heinz system share this *property of nondegeneracy*.

Theorem 6.4. (Nondegeneracy result by H. Lewy, P. Berg, E. Heinz)
An arbitrary solution of the Lewy-Heinz system

$$\mathbf{z}(u, v) = (x(u, v), y(u, v)) \in \Gamma_*(B, a_0, a_1)$$

possesses in B a positive Jacobian according to

$$J_{\mathbf{z}}(u, v) = \frac{\partial(x, y)}{\partial(u, v)}\bigg|_{(u,v)} = [\nabla x, \nabla y]\big|_{(u,v)} > 0 \quad \text{for all} \quad (u, v) \in B, \qquad (14)$$

and consequently furnishes a positive-oriented C^2-diffeomorphism from the open unit disc B onto itself.

Proof: 1.) At first, we show indirectly that $J_{\mathbf{z}}(0, 0) \neq 0$ holds true.
If this were not the case, we would have

$$\begin{vmatrix} x_u(0,0) & y_u(0,0) \\ x_v(0,0) & y_v(0,0) \end{vmatrix} = 0.$$

From Proposition 6.2 we know that the Lewy-Heinz class is invariant under rotations, such that we can realize the system of linear equations

$$\begin{pmatrix} x_u(0,0) & y_u(0,0) \\ x_v(0,0) & y_v(0,0) \end{pmatrix} \circ \begin{pmatrix} 0 \\ 1 \end{pmatrix} = \begin{pmatrix} 0 \\ 0 \end{pmatrix}$$

by the transition to an adequately rotated map. With the aid of Proposition 6.3, we find a function

$$f(x, y) = y - g(x) : K_\epsilon \to \mathbb{R} \in C^2(K_\epsilon) \text{ with } f(0, 0) = 0 \text{ and } \nabla f(0, 0) = (0, 1)$$

such that the nonlinear combination

$$\Phi(u, v) := f(x(u, v), y(u, v)), \ (u, v) \in K_\epsilon$$

is subject to the differential inequality (8). Furthermore, $\Phi(0,0) = 0$ holds true and we deduce

$$\nabla\Phi(0,0)$$
$$= \Big(f_x(0,0)x_u(0,0) + f_y(0,0)y_u(0,0), f_x(0,0)x_v(0,0) + f_y(0,0)y_v(0,0) \Big)$$
$$= \Big(y_u(0,0), y_v(0,0) \Big) = (0,0).$$

According to the central result [HW] by P. Hartman and A. Wintner (compare the Theorem 1.2 in Section 1 of Chapter 9), there exists a constant $\hat{c} = \hat{a} - i\hat{b} \in \mathbb{C} \setminus \{0\}$ and an integer $m \in \mathbb{N}$ with $m \geq 2$, such that the following asymptotic expansion holds true:

$$\Phi(u,v) = \text{Re}\Big(\hat{c}\,w^m\Big) + o(|w|^m), \quad w = u + iv \in K_\epsilon,\, w \to 0. \qquad (15)$$

We complement our function $f = f(x,y) : K_\epsilon \to \mathbb{R}$ with the function $g(x,y) := -x$. Then we obtain the positive-oriented, origin-preserving plane mapping

$$h(x,y) := \Big(f(x,y), g(x,y) \Big) : K_\epsilon \to \mathbb{R}^2 \text{ with } J_h(0,0) = +1\,,\, h(0,0) = (0,0).$$

Now we consider *the locally bent mapping* near the origin

$$\hat{z}(u,v) := h(z(u,v)) = \Big(f(x(u,v),y(u,v)), g(x(u,v),y(u,v)) \Big)$$
$$= \Big(\Phi(u,v), -x(u,v) \Big) = \Big(\text{Re}(\hat{c}w^m) + o(|w|^m), -x(u,v) \Big), \qquad (16)$$
$$w = u + iv \in K_\epsilon,\, w \to 0,$$

which is positive-oriented and topological. Therefore, the integer in (16) must be $m = 1$, since the index of this mapping at the origin is exactly m. This gives a contradiction to the condition $m \geq 2$ from above, and the statement $J_z(0,0) \neq 0$ is proved.

2.) Secondly, we show that $J_z(u,v) \neq 0$ for all $(u,v) \in B$ holds true.

When we derived part 1.), we utilized the properties for z in the class $\Gamma_*(B, a_0, a_1)$; however, the image property $z(\overline{B}) = \overline{B}$ has *not yet* been employed also in the basic Propositions 6.2 and 6.3! This allows a translation of our plane mapping without changing those assumptions, which are necessary to show the nonvanishing of the Jacobian!

Let $(u_0, v_0) \in B$ be an arbitrary interior point of the unit disc, with the positive distance $r \in (0,1]$ to the boundary ∂B. With $z_0 := z(u_0, v_0) \in B$ we regard *the translated mapping*

$$z'(u,v) := z\Big(u_0 + r(u - u_0), u_0 + r(v - v_0) \Big) - z_0, \quad w = u + iv \in \overline{B}, \quad (17)$$

which is topological, positive-oriented, and origin-preserving. Moreover, our translated mapping $\mathbf{z}'(u,v)$ fulfills the Lewy-Heinz system (1) with certain modified coefficient functions $h_1', \ldots, h_4'$ and $\tilde{h}_1', \ldots, \tilde{h}_4'$, which preserve the structure conditions (2) and (3). We apply the arguments in part 1.) and see that the Jacobian of the mapping $\mathbf{z}' : \overline{B} \to \mathbb{R}^2$ does not vanish at the origin. Consequently, our second statement is established.

3.) Now we take into account the boundary condition, namely that $\mathbf{z} : \partial B \to \partial B$ represents a positive-oriented topological map. The sign of the determinant of the mapping $\mathbf{z}$ can only be ± 1, and for continuity reasons must be identical to $+1$ or -1 in the whole disc B. We invoke an index-sum formula (see Theorem 1.17 from Section 1 in Chapter 3) and exclude the case that $J_{\mathbf{z}}(u,v) < 0$ might occur at an interior point $(u,v) \in B$. This completes the proof of our nondegeneracy theorem.   q.e.d.

Remarks: This nondegeneracy result has been established by H. Lewy for systems with real-analytic coefficients, by P. Berg for solutions of non-analytic linear systems, and by E. Heinz for the nonlinear systems considered here. The theorem remains true for the slightly extended class $\overline{\Gamma}_*(B, 0, a_1, N)$ explained in Definition 6.6 below.

We recall the class of mappings $\Gamma(B, a, b, N)$ in Definition 5.3 of Section 5 from Chapter 12 and formulate the following

Definition 6.5. *We extend the class of mappings $\Gamma(B, a, b, N)$ from Definition 5.3 of Section 5 in Chapter 12 to the extended class $\overline{\Gamma}(B, a, b, N)$ by replacing the condition i) there with the following weaker assumption:*

The function $\mathbf{z}(w) = (x(u,v), y(u,v)) : \overline{B} \to \mathbb{R}^2 \in C^2(B) \cap C^0(\overline{B})$

maps the circumference ∂B weakly monotonically, in a positive-oriented way,

with the winding number $+1$, onto the circular line ∂B.

$$(18)$$

In the same way we formulate

Definition 6.6. *We extend the class of mappings $\Gamma_*(B, a_0, a_1, N)$ from Definition 6.1 above to the extended class $\overline{\Gamma}_*(B, a_0, a_1, N)$ by replacing the condition i) there with the following weaker assumption:*

-i) The function $\mathbf{z}(w) = (x(u,v), y(u,v)) : \overline{B} \to \mathbb{R}^2 \in C^2(B) \cap C^0(\overline{B})$ maps the open disc B homeomorphically and positive-oriented onto B, whereas the boundary mapping $\mathbf{z} : \partial B \to \partial B$ satisfies the condition (18) from above.

We immediately infer the following inclusion

$$\overline{\Gamma}_*(B, a_0, a_1, N) \subset \overline{\Gamma}(B, 4a_0, 0, N) \qquad (19)$$

from Theorem 6.4 above. Theorem 5.4 in Section 5 of Chapter 12 for the class of mappings $\Gamma(a, b, N)$ pertains to the class $\overline{\Gamma}(a, b, N)$ with weakly monotonical boundary values, as an inspection of the proof there shows. Consequently, those statements are applicable to the functions of the extended Lewy-Heinz class $\overline{\Gamma}_*(B, a_0, b_0, N)$, and we can estimate $|\nabla z|$ in the interior of the unit disc from above and from below. We shall use the first estimate to prove a compactness theorem for Lewy-Heinz diffeomorphisms in the subsequent Theorem 6.8. This is achieved within the smaller class $\overline{\Gamma}_*(B, a_0, a_1, a_2, \alpha, N)$ of Definition 6.7 below, requiring additionally a bound on the Hölder-norm of the coeffient functions.

Definition 6.7. *With the constants $a_0, a_1, a_2 \in [0, +\infty)$, $\alpha \in (0, 1)$ and $N \in (0, +\infty)$ being prescribed, we denote the following extended* **Lewy-Heinz class of mappings** *by the symbol $\overline{\Gamma}_*(B, a_0, a_1, a_2, \alpha, N)$:*

i) *The function $\mathbf{z}(w) = (x(u, v), y(u, v)) : \overline{B} \to \mathbb{R}^2$ belongs to the class $\overline{\Gamma}_*(B, a_0, a_1, N)$ from Definition 6.6;*

ii) *The Jacobian of the mapping $\mathbf{z}$ satisfies $J_{\mathbf{z}}(u, v) > 0$ for all $(u, v) \in B$, and consequently maps the open disc B diffeomorphically and positive-oriented onto B;*

iii) *The coefficient functions in the Lewy-Heinz system (1) satisfy the following Hölder conditions*

$$|h_l(z_1) - h_l(z_2)| \leq a_2 |z_1 - z_2|^\alpha \quad \text{for all} \quad z_1, z_2 \in \overline{B} \text{ and } l = 1, 2, 4;$$

$$|\tilde{h}_l(z_1) - \tilde{h}_l(z_2)| \leq a_2 |z_1 - z_2|^\alpha \quad \text{for all} \quad z_1, z_2 \in \overline{B} \text{ and } l = 2, 3, 4. \tag{20}$$

Consequently, the $C^\alpha(\overline{B})$-norms of all coefficient functions $h_1, \ldots, \tilde{h}_4$ are bounded by a_1 and a_2.

An arbitrary element $\mathbf{z} \in \overline{\Gamma}_*(B, a_0, a_1, a_2, \alpha, N)$ is addressed as a **diffeomorphism of the Lewy-Heinz system.**

We can now prove the following

Theorem 6.8. (Compactness result for Lewy-Heinz diffeomorphisms)
For the given quantities $a_0, a_1, a_2, N \in [0, +\infty)$ and $\alpha \in (0, 1)$, the class of diffeomorphisms $\overline{\Gamma}_*(B, a_0, a_1, a_2, \alpha, N)$ is compact in the following sense:

Each sequence of diffeomorphisms

$$\{\mathbf{z}^k\}_{k=1,2,\ldots} \subset \overline{\Gamma}_*(B, a_0, a_1, a_2, \alpha, N)$$

admits subsequence $\mathbf{z}^{k_j}$, $j = 1, 2, \ldots$ which converges within $C^{2+\alpha}(B) \cap C^0(\overline{B})$ to a diffeomorphism $\mathbf{z} \in \overline{\Gamma}_*(B, a_0, a_1, a_2, \alpha, N)$, briefly

$$\mathbf{z}^{k_j} \to \mathbf{z} \quad \text{for} \quad j \to \infty \quad \text{in} \quad C^{2+\alpha}(B) \cap C^0(\overline{B}).$$

Furthermore, the following **minimal quantity for the Jacobian** *is positive for all* $0 < r < 1$, *namely*

$$\mathcal{T}(a_0, a_1, a_2, \alpha, N; r) :$$

$$= \inf \left\{ J_{\mathbf{z}}(w) \,\middle|\, w \in B \text{ with } |w| \leq r, \, \mathbf{z} \in \Gamma_*(B, a_0, a_1, a_2, \alpha, N) \right\} > 0. \tag{21}$$

Proof: 1.) Each diffeomorphism $\mathbf{z}^k$ solves the Lewy-Heinz system (1), where the coefficient functions $h_1^k, \ldots, \tilde{h}_4^k$ are given for $k = 1, 2, \ldots$. Due to the Hölder-bound (20), together with the structure conditions (2) and (3), we can select a subsequence such that these coefficient functions converge in the $C^\mu(\overline{B})$-norm to the limit functions $h_1, \ldots, \tilde{h}_4$ for all $0 < \mu < \alpha$. Therefore, we assume $h_1^k \to h_1, \ldots, \tilde{h}_4^k \to \tilde{h}_4$ within $C^\alpha(\overline{B})$ for $k \to \infty$. Due to Theorem 5.4 in Section 5 of Chapter 12, we can select a subsequence from the associate diffeomorphisms $\mathbf{z}^k$, which is uniformly convergent on $\overline{B}$; here again we make the transition to the subsequence tacitly. Since we have a joint estimate for the gradient of $\mathbf{z}^k$ in the interior of B, the Lewy-Heinz system yields a joint estimate of the $C^{1+\beta}(\Omega)$-norm for all domains $\Omega \subset\subset B$ and all exponents $0 < \beta < 1$. From the potential theory we can ascertain the convergence of our subsequence $\mathbf{z}^{k_j}$, $j = 1, 2, \ldots$ within $C^{2+\alpha}(\Omega) \cap C^0(\overline{B})$ to a solution $\mathbf{z}$ of the Lewy-Heinz system for the coefficients $h_1, \ldots, \tilde{h}_4$.

2.) We have still to show that $J_{\mathbf{z}}(w) \neq 0$ for all $w \in B$ holds true for our limit mapping. As in part 1.) of the proof for Theorem 6.4 we assume, on the contrary, that $J_{\mathbf{z}}(0,0) = 0$ were true. Then we consider the locally bent mapping from (16), namely

$$\widehat{\mathbf{z}}(w) := h(\mathbf{z}(w)), \, w \in K_\epsilon,$$

which has an index ≥ 2 at the origin. Since our diffeomorphisms converge uniformly according to

$$\mathbf{z}^k(w) \to \mathbf{z}(w), \, w \in \overline{B} \quad \text{for} \quad k \to \infty,$$

the approximating mapping

$$\widehat{\mathbf{z}^k}(w) := h(\mathbf{z}^k(w)), \, w \in K_\epsilon$$

must have an index ≥ 2 at the origin for $k \geq k_0$ as well, with a sufficiently large k_0. On account of $J_h(0,0) = 1$, the mapping $\widehat{\mathbf{z}^k}$ is positive-oriented and topological, and consequently possesses the index $+1$ for $k \geq k_0$. This contradiction shows that the limit mapping has a nonvanishing Jacobian. With part 3.) of the proof for Theorem 6.4, we deduce the diffeomorphic character of $\mathbf{z}$ from B onto itself.

3.) Now we establish that the quantity (21) is positive: If the quantity $\mathcal{T}(a_0, a_1, a_2, \alpha, N; r)$ vanished, we could find a sequence of points $w^k \in B$

with $|w^k| \le r$ and functions $\mathbf{z}^k \in \Gamma_*(B, a_0, a_1, a_2, \alpha, N)$ for $k = 1, 2, \ldots$ such that

$$J_{\mathbf{z}^k}(w^k) \to 0 \quad \text{for} \quad k \to \infty$$

is correct. Then we can select a subsequence within the diffeomorphisms $\{\mathbf{z}^k\}_{k=1,2,\ldots}$ converging to a diffeomorphism $\mathbf{z}$ again, more precisely

$$\mathbf{z}^{k'} \to \mathbf{z} \quad \text{for} \quad k' \to \infty \quad \text{in} \quad C^{2+\alpha}(B) \cap C^0(\overline{B}).$$

Moreover, we select a further subsequence such that $w^{k''} \to w^*$ for $k'' \to \infty$ holds true with $|w^*| \le r$. With the identity

$$0 = \lim_{k'' \to \infty} J_{\mathbf{z}^{k''}}(w^{k''}) = J_{\mathbf{z}}(w^*)$$

we arrive at a contradiction to the fact that $\mathbf{z} \in \overline{\Gamma}_*(B, a_0, a_1, a_2, \alpha, N)$ holds true.   q.e.d.

Some remarks to Dirichlet's boundary value problem for Lewy-Heinz systems:

Let us prescribe a weakly monotonically increasing, continuous function

$$\theta = \theta(\varphi) : [0, 2\pi] \to [0, 2\pi] \quad \text{with} \quad \theta(0) = 0 \quad \text{and} \quad \theta(2\pi) = 2\pi. \tag{22}$$

Now we want to solve the Lewy-Heinz system (1) *under the Dirichlet boundary conditions*

$$\mathbf{z}(\cos\varphi, \sin\varphi) = \Big(\cos\theta(\varphi), \sin\theta(\varphi)\Big) \quad \text{for} \quad 0 \le \varphi \le 2\pi \tag{23}$$

within the class of diffeomorphisms $\overline{\Gamma}_*(B, a_0, a_1, a_2, \alpha, N)$. Then it seems natural to embed this problem into the following family of problems $\mathcal{P}(\tau)$ with the deformation parameter $0 \le \tau \le 1$, namely

$$\mathcal{P}(\tau): \quad \mathbf{z}(u, v; \tau) = (x(u, v; \tau), y(u, v; \tau)) \in \overline{\Gamma}_*(B, a_0, a_1, a_2, \alpha, N) \quad \text{satisfies}$$

$$\Delta x(u, v; \tau) = \tau h_1(x(u, v; \tau), y(u, v; \tau))|\nabla x(u, v; \tau)|^2$$

$$+\tau h_2(x, y)(\nabla x \cdot \nabla y)\Big|_{(u, v; \tau)} + \tau h_3(x, y)|\nabla y|^2\Big|_{(u, v; \tau)} + h_4(x, y)\frac{\partial(x, y)}{\partial(u, v)}\Big|_{(u, v; \tau)},$$

$$\Delta y(u, v; \tau) = \tau \tilde{h}_1(x, y)|\nabla x|^2\Big|_{(u, v; \tau)} + \tau \tilde{h}_2(x, y)(\nabla x \cdot \nabla y)\Big|_{(u, v; \tau)}$$

$$+\tau \tilde{h}_3(x, y)|\nabla y|^2\Big|_{(u, v; \tau)} + \tau \tilde{h}_4(x, y)\frac{\partial(x, y)}{\partial(u, v)}\Big|_{(u, v; \tau)} \quad \text{for all } w = u + iv \in B,$$

and $\mathbf{z}(\cos\varphi, \sin\varphi; \tau) = \Big(\cos((1 - \tau)\varphi + \tau\theta(\varphi)), \sin((1 - \tau)\varphi + \tau\theta(\varphi))\Big),$

for all $0 \le \varphi \le 2\pi$;   with the parameter $0 \le \tau \le 1$.

We see that $\mathcal{P}(0)$ is uniquely solved by the identity map

$$\mathbf{z}(u, v; 0) := (u, v), \ (u, v) \in \overline{B}.$$

We infer from Theorem 6.8 that the set

$$T := \{\tau \in [0, 1] : \mathcal{P}(\tau) \text{ possesses a solution}\}$$

is a closed, nonvoid set. More precisely, if we have for $k = 1, 2, \ldots$ solutions $\mathbf{z}(., \tau_k)$ of the problem $\mathcal{P}(\tau_k)$ with $\tau_k \in [0, 1]$ satisfying $\tau_k \to \tau_*$ for $k \to \infty$, we can select a subsequence such that $\mathbf{z}(., \tau_{k'})$, $k' = 1, 2, \ldots$ converges to a solution $\mathbf{z}(., \tau_*)$ of $\mathcal{P}(\tau_*)$, because of our compactness result above. Now the decisive question remains: For which class of coefficient functions $h_1, \ldots, \tilde{h}_4$ is the set T open? This would imply $T = [0, 1]$ and the problem $\mathcal{P}(1)$ would be solved.

This again raises the question of stability for the boundary value problem above. To my knowledge, even the uniqueness question for this system with Dirichlet boundary values is completely open! To establish a maximum principle similar to W. Jäger's result from Section 1 in Chapter 12 is highly desirable! Even conditions on the Darboux system (17) within the next section, associate to the Weyl embedding problem, which guarantee uniqueness and continuous dependence from the boundary values via a maximum principle, are not known to this day!

In this context we pose the intricate question, for which class of Lewy-Heinz systems does the topological boundary behavior enforce the nonvanishing of the Jacobian in the interior of the unit disc – without prior knowledge of the one-to-one character for the mapping? Here the only positive answer known has been given by H. Kneser in his paper [Kn] for harmonic mappings with the coefficients $h_1 = \ldots = \tilde{h}_4 = 0$. Only in the case of harmonic mappings with respect to Riemannian metrics, where the uniqueness of the Dirichlet problem has been established by W. Jäger and H. Kaul, do the investigations by J. Jost cited above provide a result.

These circumstances render it even more admirable that E. Heinz managed to show the existence of diffeomorphisms of the system (1) under Dirichlet boundary conditions via the Leray-Schauder degree of mapping in Banach spaces (see Chapter 7). For these topics we would like to recommend the study of the following treatise by

E. HEINZ: *Existence theorems for one-to-one mappings associated with elliptic systems of second order.* Part I: Journ. d'Analyse Math. **15**, 325-352 (1962); Part II: Journ. d'Analyse Math. **17**, 145-184 (1966).

Moreover, we refer to the survey article [H11] (see the next paragraph) with highly interesting open problems.

Finally, we mention that one can exclude the vanishing of the Jacobian for the systems (1) on the boundary, having controlled the differentiable boundary behavior before, via the boundary point lemma by E. Hopf. Here we refer the reader to the proof of Theorem 1.7 in Section 1, where a comparable situation is treated.

For mappings in the class $\overline{\Gamma}_*(B, a_0, a_1, a_2, \alpha, N)$, we know from (21) that the Jacobian remains positive in the interior – but we do not yet have a common explicit bound, depending only on the given quantities! We shall strengthen this statement (21) to an a priori estimate of the Jacobian from below, which is independent of the quantities a_2 and α and consequently without Hölder-bounds on the coefficients. This feature is decisive in order to gain valuable a priori estimates for Monge-Ampère equations in Section 7. On account of the inclusion (19), one is tempted to transform the estimate for the gradient from below within the class $\overline{\Gamma}(B, 4a_0, 0, N)$ to an estimate for the Jacobian from below. This strategy was possible in Section 7 and Section 8 from Chapter 12 for conformal mappings with respect to a Riemannian metric, since we could build upon the conformality relations there. In the case of the Lewy-Heinz mappings, *no* conformality relations are at our disposition – and the estimate for the Jacobian from below has to be achieved independently! With the aid of the similarity principle by L. Bers and I. N. Vekua, we shall now prove the following stupendous result [H9], which we again owe to E. Heinz.

Theorem 6.9. (A priori estimate of the Jacobian)
For the given quantities $a_0, a_1, N \in [0, +\infty)$ and all radii $0 < r < 1$, there exists a constant $\mathcal{T}_*(a_0, a_1, N, r) \in (0, \infty)$, such that the following estimate is valid:

$$J_{\mathbf{z}}(u, v) = \frac{\partial(x, y)}{\partial(u, v)}\bigg|_{(u,v)} \geq \mathcal{T}_*(a_0, a_1, N, r)$$
$$\text{for all} \quad (u, v) \in B \quad \text{with} \quad u^2 + v^2 \leq r^2 \tag{24}$$

and for all diffeomorphisms $\mathbf{z} = \mathbf{z}(u, v) \in \Gamma_*(B, a_0, a_1, N)$.

Proof: 1.) At first, we utilize the inclusion (19) and follow the parts 1. – 3. in the proof of Theorem 5.4 from Section 5 of Chapter 12, quoting the formulas there with the superscript *. Thus we choose the radius $r \in (0, 1)$ sufficiently large and the quantity

$$\delta = \delta(r) := \frac{1 - r}{2} \in \left(0, \frac{1}{4}\right)$$

correspondingly small, such that (34)* is satisfied. From (37)* we obtain the gradient estimate

$$|\nabla \mathbf{z}(u, v)| \leq \Lambda(4a_0, 0, N, r), \ (u, v) \in B_{r+\delta} := \{w = u + iv \in B \big| |w| < r + \delta\}. \tag{25}$$

Now we choose $\delta(r, N) \in (0, \frac{1}{8})$ sufficiently small, such that the inequality (38)* holds true. Consequently, the covering property $B_{\frac{1}{2}} \subset \mathbf{z}(B_r)$ is realized, which implies the estimate

$$\int\int_{B_r} J_{\mathbf{z}}(u,v)\,du\,dv \geq \frac{\pi}{4} \tag{26}$$

for the integral of the Jacobian. Therefore, we arrive at the statement:

$$J_{\mathbf{z}}(w_*) > \frac{1}{4} \quad \text{is valid for one point} \quad w_* \in \overline{B_r} \quad \text{at least.} \tag{27}$$

If the latter were false, we could estimate

$$\int\int_{B_r} J_{\mathbf{z}}(u,v)\,du\,dv \leq \int\int_{B_r} \frac{1}{4}\,du\,dv = \frac{\pi r^2}{4} < \frac{\pi}{4}, \tag{28}$$

in contradiction to (27).

2.) Diminishing $\delta = \delta(r, N, a_0, a_1) = \dfrac{1-r}{2} > 0$ in the subsequent part 4.) for the last time, we derive **a Harnack-type inequality for the Jacobian with the a priori constants** $\lambda(a_0, a_1, N, r) > 0$ and $\nu(a_0, a_1, N, r) > 0$ in the following estimate:

$$J_{\mathbf{z}}(u_0, v_0) \geq \Big(\lambda(a_0, a_1, N, r)\, J_{\mathbf{z}}(u, v)\Big)^{\nu(a_0, a_1, N, r)} \tag{29}$$

for all $(u,v) \in B$ with $(u - u_0)^2 + (v - v_0)^2 \leq \delta^2$ and all $(u_0, v_0) \in \overline{B_r}$.

With the aid of this Harnack-type inequality, we complete the proof of the estimate (24) as follows: We join an arbitrary point $w \in \overline{B_r}$ by the straight line

$$\mathbf{w} = w + t(w_* - w),\ 0 \leq t \leq 1$$

with the point w_* from (27) within the convex set $\overline{B_r}$. Then we choose the integer $n = n(r, N, a_0, a_1) \in \mathbb{N}$ so large that the equidistant points

$$w_k := w + \frac{k}{n}(w_* - w),\ k = 0, \ldots, n$$

satisfy the inequalities

$$|w_k - w_{k-1}| \leq \delta(r, N, a_0, a_1),\ k = 1, \ldots, n.$$

With the aid of (29) and (27), we estimate as follows

$$J_{\mathbf{z}}(w) = J_{\mathbf{z}}(w_0) \geq \Big(\lambda(a_0, a_1, N, r)\, J_{\mathbf{z}}(w_1)\Big)^{\nu(a_0, a_1, N, r)}$$

$$\geq \lambda(a_0, a_1, N, r)^{2\nu(a_0, a_1, N, r)}\, J_{\mathbf{z}}(w_2)^{\nu^2(a_0, a_1, N, r)} \geq \ldots$$

$$\geq \lambda(a_0, a_1, N, r)^{n\nu(a_0, a_1, N, r)}\, J_{\mathbf{z}}(w_n)^{\nu^n(a_0, a_1, N, r)} \tag{30}$$

$$= \lambda(a_0, a_1, N, r)^{n\nu(a_0, a_1, N, r)}\, J_{\mathbf{z}}(w_*)^{\nu^n(a_0, a_1, N, r)}$$

$$\geq 4^{-\nu^n(a_0, a_1, N, r)}\lambda(a_0, a_1, N, r)^{n\nu(a_0, a_1, N, r)} =: T_*(a_0, a_1, N, r) > 0.$$

3.) In order to establish (29), we only have to consider the point

$$w_0 = (u_0, v_0) = (0, 0),$$

due to the invariance of the following arguments w.r.t. the necessary translations. We shall refine part 1.) in the proof of Theorem 6.4 and consider the auxiliary function

$$\gamma(\alpha, \beta) := \left| \begin{pmatrix} x_u(0,0) \ y_u(0,0) \\ x_v(0,0) \ y_v(0,0) \end{pmatrix} \circ \begin{pmatrix} -\beta \\ \alpha \end{pmatrix} \right|^2 > 0, \ (\alpha, \beta) \in S^1.$$

Obviously, we can select a direction $(\alpha, \beta) \in S^1$ such that

$$\gamma(\alpha, \beta) \le J_z(0, 0)$$

is fulfilled. According to Proposition 6.2, we perform a rotation with the given directional vector (α, β) and can assume

$$0 < |\nabla y(0, 0)|^2 \le J_z(0, 0). \tag{31}$$

4.) Our Proposition 6.3 provides a function, with the radius $\epsilon = \epsilon(a_0, a_1) > 0$ for its disc of existence, namely

$$f(x, y) = y - g(x) : K_\epsilon \to \mathbb{R} \in C^2(K_\epsilon) \text{ with } f(0,0) = 0 \text{ and } \nabla f(0,0) = (0, 1),$$

such that the nonlinear combination

$$\Phi(u, v) := f(x(u, v), y(u, v)), \ (u, v) \in K_\epsilon$$

is subject to the differential inequality (8) and satisfies $\Phi(0, 0) = 0$. Therefore, its nonvanishing gradient

$$\Phi_w(u, v) = y_w(u, v) - g'(x(u, v))x_w(u, v) \ne 0, \ (u, v) \in K_\epsilon$$

yields a pseudoholomorphic function. From Theorem 5.1 in Section 5 of Chapter 12 we find a constant $\mu(r, a_0, a_1, N) > 0$ and an exponent $\nu(a_0, a_1, N, r) > 0$ such that the following a priori estimate is valid:

$$|\Phi_w(u, v)| \le \mu(r, a_0, a_1, N) \, |\Phi_w(0, 0)|^{2\nu^{-1}(a_0, a_1, N, r)}, \quad (u, v) \in K_\epsilon. \tag{32}$$

5.) Now we consider the locally bent mapping $\hat{z}(u, v)$ from formula (16) in the proof of Theorem 6.4, whose determinant is estimable via a constant $c(r, a_0, a_1, N) > 0$ equivalently. With the aid of (25), (31), and (32), we deduce the following inequality:

$$J_{\mathbf{z}}(u,v) \leq c(r,a_0,a_1,N)\, J_{\hat{\mathbf{z}}}(u,v)$$

$$= c(r,a_0,a_1,N)\,[\nabla\Phi(u,v),\, -\nabla x(u,v)]$$

$$\leq 2c(r,a_0,a_1,N)\,|\Phi_w(u,v)| \cdot |\nabla x(u,v)|$$

$$\leq 2c(r,a_0,a_1,N)\,\mu(r,a_0,a_1,N)\,|\Phi_w(0,0)|^{2\nu^{-1}(a_0,a_1,N,r)} \cdot \Lambda(4a_0,0,N,r)$$

$$= 2c(r,a_0,a_1,N)\,\mu(r,a_0,a_1,N)|y_w(0,0|^{2\nu^{-1}(a_0,a_1,N,r)} \cdot \Lambda(4a_0,0,N,r)$$

$$\leq \lambda^{-1}(a_0,a_1,N,r) \cdot J_{\mathbf{z}}(0,0)^{\nu^{-1}(a_0,a_1,N,r)} \quad \text{for all} \quad (u,v) \in K_\epsilon.$$

(33)

Here we have introduced the constant

$$\lambda(a_0,a_1,N,r):$$

$$= \left\{ 2c(r,a_0,a_1,N)\,\mu(r,a_0,a_1,N)\,4^{-\nu^{-1}(a_0,a_1,N,r)}\,\Lambda(4a_0,0,N,r) \right\}^{-1}.$$

(34)

Thus we have established the Harnack-type inequality (29), and the proof of our theorem is complete.   q.e.d.

7 On Weyl's Embedding Problem and Elliptic Monge-Ampère Equations

Applications of the theory for nonlinear elliptic systems to *the Weyl embedding problem* and the associate *Monge-Ampère equations* opens a new domain. Here we refer the reader to the very valuable lecture notes [Sc] by Friedmar Schulz. Furthermore, we recommend the excellent survey article [H11] by E. Heinz. This lecture was held in connection with the *Cantor Medal* awarded to Professor Erhard Heinz by the *Deutsche Mathematiker-Vereinigung* on the *DMV-Jahrestagung 1995* at the Universität Ulm.

Let us consider the following partial differential equation *in the Monge notation* $p = z_x$, $q = z_y$, $r = z_{xx}$, $s = z_{xy}$, $t = z_{yy}$, namely

$$0 = F(x,y,z,p,q,r,s,t) := rt - s^2 + A(x,y,z,p,q)r$$
$$+2B(x,y,z,p,q)s + C(x,y,z,p,q)t - E(x,y,z,p,q).$$

(1)

This **Monge-Ampère equation** (1) is **elliptic for a given solution**

$$z = z(x,y),\ (x,y) \in \Omega \quad \text{of the class} \quad C^2(\Omega)$$

on the bounded domain $\Omega \subset \mathbb{R}^2$ of the topological type of the disc, if the following condition for *the discriminant* is satisfied:

$$D(x,y) := F_r F_t - \tfrac{1}{4} F_s^2$$

$$= (A+t)(C+r) - (B-s)^2$$

$$= Ar + 2Bs + Ct + (rt - s^2) + AC - B^2 \tag{2}$$

$$= AC - B^2 + E > 0, \ (x,y) \in \Omega.$$

Now we introduce *isothermal parameters into the Riemannian metric*

$$ds^2 \Big|_{(x(u,v),y(u,v))} = \left\{ F_t dx^2 - F_s dxdy + F_r dy^2 \right\} \Big|_{(x(u,v),y(u,v))}$$

$$= \left\{ (C+r)dx^2 - 2(B-s)dxdy + (A+t)dy^2 \right\} \Big|_{(x(u,v),y(u,v))} \tag{3}$$

$$= \Lambda(u,v)(du^2 + dv^2), \ (u,v) \in B,$$

$$\text{where} \quad \Lambda(u,v) > 0, \ (u,v) \in B \quad \text{holds},$$

with the following *uniformizing mapping*

$$\mathbf{z}(u,v) = x(u,v) + iy(u,v) : B \to \Omega. \tag{4}$$

As is usual, we take the symbols B for the interior and $\overline{B}$ for the closed unit disc, without fearing any confusion with the coefficient function $B = B(\ldots)$. In Proposition 9.4.2 from Section 9.4 of the interesting book [Sc] by Friedmar Schulz, the following plane elliptic system (5), (6) has been derived for the uniformizing mapping:

$$\Delta x(u,v) = h_1(x(u,v),y(u,v))|\nabla x(u,v)|^2 + h_2(x,y)\nabla x \cdot \nabla y(u,v)$$

$$+ h_3(x(u,v),y(u,v))|\nabla y(u,v)|^2 + h_4(x(u,v),y(u,v))\frac{\partial(x,y)}{\partial(u,v)}, \tag{5}$$

$$\Delta y(u,v) = \tilde{h}_1(x(u,v),y(u,v))|\nabla x(u,v)|^2 + \tilde{h}_2(x,y)\nabla x \cdot \nabla y(u,v)$$

$$+ \tilde{h}_3(x(u,v),y(u,v))|\nabla y(u,v)|^2 + \tilde{h}_4(x(u,v),y(u,v))\frac{\partial(x,y)}{\partial(u,v)}. \tag{6}$$

Here the coefficient functions are defined by the identities

$$k_1(x,y,z,p,q) := B_q - \frac{1}{2D}(D_x + D_z p - D_p C + D_q B)$$

$$k_2(x,y,z,p,q) := -A_q - B_p - \frac{1}{2D}(D_y + D_z q + D_p B - D_q A)$$

$$k_3(x,y,z,p,q) := A_p \qquad\qquad k_4(x,y,z,p,q): \tag{7}$$

$$= \frac{1}{\sqrt{D}}\left(A_x + B_y + A_z p + B_z q - A_p C + (A_q + B_p)B - B_q A - \frac{1}{2}D_p \right)$$

and

$$\tilde{k}_1(x,y,z,p,q) := C_q$$

$$\tilde{k}_2(x,y,z,p,q) := -B_q - C_p - \frac{1}{2D}(D_x + D_z p - D_p C + D_q B)$$

$$\tilde{k}_3(x,y,z,p,q) := B_p - \frac{1}{2D}(D_y + D_z q + D_p B - D_q A) \tag{8}$$

$$\tilde{k}_4(x,y,z,p,q):$$

$$= -\frac{1}{\sqrt{D}}\left(C_y + B_x + C_z q + B_z p - B_p C + (B_q + C_p)B - C_q A - \frac{1}{2}D_q\right),$$

when we insert the given solution

$$h_l(x,y) := k_l\Big(x,y,z(x,y),p(x,y),q(x,y)\Big) \qquad \text{and}$$

$$\tilde{h}_l(x,y) := \tilde{k}_l\Big(x,y,z(x,y),p(x,y),q(x,y)\Big) \quad \text{for} \quad l = 1\ldots 4. \tag{9}$$

For later use we note the following identities:

$$\tilde{k}_1 = C_q, \quad k_1 - \tilde{k}_2 = C_p + 2B_q, \quad \tilde{k}_2 - \tilde{k}_3 = -A_q - 2B_p, \quad \tilde{k}_3 = A_p. \tag{10}$$

Due to the Proposition 9.4.2 from [Sc], the second derivatives r, s, t of the Monge-Ampère equation have the following representation:

$$\frac{A+t}{\sqrt{D}} = \frac{|\nabla x|^2}{J_{\mathbf{z}}(u,v)}, \quad \frac{B-s}{\sqrt{D}} = \frac{\nabla x \cdot \nabla y}{J_{\mathbf{z}}(u,v)}, \quad \frac{C+r}{\sqrt{D}} = \frac{|\nabla y|^2}{J_{\mathbf{z}}(u,v)}$$

$$\text{with the Jacobian} \quad J_{\mathbf{z}}(u,v) := \frac{\partial(x,y)}{\partial(u,v)} = [\nabla x, \nabla y], \ (u,v) \in B; \tag{11}$$

here we see the determinant as the alternating bilinear form in the arguments of the bracket $[.,.]$ again. When we want to control the second derivatives for solutions of the p.d.e. (1) via the uniformization method in the subsequent Theorem 7.1, we have to prevent the Jacobian for the uniformizing mapping (4) from vanishing and to estimate the Jacobian from below.

The **Weyl embedding problem** requires the determination of an isometric embedding $\mathbf{x} : S^2 \to \mathbb{R}^3$ for an elliptic metric

$$ds^2 = E(u,v)du^2 + 2F(u,v)dudv + G(u,v)dv^2 \quad \text{on the unit sphere} \quad S^2 \tag{12}$$

of positive Gaussian curvature $K(u,v) > 0$ as a closed convex surface. Therefore, one has to solve the following *nonlinear system of first order*

$$|\mathbf{x}_u(u,v)|^2 = E(u,v), \ \mathbf{x}_u(u,v) \cdot \mathbf{x}_v(u,v) = F(u,v), \ |\mathbf{x}_v(u,v)|^2 = G(u,v) \text{ on } S^2. \tag{13}$$

Equivalent to this first order system (13), the function $\rho(u,v) := \frac{1}{2}\mathbf{x}(u,v)^2$ satisfies **the Darboux differential equation**

$$\left(\rho_{uu} - \Gamma_{11}^1 \rho_u - \Gamma_{11}^2 \rho_v - E\right)\left(\rho_{vv} - \Gamma_{22}^1 \rho_u - \Gamma_{22}^2 \rho_v - G\right)$$

$$-\left(\rho_{uv} - \Gamma_{12}^1 \rho_u - \Gamma_{12}^2 \rho_v - F\right)^2 \tag{14}$$

$$= K(u,v)\left\{2\rho(EG - F^2) - (G\rho_u^2 - 2F\rho_u\rho_v + E\rho_v^2)\right\} \quad \text{on } S^2.$$

Here *the Christoffel symbols* Γ_{ij}^k appear (see [K], Section 3.8 or [BL]), which depend only on the coefficients E, F, G of the given first fundamental form ds^2 and their first derivatives. However, the Gaussian curvature K additionally depends on the second derivatives of the given metric. This p.d.e. (14) represents an elliptic Monge-Ampère equation in the normal form (1), substituting $x = u$, $y = v$, $z = \rho$, $p = \rho_u$, $q = \rho_v$. This equation possesses coefficients which are linear in p and q as well as independent of z, and a right-hand side involving the Gaussian curvature as follows:

$$A(u,v,z,p,q) = -\Gamma_{22}^1(u,v)p - \Gamma_{22}^2(u,v)q - G(u,v),$$

$$B(u,v,z,p,q) = \Gamma_{12}^1(u,v)p + \Gamma_{12}^2(u,v)q + F(u,v),$$

$$C(u,v,z,p,q) = -\Gamma_{11}^1(u,v)p - \Gamma_{11}^2(x,y)q - E(u,v),$$

$$E(u,v,z,p,q) := K(u,v)\left\{2z\Big(E(u,v)G(u,v) - F(u,v)^2\Big)\right. \tag{15}$$

$$\left. -\Big(G(u,v)p^2 - 2F(u,v)pq + E(u,v)q^2\Big)\right\} - (AC - B^2)\Big|_{(u,v,z,p,q)}.$$

The ellipticity condition (2) reduces to the inequality

$$D(u,v) = K(u,v)\Big(\mathbf{x}(u,v), \mathbf{x}_u(u,v), \mathbf{x}_v(u,v)\Big) > 0 \quad \text{on} \quad S^2, \tag{16}$$

where the origin of our coordinate system lies within the closed convex surface.

The characteristic metric (3), associated with the Darboux equation (14), stands in a conformal relation with the second fundamental form of the surface $\mathbf{x}(u,v)$. Thus we introduce as characteristic parameters the so-called *conjugate isothermal parameters* (α, β), rendering the second fundamental form of $\mathbf{x}$ isothermal. The uniformizing mapping solves **the Darboux system**

$$\Delta u(\alpha, \beta) = -\left(\Gamma_{11}^1 + \frac{K_u}{2K}\right)|\nabla u|^2 - \left(2\Gamma_{12}^1 + \frac{K_v}{2K}\right)\nabla u \cdot \nabla v - \Gamma_{22}^1|\nabla v|^2,$$

$$\Delta v(\alpha, \beta) = -\Gamma_{11}^2|\nabla u|^2 - \left(2\Gamma_{12}^2 + \frac{K_u}{2K}\right)\nabla u \cdot \nabla v - \left(\Gamma_{22}^2 + \frac{K_v}{2K}\right)|\nabla v|^2. \tag{17}$$

Remarks:

a) This is a special case of the system (4) – (6), replacing (x, y) with (u, v) and (u, v) with (α, β). Here the coefficient functions $k_1(u, v)$, $k_2(u, v)$,

$k_3(u,v)$, $\tilde{k}_1(u,v)$, $\tilde{k}_2(u,v)$, $\tilde{k}_3(u,v)$ are independent of the variables z, p, q and $k_4(u,v)$ as well as $\tilde{k}_4(u,v)$ vanish. Therefore, the combinations from (10) depend only on (u,v) and *not* on z, p, q, which can also be deduced from the identities (15) for the Darboux equation.

b) For these profound results, we refer our readers to the article [H11] by E. Heinz and the underlying publications. Here we recommend the classical book [B] by L. Bianci and the comprehensive four volumes [D] by G. Darboux. They both show the excellent level that differential geometry enjoyed at the turn of the 20th century!

c) Moreover, we recommend the study of Chapter 10 in the book [Sc] by F. Schulz. There in section 10.3, an equation applicable only to graphs is called a Darboux equation; in Proposition 10.4.1 the Darboux system differs from the system (17) above.

d) For surfaces with positive Gaussian curvature $K(\alpha, \beta) > 0$ in conjugate-isothermal parameters (α, β), Franz Rellich has found an elliptic system in divergence form for the position vector $\mathbf{x}(\alpha, \beta)$ in the treatise [R], namely

$$\frac{\partial}{\partial \alpha}\left\{ \sqrt{K(\alpha, \beta)}\,\mathbf{x}_\alpha(\alpha, \beta)\right\} + \frac{\partial}{\partial \beta}\left\{ \sqrt{K}\,\mathbf{x}_\beta\right\}\Big|_{(\alpha,\beta)} = 2\sqrt{K}\,\mathbf{x}_\alpha \times \mathbf{x}_\beta\Big|_{(\alpha,\beta)}. \quad (18)$$

These p.d.e.s are similar to the well-known H-surface system; however, they belong to the inner geometry of the surface $\mathbf{x}$. Both systems were invented by the academic teacher of E. Heinz, G. Hellwig, J. Moser, E. Wienholtz and many others.

A general result for Monge-Ampère equations [Ni] by L. Nirenberg implies that one can derive a Hölder-estimate for the second derivatives of their strictly elliptic solutions from a bound on their second derivatives. This interesting theorem has been derived in Section 3.4 and Theorem 3.4.3 of the book [Sc] by F. Schulz via a technique due to S. Campanato.

Now we shall only require a C^1-bound on the solution and thus estimate the second derivatives – and consequently their Hölder norm as well: Then an intricate structure condition appears additionally, which selects those strictly elliptic solutions of the Monge-Ampère equation estimable in that sense!

On the basis of Section 6 this profound Theorem 7.1 shall be proved by the uniformization method – and the Hölder estimates for the second derivatives follow via potential-theoretic arguments.

Theorem 7.1. (Interior estimates for elliptic Monge-Ampère equations due to E. Heinz)

Assumptions:

i) *Given the bounded domain $\Omega \subset \mathbb{R}^2$, the coefficient functions $A = A(x,y,z,p,q)$, $B = B(x,y,z,p,q)$, $C = C(x,y,z,p,q)$ of the p.d.e. (1) belong to the class $C^{1+\alpha}(\overline{\Omega} \times \mathbb{R}^3)$, with the arbitrary Hölder coefficient $\alpha \in (0,1)$, and $E = E(x,y,z,p,q)$ to the class $C^1(\overline{\Omega} \times \mathbb{R}^3)$. Moreover, they satisfy the growth condition*

$$|A|, |B|, |C|, |E|; |A_x|, \dots, |E_q| \le \lambda(R)\, for\, (x,y) \in \overline{\Omega}\, and\, |z|, |p|, |q| \le R$$

$$with\, 0 < R < +\infty,\, where\, \lambda = \lambda(R) : (0, +\infty) \to (0, +\infty)\, is\, monotone.$$

$$(19)$$

ii) *The solution* $z = z(x,y) : \Omega \to \mathbb{R} \in C^{3+\alpha}(\Omega)$ *of the p.d.e. (1) is* bounded inclusive of its first derivatives *according to*

$$\|z\|_{C^1(\Omega)} := \sup_{(x,y) \in \Omega} |z(x,y)| + \sup_{(x,y) \in \Omega} |\nabla z(x,y)| \le \gamma\, with\, the\, constant\, \gamma > 0$$

$$(20)$$

and is strictly elliptic for the Monge-Ampère equation *in the sense that*

$$D(x,y) = \{AC - B^2 + E\}\Big|_{(x,y,z(x,y),p(x,y),q(x,y))} \ge \mu \qquad (21)$$

$$for\, all\quad (x,y) \in \Omega\quad with\, a\, constant\quad \mu > 0.$$

iii) *For the combinations*

$$\omega_1(x,y) := \tilde{k}_1\Big|_{(x,y,z(x,y),p(x,y),q(x,y))} = C_q(x,y,z(x,y),p(x,y),q(x,y)),$$

$$\omega_2(x,y) := (k_1 - \tilde{k}_2)\Big|_{(x,y,z(x,y),p(x,y),q(x,y))}$$
$$= C_p(x,y,z(x,y),p(x,y),q(x,y)) + 2B_q(x,y,z(x,y),p(x,y),q(x,y)),$$

$$\omega_3(x,y) := (k_2 - \tilde{k}_3)\Big|_{(x,y,z(x,y),p(x,y),q(x,y))}$$
$$= -A_q(x,y,z(x,y),p(x,y),q(x,y)) - 2B_p(x,y,z(x,y),p(x,y),q(x,y)),$$

$$\omega_4(x,y) := k_3\Big|_{(x,y,z(x,y),p(x,y),q(x,y))} = A_p(x,y,z(x,y),p(x,y),q(x,y)),$$

$$(22)$$

we require the structure conditions

$$|\omega_l(z_1) - \omega_l(z_2)| \le \sigma\, |z_1 - z_2|\quad for\, all\quad z_1, z_2 \in \overline{B}$$

$$(23)$$

$$and\quad l = 1, \dots, 4\quad with\, the\, constant\quad \sigma > 0.$$

Statement: For sufficiently small $d > 0$, *we can estimate the* $C^{2+\nu}$-*norm in* the interior set

$$\Omega[d] := \Big\{ (x,y) \in \Omega : dist\big((x,y), \partial\Omega\big) > d \Big\}$$

for all exponents $0 < \nu < 1$ *by an a priori constant* $\Theta(\dots)$ *as follows:*

$$\|z\|_{C^{2+\nu}(\Omega[d])} \le \Theta(\lambda(\gamma), \sigma, \nu, \gamma, d) < \infty. \qquad (24)$$

Proof:

1.) We consider the standard domain $\Omega := \{(x,y) \in \mathbb{R}^2 | \, x^2 + y^2 < 1\}$ and then obtain the estimate (24) via a localization argument. For our solution $z(x,y) : \Omega \to \mathbb{R} \in C^{3+\alpha}(\Omega)$ of the p.d.e. (1) we introduce isothermal parameters into the associate $C^{1+\alpha}$–metric (3); please see Section 7 and Section 8 in Chapter 12. The uniformizing mapping $\mathbf{z}$ from (4) with $f(0,0) = (0,0)$ satisfies the conditions (5) – (9) and constitutes a Lewy-Heinz system in the sense of Definition 6.1 from Section 6. We shall find out admissible constants a_0, a_1, N such that $f \in \Gamma_*(B, a_0, a_1, N)$ is valid.

2.) We look at our coefficients $h_1(x,y), \ldots, \tilde{h}_4(x,y)$ from (9) for the Lewy-Heinz system (5) and (6), which are generated by the functions $k_1(x,y,z,p,q)$, $\ldots, \tilde{k}_4(x,y,z,p,q)$ from (7) and (8). Taking (19), (20), (21) into account, we find a constant $\hat{a}_0 = a_0(\lambda(\gamma), \mu) > 0$, such that we have the following estimate:

$$|h_1(x,y)|, \ldots, |\tilde{h}_4(x,y)| \leq \hat{a}_0, \quad (x,y) \in \Omega. \tag{25}$$

The structure condition (3) in Definition 6.1 of Section 6 is fulfilled with $\hat{a}_1 := \sigma$, as we infer from (22) and (23) immediately.

Finally, we estimate Dirichlet's integral for the uniformizing mapping. From the formulas (11) we deduce the identity

$$|\nabla x(u,v)|^2 + |\nabla y(u,v)|^2 = \frac{(A+t) + (C+r)}{\sqrt{D}} \cdot J_{\mathbf{z}}(u,v), \; (u,v) \in B. \tag{26}$$

Then we estimate, without explicitly noting the double integrals, as follows:

$$\int_B |\nabla \mathbf{z}(u,v)|^2 \, du dv = \int_B \frac{A+C}{\sqrt{D}} \cdot J_{\mathbf{z}}(u,v) \, du dv + \int_B \frac{t+r}{\sqrt{D}} \cdot J_{\mathbf{z}}(u,v) \, du dv$$

$$\leq \frac{1}{\sqrt{\mu}} \left\{ \int_B (A+C) \cdot J_{\mathbf{z}}(u,v) \, du dv + \int_B (t+r)\Big|_{\mathbf{z}(u,v)} \cdot J_{\mathbf{z}}(u,v) \, du dv \right\}$$

$$\leq \frac{1}{\sqrt{\mu}} \left\{ 2\lambda(\gamma) \int_B J_{\mathbf{z}}(u,v) \, du dv + \int_\Omega \Big(z_{xx}(x,y) + z_{yy}(x,y) \Big) \, dx dy \right\}$$

$$= \frac{1}{\sqrt{\mu}} \left\{ 2\pi\lambda(\gamma) + \int_{\partial\Omega} \frac{\partial z}{\partial \nu}\Big|_{(x,y)} \, d\sigma(x,y) \right\} \leq \hat{N}(\lambda(\gamma), \mu)$$

$$\tag{27}$$

with the a priori constant

$$\hat{N}(\lambda(\gamma), \mu, \gamma) := \frac{1}{\sqrt{\mu}} \left\{ 2\pi\lambda(\gamma) + 2\pi\gamma \right\}.$$

Here we have used the derivative w.r.t. the exterior normal $\nu = \nu(x,y)$ to the domain Ω and the line element $d\sigma(x,y)$ on $\partial\Omega$. The necessary exhaustion of the disc is left as an exercise for the reader.

3.) With the aid of (25) and (27), we can estimate $|\nabla \mathbf{z}(u,v)|$ in the disc $B_{r(d)}$ of an adequate radius $r = r(d) \in (0,1)$ from above by a constant, depending only on $\lambda(\gamma), \mu, \gamma$ and d. This result is contained in Theorem 5.4

of Section 5 from Chapter 12. When we require the structure conditions (22) and (23) additionally, we can estimate the Jacobian $J_{\mathbf{z}}(u, v)$, $(u, v) \in B_{r(d)}$ from below by a positive constant, depending only on $\lambda(\gamma)$, μ, d, γ and σ. Invoking the formulas (11) for the second derivatives of our solution, we attain the fundamental estimate

$$|r(x, y)|, \ |s(x, y)|, \ |t(x, y)| \leq \theta(\lambda(\gamma), \sigma, \mu, \gamma, d) \quad \text{for all} \quad (x, y) \in \Omega[d] \quad (28)$$

with the a priori constant $\theta(\ldots)$.

4.) Disposing of a modulus of continuity for the uniformizing mapping (4) on the closed disc $\overline{B}$, we can estimate this mapping via Theorem 2.7 in Section 2 of Chapter 12 for all $0 < \nu < 1$ as follows:

$$\|\mathbf{z}\|_{C^{1+\nu}(B_{r(d)})} \leq c(\lambda(\gamma), \nu, \mu, \gamma, d). \tag{29}$$

With the aid of our estimate for the Jacobian from below, a similar estimate holds true for the inverse mapping $\mathbf{w}(x, y) := (u(x, y), v(x, y)) : \overline{\Omega} \to \overline{B}$, namely

$$\|\mathbf{w}\|_{C^{1+\nu}(\Omega[d])} \leq c(\lambda(\gamma), \nu, \mu, \gamma, \sigma d); \tag{30}$$

here the quantity σ naturally appears! On account of (28) with the Assumptions i) and ii) above, the coefficient functions

$$A(x, y) := A(x, y, z(x, y), z_x(x, y), z_y(x, y)), \quad B(x, y) := B(\ldots),$$

$$C(x, y) := C(x, y, z(x, y), z_x(x, y), z_y(x, y)), \quad \text{and} \quad D(x, y)$$

have the following interior C^1-bounds:

$$\|A(x, y)\|_{C^1(\Omega[d])}, \ldots, \|D(x, y)\|_{C^1(\Omega[d])} \leq \widehat{\theta}(\lambda(\gamma), \sigma, \mu, \gamma, d). \tag{31}$$

We return to formulae (11), and we rewrite them into the following *resolving identities for the second derivatives* within $\Omega[d]$, namely

$$r(x, y) = \frac{|\nabla y|^2}{J_{\mathbf{z}}(u, v)}\bigg|_{u=u(x,y),\, v=v(x,y)} \cdot \sqrt{D(x, y)} - C(x, y),$$

$$s(x, y) = -\frac{\nabla x \cdot \nabla y}{J_{\mathbf{z}}(u, v)}\bigg|_{u=u(x,y),\, v=v(x,y)} \cdot \sqrt{D(x, y)} + B(x, y), \tag{32}$$

$$t(x, y) = \frac{|\nabla x|^2}{J_{\mathbf{z}}(u, v)}\bigg|_{u=u(x,y),\, v=v(x,y)} \cdot \sqrt{D(x, y)} - A(x, y).$$

Now the Hölder estimate for the second derivatives follows from these identities (32) in combination with (29) – (31). Thus we arrive at the a priori estimate (24) with the constant $\Theta(\lambda(\gamma), \sigma, \nu, \gamma, d)$ stated in our theorem. q.e.d.

Remarks:

a) The class of Monge-Ampère equations with real-analytic coefficients, where
    the functions $k_1, \ldots, k_4$, $\tilde{k}_1, \ldots, \tilde{k}_4$ from (7) and (8) only depend on x, y,
    has been treated by H. Lewy. Then E. Heinz considered differentiable coef-
    ficients and transformed the compactness results of H. Lewy to these more
    profound a priori estimates. Moreover, the peculiar structure conditions
    from (22) and (23) were discovered by E. Heinz for this general class of
    Monge-Ampère equations. This requires prior intensive study for the class
    $\Gamma_*(B, a_0, a_1, N)$ in the Section 6. H. Lewy had already given an example
    (see [Sc] Section 9.5) that the estimate of the second derivatives is *not*
    possible by the solely use of the ellipticity condition!
b) In the Darboux equation (14), the coefficients A, B, C from (15) depend on
    p and q linearly. Therefore, the structure conditions (22) and (23) involve
    neither the unknown solution $z = z(x, y)$ nor its derivatives $p(x, y)$ and
    $q(x, y)$. This is in accordance with the Darboux system (17), where the
    coefficients depend only on the characteristic parameters – and neither
    the solution u nor the gradient ∇u appear there.
c) Prescribing a Riemannian metric (12) of the class $C^3(S^2)$ on the sphere,
    Theorem 7.1 yields a compactness result for the realizations (13) of this
    metric by strictly convex surfaces via the Darboux equation. Here we ob-
    serve that the growth condition i) is satisfied for the coefficients (15) of
    the Darboux equation (14), since the Gaussian curvature K still possesses
    bounded derivatives. Via Remark b) we see that the structure condition
    iii) is valid for the Darboux equation as well in this situation of a C^3-
    metric. When we consider the Darboux system (17) directly, we observe
    that the terms involving the derivatives of K cancel in the structure con-
    ditions! Thus one can solve the Weyl embedding theorem by a continuity
    and approximation method.
d) The Weyl embedding problem under the optimal regularity condition
    $ds^2 \in C^{2+\alpha}$ has been independently solved by F. Schulz and I. Sabitov.
    Finally, we would like to recommend for further studies the inspiring sur-
    vey article [H11], in which the boundary behavior and boundary estimates
    [H10] by E. Heinz are presented as well. We hope that these investiga-
    tions will encourage further research of the associate geometric embedding
    problems.

8 Some Historical Notices to Chapter 13

B. Riemann was the first to solve Plateau's problem for a quadrilateral in
1867, treating a very special Riemann-Hilbert problem. In 1866 and 1887,
K. Weierstraß elaborated the close relationship between the theories of holomorphic functions and minimal surfaces, respectively.

In the case $H = 0$, Plateau's problem has been solved independently by
T. Radó and J. Douglas, and later R. Courant created the approach using
Dirichlet's principle. R. Courant's book on *Dirichlet's Principle*, from 1950,

was very influential in this context; Courant personally built the bridge between Germany and the United States of America – in mathematics and beyond. The treatise above, inspired already by D. Hilbert, gives a simplified solution of Plateau's problem by J. Douglas and T. Radó, the first Fields medalists, from 1930.

Furthermore, we would like to emphasize the great influence of J. C. C. Nitsche with his 1975 monograph *Vorlesungen über Minimalflächen* and his inspiring lectures in the *Oberwolfach conferences on the Calculus of Variations*, which initiated a new flourishing of the classical minimal surface theory.

The pioneering solution of Plateau's problem for branched immersions of constant mean curvature by E. Heinz in 1954 was followed by a quantitative improvement of H. Werner. In 1966, S. Hildebrandt investigated the behavior of minimal surfaces at the nonanalytic boundary. Moreover, Plateau's problem for prescribed variable mean curvature was solved by S. Hildebrandt in 1969/70.

The Dirichlet problem, for the nonparametric equation of prescribed mean curvature H, was originally treated by T. Radó in 1930 for the case $H = 0$ and by J. Serrin for variable H in 1967. The latter result holds even in arbitrary dimensions and is described in the well-known book [GT] by D. Gilbarg and N. Trudinger.

The isoperimetric problem shows that uniqueness questions might be much more profound and intricate than existence questions: Although we have known since antiquity that the sphere solves the isoperimetric problem, answers to the associate classification problem were given throughout the last century: H. Liebmann in 1900, H. Hopf in 1950, A. D. Alexandrov in 1958, J. L. Barbosa and M. do Carmo in 1984, H. C. Wente in 1986 made their interesting contributions from Section 5 in this context.

Weyl's embedding problem, realizing the first fundamental form, asks for the solution of the most natural system of p.d.e.s; however, this problem leads to the most profound nonlinear partial differential equation, the Monge-Ampère equation. The admirable contributions by H. Weyl, H. Lewy, L. Nirenberg, A. V. Pogorelov, E. Heinz, F. Schulz and others laid the foundations for the entire theory. The really important open questions concern the uniqueness and stability problems again! The duality between analytic and geometric aspects, which this theory shares with the minimal surfaces, possesses a high potential for future research.

Figure 2.11   MINIMAL SURFACES SPANNING VARIOUS CONTOURS
taken from the title-page of the 1975 monograph Vorlesungen über Minimalflächen by J. C. C. Nitsche within the Springer-Grundlehren.

References

[BS]   H. Behnke, F. Sommer: *Theorie der analytischen Funktionen einer komplexen Veränderlichen.* Grundlehren der math. Wissenschaften **77**, Springer-Verlag, Berlin ..., 1955.

[B]   L. Bianci: *Vorlesungen über Differentialgeometrie.* Dt. Übersetzung von Max Lukat. Verlag B. G. Teubner, Leipzig, 1899.

[BL]   W. Blaschke, K. Leichtweiss: *Elementare Differentialgeometrie.* Grundlehren der math. Wissenschaften **1**, 5. Auflage, Springer-Verlag, Berlin ..., 1973.

[CH]   R. Courant, D. Hilbert: *Methoden der mathematischen Physik I, II.* Heidelberger Taschenbücher, Springer-Verlag, Berlin ..., 1968.

[D]   G. Darboux: *Leçons sur la théorie générale des surfaces I – IV.* Première édition: Gauthier-Villars, Paris, 1887 – 1896. Reprint: Chelsea Publ. Co., New York, 1972.

[De]   K. Deimling: *Nichtlineare Gleichungen und Abbildungsgrade.* Hochschultext, Springer-Verlag, Berlin ..., 1974.

[DHKW]   U. Dierkes, S. Hildebrandt, A. Küster, O. Wohlrab: *Minimal surfaces I, II.* Grundlehren der math. Wissenschaften **295**, **296**, Springer-Verlag, Berlin ..., 1992.

[DHS]   U. Dierkes, S. Hildebrandt, F. Sauvigny: *Minimal surfaces.* Grundlehren der math. Wissenschaften **339**, Springer-Verlag, Berlin ..., 2010.

[DHT1]   U. Dierkes, S. Hildebrandt, A. Tromba: *Regularity of minimal surfaces.* Grundlehren der math. Wissenschaften **340**, Springer-Verlag, Berlin ..., 2010.

[DHT2]   U. Dierkes, S. Hildebrandt, A. Tromba: *Global analysis of minimal surfaces.* Grundlehren der math. Wissenschaften **341**, Springer-Verlag, Berlin ..., 2010.

[E]   L. C. Evans: *Partial Differential Equations.* AMS-Publication, Providence, RI., 1998.

[G]   P. R. Garabedian: *Partial Differential Equations.* Chelsea, New York, 1986.

[Gi]   M. Giaquinta: *Multiple integrals in the calculus of variations.* Annals Math. Stud. **105**. Princeton University Press, Princeton N.J., 1983.

[GT]   D. Gilbarg, N. S. Trudinger: *Elliptic Partial Differential Equations of Second Order.* Grundlehren der math. Wissenschaften **224**, Springer-Verlag, Berlin ..., 1983.

[Gr]   H. Grauert: *Funktionentheorie I.* Vorlesungsskriptum an der Georg-August-Universität Göttingen im Wintersemester 1964/65.

[GF]   H. Grauert, K. Fritzsche: *Einführung in die Funktionentheorie mehrerer Veränderlicher.* Hochschultext, Springer-Verlag, Berlin ..., 1974.

[GL]   H. Grauert, I. Lieb: *Differential- und Integralrechnung III.* 1. Auflage, Heidelberger Taschenbücher, Springer-Verlag, Berlin ..., 1968.

[GuLe]   R. B. Guenther, J. W. Lee: *Partial Differential Equations of Mathematical Physics and Integral Equations.* Prentice Hall, London, 1988.

[HW]   P. Hartman, A. Wintner: *On the local behavior of solutions of nonparabolic partial differential equations.* American Journ. of Math. **75** (1953), 449–476.

[H1]   E. Heinz: *Differential- und Integralrechnung III.* Ausarbeitung einer Vorlesung an der Georg-August-Universität Göttingen im Wintersemester 1986/87.

[H2]   E. Heinz: *Partielle Differentialgleichungen.* Vorlesung an der Georg-August-Universität Göttingen im Sommersemester 1973.

[H3]   E. Heinz: *Lineare Operatoren im Hilbertraum I.* Vorlesung an der Georg-August-Universität Göttingen im Wintersemester 1973/74.

[H4]   E. Heinz: *Fixpunktsätze.* Vorlesung an der Georg-August-Universität Göttingen im Sommersemester 1975.

[H5]   E. Heinz: *Hyperbolische Differentialgleichungen.* Vorlesung an der Georg-August-Universität Göttingen im Wintersemester 1975/76.

[H6]   E. Heinz: *Elliptische Differentialgleichungen.* Vorlesung an der Georg-August-Universität Göttingen im Sommersemester 1976.

[H7]   E. Heinz: *On certain nonlinear elliptic systems and univalent mappings.* Journal d'Analyse Math. **5** (1956/57), 197–272.

[H8]   E. Heinz: *An elementary analytic theory of the degree of mapping.* Journal of Math. and Mechanics **8** (1959), 231–248.

[H9]   E. Heinz: *Zur Abschätzung der Funktionaldeterminante bei einer Klasse topologischer Abbildungen.* Nachr. Akad. Wiss. in Göttingen, II. Math.-Phys. Klasse, Heft Nr.9, 1968.

[H10]   E. Heinz: *Lokale Abschätzungen und Randverhalten von Lösungen elliptischer Monge-Ampèrescher Gleichungen.* Journ. reine angew. Math. **438** (1993), 1–29.

[H11]   E. Heinz: *Monge-Ampèresche Gleichungen und elliptische Systeme.* Jahresbericht der Deutschen Mathematiker-Vereinigung **98** (1996), 173–181.

[He1]   G. Hellwig: *Partielle Differentialgleichungen.* Teubner-Verlag, Stuttgart, 1960.

[He2]   G. Hellwig: *Differentialoperatoren der mathematischen Physik.* Springer-Verlag, Berlin ..., 1964.

[He3]   G. Hellwig: *Höhere Mathematik I – IV.* Vorlesungen an der Rheinisch-Westfälischen Technischen Hochschule Aachen; Mitschrift aus dem Wintersemester 1978 bis zum Sommersemester 1980.

[Hi1]   S. Hildebrandt: *Analysis 1.* Springer-Verlag, Berlin ..., 2002.

[Hi2]   S. Hildebrandt: *Analysis 2.* Springer-Verlag, Berlin ..., 2003.

[HS]   F. Hirzebruch und W. Scharlau: *Einführung in die Funktionalanalysis.* Bibl. Inst., Mannheim, 1971.

[HC]   A. Hurwitz, R. Courant: *Funktionentheorie.* Grundlehren der math. Wissenschaften **3**, 4. Auflage, Springer-Verlag, Berlin ..., 1964.

[J]   F. John: *Partial Differential Equations*. Springer-Verlag, New York ...,
         1982.

[Jo]   J. Jost: *Partielle Differentialgleichungen. Elliptische (und parabolische)
         Gleichungen*. Springer-Verlag, Berlin ..., 1998.

[K]   W. Klingenberg: *Eine Vorlesung über Differentialgeometrie*. Heidelberger
         Taschenbücher, Springer-Verlag, Berlin ..., 1973.

[Kn]   H. Kneser: *Lösung der Aufgabe 42*. Jahresber. Dt. Math.-Vereinigung **35**
         (1926), 123–124.

[M]   Claus Müller: *Spherical Harmonics*. Lecture Notes in Math. **17**, Springer-
         Verlag, Berlin ..., 1966.

[Mu]   Frank Müller: *Funktionentheorie und Minimalflächen*. Vorlesung an
         der Universität Duisburg-Essen, Campus Duisburg, im Sommersemester
         2011.

[N]   J. C. C. Nitsche: *Vorlesungen über Minimalflächen*. Grundlehren der
         math. Wissenschaften **199**, Springer-Verlag, Berlin ..., 1975.

[Ni]   L. Nirenberg: *On nonlinear elliptic partial differential equations and
         Hölder continuity*. Comm. Pure Appl. Math. **6** (1953), 103–156.

[R]   F. Rellich: *Die Bestimmung einer Fläche durch ihre Gaußsche Krüm-
         mung*. Math. Zeitschrift **43** (1938), 618–627.

[Re]   R. Remmert: *Funktionentheorie I*. Grundwissen Mathematik **5**, 2. Auflage,
         Springer-Verlag, Berlin ..., 1989.

[Ru]   W. Rudin: *Principles of Mathematical Analysis*. McGraw Hill, New York,
         1953.

[S1]   F. Sauvigny: *Einführung in die reelle und komplexe Analysis mit ihren
         gewöhnlichen Differentialgleichungen 1*. Vorlesungsskriptum an der BTU
         Cottbus im Wintersemester 2006/07.

[S2]   F. Sauvigny: *Einführung in die reelle und komplexe Analysis mit ihren
         gewöhnlichen Differentialgleichungen 2*. Vorlesungsskriptum an der BTU
         Cottbus im Sommersemester 2007.

[S3]   F. Sauvigny: *Flächen vorgeschriebener mittlerer Krümmung mit einein-
         deutiger Projektion auf eine Ebene* Math. Zeitschrift **180** (1982), 41–67.

[S4]   F. Sauvigny: *On immersions of constant mean curvature: Compactness
         results and finiteness theorems for Plateau's problem*. Archive for Rational
         Mechanics and Analysis **110** (1990), 125–140.

[S5]   F. Sauvigny: *Deformation of boundary value problems for surfaces with
         prescribed mean curvature*. Analysis **21** (2001), 157–169.

[S6]   F. Sauvigny: *Un problème aux limites mixte des surfaces minimales avec
         une multiple projection plane et le dessin optimal des escaliers tournants*.
         Annales de l'Institut Henri Poincaré – Analyse Non Linéaire **27** (2010),
         1247–1270.

[Sc]   F. Schulz: *Regularity theory for quasilinear elliptic systems and Monge-
         Ampère equations in two dimensions*. Lecture Notes in Math. **1445**,
         Springer-Verlag, Berlin ..., 1990.

[V]   I. N. Vekua: *Verallgemeinerte analytische Funktionen*. Akademie-Verlag,
         Berlin, 1963.

Index

F. Sauvigny, *Partial Differential Equations 2*, Universitext,
DOI 10.1007/978-1-4471-2984-4, © Springer-Verlag London 2012